T0309796

GLACIAL SEDIMENTARY PROCESSES AND PRODUCTS

Other publications of the International Association of Sedimentologists

Medial moraines on a tributary of Kaskawulsh Glacier, Icefield Ranges, Yukon, Canada. (Photograph by M.J. Hambrey)

SPECIAL PUBLICATION NUMBER 39 OF THE INTERNATIONAL ASSOCIATION OF SEDIMENTOLOGISTS

Glacial Sedimentary Processes and Products

EDITED BY

Michael J. Hambrey
Centre for Glaciology, Aberystwyth University, Wales, UK

Poul Christoffersen
Scott Polar Research Institute, University of Cambridge, UK

Neil F. Glasser and Bryn Hubbard
Centre for Glaciology, Aberystwyth University, Wales, UK

SERIES EDITOR

Isabel Montanez
University of California, Davis

© 2007 International Association of Sedimentologists
and published for them by
Blackwell Publishing Ltd

BLACKWELL PUBLISHING
350 Main Street, Malden, MA 02148–5020, USA
9600 Garsington Road, Oxford OX4 2DQ, UK
550 Swanston Street, Carlton, Victoria 3053, Australia

The right of Michael J. Hambrey, Poul Christoffersen, Neil F. Glasser, and Bryn Hubbard
to be identified as the Authors of the Editorial Material in this Work has been asserted in
accordance with the UK Copyright, Designs, and Patents Act 1988.

All rights reserved. No part of this publication may be reproduced, stored in a
retrieval system, or transmitted, in any form or by any means, electronic, mechanical,
photocopying, recording or otherwise, except as permitted by the UK Copyright,
Designs, and Patents Act 1988, without the prior permission of the publisher.

First published 2007 by Blackwell Publishing Ltd

1   2007

Library of Congress Cataloging-in-Publication Data

International Conference on 'Glacial Sedimentary Processes and Products' (2005 : University of Wales)
    Glacial sedimentary processes and products / edited by Michael J. Hambrey . . . [*et al.*].
      p. cm. — (International Association of Sedimentologists special publications ; 39)
    Papers presented at the International Conference on 'Glacial Sedimentary Processes and Products'
    held at the University of Wales, Aberystwyth in August 2005.
    Includes bibliographical references and index.
    ISBN 978-1-4051-8300-0 (pbk. : alk. paper)
    1. Drift—Congresses.   2. Glacial landforms—Congresses.   I. Hambrey, M.J.   II. Title.

QE579.1569 2005
551.31'5—dc22

                                                                                      2007038525

A catalogue record for this title is available from the British Library.

Set in 10.5/12.5pt Palatino
by Graphicraft Limited, Hong Kong
Printed and bound in Italy
by Rotolito Lombarda S.p.A

The publisher's policy is to use permanent paper from mills that operate a sustainable
forestry policy, and which has been manufactured from pulp processed using acid-free
and elementary chlorine-free practices. Furthermore, the publisher ensures that the
text paper and cover board used have met acceptable environmental accreditation
standards.

For further information on
Blackwell Publishing, visit our website:
www.blackwellpublishing.com

Contents

Preface

This volume is the outcome of an international conference on 'Glacial sedimentary processes and products' held at the University of Wales, Aberystwyth in August 2005. The idea for such a conference dates back to the Congress of the International Association of Sedimentologists (IAS) in Johannesburg, July 2002, when following a session on '3 billion years of Earth's glacial history' it was apparent that there was little interchange of ideas between those working on modern glacial processes and those concerned with the palaeoglacial record. It was also apparent that within IAS there was a slowly growing interest in the glacial sedimentary record, manifest primarily in the rising number of papers in the journal *Sedimentology*. In the light of the successful series of IAS conferences on fluvial sediments and carbonates, it was felt that the time was ripe for IAS to sponsor its first glacial sediments conference. In view of recent developments in glaciology it was also apparent that glacial geologists and ice-sheet numerical modellers were beginning to work together to determine the scale and dynamics of past ice sheets. Thus, in addition to IAS, we brought together additional co-sponsors for the Aberystwyth meeting:

The International Glaciological Society
The International Commission of Snow and Ice (now International Association of Cryospheric Sciences)
The Quaternary Research Association
The International Union for Quaternary Research (INQUA)
The Antarctic Climate Evolution Programme of the Scientific Committee on Antarctic Research
The British Geological Survey

The conference, and hence this volume, represents the first outcome of two international working groups which are working closely together:

• The International Association of Cryospheric Sciences' Working Group on *Debris Transport in Glaciers*; and
• INQUA's Working Group on *Glacial Sedimentary Processes and Landforms*.

The principal aims of these Working Groups are:

• to promote dialogue between researchers in the fields of glacier dynamics, contemporary glacial processes, glacial sedimentology and ice sheet modellers; and
• to apply a multidisciplinary approach to enhance our understanding of debris entrainment and transfer in glaciers, and subsequent deposition at contemporary ice margins; and to use this knowledge to constrain numerical models of contemporary and former ice masses.

These aims are reflected in the contents of this volume, which are divided into four themes: 'Glacier dynamics and sedimentation', 'Modelling glaciers and ice sheets', 'Quaternary glacial systems' and 'Pre-Quaternary glacial systems'. The papers represent a cross-section of those delivered at the conference. The abstracts of all the oral and poster presentations at the conference, totalling 139 contributions are available on-line through the IUGG/CCS Working Group website (http://www.lboro.ac.uk/research/phys-geog/dtg/index.html). The papers in this volume were refereed to the standards used by *Sedimentology*, each of the editors selecting two referees. Ian Jarvis and Isabel Mõntanez ensured that the editors' work was in line with IAS policy. The amount of effort put in by the referees has been most impressive, and we acknowledge their contributions and express our thanks to them. They are acknowledged in each paper, except where they prefer to remain anonymous.

This volume is dedicated to the late W. Brian Harland (1917–2003) of Cambridge University who, as David Drewry records in his tribute on page xi, played a leading role in gathering data on and interpreting Earth's glacial record, and promoting the concept of Neoproterozoic glaciation, long before this became fashionable in the context of the 'Snowball Earth hypothesis'. His career was dominated by fieldwork in the High-Arctic archipelago of Svalbard, where modern glacial sedimentary processes are fully in evidence, and his observations there undoubtedly informed his thinking about the past glacial record. Yet his own interests spanned the full spectrum of geology, and he made major contributions to tectonic geology, stratigraphy and the geological time-scale. Furthermore, he introduced an astonishing number of undergraduates, post-graduates, post-doctoral researchers and academic staff to the delights of Svalbard (including one of the editors, MJH). The spin-off from his work is manifested in several hundred scientific papers, both those of his own and those resulting from follow-up work by geologists who were fortunate enough to join his expeditions.

We acknowledge the financial support of several corporate sponsors and a private sponsor, whose contributions facilitated the participation of post-graduates and scientists from developing countries at the conference itself. They are:

Statoil UK Ltd.
Cardigan Sand & Gravel Ltd.
Reynolds Geo-Sciences Ltd.
Sir Mark Moody Stuart

Underpinning all the work associated with running the conference and producing this volume has been the support of postgraduates and staff of the Institute of Geography and Earth Sciences at the University of Wales, Aberystwyth (especially Inka Koch). In addition, Martin Siegert of the University of Bristol (now University of Edinburgh) and Peter Janssen of the University of Stockholm, in their capacities as co-convenors, advised us on the key themes that should be explored.

Michael J. Hambrey, Poul Christoffersen,
Neil F. Glasser and Bryn Hubbard

Aberystwyth, October 2006

Foreword
Brian Harland and the Neoproterozoic 'Snowball Earth' concept

DAVID J. DREWRY

Office of the Vice-Chancellor, The University of Hull, HU6 7RX, UK (e-mail: david.drewry@hull.ac.uk)

On 8th July 2005, the G8 leaders, issued a communiqué which stated: 'Climate change is a serious and long-term challenge . . . We know that increased need and use of energy from fossil fuels . . . contribute in large part to increases in greenhouse gases associated with warming of the Earth's surface' (The Gleneagles Communiqué, 2005).

For scientists, particularly geologists, the many changes to the Earth's environment in the past are neither startling nor largely inexplicable, and the notion that human beings can have a measurable impact in raising the levels of radiatively-active gases in the atmosphere can have root in fertile ground. The explanation of Ice Ages, periods of desertification or the steady rhythmic shift of climate belts are phenomena that have engaged earth scientists from the earliest times. Indeed the *Conference on Glacial Sedimentary Processes and Products* in Aberystwyth, Wales in August 2005, was a significant contribution to such studies. But the magnitude of the rapid changes from a glacial-bound world to super-greenhouse conditions might be said to astound even the most ardent supporters of climate change research.

They certainly created colossal scepticism and disbelief amongst the geological and scientific community when first propounded some forty years ago. Yet the proposition that the Earth might have experienced a full planetary ice cover in the past involving the improbable glaciation of the tropics came not from the hallucinations of a science fiction writer or the twilight fringe of pseudo-scientific anoraks but from the pen of a well-respected academic in the Department of Geology (now Earth Sciences) at Cambridge University.

Brian Harland (Fig. 1) had first noticed the presence of tillites in rocks of late Precambrian age from his extensive work in Svalbard – a part of the Arctic that had bewitched him as a young man, seventy years ago just prior to the Second World War (and indeed has captivated many others both before

Fig. 1 Brian Harland with his wife Elisabeth on the deck of *MV Copius*, a vessel which provided support for numerous expeditions to Svalbard. (Photo: Michael Hambrey).

and since). Exposed in northeastern Spitsbergen and western Nordaustlandet is a glacial sequence, almost 1 km in thickness, sandwiched between deposits of a very different composition and origin – thick carbonate sequences – which taken together testify to some dramatic lurch in Earth's climate about 600 million years ago.

Assembling evidence from other continents, Harland spotted similar patterns. What was more astonishing was that the palaeogeographic position of these Neoproterozoic glacial deposits placed them, according to the then best available evidence, in equatorial latitudes. Within the glacigenic sequences were unequivocal ice-rafted sediments containing an abundance of distinctive drop-stones. They imputed, not mountain glaciation in these tropical regions, but extensive ice masses at sea level on the margins of what is now termed Laurentia. Debris-laden icebergs would be calving, drifting, melting and releasing their debris to the floor of the proto-Iapetus.

Brian Harland proposed at a Conference in January 1963 in Newcastle (Harland, 1963), and further propounded in *Geologische Rundschau* (Harland, 1964) and *Scientific American* (Harland & Rudwick, 1964), the radical view of a world-wide glacial event of many millions of years duration that terminated abruptly and was followed by an equally extreme warm period. This latter interval saw the deposition of thick units of carbonate rock, and importantly presaged the conditions for the Ediacaran and the great Cambrian biological explosion.

To promote such an hypothesis in the early 1960s required courage and conviction; Harland received cynical and coruscating criticism of his ideas at the time. It is important to recall that the concept of continental drift – the term 'plate tectonics' had yet to be coined and defined – was just climbing out of the slough of scientific aprobrium. Modern evidence for continental motions was being amassed steadily – Harry Hess, Bruce Heezen, and Bob Deitz were postulating the role of mid-ocean ridges and Drummond Matthews and Fred Vine were at the cusp of publishing their *Nature* paper (Vine & Matthews, 1963) on magnetic lineations in the oceans that clinched the case for sea-floor spreading. Many of the techniques which we take for granted in the geosciences were in their infancy – geomagnetism was only then establishing itself as a reliable technique (it was a vital strand in Harland's thesis, but the evidence he required was woefully inadequate), geochemical, particularly isotopic determinations were commencing, but accuracy and applications were limited. Certainly the ability to direct awesome computer power to creating simulations of Earth, ocean and atmosphere, and run coupled models was a generation away.

But Harland had built his case with typical meticulous care and logical deduction – the same qualities that he instilled in generations of undergraduates he taught in the lecture theatre, laboratory and in the field. His technique, as re-called by Peter Friend in Harland's obituary in *The Independent* newspaper in November 2003, was, '. . . to train young scientists to make simple and clear observations and then argue on the basis of these, *'ignoring all preconceptions'* (Friend, 2003).

Harland's ideas on planetary glaciation were later picked up by Joseph Kirschvink at the

California Institute of Technology (Kirschvink, 1992) and at Harvard University by Paul Hoffman and Dan Schrag (Hoffman *et al.*, 1998), and became landmark contributions to what is now termed the 'Snowball Earth' hypothesis – accessibly documented by Gabrielle Walker, a writer of popular science (Walker, 2003). These scientists, using more modern and accurate techniques, were able to develop Harland's early work and extend it, importantly providing mechanisms for initiating and sustaining the global Marinoan and earlier Sturtian glaciations and their rapid transition to a hyper-greenhouse world. The former, initiation phase, referred to the earlier and equally controversial work of Mikhail Budyko on modelling run-away ice-albedo feedback (Budyko, 1969). During this author's early years in Cambridge, Budyko came to spend a sabbatical period at the Scott Polar Research Institute, working on his ice-albedo model – I am reasonably confident that he never met Brian Harland or discussed the possible concatenation of their various ideas; but that would have been one of those tantalising scientific 'what ifs'.

The tillites that had inspired Harland continued to intrigue him in later years. He approached their study not from a need to understand the mechanisms or processes which produced them, nor their detailed petrography or geochemistry; for these he relied on other scientists, especially glaciologists and glacial geologists. Brian Harland and I often met and he was frequently challenged on some of his interpretations of glacial sequences! He would always demand clear argument and evidence, but once convinced would absorb and incorporate the new knowledge into his studies. Harland's interest in tillites continued to be in their distribution in time and space as distinctive indicants of palaeo-environmental conditions and as a seductive pointer for climate change. Convinced of the need to produce reliable time scales and assemble basic data for use by the geoscientific community he commenced to edit a new genre of publications. *The Phanerozoic Time Scale* (Harland *et al.*, 1964), *The Fossil Record* (Harland *et al.*, 1967), and *A Geological Time Scale* (Harland *et al.*, 1990) were the result, the last mentioned being produced also as a poster which decorated many a geologist's pin-board or wall.

During his searches, Harland found the unlinked descriptions and individual interpretations of rock units particularly frustrating. Thus he became a key protagonist of the International Geological Correlation Programme (IGCP) and, in the 1970s, initiator and coordinator with the able and untiring assistance of Michael Hambrey, of its Pre-Pleistocene Tillite Project (IGCP 38). This was a massive international undertaking involving 165 authors who produced 212 papers totalling 1004 pages of published material (Hambrey & Harland, 1991). Brian Harland would have been delighted to see the significant number of papers and posters being given on tillites at the Aberystwyth meeting, and particularly the continued investigation and debate concerning the Neoproterozoic glaciations.

I should conclude this reflection on Brian Harland's scientific life by returning to Svalbard – which held a special place in his heart and around which most of his professional work revolved. The Cambridge Spitsbergen Expeditions (CSE) that he initiated was a remarkable programme of investigations by boat, sledge, snow-scooter, all-terrain vehicle, occasionally helicopter and back-packing on foot and ski, operating mostly out of Ny-Ålesund and his much loved hut 'Mexico' (Fig. 2). Brian personally led 29 of the 43 summer expeditions, accompanied on 13 of these by his wife, Elisabeth (Fig. 1). These expeditions systematically covered the islands and revealed their geological story. No single person has contributed more to the geological understanding of this part of the North Atlantic than Brian Harland, and, I would posit, nor probably to any area of comparable size and complexity. More than 300 undergraduates and staff were involved with the CSE, including 50 graduate collaborators. Out of CSE emerged the Cambridge Arctic Shelf Programme in 1975 – now an independent scientific research company working on geological-related problems world-wide. *The Geology of Svalbard* (Harland, 1998) is an enduring testament to his fastidious and insightful work, but even these 500 or so pages and the 2.2 kg weight are not a true reflection of his extraordinary encyclopaedic knowledge. David Gee, in a review of this book, commented, 'It will remain the Svalbard bible for many years to come. One should remember to read it in the spirit of the author: with an inquiring mind never satisfied with conventional wisdom and present-day interpretations' (Gee, 2001).

Brian Harland was a man of great scientific foresight, with tireless intellectual and physical energy. He could be challenging and robust, stubborn and irascible, dutiful and engaging with a high moral conviction. I wonder what he would have made of the G8 concerns on climate change and President George Bush's equivocal position?

It is, therefore, with considerable pleasure and respect that the *International Conference on Glacial Sedimentary Processes and Products*, Aberystwyth, 22–27th August 2005 has been dedicated to the memory of Brian Harland.

Fig. 2 Brian Harland in Svalbard with two of his Cambridge Spitsbergen Expedition's motor boats, *Collenia* and *Salterella*, appropriately named after Svalbard fossils. In the background is the glacier Scottbreen, surrounded by mountains bearing Neoproterozoic tillites. (Photo: Michael Hambrey).

REFERENCES

Budyko, M.I. (1969) The effects of solar radiation variations on the climate of the Earth. *Tellus*, **21**, 611–619.

Friend, P.F. (2003) Walter Brian Harland 1917–2003, Obituary. *The Independent*, 14 November 2003, p. 22.

Gee, D. (2001) Review 'The Geology of Svalbard' by W.B. Harland. *Polar Record*, 37, 160–162.

Hambrey, M.J. and Harland, W.B., Eds. (1984) *Earth's pre-Pleistocene glacial record*. Cambridge University Press, Cambridge, 1004 pp. Available online at http://www.aber.ac.uk/glaciology/epgr.htm.

Harland, W.B. (1963) Evidence of late Precambrian glaciation and its significance. In: *Problems in Palaeoclimatology* (Ed. A.E.M. Nairn). Proceedings of the NATO Palaeoclimates Conference, Newcastle-upon-Tyne, January 7–12, 1963, pp. 119–149 Interscience, London.

Harland, W.B. (1964) Critical evidence for a great Infra-Cambrian Glaciation. _Geologische Rundschau_, **54**, 45–61.

Harland, W.B. (1998) _The Geology of Svalbard_. Geol. Soc. Lond. Mem., **17**, 522 pp.

Harland, W.B. and Rudwick, M.J.S. (1964) The Great Infra-Cambrian ice age. _Scientific American_, **211** (2), 28–36.

Harland, W.B., Smith, A.G. and Willcocks, B., Eds (1964) The Phanerozoic time-scale (a symposium). _Quart. J. Geol. Soc., Lond._, **120**, supplement.

Harland, W.B., Holland, C.H., House, M.R., Hughes, N.F., Reynolds, A.B., Rudwick, M.J.S., Satterthwaite, G.E., Tarlo, L.B.H. and Willey, E.C. Eds (1967) _The Fossil Record_. The Geological Society, London, 828 pp.

Harland, W.B., Armstrong, R.L., Cox, A.V., Craig, L.E., Smith, A.G. and Smith, D.G. (1990) _A Geologic Time Scale_, 1989 edition, Cambridge University Press, 265 pp.

Hoffman, P.F., Kauffman, A.J., Halverson, G.P. and Schrag, D.P. (1998) A Neoproterozoic snowball Earth. _Science_, **281**, 1342–46.

Kirschvink, J.L. (1992) Late Precambrian low-latitude global glaciation: the snowball earth. In: _The Proterozoic Biosphere: a multidisciplinary study_ (Eds J.W. Schopf, C. Klein and D. Maris, D.), pp. 51–52. Cambridge University Press, Cambridge.

The Gleneagles Communiqué (2005) UK, Foreign & Commonwealth Office website: http://www.fco.gov.uk/Files/kfile/PostG8_Gleneagles_Communique,0.pdf

Vine, F.J. and Matthews, D.H. (1963) Magnetic anomalies over oceanic ridges. _Nature_, **199**, 947–949.

Walker, G. (2003) _Snowball Earth_. Bloomsbury, London, 269 pp.

Introduction to Papers

MICHAEL J. HAMBREY*, POUL CHRISTOFFERSEN*†, NEIL F. GLASSER*
and BRYN HUBBARD*

*Centre for Glaciology, Institute of Geography & Earth Sciences, Aberystwyth University, Wales, Ceredigion SY23 3DB, UK
(e-mail: mjh@aber.ac.uk)
†Present address: Scott Polar Research Institute, University of Cambridge, Lensfield Road, Cambridge, CB2 1ER, UK

INTRODUCTION

Glacial Sedimentary Processes and Products includes twenty papers that demonstrate the breadth and dynamism of this research field. Broadly following the themes of the conference at which these papers were presented, we divide the book into four sections, as highlighted below, in which each of the papers is briefly summarised.

GLACIER DYNAMICS AND SEDIMENTATION

Three papers ranging from ice-sheet-scale investigations to studies on valley glaciers make up this section. **Siegert** *et al.* follow up recent research, in which water was hypothesized to flow from one Antarctic subglacial lake to another, to investigate the possibility of linkages between lake networks, and the consequences of such linkages for subglacial erosion and deposition. The authors use ice surface and bed DEMs to reconstruct the ice sheet's basal hydraulic potential field and use this to investigate the nature of water flows between known, located lakes (of which there are over 140). The authors conclude that most lakes in the Dome C region of the East Antarctic Ice Sheet, where there is already a relatively high density of individual lakes, are likely to be linked. Further, these water flows most probably form a connected drainage network of linked lakes with the capacity to transfer basal water and sediments from the ice sheet interior down-valley to the Southern Ocean.

On a much smaller scale, **Midgley** *et al.* describe sediment composition, morphology and structural characteristics of a moraine-mound complex in front of Midre Lovénbreen, a polythermal valley glacier in Svalbard. They identified a range of lithofacies of subglacial, supraglacial and proglacial origin, and integrate the structural attributes with a geomorphological characterisation based on digital elevation models. The integrated analysis shows that moraine-mound formation is linked to structural controls on the elevation of sediment within the glacier.

On a temperate glacier, **Rousselot** *et al.* describe the development of, and provide preliminary results from, an innovative laboratory tool designed to investigate key material properties related to subglacial ploughing. Here, the authors have applied the measurement principles of a subglacial 'ploughmeter' to construct an analogue 'instrumented tip' for laboratory use. The tip itself houses an array of strain gauges and pressure sensors, while the sediment through which the tip is dragged is additionally instrumented for pore-water pressure and bulk volume. Tests on water-saturated subglacial sediment recovered from the Unteraargletscher, Switzerland, revealed several interesting properties, including the establishment of pore-water pressure gradients within the sediment in the vicinity of the ploughing tip, indicating compression (reflected by above-hydrostatic pore-water pressure) in front of the ploughing tip and dilation (below-hydrostatic pore-water pressure) behind the tip. The authors also noted that the sediment had a 'memory' such that virgin sediment presented greater resistance to ploughing than did sediment that had been pre-ploughed.

MODELLING GLACIERS AND ICE SHEETS

Numerical modelling represents one of the most exciting and rewarding developments in the field of glacial geology at the present time, and two papers cover this topic. First, **Pollard and DeConto** present the results of a numerical model of ice-sheet flow that has been coupled both to an ice-shelf flow model and to a model of subglacial sediment deformation. Running the model along an idealised 1-D flowline over a 10 Ma period allows ice and sediment to be advected along the flow path, and for the bulk depositional style of the resulting offshore sediments to be reconstructed. Varying key system and material properties, including sediment viscosity, a sediment transport factor, and surface mass-balance (to force ice-sheet retreats every 100 ka), results in a range of offshore sediment assemblages. Importantly, these are expressed explicitly not only in terms of their bulk form, but also in terms of their internal architecture and the age of internal units. In their most sophisticated model run, the authors include a pre-existing ice shelf, allow for pelagic deposition, include sea-level variations with a cyclicity of 100 ka, and include realistic sediment geotechnical properties. The resulting offshore deposit has a complex form, comprising several internal units of markedly differing architectures, demonstrating the clear utility of this approach to guide the investigation and interpretation of offshore glacigenic sediments.

The second paper by **Siegert** reviews sediment erosion, transport and deposition by former large ice sheets from a numerical modelling perspective. Three examples are used to examine how the integration of geological data with ice dynamical models can elucidate large-scale ice-sheet evolution. The paper also contains a brief discussion of promising new trends and techniques. Many of these approaches are now being undertaken by one of the flagship programmes of the Scientific Committee on Antarctic Research, namely Antarctic Climate Evolution, that is drawing together especially the glacial geological, the geophysical and the modelling communities.

QUATERNARY GLACIAL SYSTEMS

Developments in our understanding of Quaternary glacial systems are reflected in the nine papers in this section. Commonly in the past, Quaternary geologists had little interaction with glaciologists, but in recent years the picture looks much healthier, and both scientific disciplines are working closer together than ever before. Researchers now adopt a wide range of tools and approaches to reconstruct depositional environments. The papers in this section reflect to some extent this improved situation, as a number of the authors have themselves worked in both camps.

First, working on the continental shelf of Antarctica, **Willmott** *et al.* present new data from sediment cores and seismic profiles concerning glaciomarine sediment drifts from Gerlache Strait, Antarctic Peninsula. Sediment drifts are accumulations of hemipelagic sediment contained within a depositional geometry, which is influenced by bottom currents, often showing evidence of re-mobilization of foci of accumulation and erosion by currents. The extent and palaeoenvironmental importance of these accumulations on the continental shelf of Antarctica remain poorly defined so their paper is a welcome addition to the field. Their cores reveal a variety of sediment facies, with couplets of laminated diatom ooze and sandy mud alternating with massive, bioturbated, terrigenous silt/clay units to a predominantly massive, bioturbated, silty-clay unit. Seismic profiles across the drift show an internal structure defined by wavy stratified reflectors. Willmott *et al.* use these data to suggest that the drifts formed in estuarine conditions during the Late to Middle Holocene, as a result of large fresh water input from melting ice margins, melting sea ice, warmer surface temperatures, and lack of exchange with the Antarctic Circumpolar Current.

Back on land, **Russell** *et al.* describe the structure and sedimentology of the Waterloo Moraine in southern Ontario, Canada. The moraine, which is composed mainly of stratified sediments, was deposited in two distinct depositional environments – a conduit or esker setting and a subaqueous fan setting. The implication from the sedimentology is that much of the moraine was formed rapidly during large flood discharge events and not from diurnal meltwater events. The conclusion is that the Waterloo Moraine was the depocentre of high-magnitude meltwater discharge events. This paper is an important contribution because of the debate concerning the origin of stratified moraines.

On a topic that is not explicitly glacial sedimentological, but which deals with processes in glacial materials following deposition, **Delisle** *et al.* use palaeoenvironmental records to estimate permafrost evolution in northern Germany during the Pleistocene Epoch. The study shows that the growth and decay of permafrost is controlled by climate forcing and terrestrial heat-flow variations associated with subterranean salt structures. The maximum vertical extent of permafrost is estimated at 130–170 m.

Processes at the margin of the Saalian Scandinavian ice sheet are documented by **Winsemann** *et al.* using a combination of geological and geophysical data to identify the nature of subaqueous fan-deposits from glacial Lake Rinteln in northern Germany. The study shows that gravity flows were the dominant sediment transport mechanism on the fans when meltwater discharge was relatively continuous. Sediments that are rich in ice-rafted debris characterise a phase of glacial retreat and a major subglacial discharge event is associated with a significant fall in lake-level.

Delaney provides a detailed study of glaciolacustrine sedimentation associated with the Late Devensian ice sheet of Ireland, and illustrates neatly how inferences can be made concerning the seasonal controls on the deposition of glaciolacustrine sediments. Her study is based on careful logging and microfabric analysis using a scanning electron microscope (SEM) of laminated sediments in cores from beneath two bogs west of Lough Ree in central Ireland. The glaciolacustrine sediments are primarily rhythmically laminated silts alternating with clay layers. These sediments are interpreted as representing annually laminated lake sediments (varves).

Demonstrating the benefits of using modern geophysical equipment, **Kulessa** *et al.* applied two methods to the investigation of the internal structure of a drumlin located in County Down, Northern Ireland. This drumlin was investigated by traditional (surface and borehole) logging, as well as by (a) seismic refraction imaging and (b) electrical resistivity imaging. Seismic refraction imaging was found to be particularly useful in terms of defining and tracing structural layering within the body of the drumlin. The technique, for example, allowed the authors to identify five distinct internal layers, including a highly permeable and spatially-extensive basal layer of weathered bedrock that would otherwise have been difficult to trace. Although of lower resolution, electrical resistivity imaging was also found to be useful in terms of mapping variations in moisture content and sedimentary inclusions within more varied

tills. It is clear from this pioneering study that geophysical techniques, such as those used here, have a great deal to offer to the investigation of depositional landforms. This reflects the ability of geophysical methods to (a) cover a surface large area, (b) reveal 3-D internal structure of a feature in a non-invasive manner, and (c) provide information relating to the nature of that structure.

Two papers next consider the sedimentology of eskers in glaciated areas, helping to shed new light on the origin, palaeohydrology, depositional processes and environments of deposition of these landforms. **Bennett** *et al.* describe the stratigraphy and sedimentology of a 13-km long esker system (the Newbigging esker system) in Lanarkshire, southern Scotland. The esker has a variable morphology. In some places it is a single ridge, while elsewhere it comprises a series of multiple subparallel ridges or has a complex multi-ridge structure. From exposures in quarries along the length of the esker, these authors link the sedimentology to the morphology of the esker. Both the sedimentary architecture and the morphology are consistent with an interpretation of progressive infilling of a large lake basin by successive shifts in sediment input as the ice margin receded. Bennett *et al.* conclude their paper by presenting a model that emphasises composite tunnel, subaqueous fan and supraglacial esker sedimentation.

The second paper on esker formation is by **Gale and Hoare**, who consider the age and origin of the Blakeney Esker of north Norfolk and its implications for the glaciology and glacial history of the southern North Sea Basin. In this contribution, the Blakeney Esker is interpreted as subglacial in origin. The esker developed during a glacial standstill and under steep hydraulic gradients close to the former ice sheet margin. The conduit that provided a source for the esker was incised into the base of the ice sheet, although there is localised evidence of simultaneous incision of the glacier bed. By considering the stratigraphic position of the esker in relation to surrounding glacigenic deposits, Gale and Hoare argue that it is Middle Quaternary in age.

Working in Ennerdale, a valley in the English Lake District, **Graham and Hambrey** present new information about the sediments and landforms developed during the Younger Dryas glaciation in Britain. The genesis of moraines associated with

British glaciers of Younger Dryas age has long been controversial and their paper is therefore a valuable addition to this debate. Graham and Hambrey use sedimentological and geomorphological evidence to argue that the landforms in Ennerdale developed from a combination of ice-marginal deposition and englacial thrusting. They argue that, whilst englacial thrust moraines may not be commonly associated with British Younger Dryas glaciers, under certain topographic and thermal conditions englacial thrusting is an important process in landform development.

PRE-QUATERNARY GLACIAL SYSTEMS

The papers in this section include representatives from ancient glacial records, including the earlier Cenozoic, the Ordovician and the Neoproterozoic. These contributions focus on the sedimentology of glacial successions, in which inferences about depositional processes rely less on landform characteristics than on the lithostratigraphic record. Nevertheless, well-preserved erosional landforms in some cases aid the interpretation.

The paper by **Barrett** is a comprehensive summary of the vast amount of data collected during a 3-season multi-national Cape Roberts drilling project off the coast of Victoria Land in the western Ross Sea. This work is of fundamental importance in understanding the earlier history of the East Antarctic Ice Sheet. The topic has important societal implications, since current global warming trends suggest that we are entering a phase analogous to that when the Antarctic Ice Sheet was much more dynamic and even temperate in character. The three holes drilled provide a record spanning the interval 34 Ma (earliest Oligocene) to 17 Ma (early Miocene). The strata were deposited in a rift basin and constitute a variety of facies: conglomerate, sandstone, mudstone and diamictite, with marine fossils throughout. Repetitive vertical facies associations indicate glacio-eustatic fluctuations on a wave-dominated coast that was influenced by fluctuating tidewater glaciers. Precise dating of the strata near the Oligocene/Miocene boundary has enabled an orbitally driven 40,000-year cycle to be determined. In terms of climate, recorded especially by the preservation of terrestrial pollen, a cool temperate regime is envisaged,

and cooling to the present cold polar state occurred much later.

A paper by **Fasano** *et al.* focuses on the onshore area of Victoria Land, where a tillite of indeterminate age (within the Cenozoic Era) is preserved along the inland margin of the Transantarctic Mountains, close to the ice sheet. In order to better understand the processes responsible for deposition of the tillite, the authors introduce an innovative technique that is new to glacial geology, known as μX-ray tomography. This technique enables the sediment texture to be characterized in 3-D to a level that goes beyond standard microstructural techniques, including determining the ice-flow direction. Unlike normal sequential thin-sectioning techniques, μX-ray tomography preserves the sample intact, and is particularly useful for core-analysis.

Moving back through time, two papers are presented from the late Ordovician glaciation of Africa. This relatively short-lived event (a few million years at most), was characterized by the growth of a large ice sheet over the North Gondwana Platform in North Africa. As described by **Ghienne** *et al.*, the evidence for glaciation is remarkably well-preserved, especially in Libya, where a number of cycles of sedimentation are recorded. Facies include coarse grained glaciomarine sediments, deep-marine fans, glaciofluvial sediments, outwash systems and shelf deposits. Good exposure has enabled the 3-D facies architecture to be determined, and several examples are illustrated in detailed cross-sections. Three domains are identified: a continental interior, a glaciated continental shelf, and a non-glaciated shelf. Major morphological features, notably ice-stream-generated glacial troughs, cut across these domains, and bear some similarities with features found on modern high-latitude continental shelves.

The second Ordovician paper by **Kumpulainen** documents equivalent Gondwana Platform deposits for the first time in Eritrea and Ethiopia. Several sections spaced over a distance of 200 km were measured and a range of facies documented, including diamictite, sandstone and mudstone, whilst glacially erosional surfaces provide additional palaeoenvironmental data. Facies assemblages indicate both terrestrial (in Ethiopia) and marine (in Eritrea) glacial settings, and the sequence was apparently deposited during a single phase of deglaciation at the end of the Ordovician Period. The sequence is believed to reflect recession of the Gondwana ice sheet that was centred over North Africa.

The last paper in this volume is a comprehensive critical review by **Etienne** *et al.* that deals with a particularly controversial phase in Earth history, the Neoproterozoic Era. This era is commonly regarded as having the most severe cold phase experienced by the planet, since glacial deposits are widespread on all continents. The concept of severe ice ages has become popularly know as 'Snowball Earth' and the most extreme view proposes that most of the planet, including the oceans, was ice-covered. The authors of this paper use a sedimentological approach to evaluate the evidence, using our understanding of the Cenozoic glacial sequences to inform the debate. They conclude that most Neoproterozoic glacial successions exhibit spatial and temporal variability, and have

facies associations that are typical of dynamic ice masses. Most sequences were deposited on glaciated continental margins without the need for global synchroneity or necessarily severe climatic cooling. Thus a complete hydrological shutdown, as envisaged by the most extreme 'Snowball Earth' views is not supported by the evidence from Neoproterozoic glacial successions.

In summary, this volume demonstrates through a wide range of studies (i) that our understanding of glacial processes in modern environments continues to improve; (ii) that major leaps in understanding of how major ice masses deliver sediment to continental margins are now possible using numerical modelling approaches, and (iii) that application of understanding of modern glaciers to Quaternary and pre-Quaternary successions enables us to solve major questions regarding the temporal and spatial scales of former ice masses.

Part I

Glacier dynamics and sedimentation

Debris apron comprising reworked fluvial and wind-blown sands with internal deformation including thrusting, at Wright Lower Glacier, Victoria Land, Antactica. (Photograph by M.J. Hambrey)

Hydrological connections between Antarctic subglacial lakes, the flow of water beneath the East Antarctic Ice Sheet and implications for sedimentary processes

MARTIN J. SIEGERT*, ANNE LE BROCQ† and ANTONY J. PAYNE†

*School of GeoSciences, Grant Institute, University of Edinburgh, West Mains Road, Edinburgh, EH9 3JW, UK
(e-mail: m.j.siegert@edinburgh.ac.uk)
†Centre for Polar Observation and Modelling, Bristol Glaciology Centre, School of Geographical Sciences,
University of Bristol, Bristol BS8 1SS, UK

ABSTRACT

Subglacial lakes are commonly referred to as unique environments isolated for millions of years. The recent detection of a rapid transmission of water between subglacial lakes indicates, however, that these environments may be connected hydrologically and that sporadic discharge of lake water may be an expected process. Knowledge of this flow at a continental scale is important to understanding habitats provided by subglacial lakes, the potential routes by which stored basal water can be exported to the ice margin and the development of glacial-fluvial landforms. Here, an assessment of Antarctic subglacial water *flow-paths* is presented, based on hydro-potential gradients derived from basal and ice surface topographies. The assessment reveals that most subglacial lakes around Dome C are likely to be linked. One *flow-path* in particular connects >10 lakes located within adjacent topographic valleys, including Lake Concordia and Lake Vincennes. Subglacial water at Dome C has the potential to flow to the ocean, as the ice base is warm continuously between the ice divide and the margin. Such flow is likely to be organised into distinct drainage basins, in which water is routed to the proglacial zone through only a small number of outlets, which potentially has a strong influence on the development of sedimentary landforms.

Keywords basal hydrology, ice-sheet dynamics, glacial history

INTRODUCTION

Rapid release of water from subglacial lakes, and well-organised (probably channellised) flow of this water to other lakes, have been demonstrated recently from analysis of satellite altimetric surface changes (Wingham *et al.*, 2006). Such flow may be a common means by which stored water is re-distributed beneath the East Antarctic Ice Sheet. The paths of such flow have also been shown to be predictable through consideration of the subglacial water pressure potential (Shreve, 1972). The number of measured Antarctic subglacial lakes currently stands at 145, the majority of which are located around the Dome C and Ridge B regions of East Antarctica (Siegert *et al.*, 2005a). Thermomechanical ice-sheet models predict basal

temperatures at or near the pressure-melting point in these regions (e.g. Huybrechts, 1990; Siegert *et al.*, 2005b). It is possible, therefore, that many Antarctic subglacial lakes are hydrologically connected. The aim of this paper is to provide a first assessment of the expected hydrological paths in the vicinity of known Antarctic subglacial lakes, focussing on the Dome C and Ridge B regions where there is sufficient surface and subglacial topography data to constrain the water pressure calculation to at least a gross level (i.e. ±25 km) and to investigate the potential geomorphological implications of these paths. Such assessment is important to determining the likelihood of water escaping to the ice sheet margin, explaining the onshore and offshore sedimentary record, evaluating the habitats of subglacial lakes and planning future exploration of these extreme environments (e.g. Priscu *et al.*, 2005).

CALCULATING SUBGLACIAL WATER FLOW-PATHS

Assuming that subglacial water pressure (p) is equal to the pressure of the overlying ice, the flow-paths of subglacial water can be calculated from the pressure potential (ϕ), i.e. the sum of gravitational potential energy and water pressure, following Shreve (1972): $p = \rho_i gh$, $\phi = p + (\rho_w gz_b)$ and, therefore, $\phi = \rho_i gz_s + (\rho_w - \rho_i)gz_b$, where ρ_w is the density of water (1000 kg m^{-3}), g is the gravitational acceleration (9.81 m s^{-2}), z_b (m) is the elevation of the bed, z_s (m) is the elevation of the ice surface, ρ_i is the density of ice (910 kg m^{-3}) and h (m) is the overlying ice thickness (i.e. $z_s - z_b$). Using values of z_b derived from the BEDMAP topographic dataset of Antarctica (Lythe *et al.*, 2000), and z_s from the RAMP altimetric dataset of Antarctica (Liu *et al.*, 1999), the water pressure potential for the entire ice sheet base can be calculated. Flowpaths can then be constructed assuming that water flow is perpendicular to pressure potential contours.

Although it must be noted that the BEDMAP dataset does not represent a comprehensive evaluation of subglacial topography, at some locations it is an appropriate dataset for the calculation of basal water flowpaths. At Dome C, topographic coverage is good enough to denote the large-scale topography (features with lateral extent ≥10 km) well. Importantly, roughness at a smaller scale at Dome C has been shown to be low (Siegert *et al.*, 2005b), which suggests the likelihood of substantial topographic barriers to water flow is low in this region. The accuracy of the dataset has been shown to be sufficient to predict the flowpath of water between subglacial lakes >250 km apart (Wingham *et al.*, 2006).

FLOW OF WATER BENEATH THE ICE SHEET

Antarctic subglacial water drainage is organized into a series of distinct catchments and pathways (Fig. 1). While the ice surface slope dictates to a first order the direction of subglacial water flow, basal topography has a noticeable effect on flow-paths. It is important to note that there are very few, if any, significant regions where water may pond (i.e. where there is a local minimum in the piezo surface). Hence the potential for the build up of very large subglacial lakes, such as Lake Vostok, is low (nb. Lake Vostok is not identified by this method, as BEDMAP does not account for the bathymetry of this large lake). At Dome C there are no large measurable water-pressure 'basins'. Consequently, the vast majority of the subglacial landscape will be involved in a large-scale hydrological network beginning at the ice sheet centre and terminating at the margin.

Results from thermomechanical ice sheet modeling show many regions of the Antarctic Ice Sheet at or near the pressure melting temperature (Fig. 2A), including the location of lakes around Dome C (Fig. 2B). If the regional ice sheet base is warm, it is highly likely that water will flow at the base according to Shreve (1972). If the ice base is below the pressure melting point, hydrological connections may still be possible, but additional energy will be required from the flow locally to both raise the temperature and melt ice. Such energy may be available from rapid releases (e.g. Wingham *et al.*, 2006).

There are at least three continuous warm-based pathways from the Dome C interior to the ice margin: to the Byrd Glacier within the Transantarctic Mountains; to Totten Glacier in Wilkes Land; and to the Mertz and Ninnis glaciers in George V Land. At Ridge B, there is also a continuously warm-based hydrological pathway to Byrd Glacier (Figs 2 and 3).

The association between known subglacial lakes, modelled warm basal conditions, and a predicted dendritic basal drainage network, in which smaller tributaries feed larger trunk pathways, makes it possible that subglacial lakes are connected into a hierarchical drainage system, within which lake-water mixing (and consequently biodiversity) may increase toward the ice margin. Furthermore, under this ordered system, the total water flux from lake discharges is likely to increase towards the margin. We anticipate that much of the melt water generated at the centre of the ice sheet could well pass through several connected lakes during its passage to the ice sheet margin. In fact, flow to the margin must happen unless it can be refrozen to the underside of the ice sheet in sufficient quantities to balance melt water generated. Wingham *et al.* (2006) expected the flow of lake water to be

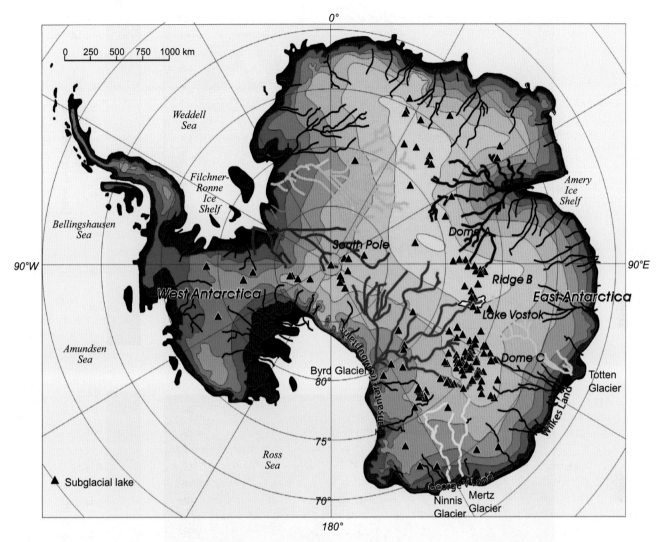

Fig. 1 The location of 145 known Antarctic subglacial lakes, the ice-sheet surface elevation (both adapted from Siegert *et al.*, 2005a), and large-scale *flow-paths* of subglacial water. Ice surface contours are provided in 500 m intervals. The largest drainage basins are coloured distinctly, although no significance is attached to these colours.

sporadic and inherently unstable. They reasoned that if a hydrological connection existed between a subglacial lake and the ocean, there would be little to stop the lake draining (as the ice margin has a far lower potential than the centre of the ice sheet base) provided there is sufficient energy available to melt the channel through which water flows, and to deform the ice (i.e. for draining and ponding of water).

SUBGLACIAL LAKE CONNECTIONS

Examination of how hydrological flow-paths link subglacial lake locations allows an indication of the potential connectivity of subglacial lakes (Fig. 3). In particular, where numerical modelling predicts warm basal conditions such connectivity is likely (Fig. 2).

Most subglacial lakes around Dome C are located within or close to smooth topographic lowlands, which are thought to be the product of ancient glacial erosion by a former smaller ice cap (Drewry, 1975; Siegert *et al.*, 2005b). The predicted *flow-paths* of subglacial water allow us to develop an understanding of possible hydrological linkages between lakes at Dome C (Fig. 3).

The majority of subglacial lakes at Dome C connect to at least one lake downstream. A large proportion of these feed into more than one

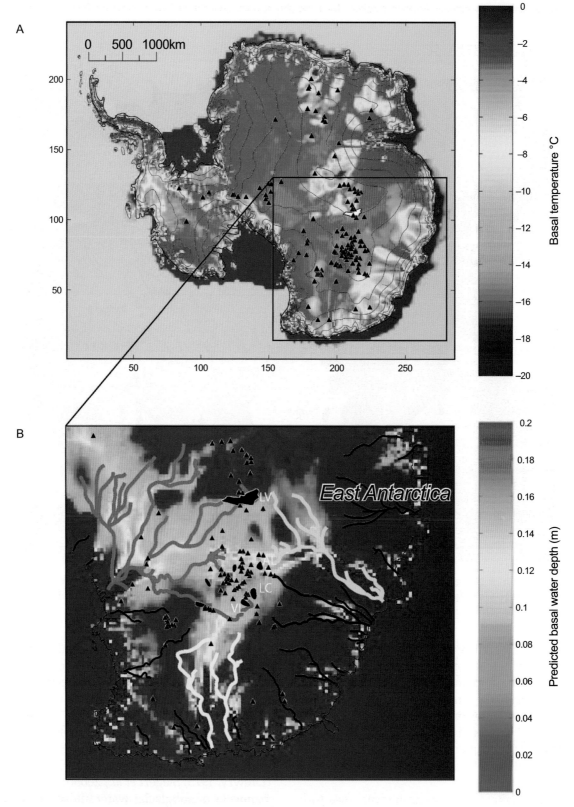

Fig. 2 (A) Basal temperature of the Antarctic Ice Sheet (in °C), derived from a thermomechanical ice sheet model (adapted from Siegert *et al.*, 2005b), and the locations of Antarctic subglacial lakes (shown as solid triangles apart from Lake Vostok, which is outlined). (B) The location of basal water (the average depth in m), predicted by the ice sheet model, around the Dome C and Dome A regions. Names of subglacial lakes are as follows: LV, Lake Vostok; AL, Aurora Lake, LC, Lake Concordia; VL; Vincennes Lake. The positions of potential water *flow-paths*, as provided in Fig. 1, are also included.

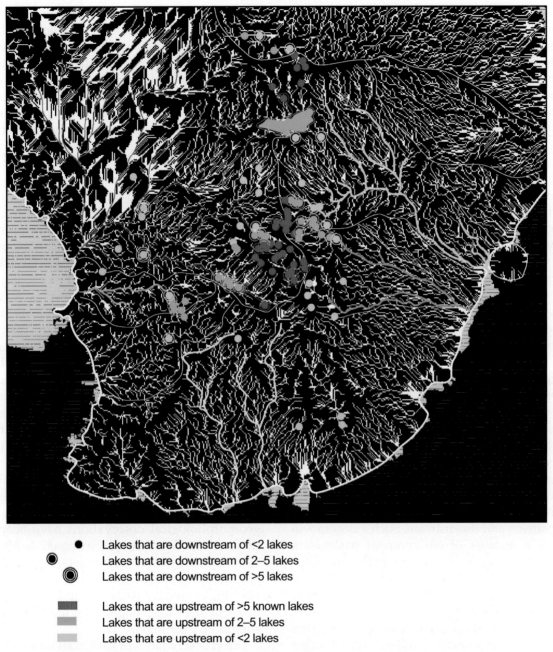

● Lakes that are downstream of <2 lakes
◎ Lakes that are downstream of 2–5 lakes
◉ Lakes that are downstream of >5 lakes

▬ Lakes that are upstream of >5 known lakes
▬ Lakes that are upstream of 2–5 lakes
▬ Lakes that are upstream of <2 lakes

Fig. 3 Complete drainage pathways of subglacial water around the Dome C and Ridge B regions of East Antarctica. The key reveals how subglacial lakes are coded to illustrate whether they are fed by, or feed into, other subglacial lakes. The largest drainage basins are coloured distinctly as in Fig. 1. Blue regions denote floating ice, while black shade represents the ocean.

downstream lake. One hydrological *flow-path* connects lakes within the Vincennes Subglacial Basin and the Adventure Subglacial Trench, where subglacial lake connectivity has been observed (Wingham *et al.*, 2006). This path involves the connection of over 20 individual lakes, including Lake Concordia and the Vincennes Lake (Tabacco *et al.*, 2003; Tikku *et al.*, 2005) (Fig. 2B). Although subglacial water flow into Lake Concordia has previously been postulated by Tikku *et al.* (2005), the flowpaths out of the lake along warm basal conditions makes it highly likely that water also escapes from Lake Concordia to, possibly, several lakes downstream.

As even small changes to the ice sheet surface have large consequences for subglacial hydrology, the present distribution of basal flowpaths is unlikely to be fixed through time. That said, the large-scale subglacial topography has a strong effect on water flow, which runs along the axes of many subglacial valleys around Dome C. Consequently, while small changes to the position of the ice divide may result in reconfiguration of the boundaries of drainage basins, it probably will not affect the flow routes of water along subglacial valleys. Hence, many of the subglacial lake connections found today may have existed in a similar stable form over the past few glacial cycles (during which time topography would probably have been largely unchanged).

IMPLICATIONS FOR SUBGLACIAL LAKE EXPLORATION AND OUTBURST FLOODING

The notion of subglacial lakes (including Lake Vostok) as isolated water bodies (e.g. Siegert, 1999; Priscu *et al.*, 2003) is clearly in need of revision. As a consequence, plans to explore subglacial lake environments must take note of the probability of a hydrological system that connects lakes both upstream and downstream. Such a system will influence a number of environment parameters of the lake system, including hydrochemistry (including solute acquisition), water residence times, biodiversity (between lakes) and water circulation. In terms of risk during exploration, it is now clear that should a lake be contaminated during *in situ* examination, other lakes located downstream may well be affected over time. Clearly this is an issue that must be considered by proposals to explore many of the subglacial lakes across Dome C, but probably not all. Across the lowlands, which have well defined flow-paths, where quite large lakes are located, these concerns will be greatest. Over the higher ground some lakes could be hydrologically isolated. Further geophysical survey on such lakes to define the basal hydrology further is a prerequisite to exploratory examination.

As ice-sheet modelling indicates continuous warm based conditions from Dome C to the coast (Fig. 2) it is likely that subglacial lake water could escape the ice sheet. The subglacial drainage system indicates only a few outlets of such water, however (Fig. 3). The majority of lakes drain into the Byrd Glacier drainage basin, which cuts across the Transantarctic Mountains and into the Ross Ice Shelf. Most other lakes at Dome C feed into the Aurora basin and the Totten Glacier, from where lake water could flow into the ocean.

The probability of substantial volumes of meltwater being issued to the ice margin by rapidly draining Antarctic subglacial lakes has important implications for active subglacial geomorphological processes. Now that subglacial lake outbursts have been identified as probably being commonplace, a logical next step is to identify and comprehend evidence of such activity in the past. For example, the Antarctic Dry Valleys contain channels of 10^3 m^2 in cross-section that attest to intense subglacial floods of considerable size (Denton & Sugden, 2005) that reached the ice sheet margin. Today, the potential path of subglacial water into the Dry Valleys comes from the Wilkes Basin, where several subglacial lakes are known to exist. As noted by Denton & Sugden (2005), expansion of the ice sheet in this area would reduce the ice-surface slope over the Wilkes Basin, so allowing water to pond and be issued in potentially large quantities to the ice margin and the ocean. Similar geomorphological observations have been measured on the Soya Coast (Sawagaki & Hirakawa, 1997). Investigation of surface geomorphological and sedimentary records for further evidence of past flooding events is clearly warranted to establish, if possible, the rates, magnitudes and frequencies of subglacial lake floods.

Direct observations of subglacial water 'outbursts' are clearly critical to evaluating the dynamics of the process, but are currently restricted to a single event in 1985 at Casey Station in East Antarctica, which lasted six months (Goodwin, 1988). This single observation is testament to the possibility of flooding from beneath polar ice masses, where little or no melting occurs on the surface. According to Goodwin (1988) the origin of the flood was stored subglacial water, which collected slowly over time until a threshold was reached, upon which rapid unstable outflow occurred. The process witnessed at Casey Station may, therefore, be applicable to other regions of stored subglacial water such as at Dome C.

IMPLICATIONS FOR ICE DYNAMICS AND GLACIAL SEDIMENTARY PROCESSES IN EAST ANTARCTICA

The collection and transfer of water from large upstream subglacial catchments to discrete regions of the ice sheet margin has potential implications for ice stream dynamics. It is highly likely that the bases of at least some East Antarctic ice streams are underlain by sedimentary material (e.g. Bamber *et al.*, 2006). Such sediments, to become weak to the point of deforming to facilitate ice flow, need to be saturated with water. The results presented here indicate that the bulk of water emanating from the Dome C region can be routed into only a few ice-marginal outlets. The consequence is that any sedimentary material beneath the Byrd, Totten, Mertz and Ninnis outlet glaciers is likely to be fed with water. Unless this water escapes to the margin *via* well defined channels, it will act to increase pore-water pressures and so reduce the strength of sediments, which will increase ice velocity. The deformation of sediments beneath ice streams will drive material in the direction of ice flow to the ice margin and, given that the supply of water to these outlets is likely to have occurred over a considerable period (the LGM ice sheet configuration of East Antarctica is quite similar to that of today; Huybrechts, 2002), the build up of very large proglacial sedimentary sequences, possibly developed over millions of years, should be expected. If no subglacial sediment is present beneath East Antarctic ice streams (i.e. the glaciers rest on bedrock), water from upstream catchments may still influence ice flow provided distributed subglacial hydrological systems operate (Kamb, 1987).

The rapid transfer of subglacial water may also lead to glaciofluvial sedimentary processes, similar to that proposed in huge magnitude by Shaw *et al.* (1989) and Shaw (1996), and on a lesser scale by Shoemaker (1992), for the formation of drumlin fields across southern Canada during the last glacial phase. The distribution of subglacial water beneath the Antarctic ice sheet, involving episodic discharges from subglacial lakes, is likely to be a common process (Wingham *et al.*, 2006). It is therefore likely that subglacial sedimentary landforms, moulded by the occasional rapid supplies of significant quantities of water, are actively being developed in Antarctica. The implication of this, for interpretation of the sedimentary geological record, is that many landforms may originate from short-term fluvial processes operating beneath large ice sheets as well as longer-term processes such as ice flow. The subsequent task is to identify the degree to which glacial landforms exhibit signs of subglacial fluvial activity in order to ascertain the influence of subglacial discharges on past and present ice-sheet dynamics.

ACKNOWLEDGEMENTS

Funding for this work was provided by the UK Natural Environment Research Council's Centre for Polar Observation and Modelling (CPOM). ALB acknowledges funding of a UK NERC PhD studentship (no. NER/S/D/2003/11902).

REFERENCES

Bamber, J.L., Ferraccioli, F., Shepherd, T., Rippin, D.M., Siegert, M.J. and Vaughan, D.G. (2006) East Antarctic ice stream tributary underlain by major sedimentary basin. *Geology*, **34**, 33–36.

Denton, G.E. and Sugden, D.E. (2005) Meltwater features that suggest Miocene ice-sheet overriding of the Transantarctic Mountains in Victoria Land, Antarctica. *Geogr. Ann.*, **87 A**, 67–85.

Drewry, D.J. (1975) Initiation and growth of the East Antarctic ice sheet. *J. Geol. Soc. London*, **131**, 255–273.

Goodwin I.D. (1988) The nature and origin of a jökulhlaup near Casey Station, Antarctica. *J. Glaciol.*, **34**, 95–101.

Huybrechts, P. (1990) A 3-D model for the Antarctic Ice Sheet: a sensitivity study on the glacial-interglacial contrast. *Climate Dynamics*, **5**, 79–92.

Huybrechts, P. (2002) Sea-level changes at the LGM from ice-dynamic reconstructions of the Greenland and Antarctic ice sheets during the glacial cycles. *Quatern. Sci. Rev.*, **21**, 203–231.

Kamb, B. (1987) Glacier surge mechanism based on linked cavity configuration of the basal water conduit system. *J. Geophys. Res.*, **92**, 9083–9100.

Liu, H., Jezek, K.C. and Li, B. (1999) Development of an Antarctic digital elevation model by integrating cartographic and remotely sensed data: A geographic information system based approach. *J. Geophys. Res.*, **104**, 23199–23213.

Lythe, M.B., Vaughan, D.G. and the BEDMAP consortium (2000) BEDMAP – bed topography of the Antarctic. 1:10,000,000 map, BAS (Misc.), 9. Cambridge: British Antarctic Survey.

Priscu, J.C., Bell, R.E., Bulat, S.A., Ellis-Evans, J.C., Kennicutt, M.C. II., Lukin, V.V., Petit, J.R., Powell, R.D., Siegert, M.J. and Tabacco, I.E. (2003) An international plan for Antarctic subglacial lake exploration. *Polar Geogr.*, **27**, 69–83.

Sawagaki, T. and Hirakawa, K. (1997) Erosion of bedrock by subglacial meltwater, Soya Coast, East Antarctica. *Geogr. Ann.*, **79**, 223–238.

Shaw, J. (1996) A meltwater model for Laurentide subglacial landscapes. In: *Geomorphology Sans Frontiers.* (Eds McCann, S.B. and Ford, D.C.). London, Wiley, 181–236.

Shaw, J., Kvill, D. and Rains, R.B. (1989) Drumlins and catastrophic subglacial floods. *Sed. Geol.*, **62**, 177–202.

Shreve, R.L. (1972) Movement of water in glaciers. *J. Glaciol.*, **11**, 205–214.

Shoemaker, E.M. (1992) Water sheet outburst floods from the Laurentide Ice Sheet. *Can. J. Earth Sci.*, **29**, 1250–1264.

Siegert, M.J. (1999) Antarctica's Lake Vostok. *American Scientist*, **87**, 510–517.

Siegert, M.J., Carter, S., Tabacco, I.E., Popov, S. and Blankenship, D.D. (2005a) A revised inventory of Antarctic subglacial lakes. *Antarct. Sci.*, **17**, 453–460.

Siegert, M.J., Taylor, J. and Payne, A.J. (2005b) Spectral roughness of subglacial topography and implications for former ice-sheet dynamics in East Antarctica. *Global Planet. Change*, **45**, 249–263.

Tabacco, I.E., Forieri, A., Vedova, A.D., Zirizzotti, A., Bianchi, C., De Michelis, P. and Passerini, A. (2003) Evidence of 14 new subglacial lakes in the Dome C-Vostok area. *Terra Antarct.*, **8**, 175–179.

Tikku, A.A., Bell, R.E., Studinger, M., Clarke, G.K.C., Tabacco, I. and Ferraccioli, F. (2005) Influx of meltwater to subglacial Lake Concordia, East Antarctica. *J. Glaciol.*, **51**, 96–104.

Wingham, D.J., Siegert, M.J., Shepherd, A.P. and Muir, A.S. (2006) Rapid discharge connects Antarctic subglacial lakes. *Nature*, **440**, 1033–1036.

Sedimentology, structural characteristics and morphology of a Neoglacial high-Arctic moraine-mound complex: Midre Lovénbreen, Svalbard

NICHOLAS G. MIDGLEY*, NEIL F. GLASSER† and MICHAEL J. HAMBREY†

*School of Animal, Rural and Environmental Sciences, Nottingham Trent University, Brackenhurst, Southwell, Nottinghamshire, NG25 0QF, UK (e-mail: nicholas.midgley@ntu.ac.uk)
†Centre for Glaciology, Institute of Geographical and Earth Sciences, Aberystwyth University, Wales, Ceredigion, SY23 3DB, UK

ABSTRACT

Detailed analyses of the sediments, structural characteristics and morphology of a Neoglacial moraine-mound complex at a high-Arctic, polythermal, land-based glacier – midre Lovénbreen, Svalbard – have been undertaken in order to evaluate their mode of formation and subsequent modification. Ten moraine mound lithofacies were identified: four types of diamicton, four types of gravel, sand and mud. A buried debris-bearing basal ice facies was also identified following excavation of a single moraine mound. The moraine mound lithofacies and buried basal ice reflect subglacial, glaciofluvial and basal ice layer processes. Structural attributes of the moraine mounds include faulting, folding and shearing. Three broad moraine mound morphological types are identified and characterised using Digital Elevation Models. Moraine mound formation is linked to elevation of sediment within the ice, associated with englacial thrusting. The recognition of buried ice in a moraine mound implies that on melting, the landform will become degraded. Hence the preservation potential of thrust-related features within these moraine mounds is low, unless the debris content of the ice is high.

Keywords Hummocky moraine, moraine-mound complexes, glaciotectonics, polythermal glacier, midre Lovénbreen, Svalbard.

INTRODUCTION

The aim of this paper is to integrate the evaluation of sediment, structure and morphology of moraine mounds (commonly described as 'hummocky moraine') at the margin of a modern high-Arctic polythermal glacier. This paper focuses on Neoglacial moraine mound development at a land-based valley glacier, midre Lovénbreen, which is approximately 5 km from Ny-Ålesund on the peninsula named Brøggerhalvøya in NW Spitsbergen, Svalbard (Figs 1A and B). This work is significant for understanding ice-marginal processes because it combines detailed sediment analysis, structural characteristics and accurate three-dimensional characterisation of moraine mound morphology – such an approach has not previously been undertaken at these Svalbard moraine-mound complexes.

Previous work on Svalbard glaciers has demonstrated the importance of englacial thrusting in the development of moraine-mound complexes (e.g. Hambrey and Huddart, 1995; Huddart and Hambrey, 1996; Bennett et al., 1996; Hambrey et al., 1997; Glasser et al., 1998; Bennett et al., 1998; Bennett, 2001). However, this model of moraine development by englacial thrusting is contentious (Evans, 2000), and the process has also been controversially reinterpreted as the reorientation of sediment-filled crevasses at Kongsvegen, Svalbard (Woodward et al., 2002; Glasser et al., 2003; Woodward et al., 2003). An improved understanding of these contemporary moraines will aid the interpretation of formerly glaciated areas, since morphological similarity has been suggested between some Svalbard and British Younger Dryas moraine mounds (Hambrey et al., 1997; Bennett et al., 1998;

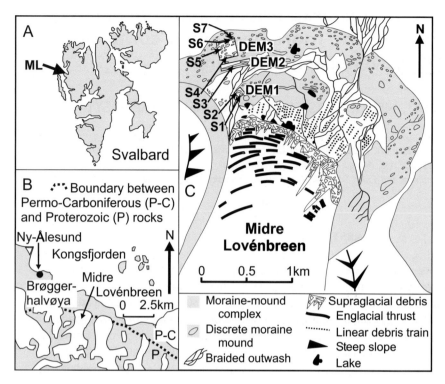

Fig. 1 (A) Svalbard, Norwegian high-Arctic. (B) Location of midre Lovénbreen (ML) on Brøggerhalvøya. (C) Geomorphological map of the midre Lovénbreen proglacial area (localities S1 to S7 are the excavated sections shown in Figures 3b and 4, and the locations of DEMs shown in Figure 5 are indicated by boxes DEM 1 to DEM 3 (geomorphological map modified from Hambrey *et al.*, 1997).

Graham and Midgley, 2000; Midgley, 2001; Graham, 2002). Whilst the englacial thrusting model of moraine formation may be appropriate for some British Younger Dryas moraine sites, it does not apply universally (e.g. Lukas, 2005).

GEOLOGICAL AND GLACIOLOGICAL SETTING OF MIDRE LOVÉNBREEN

The bedrock of Brøggerhalvøya is composed of three groups consisting of Proterozoic metamorphic rocks (schist, marble and amphibolite), Permo-Carboniferous cherty limestone and Tertiary sandstone and coal measures (Fig. 1B; Hjelle, 1993; Harland, 1997). During the Neoglacial maximum the terminal zone of midre Lovénbreen was underlain by the younger sedimentary rocks, but due to subsequent recession the glacier is now underlain and surrounded by the metamorphic rocks.

Ground-penetrating radar investigations by Björnsson *et al.* (1996) indicate that midre Lovénbreen is cold-based at the snout and along the lateral edges, but has a warm-based inner basal layer. Mass balance data for midre Lovénbreen from 1968 to 2001 indicates a variable annual mass

balance with an overall negative trend (Haeberli *et al.*, 2003). Midre Lovénbreen has been classified as a non-surge type glacier (e.g. Jiskoot *et al.*, 2000), but it may have changed from a surge-type to a non-surge type glacier (Hansen, 2003). Midre Lovénbreen has receded by approximately 1 km from its Neoglacial maximum, as defined by the outer limits of the moraine-mound complex (Fig. 1C). The structures of midre Lovénbreen, e.g. folding, foliation and thrusts (Hambrey & Glasser, 2003; Hambrey *et al.*, 2005) to a large extent control delivery of sediment to the proglacial area. In particular, the snout of midre Lovénbreen exhibits a range of structures revealed by surface debris accumulations including numerous englacial thrusts transverse to glacier flow (Hambrey *et al.*, 1997; Glasser and Hambrey, 2001).

METHODOLOGY

Analysis of sediment lithofacies was undertaken by means of: textural description; particle-size analysis using c. 100 g of the sand, silt and clay component (<2 mm) with sieve stacks and a SediGraph; and analysis of clast attributes such as shape and roundness. Textural description of sediments was undertaken using a classification for unsorted sediments that was adapted from Hambrey (1994). Clast morphology was described by the shape and roundness of 50 randomly chosen small pebble- to cobble-sized clasts of mixed lithology, assessed by measurement of the a, b and c-axis and visually using Powers (1953) roundness categories. Co-variant plots of RA (% of very angular and angular clasts) against C_{40} (% of clasts with c : a axis ratio ≤ 0.4) were plotted for each moraine lithofacies and representative control samples, including scree, fractured bedrock, glaciofluvial sediment and a basal till (Benn and Ballantyne, 1994; Bennett *et al.*, 1997). Structural characteristics of moraine mounds were identified from shallow (<1.5 m depth) excavations through the ridge crests and a single stream-cut section. A Geodimeter Total Station was used to collect topographic data, with contoured Digital Elevation Models produced using Leica Liscad software with a contour interval of 1 m. Point density for the survey of topography was highest on the moraine mounds and lowest on the adjacent areas, where accurate representation of topography was less important.

MORAINE MOUND FACIES

On the basis of shallow (<1.5 m) excavations and a single stream-cut section in the moraine-mound complex, ten lithofacies are defined. Moraine mound facies include: clast-poor intermediate diamicton, clast-rich sandy diamicton, clast-rich intermediate diamicton, clast-rich muddy diamicton, sandy gravel, two types of gravel, gravel with sand, sand and mud. For each facies, clast roundness data are shown in Fig. 2A, RA : C_{40} data are shown in Fig. 2B and a ternary diagram of sand, silt and clay components for each facies is shown in Fig. 2C. All facies lack internal sedimentary structure, with the exception of the sand and mud facies, which displays evidence of sedimentary structure including parallel lamination, cross lamination and bedding. An ice facies found buried within a single moraine mound is also identified. The buried ice facies exhibits mud laminae, clotted ice and clear to bubbly ice lamination (Fig. 3A). The mud laminae are less than 10 mm thick, intercalated with up to 15 mm thick layers of clear ice.

The clotted ice exhibits dispersed mud accumulations (<5 mm length) within a clean ice surround. The alternating clear and bubbly ice laminae are approximately 10 mm thick and the bubbles exhibit a-axis elongation.

INTERPRETATION OF MORAINE MOUND FACIES

The diamicton facies all display evidence of subglacial transport with dominant subangular and high subrounded clast categories developed by subglacial crushing and abrasion (Croot and Sims, 1996; Kjær, 1999). The wide range of particle-size fractions >0 Ø indicates an 'extremely poorly sorted' nature in common with basal abrasion products (Croot and Sims, 1996). RA : C_{40} analysis of the diamicton facies shows that the clasts are distinct from samples whose depositional process is known, such as scree, fractured bedrock, supraglacial and glaciofluvial clasts, but are comparable to the basal till clast sample. The diamicton facies are all interpreted as subglacial sediments. The position of the clast-poor intermediate diamicton adjacent to the buried ice, which has characteristics of basal glacier ice (discussed below), indicates that this facies represents a debris band associated with the basal ice layer. Debris incorporation into the basal ice layer can result from regelation, bulk freeze-on and thrusting of subglacial sediment (Alley *et al.*, 1998). Thrusting and folding can subsequently elevate this incorporated debris (Knight, 1997). The inward migration of the thermal boundary between outer cold-based ice and inner warm-based ice can also provide effective debris incorporation by freeze-on of subglacial sediment (Alley *et al.*, 1998). The longitudinal compression associated with the transitional from warm-based ice to cold-based ice may result in thrusting near the thermal boundary (Alley *et al.*, 1998). Thrusting developed in response to longitudinal compression may result in the incorporation of subglacial sediment, and also elevate debris within the basal ice layer.

The sandy gravel facies has clast characteristics, notably the predominance of subangular and subrounded clasts, which indicate evidence of subglacial transport. The RA : C_{40} plot shows that the clast component is indistinguishable from that of the diamicton facies interpreted as

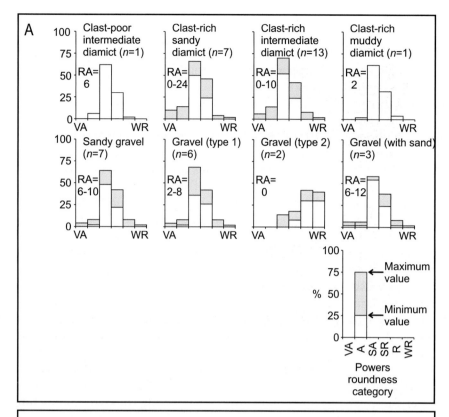

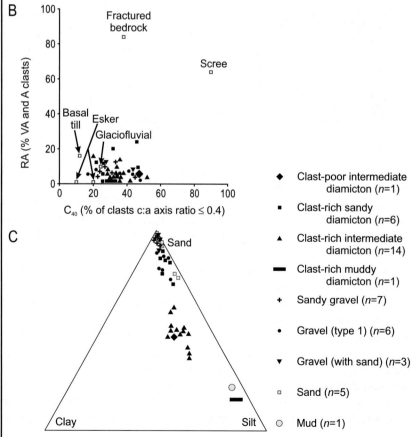

Fig. 2 (A) Co-variant plots of clast roundness from moraine mound samples (*n* = number of samples each consisting of 50 clasts). (B) Co-variant plot of RA (index of very angular and angular clasts) against C_{40} (index of clasts with c : a axis ratio ≤0.4) from moraine mounds and control samples (each symbol represents a 50 clast sample). (C) Ternary plot of sand, silt and clay components from moraine mound lithofacies samples.

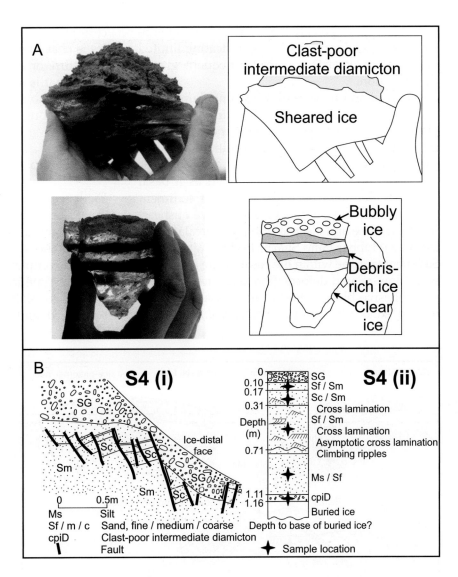

Fig. 3 (A) Photographs and interpretations of samples of buried ice facies obtained from excavation at Section S4, midre Lovénbreen. (B) Examples of sedimentary facies and structures in moraine mounds at midre Lovénbreen: S4i (transverse to ridge crest) showing normal faulting in bedded sands, and log of Section S4ii (longitudinal to ridge crest) showing the stratigraphy of the ridge crest.

subglacial in origin. The matrix component has a higher degree of sorting than the diamicton facies. The sandy gravel facies is interpreted as a glacio-fluvial deposit on the basis of the low silt and clay content. The greater matrix sorting indicates that this facies develops as a lag deposit from sub-glacial sediment with the fines content removed by fluvial action (Glasser and Hambrey, 2001). The proximity to the subglacial sediment source results in clast characteristics that are both inherited and indistinguishable from active subglacial transport (Bennett *et al.*, 1997).

The gravel (type 1) is interpreted as a high-energy glaciofluvial deposit because of size-selective sorting of clasts, stratification and a lack of matrix component. Clast roundness data indicates low

transport distance from the original basal sediment source, as clast roundness and RA : C_{40} are similar to those for basal glacial clasts. The gravel (type 2) is found in both the outer moraine-mound complex and emerging from englacial thrusts in one location at the current ice margin. This type of clast facies has previously been termed 'egg gravel' (Bennett *et al.*, 1997), which has been interpreted as a reworked raised beach deposit and also linked to thrusting in association with meltwater at locations above the former marine limit (Huddart *et al.*, 1998; Glasser *et al.*, 1999). The gravel (with sand) facies results from glaciofluvial deposition, as indicated by size-selective sorting of the clasts and a minor sand and mud component. The clast characteristics resemble those typical of basal glacial transport; notably the predominance of sub-angular and subrounded clasts, indicating limited modification of clasts by the subsequent glaciofluvial action.

The sand and mud exhibit bedding, lamination and a relatively high degree of sorting, indicating glaciofluvial deposition. The sand and mud result from low energy glaciofluvial deposition following short-distance transport.

The buried ice facies is interpreted as a basal ice layer identified by the presence of sediment laminae, anatomising ice layers indicative of shearing, clotted ice and clear to bubbly intercalated ice laminae (Hubbard and Sharp, 1995; Knight, 1997). Sediment laminae within the basal ice are associated with 'flow-related deformation and attenuation of non-cohesive blocks of [stratified basal ice facies termed] solid sub-facies ice' (Waller *et al.*, 2000). The

clear ice indicates evidence of regelation. The bubbly ice is characteristic of 'true' glacier ice originating from firn, or is the product of basal adfreezing of water that is supersaturated in gases. The bubbly ice also shows evidence of deformation, in the form of elongation of the bubbles.

SEDIMENT DEFORMATION

Evidence of faulting, folding and shearing of sediment slabs was observed in seven sections through moraine mounds (sections S1 to S7 are shown in Figs 3B and 4, within the context of the moraine-mound complex in Fig. 1C and in relation to each moraine type in Fig. 5).

Inclined slabs of sediment are illustrated in section S1 and are defined by a clast-rich sandy

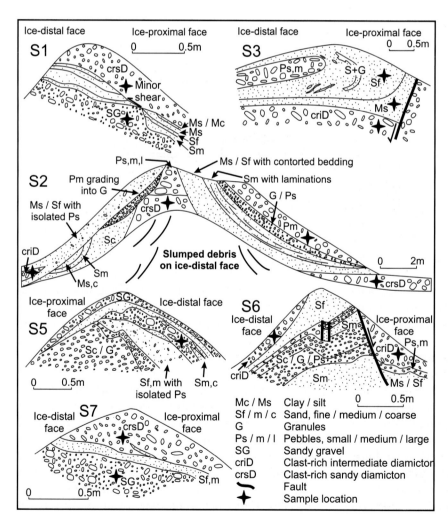

Fig. 4 Examples of sedimentary facies and structures in moraine mounds at midre Lovénbreen: section S1 showing tilted sedimentary slabs and minor shear of sand and mud facies, section S2 showing tilted sedimentary slabs of sands and gravels, section S3 showing normal faulting within the clast-rich intermediate diamicton, section S5 showing folding of sedimentary facies, section S6 showing faulted clast-rich intermediate diamicton against sands and gravels, section S7 showing tilted sand bed between sandy gravel and clast-rich sandy diamicton.

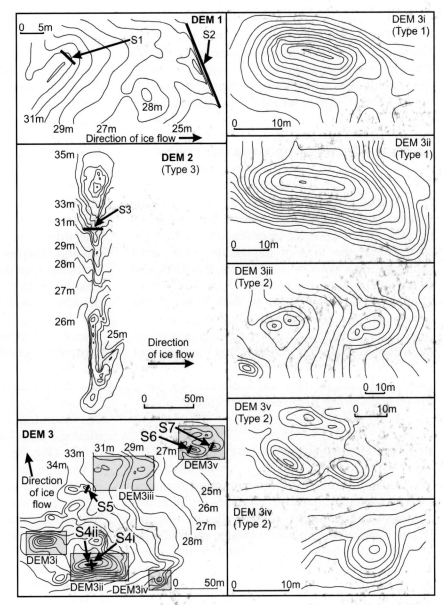

Fig. 5 Digital Elevation Models (DEMs) of moraine mounds in front of midre Lovénbreen. DEM 1 (*n* = 583) Type A (rectilinear ice-proximal face) moraine mounds. DEM 2 (*n* = 2014) Type E (highly elongated) moraine mound (0.5 m contour interval). DEM 3 (*n* = 5381), comprising DEM 3i Type A (rectilinear ice-proximal face) moraine mound; DEM 3ii Type B (curvilinear ice-proximal face) moraine mound (0.5 m contour interval); DEM 3iii Type C (low relief rounded mound) moraine mounds (0.5 m contour interval); DEM 3iv Type C (high relief rounded mound) moraine mounds (0.5 m contour interval); and DEM 3v Type D (convex-crested ridge) moraine mounds (0.5 m contour interval).

diamicton and sandy gravel, separated by sand and mud facies (Fig. 4). The strike and dip of the inclined sand and mud facies reflects the ice-proximal slope angle of 31° and moraine mound crest orientation of 307°. Inclined slabs of bedded sand and gravel are also found in section S2 (Fig. 4). The surface of the pebble gravel facies forms a rectilinear slope on the up-glacier side of the moraine mound. The base of section S3 shows a normal fault in a clast-rich intermediate diamicton (Fig. 4). Vertical displacement of 0.9 m is shown by the sharp contact, inclined at 80°, between the

clast-rich intermediate diamicton and the overlying silt facies. Normal faulting is seen in section S4i with beds of medium and coarse sand showing vertical displacement of 0.05 to 0.5 m (Fig. 3B). Below the bedded sand is a thin (~0.05 m) horizon of clast-poor intermediate diamicton and the buried ice facies of undetermined thickness, shown at the base of section S4ii (Fig. 3B). Folding of sedimentary structures is shown in section S5 resulting in an undulating bed surface (Fig. 4). Faulting is indicated by the displaced contact in section S6 of a clast-rich intermediate diamicton with bedded sands and gravels (Fig. 4). Section S7 shows an inclined bed of fine to medium sand between a sandy gravel and clast-rich sandy diamicton (Fig. 4). The strike of the sand bed is equivalent to the ridge crest orientation at 207°, but the dip of the bed is 15°, whereas the dip of the ice-proximal slope is 32°.

INTERPRETATION OF DEFORMATION STYLES

The normal faulting shown in sections S3, S4 and S6 is associated with extensional brittle deformation. Extensional deformation may have been achieved by ablation of buried ice, or of ice surrounding the sediment at the time of deposition. Folding of sediment in sections S5 and S6 may also be linked to the ablation of buried ice and minor slumping. Tilting of sediment slabs is indicated by the orientation of original depositional structures, as in sections S1, S2 and S7. The retention of original depositional structures indicates that the sediment slabs were highly competent during displacement, and could indicate that the sediment was in a frozen state whilst displacement occurred.

MORAINE MOUND MORPHOLOGY

The midre Lovénbreen moraine-mound complex forms an arcuate band of moraines terminating around 1 km from the current glacier margin. Detailed characterisation of moraine morphology – using contoured Digital Elevation Models – was undertaken along a transect from the glacier margin to the outer moraine-mound complex. Three broad moraine mound morphological types are characterised and include: rectilinear and curvilinear ice-proximal face ridge, rounded mounds and ridges and a highly elongated ridge. The coverage within the moraine-mound complex of each Digital Elevation Model is shown in Fig. 1C.

Rectilinear and curvilinear ice-proximal face (n = 4)

The rectilinear ridge type is characterised by a rectilinear ice-proximal face. The ice-proximal faces have slope angles of 31, 33 and 40°, ridge-crest orientations that vary from 237 to 307° and have sharp ridge crest between 8 m and 14 m in length (Fig. 5). Active slumping occurs on parts of upper ice-distal faces in association with steep upper slope angles (~48°), and slumped sediment accumulates on gently inclined lower slopes (~30°). Sediment facies incorporated in this moraine type include clast-rich sandy diamicton, clast-rich intermediate diamicton, gravel (type 1), sandy gravel, sand and mud. Structural characteristics include tilting of intact sediment slabs and extensional shearing demonstrated in sections S1 and S2. The curvilinear ridge type possesses a curving rectilinear ice-proximal slope of 34° and a more irregular ice-distal slope of around 25° (Fig. 5). The ridge-crest orientation varies along the ridge from 240 to 220°. The ridge crest has a rounded longitudinal profile and tapering ridge-crest termination. Sediment facies incorporated within this moraine type include clast-poor intermediate diamicton, clast-rich sandy diamicton, sandy gravel, gravel with sand, sand and the buried basal ice facies. Structural characteristics found in this moraine type are normal faults, shown in section S4.

Rounded mound and ridge (n = 8)

The rounded mound type has a conical morphology with low elongation of the longitudinal profile (Fig. 5). The relief of the rounded mound is variable and ranges from 1 to 4 m. Sediments incorporated in this moraine type include gravel (type 1), gravel (type 2) and gravel with sand. Limited excavation did not reveal any structural characteristics within this moraine mound type. The rounded ridge type exhibits variable longitudinal section slope angles, resulting in a rounded convex longitudinal profile. Ridge-crest orientation varies from 207 to 258° and relief varies from 2 m to 4 m (Fig. 5). Ice-proximal slopes vary from 17 to 33° and ice-distal slopes vary from 25 to 37°. In all examples, the ice-proximal slope is at a shallower angle than the ice-distal slope. Sediments incorporated in this moraine type include clast-rich sandy diamicton, clast-rich intermediate diamicton, gravel (type 1), sandy gravel, sand and mud. Structural characteristics of this moraine type include faulting and folding, shown in sections S5, S6 and S7.

Highly elongated ridge (n = 1)

This ridge type has a longitudinal profile of 290 m with an overall orientation of 255°. The ridge is around 15 to 20 m across, and along the length of the ridge a number of local topographic high-points give heights between 2.5 and 5 m above the surrounding area (Fig. 5). The ridge is somewhat dissected around midway along the ridge, as the height above the surrounding area at this point is only around 0.5 m. A subsidiary ridge 90 m in length with a crest-line orientation of 290° intersects the main ridge at the east end. Sediments incorporated in this moraine type include clast-rich sandy diamicton, clast-rich intermediate diamicton, clast-rich muddy diamicton, gravel (type 1), sand and mud. Structural characteristics of this moraine type include normal faulting, shown in section S3.

INTERPRETATION OF MORAINE MOUNDS

Rectilinear and curvilinear ice-proximal face

The inclusion of sandy gravel and diamicton within this moraine mound type indicates the elevation of both glaciofluvial sediments that have been overridden during glacier advance and subglacial debris. This moraine mound type also shows a close association between morphology and internal structural character. Sections S1 and S2 show moraine mound development by inclusion of inclined sediment slabs that are reflected in the rectilinear ice-proximal face. The identification of rectilinear ice-proximal faces found on this moraine mound type has been suggested to mirror the orientation of linear englacial thrusts (Bennett *et al.*, 1999). The inclusion of inclined sediment slabs that are relatively undeformed probably indicates a subglacial freeze-on mechanism and elevation above an englacial thrust plane, rather than actually along the thrust plane (Bennett *et al.*, 1996; Hambrey *et al.*, 1996).

The characteristic curvilinear ice-proximal face of this moraine mound morphology is associated with arcuate englacial thrusting of sediment (Bennett *et al.*, 1999). The *in situ* fluvial deposition along the ridge crest indicates the former presence of ice surrounding the ridge with ablation of the surrounding ice resulting in topographic inversion (Boulton, 1972). The ablation of the buried basal ice to give faulting and minor slumping of the ice-distal face indicates that future ablation of buried basal ice may further alter the moraine mound

morphology. Preservation of moraine mound ice-cores in permafrost regions such as Svalbard results from the limited depth of the active layer in summer (e.g. Sollid and Sørbel, 1988). Any subsequent amelioration in climate that extends the summer active layer beyond the protective debris layer over the ice-core will increase ablation of the buried ice and modify the present moraine mound morphology.

Rounded mound and ridge

The low-relief rounded mounds contain facies that display evidence of glaciofluvial deposition, whereas the high-relief rounded mounds consist of diamicton facies that indicate a subglacial origin. The low-relief rounded mounds are interpreted as ice-marginal glaciofluvial features. Mounds and ridges of fluvial sediment are developed by a process of topographic inversion by deposition around ablating glacier ice (Boulton, 1972). The low-relief rounded mounds may therefore have a similar mode of formation to the convex-crested ridges, which also show evidence of fluvial deposition. The high-relief rounded mounds include diamicton facies that show evidence of subglacial transport, and are interpreted as part of a continuum of moraine mound morphologies related to the incorporation and elevation of basal sediment by thrusting.

The convex-crested moraine mounds show evidence of glaciofluvial deposition through the inclusion of bedded sediments that exhibit limited sorting. Boulton (1972) suggests that moraines that include bedded and sorted sediments can indicate an englacial, subglacial or ice-marginal position of formation, in associated with the meltout of glacier ice. The presence of buried ice would be indicative of an englacial origin according to the Boulton (1972) model, but was not identified in excavations of around 1 m depth through this moraine mound type at midre Lovénbreen. Section S6 shows a faulted drape of clast-rich intermediate diamicton interpreted as a subglacial facies adjacent to fluvial sediments. This could indicate development in a subglacial position or an ice-marginal position by a process of meltout of an elevated subglacial facies from the ablating glacier ice (Boulton, 1972). An englacial debris component of the glacier ice is suggested by evidence of topographic inversion with restricted ablation and down-wasting of the glacier ice. Section S5 shows an absence of faulting, but includes limited folding of bedded sediments. This moraine mound is more indicative of formation in an ice-marginal position, since incorporated subglacial sediment is absent. Subglacial and englacial formation of this moraine mound is less likely because such conduits commonly form long sinuous ridges, whereas these moraine mounds are only around 5 to 10 m in length (e.g. Huddart *et al.*, 1999). Overall, this moraine mound type is likely to indicate ice-marginal deposition.

Highly elongated ridge

The highly elongated ridge includes diamicton facies of subglacial origin and gravel, sand and mud that show evidence of glaciofluvial deposition. The normal fault in section S3 indicates that the diamicton may overlie buried ice that has experienced ablation. Sedimentological characteristics of each peak are varied and evidence of *in situ* fluvial deposition is found in the form of sand and silt along the ridge crest. This indicates fluvial activity confined by the ablating glacier ice, the recession of which may have developed the extensional faulting.

LANDFORM DEVELOPMENT BY ENGLACIAL THRUSTING AT CONTEMPORARY AND FORMER GLACIERS

This work demonstrates that a high englacial debris load can contribute directly to moraine development by the meltout of debris that has been incorporated subglacially and elevated to an englacial position. This high debris load can also contribute to moraine development by restricting ablation at the ice margin, which results in glaciofluvial deposition with subsequent topographic inversion following ablation of the surrounding ice. The identification of these diagnostic landforms may also provide evidence of englacial thrusting in former glaciated areas.

However, the long-term preservation potential of moraine mounds formed through the process of englacial thrusting is questioned by the identification of a buried ice component within the curvilinear ice-proximal face moraine mound at midre

Lovénbreen. The identification of buried ice is not isolated to midre Lovénbreen, as it has also been identified below thrust sediments at other moraine mound sites in Svalbard (Bennett *et al.*, 1999; Sletten *et al.*, 2001). Clean ice rapidly ablates, whereas ice covered by a thin sediment layer can experience little or no ablation (under current climatic conditions in Svalbard) and will therefore form a stable landform component until a stage where climatic amelioration removes permafrost conditions. The importance of buried ice included in these moraine mounds is related to the characteristics of the original englacial thrust. These characteristics include englacial thrust spacing and angle, content and thickness of incorporated debris slabs and the amount of elevation the debris slab experiences (Bennett, 2001).

The issue of incorporated ice within the Svalbard moraine mounds has important implications for moraine mounds development by englacial thrusting at formerly glaciated sites because complete ablation of any incorporated ice will have occurred. Using the Svalbard englacial thrusting model, a number of British Younger Dryas examples have already been suggested (Bennett *et al.*, 1998; Graham and Midgley, 2000; Midgley, 2001; Graham, 2002).

CONCLUSIONS

This comprehensive investigation of the morphology, sedimentology and structure of the moraine-mound complex in the proglacial area of a land-based, high-Arctic polythermal glacier has yielded information that may be applicable to many comparable glaciological settings:

1 Ten lithofacies, ranging from mud to well-sorted gravel, are identified within the moraine-mound complex, of which the clast-rich intermediate diamicton is dominant.
2 A buried basal ice facies has been identified within a single moraine mound, which demonstrates that the morphology is likely to change under conditions of climatic amelioration.
3 A range of deformation styles is revealed in excavations through the crests of moraine mounds including normal faulting, folding, tilting of intact sediment slabs and extensional deformation.

4 Three broad moraine mound morphological types are characterised that demonstrate development associated with englacial thrusting and glaciofluvial deposition at the ice margin and reflect the high englacial debris load.

5 The application of the englacial thrusting model to formerly glaciated areas should be undertaken with care until the full extent of buried ice at contemporary sites, and the role that ablation of this ice plays in landform modification, have been determined.

ACKNOWLEDGEMENTS

This research was undertaken whilst NGM was in receipt of a studentship at Liverpool John Moores University, during which he received funding for fieldwork from the Quaternary Research Association and British Geomorphological Research Group. This work represents a spin-off from a larger scale glaciological project funded by the UK Natural Environment Research Council under the ARCICE Thematic Programme (Grant number GST022192). Nick Cox, Manager of the NERC Arctic Research Station in Ny-Ålesund is thanked for logistical support. David Graham, Richard Matthews, Andrew Mellor and Phillippa Noble are thanked for assistance in the field. Matthew Bennett and Kurt Kjær are also thanked for providing constructive comments on an earlier version of this work, along with two anonymous referees.

REFERENCES

Alley, R.B., Lawson, D.E., Evenson, E.B., Strasser, J.C. and Lawson, D.E. (1998) Glaciohydraulic supercooling: a freeze-on mechanism to create stratified, debris-rich ice: II. Theory. *J. Glaciol.*, **44**, 563–569.

Benn, D.I. and Ballantyne, C.K. (1994) Reconstructing the transport history of glacigenic sediments: a new approach based on the co-variance of clast form indices. *Sed. Geol.*, **91**, 215–227.

Bennett, M.R. (2001) The morphology, structural evolution and significance of push moraines. *Earth-Sci. Rev.*, **53**, 197–236.

Bennett, M.R., Huddart, D., Hambrey, M.J. and Ghienne, J.F. (1996) Moraine development at the high-arctic valley glacier Pedersenbreen, Svalbard. *Geogr. Ann.*, **78A**, 209–222.

Bennett, M.R., Hambrey, M.J. and Huddart, D. (1997) Modification of clast shape in High-Arctic glacial environments. *J. Sed. Petrol.*, **67**, 550–559.

Bennett, M.R., Hambrey, M.J., Huddart, D. and Glasser, N.F. (1998) Glacial thrusting and moraine-mound formation in Svalbard and Britain: the example of Coire a' Cheud-chnoic (Valley of Hundred Hills), Torridon Scotland. *Quaternary Proceedings*, **6**, 17–34.

Bennett, M.R., Hambrey, M.J., Huddart, D., Glasser, N.F. and Crawford, K. (1999) The landform and sediment assemblage produced by a tidewater glacier surge in Kongsfjorden, Svalbard. *Quatern. Sci. Rev.*, **18**, 1213–1246.

Björnsson, H., Gjessing, Y., Hamran, S-E., Hagen, J.O., Liestøl, O., Pálsson, F. and Erlingsson, B. (1996) The thermal regime of sub-polar glaciers mapped by multi-frequency radio-echo sounding. *J. Glaciol.*, **42**, 23–32.

Boulton, G.S. (1972) Modern Arctic glaciers as depositional models for former ice sheets. *J. Geol. Soc. London*, **128**, 361–393.

Croot, D.G. and Sims, P.C. (1996) Early stages of till genesis: an example from Fanore, County Clare, Ireland. *Boreas*, **25**, 37–46.

Evans, D.J.A. (2000) Glaciers. *Prog. Phys. Geogr.*, **24**, 579–589.

Glasser, N.F. and Hambrey, M.J. (2001) Styles of sedimentation beneath Svalbard valley glaciers under changing dynamic and thermal regimes. *J. Geol. Soc. London*, **158**, 697–707.

Glasser, N.F., Hambrey, M.J., Crawford, K., Bennett, M.R. and Huddart, D. (1998) The structural glaciology of Kongsvegen, Svalbard and its role in landform genesis. *J. Glaciol.*, **44**, 136–148.

Glasser, N.F., Bennett, M.R. and Huddart, D. (1999) Distribution of glaciofluvial sediment within and on the surface of a High Arctic valley glacier: Marthabreen, Svalbard. *Earth Surf. Proc. Land.*, **24**, 303–318.

Glasser, N.F., Hambrey, M.J., Bennett, M.R. and Huddart, D. (2003) Comment: Formation and reorientation of structure in the surge-type glacier Kongsvegen, Svalbard. *J. Quatern. Sci.*, **18**, 95–97.

Graham, D.J. (2002) Moraine mound formation during the Younger Dryas in Britain and the Neoglacial in Svalbard. Unpublished PhD thesis, University of Wales.

Graham, D.J. and Midgley, N.G. (2000) Moraine-mound formation by englacial thrusting: the Younger Dryas moraines of Cwm Idwal, North Wales. In: Maltman, A.J., Hubbard, B. and Hambrey, M.J. (Eds) Deformation of Glacial Materials. *The Geological Society of London, Special Publications*, **176**, 321–336.

Haeberli, W., Frauenfelder, R., Hoelzle, M. and Zemp, M. (2003) Glacier Mass Balance Bulletin No. **7**, *World Glacier Monitoring Service*, Zurich, 94 pp.

Hambrey, M.J. 1994 *Glacial Environments*. University College Press, London, 296 pp.

Hambrey, M.J. and Glasser, N.F. (2003) The role of folding and foliation development in the genesis of medial moraines: examples from Svalbard glaciers. *J. Geol.*, **111**, 471–485.

Hambrey, M.J. and Huddart, D. (1995) Englacial and proglacial glaciotectonic processes at the snout of a thermally complex glacier in Svalbard. *J. Quatern. Sci.*, **10**, 313–326.

Hambrey, M.J., Dowdeswell, J.A., Murray, T. and Porter, P.R. (1996) Thrusting and debris entrainment in a surging glacier: Bakaninbreen, Svalbard. *Ann. Glaciol.*, **22**, 241–248.

Hambrey, M.J., Bennett, M.R., Huddart, D. and Glasser, N.F. (1997) Genesis of 'hummocky moraine' by thrusting in glacier ice: evidence from Svalbard and Britain. *J. Geol. Soc. London*, **154**, 623–632.

Hambrey, M.J., Murray, T., Glasser, N.F., Hubbard, A., Hubbard, B., Stuart, G., Hansen, S. and Kohler, J. (2005) Structure and changing dynamics of a polythermal valley glacier on a centennial time-scale: midre Lovénbreen, Svalbard. *J. Geophys. Res., Earth Surface.* F010006, doi: 10.1029/2004JF000128.

Hansen, S. (2003) From surge-type to non-surge-type glacier behaviour: midre Lovénbreen, Svalbard. *Ann. Glaciol.*, **36**, 97–102.

Harland, W.B. (1997) *The Geology of Svalbard*. The Geological Society of London, Memoir **17**.

Hjelle, A. (1993) *Geology of Svalbard*. Norsk Polarinstitutt Polarhåndbok 7. Norsk Polarinstitutt, Oslo.

Hubbard, B. and Sharp, M.J. (1995) Basal ice facies and their formation in the Western Alps. *Arctic Alpine Res.*, **27**, 301–310.

Huddart, D. and Hambrey, M.J. (1996) Sedimentary and tectonic development of a high-arctic, thrust-moraine complex: Comfortlessbreen, Svalbard. *Boreas*, **25**, 227–243.

Huddart, D., Bennett, M.R., Hambrey, M.J., Glasser, N.F. and Crawford, K.R. (1998) Origin of well rounded gravels in glacial deposits from Brøggerhalvøya,

northwest Spitsbergen: potential problems caused by sediment reworking in the glacial environment. *Polar Res.*, **17**, 61–69.

Huddart, D., Bennett, M.R. and Glasser, N.F. (1999) Morphology and sedimentology of a high-arctic esker system: Vegbreen, Svalbard. *Boreas*, **28**, 253–273.

Jiskoot, H., Murray, T. and Boyle, P. (2000) Controls on the distribution of surge-type glaciers in Svalbard. *J. Glaciol.*, **46**, 412–422.

Kjær, K.H. (1999) Mode of subglacial transport deduced from till properties, Mýrdalsjökull, Iceland. *Sed. Geol.*, **128**, 271–292.

Knight, P.G. (1997) The basal ice layer of glaciers and ice sheets. *Quatern. Sci. Rev.*, **16**, 975–993.

Lukas, S. (2005) A test of the englacial thrusting hypothesis of 'hummocky' moraine formation: case studies from the northwest Highlands, Scotland. *Boreas*, **34**, 287–307.

Midgley, N.G. (2001) Moraine-Mound Development in Britain and Svalbard – The Development of 'Hummocky Moraine'. Unpublished PhD thesis, Liverpool John Moores University.

Powers, M.C. (1953) A new roundness scale for sedimentary particles. *J. Sed. Petrol.*, **23**, 117–119.

Sollid, J.L. and Sørbel, L. (1988) Influence of temperature conditions in formation of end moraines in Fennoscandia and Svalbard. *Boreas*, **17**, 553–558.

Sletten, K., Lyså, A. and Lønne, I. (2001) Formation and disintegration of a high-arctic ice-cored moraine complex, Scott Turnerbreen, Svalbard. *Boreas*, **30**, 272–284.

Waller, R.I., Hart, J.K. and Knight, P.G. (2000) The influence of tectonic deformation on facies variability in stratified debris-rich basal ice. *Quatern. Sci. Rev.*, **19**, 775–786.

Woodward, J., Murray, T. and McCaig, A. (2002) Formation and reorientation of structure in the surge-type glacier Kongsvegen, Svalbard. *J. Quatern. Sci.*, **17**, 201–209.

Woodward, J., Murray, T. and McCaig, A. (2003) Reply: Formation and reorientation of structure in the surge-type glacier Kongsvegen, Svalbard. *J. Quatern. Sci.*, **18**, 99–100.

A new laboratory apparatus for investigating clast ploughing

MARIE ROUSSELOT, URS H. FISCHER *and* MICHAEL PFISTER

Laboratory of Hydraulics, Hydrology and Glaciology, ETH Zentrum, CH-8092 Zürich, Switzerland (e-mail: m.rousselot@caramail.com)

ABSTRACT

A significant portion of the basal motion of soft-bedded glaciers can be attributed to 'ploughing'. This term designates the transitional state between sliding and bed deformation which occurs when clasts that protrude into the glacier sole are dragged through the upper layer of the sediment. This process may cause pore pressures in excess of the hydrostatic value that could weaken the sediment downglacier from ploughing clasts and thus affect the strength of the ice–bed coupling. A large laboratory apparatus was developed and constructed to study, systematically and under glacially relevant conditions, the influence of sediment properties, ploughing velocity and effective pressure on excess pore-pressure generation and sediment strength. In this device, an instrumented tip is dragged at different velocities through a water-saturated sediment bed subject to different effective normal stresses. The drag force on the tip and the pore-water pressure in the adjacent sediment are measured simultaneously. In preliminary experiments performed with subglacial sediment from Unteraargletscher, Switzerland, the sediment diffusivity was estimated from consolidation records. During ploughing, pore-pressure gradients developed rapidly around the tip. Excess pore pressures were due to sediment compression in front of the tip whereas pore pressures below the hydrostatic value resulted from dilatant shearing and a wake devoid of sediment that was left behind the tip. A zone of compressed sediment formed in front of the tip. The absolute magnitude of the pore pressure changes was small relative to the effective normal stress, so that the pore pressures did not significantly influence the resistance to ploughing. Rather, the drag force on the ploughing tip was influenced by the properties of the sediment in front of the compression zone, with a greater magnitude in a virgin sediment than in one that has been ploughed before.

Keywords Glacier, ice–bed coupling, laboratory device, ploughing, pore water, subglacial sediment.

INTRODUCTION

For the last 20 years, much research has been directed toward improving our understanding of the mechanisms of basal motion beneath soft-bedded glaciers and their implications for ice flow dynamics. In particular, ice flow instabilities such as glacier surging or fast ice streaming may be caused by water-saturated and highly pressurized subglacial sediment that lubricates the glacier base (e.g. Clarke, 1987; Raymond, 1987; Alley, 1989). The extent to which bed deformation contributes to basal motion depends primarily on the rheology of subglacial sediment and the amount of shear stress transferred across the ice–bed interface.

However, the factors that control the behavior of sediment-water mixtures and the strength of ice–bed coupling have not yet been fully elucidated.

Field (e.g. Iverson et al., 1995; Hooke et al., 1997; Fischer et al., 2001) and laboratory studies (e.g. Iverson et al., 1998; Tulaczyk et al., 2000; Kamb, 2001) suggest that subglacial sediment behaves essentially as a Coulomb-plastic material, a finding which is in agreement with classical soil mechanics (e.g. Lambe & Whitman, 1979). However, transient pore-water flow induced by sediment deformation may complicate the sediment behaviour. A case of special interest arises when the sediment is deformed by ploughing (Brown et al., 1987; Alley, 1989), when clasts that protrude into the glacier sole are dragged through the upper layer of the bed. Iverson (1999) suggested that the compression of sediment in front of ploughing clasts may lead to the generation of excess pore-water pressures that have the potential to substantially weaken the sediment. By reducing ice–bed coupling, ploughing may thus have fundamental implications for glacier basal motion (Iverson *et al.*, 1994; Iverson, 1999; Fischer *et al.*, 2001; Hooyer & Iverson, 2002; Rousselot & Fischer, 2005). However, a quantitative description of this mechanism in glacier flow models is hampered by the paucity of data. In situ studies of the force on objects dragged through subglacial sediment are difficult to perform and the interpretation of the results is complicated by the variability of parameters such as subglacial water pressure, sediment texture and sliding velocity. Also, results from laboratory studies of the drag force on objects moving through granular materials (e.g. Wieghardt, 1975; Albert *et al.*, 1999; Chehata *et al.*, 2003; Zhou & Advani, 2004) are not directly applicable to clast ploughing because the experiments were performed at velocities greater than typical glacial velocities and using dry materials, with grains smaller than a few millimeters. Thomason & Iverson (2003) dragged hemispheres through water-saturated sediment using a ring-shear device. The authors reported a decrease in the drag force on the hemispheres with increasing velocities as a consequence of excess pore-pressure generation.

To further investigate clast ploughing and its influence on ice–bed coupling, a new laboratory device, the 'rotary ploughing device', was designed and constructed. This device is used to conduct experiments by dragging an instrumented tip through sediment under glacially relevant conditions. The aim is to systematically investigate the effects of parameters such as ploughing velocity and effective pressure on excess pore-pressure generation and sediment strength. In the present paper, the rotary ploughing device and the experimental procedure are described in detail and preliminary results of ploughing experiments conducted using sediments from Unteraargletscher, Switzerland, are discussed.

APPARATUS

The device consists of a cylinder containing the sediment, a combined lever arm/pulley system

for applying a stress to the sediment normal to the ploughing direction and a drive mechanism for dragging an instrumented object through the sediment (Fig. 1A). These components are supported by a frame of 2.6 m length, 1.7 m width and 3 m height that rests on a base slab of concrete.

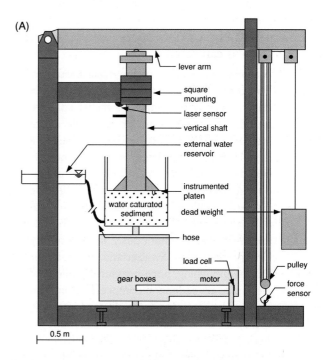

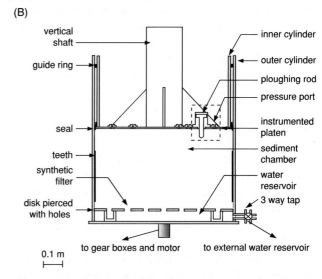

Fig. 1 Schematic diagrams of the rotary ploughing device. (A) Side view showing the support frame (dark grey), lever arm/pulley system (mid-tone grey), drive mechanism (light grey) and cylinder containing the sediment. (B) Details of the sediment chamber. A detailed view of the dashed box is shown in Fig. 2A.

The steel cylinder of 0.58 m diameter, 0.7 m height and 6 mm wall thickness is divided into a sediment chamber and a water reservoir by a steel disc welded 50 mm from the bottom of the cylinder (Fig. 1B). The disc is pierced with 8 mm diameter holes and covered with a synthetic filter to enable water-flow between the chamber and the reservoir. The water-saturated sediment contained in the sediment chamber is gripped laterally by teeth. Its initial thickness can reach 0.3 m, corresponding to a volume of about 0.08 m³, which is significantly larger than in laboratory devices commonly used in soil testing (e.g. Bowles, 1978).

The sediment is subjected to a normal stress by suspending dead weights from a lever arm by a system of pulleys. The downward force acting on the lever arm is recorded by a force sensor (Interface SSM-AJ, 20 kN) (Fig. 1A) and is transmitted to a platen of 0.565 m diameter and 8 mm thickness through a vertical shaft that sits in a square mounting to inhibit rotation while still allowing vertical motion. With this system, the sediment can be subjected to uniform normal stresses up to 400 kPa, a range that is typical for effective stresses under glaciers. The applied normal stress is known with an accuracy of ±3 kPa.

To prevent the platen from jamming, it is mounted to the bottom of a 0.3 m long and 0.55 m diameter steel cylinder which helps guide it into the sediment chamber (similar to a piston). A seal at the base of this cylinder prevents sediment and water loss during an experiment (Fig. 1B).

The platen is free to move vertically to accommodate changes in sediment volume, ensuring a constant normal stress during an experiment. Its vertical displacement is recorded by a laser sensor (Baumer electric, OADM2016441/S14F, 30–70 mm) installed on the square mounting, that measures the distance to a horizontal surface fixed on the vertical shaft with an accuracy of ±0.06 mm (Fig. 2A).

The water reservoir beneath the sediment chamber is connected via a hose and a 3-way tap to an external water reservoir that is open to the atmosphere (Fig. 1A). If the tap is open toward the external reservoir, water can flow out of or into the sediment chamber through the intermediate bottom, maintaining drained conditions during an experiment. In this case, the sediment pore water is open to the atmosphere, and the normal stress

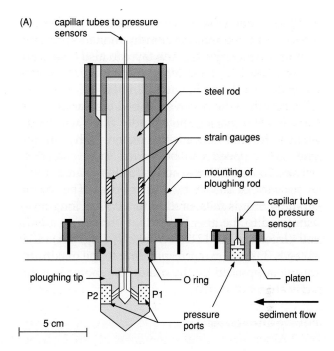

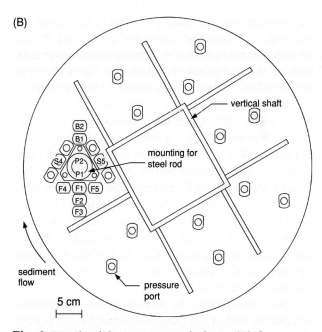

Fig. 2 Details of the instrumented platen. (A) Cross sectional view of the ploughing tip and a pressure port on the platen. (B) Map view showing the positions of the pressure ports on the platen. During ploughing experiments, the pore-water pressure is measured at the pressure ports labeled P1, P2 (on the tip) and F1, F2, F3, F4, F5, B1, B2, S4, S5 (on the platen), while the other pressure ports are closed hermetically by a plastic cap.

applied on the sediment corresponds to the effective stress, in the absence of non-hydrostatic pressure induced by ploughing. The tap can also be closed, or connected to a waste pipe to drain the water contained in the sediment chamber.

To simulate the relative ice–bed displacement at the base of a glacier, the cylinder filled with sediment is rotated beneath the platen with a motor. The motor speed can be varied by a computer between 0.7 – 5000 rpm and is further geared down by a factor of 463699 by 2 gear boxes. The torque on the motor is measured with a 0.97 m long lever arm fixed to the gear boxes that presses on a load cell (Interface M 1211-EX, 50 kN) (Fig. 1A). If it exceeds 33.6 kN m, corresponding to the maximum torque supported by the motor, the motor is automatically shut down.

Instrumented platen

A 140 mm long and 32 mm diameter steel rod terminated by a conical tip is set 110 mm from the cylinder wall into the platen and protrudes 40 mm into the sediment (Fig. 2). In response to the rotation of the sediment chamber, the tip is dragged through the sediment, following a circular path of 1.1 m circumference for a full rotation. Its relative velocity can be adjusted between 2 mm day^{-1} and 16.6 m day^{-1}, values that cover typical glacial velocities. The ploughing rod is permitted to flex along its entire length below the point at which it is held at the top of its mount. The elastic bending of the rod is measured with strain gauges (Fig. 2A) and converted into a force and azimuth with the use of a laboratory calibration, following the description of Fischer & Clarke (1994). The force applied on the ploughing tip is known with an accuracy of ±150 N.

Pore pressures in the sediment 20 mm beneath the platen and at the platen-sediment interface are measured through pressure ports situated on the lee and stoss sides of the tip and on the platen, respectively (Fig. 2). These pressure ports are connected to pressure sensors (Keller 23S, 50 kPa) via capillar tubes. Others not used for measurements are closed hermetically by plastic caps. Pore-water pressures are expressed with an accuracy of ±50 Pa relative to the hydrostatic pressure corresponding to the water level in the external reservoir (Fig. 1A).

SEDIMENT PROPERTIES

Experiments with the rotary ploughing device were performed using subglacial sediment from the recently deglaciated forefield of Unteraargletscher, Switzerland. To ensure collection of sediment as unaltered as possible by wind erosion or winnowing by water, a 3 cm thick layer was first removed from the ground surface. The Unteraargletscher sediment is coarse grained, and made up of 2% clay, 10% silt, 58% sand and 30% gravel (Fig. 3). During ploughing, the magnitude of the force on the ploughing object and the tendency for excess pore-pressure generation

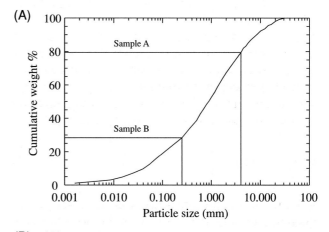

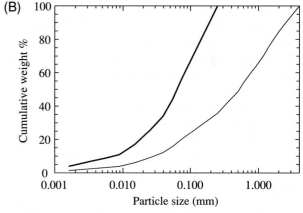

Fig. 3 (A) Grain-size distribution of Unteraargletscher sediment. Sample A, which was obtained by removing the particles larger than 4 mm from Unteraargletscher sediment, was used in the oedometer and triaxial tests. Sample B, which was obtained by removing particles larger than 0.25 mm from Unteraargletscher sediment, was used in the ploughing experiments. (B) Grain-size distribution of Sample A (thin line) and Sample B (thick line).

is influenced by the hydromechanical properties of the sediment. Therefore, the compressive, diffusive and frictional characteristics of the Unteraargletscher sediment were investigated by means of a series of consolidation, permeability and triaxial tests (standard procedures are described in Bowles, 1978). Because of the small dimensions of typical laboratory apparatus, the tests were conducted on remoulded samples of grain size <4 mm, a fraction representing about 80% of the total weight (Sample A in Fig. 3).

Compression test

Sediment compressibility depends on the stress magnitude and on the stress history (Lambe & Whitman, 1979). Thus, a sediment can be either in a virgin or normally consolidated state if it has never experienced a normal stress larger than the current one, or in an overconsolidated state, if it has been subjected to larger normal stresses in the past. For a sediment subjected to loading and unloading, the changes in void ratio e can be approximated by

$$e = e_0 - C \log\left(\frac{\sigma_n}{\sigma_0}\right), \qquad (1)$$

where e_0 and σ_0 are the initial void ratio and initial normal stress, respectively, σ_n is the normal stress and C is the dimensionless compression index, with subscript c for a normally consolidated sediment and s for an overconsolidated sediment. Hence, on a $e - \log \sigma_n$ diagram, C_c is the slope of the normal consolidation line (NCL) related to the virgin state, and C_s is the slope of the overconsolidation lines (OCL), which are the branches originating from the NCL (e.g. Lambe & Whitman, 1979).

The compressibility characteristics of Unteraargletscher sediment were determined by a confined consolidation test performed with an oedometer on a sediment sample of initial void ratio $e_0 = 0.35$. Results from this test and the corresponding best fit NCL and OCL are shown in Figure 4. The values of $C_c = 0.022$ for $25 < \sigma_n < 200$ kPa and $C_c = 0.033$ for $200 < \sigma_n < 300$ kPa were determined by locally fitting straight lines to the data points and $C_s = 0.004$ is the mean of OCL$_1$ and OCL$_2$. These values are characteristic for sediments of low compressiblity and typical for fine sands (Mitchell, 1993).

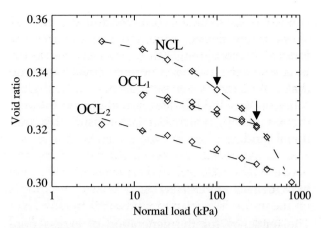

Fig. 4 Compressibility of Unteraargletscher sediment in normally consolidated (NCL) and overconsolidated (OCL) states derived from tests performed with the oedometer.

Two falling head permeameter tests were combined with the consolidation test at different normal stresses to determine the hydraulic conductivity k at the corresponding void ratios (see arrows in Figure 4). The tests yield $k = 8.4 \times 10^{-7}$ m s^{-1} for $e = 0.33$ and $k = 6.9 \times 10^{-7}$ m s^{-1} for $e = 0.32$. These values are typical for tills (10^{-12}–10^{-6} m s^{-1}; Freeze & Cherry, 1979).

Values of D were computed from values of C and k following

$$D = \frac{k(1 + e_0)\sigma_m}{0.435\rho_w g C}, \qquad (2)$$

(e.g. Lambe & Whitman, 1979) where σ_m is the value of the mean normal stress for which C has been determined, ρ_w is the density of water and g is the acceleration due to gravity. Values of $D = 1.2 \times 10^{-3}$ to 1.9×10^{-3} m^2 s^{-1} were found for void ratios between $e = 0.32$ and $e = 0.33$. These values are well above those previously reported for Unteraargletscher sediment (10^{-6} m^2 s^{-1}; Fischer *et al.*, 2001), suggesting that the sediment from the forefield has been partially winnowed by water.

Triaxial tests

Following Coulomb's theory (e.g. Lambe & Whitman, 1979), the sediment shear strength at failure τ_f is expressed as

$$\tau_f = c + \sigma' \tan \phi_r, \qquad (3)$$

where c is the cohesion, σ' is the effective normal stress on the failure plane and ϕ_r is the angle of internal friction. Results from two triaxial drained tests and two triaxial undrained tests indicate that Unteraargletscher sediment has a negligible cohesion ($c = 10$ kPa) and an angle of internal friction of 39°, which is characteristic of a frictional sand.

PLOUGHING EXPERIMENTS

The tendency for the generation of excess pore pressure downglacier from ploughing objects can be assessed by considering a dimensionless parameter R

$$R = \frac{D}{v\delta'} \qquad (4)$$

(adapted from Iverson & LaHusen, 1989; Iverson *et al.*, 1994; Fischer *et al.*, 2001) where v is the velocity of the ploughing object (i.e. the ploughing tip) through the sediment and δ is the characteristic length of the sediment compressed in front of the tip. This parameter relates the time-scale for excess pore-pressure generation (δ/v) to the time-scale for diffusive pore-pressure equilibration across δ (δ^2/D). Using typical subglacial ploughing velocities, a compression length of one tip diameter and the diffusivity determined above for Unteraargletscher sediment, we find $R \sim 10^5$, a case where generation of excess pore pressure is unlikely (Iverson *et al.*, 1994). To increase the tendency to generate excess pore pressure, ploughing experiments were performed using a less diffusive sediment that was obtained by removing grains larger than 0.25 mm from Unteraargletscher sediment (Sample B in Figure 3). Its diffusivity and the corresponding value of R were estimated as outlined below.

Experimental procedure

Prior to an experiment, water-saturated sediment is poured into the sediment chamber with the tap closed. Sediment lumps are crushed to achieve an even texture. When the desired initial thickness is reached, the sediment is covered with a few

centimeters of water. The upper platen, with all pressure ports open, is subsequently lowered into the sediment chamber. As soon as water sprays out through the pressure ports, they are closed. A maximum normal stress of 50 kPa is then imposed by the lever arm/pulley system. The pressure ports are connected to pressure sensors. The laser sensor is installed and measurements of normal stress, vertical displacement of the platen, pore-water pressures and force on the tip are started.

The standard procedure followed for a ploughing experiment is illustrated in Figure 5, which shows the general behaviour of the vertical displacement of the platen, the pore pressure, the azimuth and the drag force. In the first stage of an experiment (Stage I), the tap to the reservoir is opened such that water can drain from the sediment. Dead weights can be added to achieve the desired normal consolidation stress (not done in the experiment shown in Figure 5). This stage lasts until the sediment is fully consolidated, i.e. when the upper platen is stationary and the pore pressure is at hydrostatic level. In the second stage of the experiment (Stage II), the sediment is set in motion by rotating the cylinder, while corresponding changes in sediment volume, pore pressure, amplitude and azimuth of the drag force are recorded. After the tip has been dragged through the sediment over the required distance, sediment motion is stopped (Stage III). The measurements are terminated when the pore pressure is in equilibrium with the hydrostatic pressure, usually after a few hours. To conclude an experiment, the lever arm/pulley system and the instrumented platen are removed, and deformation features in the sediment are inspected visually.

Results

In the first stage of an experiment (Stage I), the diffusivity of the sediment can be determined from the rate at which the water is squeezed out of the sediment. In the experiment shown in Figure 6, the sediment is initially covered with a ~79 mm thick water layer and a normal stress of 37 kPa is applied to the platen. As the tap is opened ($t = 0$ minute), water flows out of the cylinder leading to a corresponding settling of the platen (note that in Figure 6 an increase in the distance recorded by the laser corresponds to a downward motion of

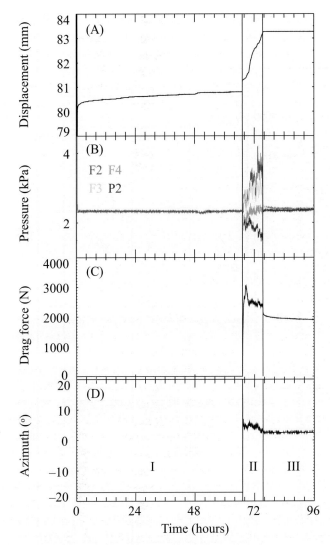

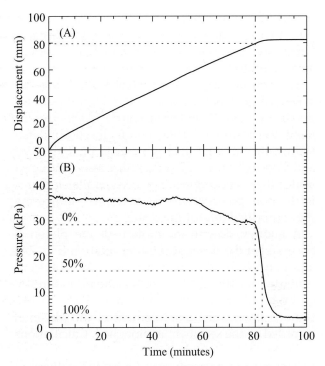

Fig. 6 Records of (A) downward displacement of platen and (B) pore-water pressure measured on the platen as a function of time during the equilibration stage of a ploughing experiment. The sediment is initially covered with a ~79 mm thick water layer which starts to drain at $t = 0$ minutes. Sediment consolidation starts at $t = 80$ minutes. A 50% consolidation is reached at $t = 83$ minutes. See text for details.

Fig. 5 General behaviour of (A) downward displacement of the platen, (B) pore pressure, (C) drag force and (D) azimuth as a function of time during equilibration (Stage I), sediment ploughing (Stage II) and after ploughing (Stage III). The vertical dashed lines correspond to the start and termination of cylinder rotation, respectively. Parameters for this experiment were $\sigma_n = 50$ kPa and $v = 10$ m day^{-1}. Positions of pressure ports F2, F3, F4 and P2 are shown in Figure 2.

the platen). At first, the normal stress is entirely supported by the water such that the water pressure is nearly constant and corresponds roughly to the applied normal stress (Fig. 6b). Subsequently, as the tip penetrates the sediment, the water pressure drops somewhat. Sediment consolidation starts when the platen touches the sediment after 79 mm of platen lowering ($t = 80$ minutes). This time can

be determined by locating the point where the rate of platen lowering starts to decrease (Fig. 6a). Consolidation theory indicates that

$$D = \frac{T_v d^2}{t}, \tag{5}$$

(e.g. Lambe & Whitman, 1979) where d is the length of the drainage path and T_v is a dimensionless time factor which depends on the degree of consolidation and on the boundary conditions. In the present case, $d = 19 \pm 0.5$ cm corresponds to the initial sediment thickness and $T_v = 0.3$ for 50% consolidation (Lang *et al.*, 2002). This degree of consolidation is reached at $t = 83$ minutes, as indicated by the pore pressure record (Fig. 6b). Accounting for uncertainties in the thickness of the sediment and the time when consolidation starts (± 1 minute), Eq. 5 yields $D = 7 \pm 2.5 \times 10^{-5}$ m^2 s^{-1}. By substituting this diffusivity and $v = 10$ m day^{-1} in Eq. 4, we find that $R < 25$, indicating that at this velocity, excess pore pressure may develop during ploughing of Unteraargletscher sediment containing only grains smaller than 0.25 mm.

The second stage of an experiment (Stage II) is started by rotating the cylinder. Figure 7 shows details of Stage II shown in Figure 5. This experiment was performed using a normally consolidated sediment with a ploughing velocity of 10 m day^{-1} and a normal stress of 50 kPa. Other tests conducted under these conditions displayed similar features in the pore pressure and force records, indicating the reproducibility of the results.

A sudden downward motion of the platen is observed at the onset of cylinder rotation (Fig. 7A) and may occur as friction, which is probably concentrated mostly in the seal, the square mounting and the pulley system, is overcome. Further continuous downward displacement of the platen is recorded as the sediment is ploughed. Calculations indicate that about half of this downward displacement can be accounted for by the volume of sediment that may have escaped the sediment chamber and entered into a small gap within the seal between the two cylinder walls. Similar problems have been described in other laboratory studies (Iverson *et al.*, 1997). Additional platen settling can be explained by a decrease in sediment thickness. This behaviour is characteristic of normally consolidated sediments that tend to decrease their volume during initial shear deformation (e.g. Lambe & Whitman, 1979).

The drag force on the ploughing tip increases for the first 70 cm of displacement to reach a peak value of 3000 N and subsequently decreases rapidly toward a nearly steady value of 2450 N, roughly 40 cm before one complete rotation of the cylinder (Fig. 7B). The azimuth record is approximately constant and indicates that the drag force is essentially oriented in the flow direction (Fig. 7C). The post-peak reduction in the drag force seems to be uncorrelated with the pore pressure evolution (Fig. 7D). Instead, we propose that a zone of compressed sediment forms in front of the ploughing tip such that the force recorded by the tip decreases as the compressed sediment enters the sediment that has been ploughed before. Support for this suggestion stems from the observation of dense sediment packed in front of the tip when the instrumented platen is removed from the cylinder

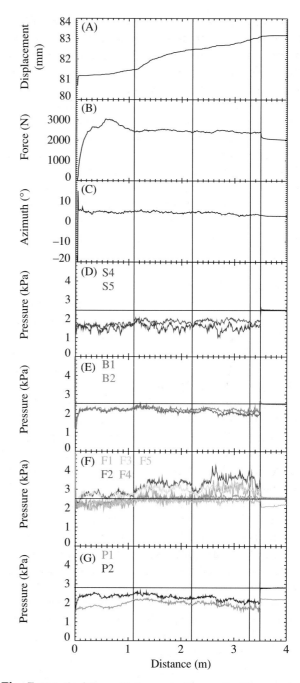

Fig. 7 Detail of Stage II shown in Figure 5 with records of (A) downward displacement of platen, (B) drag force, (C) azimuth and (D)–(G) pore-water pressures as a function of the distance ploughed by the tip. One rotation corresponds to 1.1 m (vertical dashed lines). The vertical solid line indicates rotation stop. Pore-water pressures are initially at the hydrostatic value (horizontal dashed lines in (D)–(G)), which corresponds to ~2.6 kPa at the pressure ports on the platen and ~2.8 kPa at those on the tip. The positions of the pressure ports are shown in Fig. 2.

at the conclusion of an experiment. As a result of this zone of compressed sediment in front of the tip, the zone of influence of the tip is greater than usually assumed in other works (~1 diameter of the ploughing object) (Iverson *et al.*, 1994; Fischer *et al.*, 2001; Iverson & Hooyer, 2004) but is in agreement with a modelling study of pile insertion, which predicts zones of influence extending up to 20 pile diameters in dense sand (Ladanyi & Foriero, 1998). The small amplitude in the subsequent drag force variations is likely to be indicative of a well-developed, steady flow of the sediment around the tip. At the termination of the cylinder rotation, a small decrease in the drag force is observed, probably caused by elastic relaxation within the sediment and the device (Fig. 7B).

Pore-pressure gradients develop within a few seconds in the vicinity of the tip, with values in excess of (up to 1.7 kPa at F2) and below (down to 1.4 kPa at S4) the hydrostatic pressure (Figs. 7D–G). In particular, excess pore pressures develop in front of the tip (F1, F2, F3), with a maximum at a distance of 5 cm (F2) and pore pressures below the hydrostatic value are observed on the tip itself (P1, P2), on the sides (S4, S5) and at the back (B1, B2) of the tip (Figs. 7D–G). The pressures recorded by the other sensors (F4, F5) fluctuate around the hydrostatic value. Pore pressures in excess of the hydrostatic value recorded by F1, F2 and F3 result as the sediment in front of the tip is compressed at a rate which is faster than that at which pore pressure can diffuse. This compression of sediment in front of the tip also causes the sediment to be locally overconsolidated there. Under shearing, overconsolidated sediments of low compressibility are commonly dilatant such that increasing pore volume leads to a local reduction in pore pressure. For example, in studies of pile penetration in a stiff overconsolidated clay, negative pore pressures were observed over most of the penetration length of the pile during driving (Bond & Jardine, 1991). Similarly, the locally overconsolidated sediment that flows around the tip may be sheared and the pore pressures below the hydrostatic value recorded at S4, S5 and P1 may therefore result from shear-induced dilation of the sediment there. Because of the circular geometry of the device, the shear rate, and thus the rate of sediment dilation, is greater at S4 than at S5 so that the pore pressure recorded at S4 is less than that recorded at S5 (Fig. 7D).

Fig. 8 View from above into the sediment chamber showing the wake devoid of sediment that is observed after a ploughing experiment. The direction of ploughing is counterclockwise.

The pore pressures below the hydrostatic level recorded at P2, B1 and B2 at the back of the tip can be explained by the formation of a wake devoid of sediment immediately behind the tip as it is dragged through the sediment, as seen after an experiment (Fig. 8). Similar trenches extending behind ploughing clasts have been observed in the geological record (Iverson & Hooyer, 2004).

The sediment yield strength can be estimated from the force record using a geotechnical model of cone penetration (Senneset & Janbu, 1985). This model describes the resistive force on objects pushed vertically through sediment as a function of the effective stress and the sediment frictional properties. This formulation, when adapted to the ploughing tip, yields the force per unit length on the leading edge of the tip (Fischer *et al.*, 2001). In close analogy to Fischer *et al.* (2001), by equating the bending moment exerted by this force on the ploughing tip to the bending moment recorded on the tip, we derive from the ploughing experiment a sediment yield strength ranging from 34 to 46 kPa. This value provides an upper estimate of the yield strength of Unteraargletscher sediment used in this experiment (Sample B in Figure 3) because it was calculated using the angle of internal friction characteristic of the Sample A sediment (Fig. 3).

The yield stress of a Coulomb sediment scales linearly with the effective stress. Moreover, for a constant applied normal stress, the changes in effect-ive stress are directly related to the pore-pressure variations. In the present case, the pore-pressure records indicate that the effective stress departed from the applied normal stress by absolute values up to 1.7 kPa, leading to maximum changes of 3.5% in the yield strength and in the drag force. Variations in pore pressures, therefore, did not significantly influence the drag force in this experiment.

CONCLUDING DISCUSSION

A new laboratory device of large dimensions was developed and constructed to systematically study the influence of effective stress and ploughing velocity on excess pore-pressure generation and on sediment strength. This device enables simultane-ous measurements of the drag force on a plough-ing tip and the distribution of pore pressure in the adjacent sediment. Significantly, according to standard soil-testing convention (Head, 1989), it is possible to conduct experiments with sediment containing clasts up to 1 cm, which corresponds to one-tenth of the smallest dimension of the sediment chamber (i.e. 11 cm, distance between the cylinder wall and the tip).

From the rate at which pore water is squeezed out of the sediment during compression, a hydraulic diffusivity was estimated for a glacial sediment that was collected in the forefield of Unteraargletscher. A difference of nearly two orders of magnitude was obtained for the diffusivities of two samples of the same origin but containing different frac-tions of fine particles. On a cautionary note, we acknowledge that the diffusivity estimates for the two samples were obtained using different methods. Nevertheless, since the proportion of fines in the upper layer of the substrate may vary spatially and temporally beneath glaciers as a consequence of eluviation, transport and deposition by subglacial water, this result illustrates that the hydromechan-ical properties of a sediment bed can be highly heterogeneous.

During sediment ploughing, pore-pressure gra-dients developed rapidly around the tip. Values below and in excess of the hydrostatic pressure were due to sediment shearing and compression, respectively. These observations suggest that, when a dilatant sediment is ploughed, the mean pore-pressure response may be governed by the relative volumes of sheared and compressed sediment. These volumes, in turn, probably depend on the ambient effective stress and on the size and shape of the ploughing object. If shear-induced pore pressures below the hydrostatic pressure are dominant, subglacial sediment may display a yield strength that increases with the rate of shearing. On the other hand, if excess pore pressures due to sediment compression dominate, the sediment yield strength may decrease with the rate of com-pression. This issue can be tested using the rotary ploughing device by dragging instrumented objects of various geometries through subglacial sediment at different velocities.

Water pressures below the hydrostatic value were also a consequence of a wake devoid of sediment left behind the tip. The length of such a wake may be influenced by the effective normal stress. A sys-tematic investigation of the wake length with the rotary ploughing device could therefore provide a basis for estimating the basal effective stresses of past ice sheets from the observations of ploughing structures in the bed (Iverson & Hooyer, 2004).

In the preliminary experiments, the drag force was influenced by the properties of the sediment in front of the compression zone rather than by the pore pressures in the vicinity of the tip. Specifically, the drag force was larger in a dense, virgin sedi-ment than in a loose one that had been ploughed before. Beneath glaciers, soft beds are likely to undergo deformation, independent of ploughing, at strains that result in the critical state and thus in full dilation of the sediment. Future ploughing experiments with our device will therefore be per-formed using an already ploughed sediment.

A problem with the device is the continuous decrease in sample thickness during an experiment, which is probably mainly due to frictional effects and to sediment loss in the seal. This problem can be minimized in future experiments by first over-consolidating the sediment, to eliminate part of the friction and such that the gap of the seal is filled with sediment prior to a ploughing experiment. Despite this problem, the rotary ploughing device proves useful for further exploring the effects of pore pressure changes on the drag force on a ploughing object. In particular, the results of such experiments should help in the development of constitutive equations describing sediment deformation and the corresponding basal drag during clast ploughing beneath glaciers.

ACKNOWLEDGEMENTS

This work was funded by ETH grant TH-7/01-1. We gratefully thank T. Wyder for his skillful assistance with the construction of the rotary ploughing device and with the ploughing experiments. We also thank the workshop staff of the Laboratory of Hydraulics, Hydrology and Glaciology, ETH Zürich, for the construction of the rotary ploughing device. We are indebted to S. Springman for the soil laboratory tests that were conducted at the Institute for Geotechnical Engineering, ETH Zürich, with generous assistance of D. Bystricky, M. Sperl and T. Ramholt. Logistical support was provided by the Swiss Military during sediment collection in the field and the Centro Stefano Franscini at ETH Zürich. We are grateful to D. Chandler, T. Hooyer and the scientific editor B. Hubbard for perceptive review comments and to N. Iverson for his review of an earlier version of this paper.

REFERENCES

Albert, R., Pfeifer, M.A., Barabási, A.-L. and Schiffer, P. (1999) Slow drag in a granular medium. *Phys. Rev. Lett.*, **82**, 205–208.

Alley, R.B. (1989) Water-pressure coupling of sliding and bed deformation: II. Velocity-depth profiles. *J. Glaciol.*, **35**, 119–129.

Bond, A.J. and Jardine, R.J. (1991) Effects of installing displacement piles in a high OCR clay. *Géotechnique*, **41**, 341–363.

Bowles, J.E. (1978) *Engineering Properties of Soils and their Measurement*. 2nd ed. McGraw-Hill, New York.

Brown, N.E., Hallet, B. and Booth, D.B. (1987) Rapid soft bed sliding of the Puget glacial lobe. *J. Geophys. Res.*, **92**, 8985–8997.

Chehata, D., Zenit, R. and Wassgren, C.R. (2003) Dense granular flow around an immersed cylinder. *Physics of Fluids*, **15**, 1622–1631. doi: 10.1063/1.1571826.

Clarke, G.K.C. (1987) Fast Glacier Flow: Ice Streams, Surging, and Tidewater Glaciers. *J. Geophys. Res.*, **92**, 8835–8842.

Fischer, U.H. and Clarke, G.K.C. (1994) Ploughing of subglacial sediment. *J. Glaciol.*, **40**, 97–106.

Fischer, U.H., Porter, P.R., Schuler, T., Evans, A.J. and Gudmundsson, G.H. (2001) Hydraulic and mechanical properties of glacial sediments beneath Unteraargletscher, Switzerland: implications for glacier basal motion. *Hydrol. Processes*, **15**, 3525–3540.

Freeze, R.A. and Cherry, J.A. (1979) *Groundwater*. Prentice Hall, Englewood Cliffs, NJ.

Head, K.H. (1989) *Soil Technician's Handbook*. John Wiley and Sons, New York.

Hooke, R.LeB., Hanson, B., Iverson, N.R., Jansson, P. and Fischer, U.H. (1997) Rheology of till beneath Storglaciären, Sweden. *J. Glaciol.*, **43**, 172–179.

Hooyer, T.S. and Iverson, N.R. (2002) Flow mechanism of the Des Moines lobe of the Laurentide ice sheet. *J. Glaciol.*, **48**, 575–586.

Iverson, N.R. (1999) Coupling between a glacier and a soft bed: II. Model results. *J. Glaciol.*, **45**, 41–53.

Iverson, N.R. and Hooyer, T.S. (2004) Estimating the sliding velocity of a Pleistocene ice sheet from plowing structures in the geologic record. *J. Geophys. Res.*, **109**. doi: 10.129/2004JF000132.

Iverson, N.R., Jansson, P. and Hooke, R.LeB. (1994) In situ measurement of the strength of deforming subglacial till. *J. Glaciol.*, **40**, 497–503.

Iverson, N.R., Hanson, B., Hooke, R.LeB. and Jansson, P. (1995) Flow mechanism of glaciers on soft beds. *Science*, **267**, 80–81.

Iverson, N.R., Baker, R.W. and Hooyer, T.S. (1997) A ring-shear device for the study of till deformation: tests on till with different clay contents. *Quat. Sci. Rev.*, **16**, 1057–1066.

Iverson, N.R., Hooyer, T.S. and Baker, R.W. (1998) Ring-shear studies of till deformation: Coulomb-plastic behavior and distributed strain in glacier beds. *J. Glaciol.*, **44**, 634–642.

Iverson, R.M. and LaHusen, R.G. (1989) Dynamic pore-pressure fluctuations in rapidly shearing granular materials. *Science*, **246**, 796–799.

Kamb, B. (2001) Basal zone of the West Antarctic Ice Sheet and its role in lubrication of their rapid motion. In: *The West Antarctic Ice Sheet: Behavior and Environment* (Eds R.B. Alley and R.A. Bindschadler), *Antarctic Research Series*, **77**, 157–199. American Geophysical Union.

Ladanyi, B. and Foriero, A. (1998) A numerical solution of cavity expansion problem in sand based directly on experimental stress-strain curves. *Can. Geotech. J.*, **35**, 541–559.

Lambe, T.W. and Whitman, R.V. (1979) *Soil mechanics*. SI edition. John Wiley and Sons, New York.

Lang, H.J., Huder, J. and Amann, P. (2002) *Bodenmechanik und Grundbau*. 7th ed. Springer-Verlag, Berlin.

Mitchell, J.K. (1993) *Fundamentals of Soil Behavior*. 2nd ed. John Wiley and Sons, New York.

Raymond, C.F. (1987) How do glaciers surge? A review. *J. Geophys. Res.*, **92**, 9121–9134.

Rousselot, M. and Fischer, U.H. (2005) Evidence for excess pore-water pressure generated in subglacial sediment: implications for clast ploughing. *Geophys. Res. Lett.*, **32** (L11501). doi: 10.1029/2005GL022642.

Senneset, K. and Janbu, N. (1985) Shear strength parameters obtained from static cone penetration tests. In: *Strength Testing of Marine Sediments: Laboratory and In-situ Measurements* (Eds R.C. Chaney and K.R. Demars), *American Society for Testing and Materials, Spec. Tech. Publ.*, **883**, 41–54. Philadelphia, PA.

Thomason, J.F. and Iverson, N.R. (2003) Flow mechanism of ice sheets on unlithified sediment: Plowing of clasts at the ice–bed interface. In: *GSA Abstracts, Ann. Meet. Suppl.*, **35**. Abstract 122–9.

Tulaczyk, S.M., Kamb, W.B. and Engelhardt, H.F. (2000) Basal mechanics of Ice Stream B, West Antarctica 1. Till mechanics. *J. Geophys. Res.*, **105**, 463–481.

Wieghardt, K. (1975) Experiments in granular flow. *Ann. Rev. Fluid Mech.*, **7**, 89–114. doi: 10.1146/annurev.fl.07.010175.000513.

Zhou, F. and Advani, S.G. (2004) Slow drag in granular materials under high pressure. *Phys. Rev. E*, **69**. doi: 10.1103/PhysRevE.69.061306.

Part 2

Modelling glaciers and ice sheets

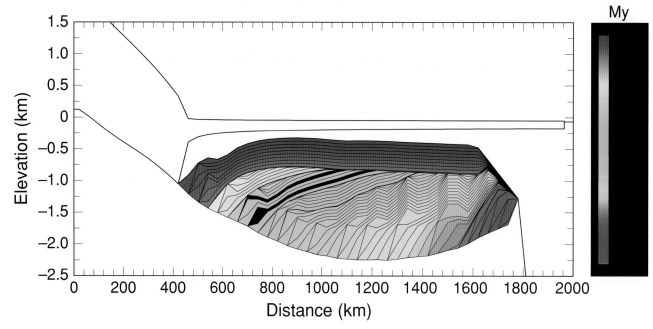

Numerical model demonstrating debris transport on an ice-sheet scale. (Image by D. Pollard and R. DeConto; see following paper)

A coupled ice-sheet/ice-shelf/sediment model applied to a marine-margin flowline: forced and unforced variations

DAVID POLLARD† *and* ROBERT M. DECONTO*

†*Earth System Science Center, Pennsylvania State University, University Park, PA 16802, USA (e-mail: pollard@essc.psu.edu)*
Department of Geosciences, University of Massachusetts, Amherst, MA 01003, USA

ABSTRACT

A standard large-scale ice-sheet model is extended by (i) adding ice stream-shelf flow using a combined set of scaled equations for sheet and shelf flow, and (ii) coupling with a deforming sediment model that predicts bulk sediment thickness. The combination of sheet and shelf flow equations is heuristic, but allows horizontal shear and longitudinal stretching without *a priori* assumptions about the flow regime, and a freely migrating grounding line. The sediment model includes bulk transport under ice assuming a weakly non-linear till rheology, and generation of till by glacial erosion. The combined model explicitly simulates off-shore sediment strata, taking one step towards the goal of direct comparisons with Cenozoic glacimarine sediments on Antarctic continental shelves.

Preliminary 1-D flowline simulations are described on a linearly sloping domain, with simple prescribed spatial and temporal variations of surface ice mass balance. Each simulation is run for 10 million years, and the various patterns of offshore sediment strata built up over the course of each run are examined. A wide variety of sediment patterns is produced depending on uncertain model parameters, including unforced irregular oscillations in the absence of external forcing variations. The results are not conclusive, but illustrate the variety of possible interactions that may play a role as more definitive 3-D models become available to link Cenozoic climate variations with the Antarctic sediment record.

Keywords Ice sheet model, ice shelf model, marine sediment, Cenozoic, Antarctica.

INTRODUCTION

A broad picture of long-term climate variations during the Cenozoic has emerged from deep sea core records (Zachos *et al.*, 2001). Overall cooling since the warm and largely ice-free early Eocene (~50 Ma) has been punctuated by major Antarctic ice growth around the Eocene-Oligocene boundary (34 Ma) and significant fluctuations through the Oligocene and Miocene. These were followed by relatively recent Northern Hemispheric glaciation in the Pliocene, leading to the current Quaternary ice ages. Beyond this broad picture, some important questions remain concerning the role of ice sheets and CO_2 forcing (Pekar and DeConto, 2006; Pekar *et al.*, 2006). Numerous large fluctuations in $\delta^{18}O$ records, especially during the Miocene,

suggest ice volume variations of roughly 30 to ~75 m equivalent sea level over time scales of several 10^5 to 10^6 years. On the face of it, that would require drastic retreats and re-growth of the East Antarctic Ice Sheet (EAIS, currently ~65 m sea-level equivalent), and/or similar amounts of Northern Hemispheric ice growth and decay for which there is scant terrestrial evidence. It is important to understand how much these $\delta^{18}O$ fluctuations are due to other factors in the composite record, or whether they imply drastic variations of EAIS since the Oligocene (Oerlemans, 2004a, 2004b; Pekar and DeConto, 2006; Pekar *et al.*, 2006).

Once a large continental-scale EAIS is formed, it is problematic to invoke drastic retreats at least of the terrestrial portions (grounded on bedrock above sea level after rebound), due to hysteresis in the ice-climate system stemming from ice-albedo and height-mass-balance feedbacks (Huybrechts, 1993, 1994; Maqueda et al., 1998; Oerlemans, 2002). Air temperatures must rise on the order of 15°C around the Antarctic terrestrial flanks before substantial melting and retreat occurs. Depending on model physics (orbital forcing or not, ice-albedo feedback or not), that corresponds to a several fold increase above present atmospheric CO_2, which is not observed in Cenozoic proxy records since the late Oligocene (Pagani et al., 2005). However, it is more plausible to invoke retreats and advances of *marine* portions of Antarctic ice, for instance in Wilkes Land (Miller and Mabin, 1998) or the West Antarctic Ice Sheet itself (WAIS, Anderson and Shipp, 2001; which may have existed since the Miocene at least, Scherer, 1991). Whether or not these have occurred and how much sea level they could account for (~6 m, current WAIS) is an open question.

There is a potential wealth of information on WAIS and EAIS variations in Cenozoic marine sediment deposits on the Antarctic continental shelf (Cooper et al., 1993; Hambrey and Barrett, 1993; Brancolini et al., 1995; De Santis et al., 1999, Anderson, 1999; Escutia et al., 2005; and others). Since the mid-Miocene at least, much of these sediments have been deposited in glacimarine or sub-ice environments, as the ice sheet grounding line has advanced and receded over the continental shelf, especially in the Ross and Weddell Seas, Prydz Bay, and off Wilkes Land and the Antarctic Peninsula. Great efforts have been made over the last several decades to study these deposits by seismic profiling, coring and sampling, and progress in inferring Cenozoic ice-sheet variations is ongoing (Anderson, 1999).

If climate and ice-sheet models can include these sediment processes explicitly and results compared directly with observed deposits, that would be a valuable tool in linking the Antarctic marine sediment record to Cenozoic ice and climate history. As a small step towards that goal, this paper extends a standard ice-sheet model to include (i) both grounded ice-sheet and floating ice-shelf flow, and (ii) sub-ice sediment production and transport, both of which are necessary to explicitly simulate shelf deposits. Preliminary results are described for experiments run over 10 million years with basic 1-D flowline geometries and simple prescribed climate forcing.

COMBINED ICE SHEET-SHELF MODEL

The combination of ice sheet and shelf flow in one efficient large-scale model is not straightforward (Vieli and Payne, 2005). Although grounded (terrestrial) ice and floating (shelf) ice have the same fundamental ice rheology, the large-scale flow regimes and scaled equations are very different. Due to basal shear stress, terrestrial ice flows mainly by vertical shear $\partial u/\partial z$ determined locally by the driving stress $\rho g H \partial h/\partial x$ (the zero-order shallow-ice approximation), whereas shelf ice deforms mainly by along-flow longitudinal stretching $\partial u/\partial x$ determined non-locally by the shelf thickness distribution. In the vicinity of the grounding line and in ice streams with very little basal stress, a combination of the two flow regimes exists. Full(er)-stress equations are available (Blatter, 1995; Pattyn, 2002; Payne et al., 2004) but are computationally expensive, and have only recently begun to be applied in 3-D (Pattyn, 2003). Other approaches have been (i) to add basal and side friction to the scaled shelf equations (MacAyeal, 1989; Dupont and Alley, 2005; Vieli and Payne, 2005), (ii) to use continuum mixtures within each grid cell (Marshall and Clarke, 1997), and (iii) to apply sets of scaled equations (sheet/stream/shelf) in pre-determined domains, with simple matching conditions at the interfaces (Huybrechts, 1990, 2002; Hulbe and MacAyeal, 1999; Ritz et al., 2001; Vieli and Payne, 2005).

The approach used here is to heuristically combine the scaled sheet and shelf equations, and is close to those used by Hubbard (1999, 2006) and Marshall et al. (2005). In these techniques, the two sets of scaled equations (for internal shear $\partial u/\partial z$ and for vertical-mean longitudinal stretching $\partial u/\partial x$) are combined into one set, which is applied at all locations with no a priori assumptions about flow regimes. Related non-rigorous sets of combined equations have also been used by Alley and Whillans (1984) and van der Veen (1985). The main difference in our approach is that the full elliptical (non-local) stretching equation is retained, which is needed for domains that include floating ice with vanishing basal stress.

The flow equations are described fully in Appendix A (page 48). A numerical iteration is performed at each timestep between the shear and stretching components that converges naturally to the appropriate scaling or to a combination of the two, depending on the magnitude of the basal drag coefficient. The combination is heuristic because neither set of equations is accurate in the transition region where both $\partial u/\partial x$ and $\partial u/\partial z$ are significant. Nevertheless, our results are reasonable in idealized tests and in 3-D modern Antarctic simulations. Work is in progress to compare them with full-stress solutions in simple flowline situations such as Huybrechts (1998). Our equations fall into type L1L2 of Hindmarsh's (2004) categorization of various approximate schemes, which is found to yield relatively accurate results. Using a similar technique to ours, Hubbard (2000) found that results for a valley glacier agree well with those using a higher-order model.

Hindmarsh (1993, 1996) and Hindmarsh and Le Meur (2001) have investigated whether the dynamics of the narrow transitional region near the grounding line matter for overall equilibrium and stability properties, but these issues remain unresolved. The general behaviour anticipated by Weertman (1974) (cf., van der Veen 1985; Huybrechts, 1998), is that the grounding line can advance relatively freely into water depths of less than a few hundred metres, but advance is inhibited and retreat can easily be induced by greater grounding-line depths; this behaviour occurs in our model with the aid of MacAyeal-Payne thermal fluctuations (Hindmarsh and Le Meur, 2001), and is important in interpreting the results described below. Recently Vieli and Payne (2005) have examined grounding-line migration in 1-D flowline sheet/stream/shelf models using fixed versus moving grids, and found suspect behaviour and grid dependencies with fixed grids. Our grid is fixed, but unlike theirs our equations combine sheet and shelf flow continuously across the grounding line. These issues will be addressed in a separate paper, but we note that the results shown here do not depend strongly on grid size (nominally 40 km), and very similar results are obtained for grid resolutions of 20, 40 and 80 km (not shown).

In addition to the ice flow equations, the ice model consists of three other standard components: (i) an ice-mass advection equation predicting ice thickness and accounting for prescribed annual surface accumulation minus ablation, (ii) an ice temperature equation including ice advection, vertical diffusion in the ice-sediment-bedrock column, and

shear heating, and (iii) a bedrock elevation equation with local relaxation towards isostatic equilibrium and elastic lithospheric deformation (Huybrechts, 1990, 2002; Ritz *et al.*, 1997, 2001).

SEDIMENT MODEL

The large-scale, long-term evolution of sediment is simulated using a predictive model of sediment thickness, including subglacial deformation by the imposed basal ice shear stress, which deforms and transports the upper few 10's of cm to meters of sediment. The sediment rheology is assumed to be weakly non-linear (Boulton and Hindmarsh, 1987; Alley, 1989; Boulton, 1996; Jenson *et al.*, 1996), in contrast to nearly plastic (Tulaczyk *et al.*, 2000; Kamb, 2001). The induced sediment velocity contributes to ice motion and effectively makes the bed much slipperier.

The deforming sediment model and equations are described in Appendix B. The main change from our previous applications (Clark and Pollard, 1998; Pollard and DeConto, 2003) is a simple parameterization of fractional exposed bedrock when the grid-mean sediment thickness is <0.5 m. This accounts for separate basal shear stresses over the sediment and bare-bedrock fractions, and assumes that the sediment patches always remain thick enough so sediment deformation never penetrates to bedrock. Where no (or little) sediment exists, quarrying or abrasion by ice on exposed bedrock erodes the bedrock profile and is a source of new sediment. No sediment deformation, basal sliding or bedrock erosion occurs when the base is below the ice pressure-melting point, when presumably there is insufficient liquid water to support most of the ice load and prevent sediment compaction.

In the marine experiments described below nothing happens to sediment in ice-free locations, except for an imposed maximum slope limit of $0.3°$ representing marine slumping down the continental slope, and an infinite sink of sediment to the deep ocean (i.e. no sediment allowed at the right hand edge of the figures below). There are no non-glacial sedimentary processes, and no pelagic deposition. Very similar ice-sediment models haves been applied to general Pleistocene cycles (Boulton, 1996) and Quaternary Barents-Sea deposits (Dowdeswell and Siegert, 1999; Howell and

Siegert, 2000). Other predictive ice sheet-sediment models have been developed by ten Brink and Schneider (1995), ten Brink *et al.* (1995), Tomkin and Braun (2002), Bougamont and Tulaczyk (2003), and Hildes *et al.* (2004).

BASIC RESULTS WITH CLIMATICALLY FORCED RETREATS

As preliminary tests of the model, 1-D flowline experiments have been run for 10 million years, with an initial linear equilibrium bedrock slope and constant sea level. Simple climatic patterns are prescribed for annual mean surface temperature T (°C) = $-20 - 0.0051 \times$ elevation (m), ice surface budget B (m yr^{-1}) = $0.15 \times 2^{(T+15)/10}$, and ocean shelf-melting $M = 0.1$ m yr^{-1}. These are constant except in experiments with forced ice retreats, in which a drastic decrease of $B = -1$ m yr^{-1} is imposed everywhere for 10^5 years once every 10^6 years causing complete wastage of all ice. As shown below, quite a wide variety of long-term ice and sediment behaviour occurs in the model depending on uncertain features such as sediment rheological parameters and initial bedrock geometry. The results are not intended to prove any particular conclusions, but just to illustrate the variety and uncertainty at this stage, and to suggest what may be possible in the future as more definitive models become available.

The most straightforward type of behaviour is shown in Fig. 1 using the nominal ice and sediment parameter values given in the Appendices. Ice sheet maximum extents do not vary much during the run, with the ice growing to nearly the same size after each forced retreat, limited to grounding line depths of <~500 m. Throughout the run fresh sediment is quarried by grounded ice acting on inland and inner-shelf bedrock, and quickly transported as deforming sediment to the grounding line where it accumulates in a wedge. This allows lengthy advances of grounded ice, as the sediment wedge reduces the bathymetry and grounding-line depths (Alley, 1991; Dahlgren *et al.*, 2002). After each forced retreat, a new wedge of sediment advances with the grounding line, resembling smaller-scale grounding-zone wedges of tidewater glaciers (Powell and Alley, 1997). The time scale of advance is on the order of 10^6 years (Fig. 1C),

controlled not by ice mass balance but by grounding line-sediment interaction. When previous deposits are reached the new sediment is draped over them, first by piling up on the inner slope and then slumping down the steep outer slope, producing the laminated strata in Fig. 1B (discussed further below). A single large triangular deposit ~1000 m thick is formed after 10 million years.

SENSITIVITY EXPERIMENTS WITH CLIMATICALLY FORCED RETREATS

The simple structure of the sediment deposits in Fig. 1 is basically due to the small amount of total sediment transported to the shelf. There is not enough transport to cause very rapid slumping down the outer slopes of earlier deposits, or conversely to build large enough deposits that form barriers to later sediment wedges (both of which produce more complex patterns as seen below). One of the more uncertain variables in the sediment model is the viscosity μ_o, which can potentially affect total transport. Our nominal model value is 1×10^{10} Pa s, and we experimented with values between 1×10^9 and 1×10^{11}. This makes little difference to the results, and a single laminated triangular deposit is formed after 10 Ma much like Fig. 1 (not shown). [*n.b.* Although total sediment transport increases with decreasing μ_o at constant basal stress τ_b (Eq. B3), it *decreases* with decreasing μ_o at constant basal (sediment-top) velocity $u_s(0)$ (seen by combining (B2) and (B3)). When μ_o is changed in our interactive model, both basal stresses and velocities change as the ice sheet reacts to the different basal slipperiness, and the net effect on sediment transport is not readily deducible from the relations in Appendix B; in fact, it decreases slightly with decreasing μ_o, at least in the range 10^9 to 10^{11} Pa s.]

The basic properties of the sediment in this context are (i) sediment-top velocity $u_s(0)$ (affecting basal slipperiness and ice flow) and (ii) total sediment transport $\int u_s dz$ (affecting shelf accumulation), and their dependence on basal shear stress. If this combination can be changed outside the constraints of the weakly non-linear model, then very different long-term interactions between ice sheets and sediment deposits can ensue. Given current uncertainty in even the basic forms of sediment rheology, hydrology and transport mechanisms

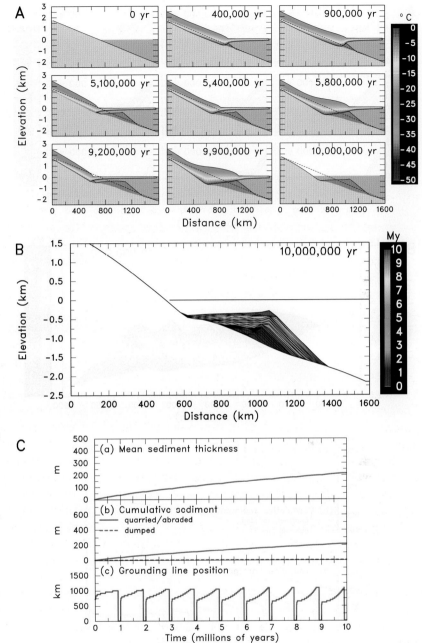

Fig. 1 Ten million year simulation with forced retreats every 1 million years, and with nominal ice and sediment model parameters. (**A**) Ice sheet temperatures (rainbow scale), bedrock (grey), sediment (brown) and ocean (blue), at various times through the run. (**B**) Isochrons of time of original sediment deposition at the end of the run. Times are relative to the start of the run. (**C**) Time series of (A) Sediment thickness averaged over the model domain; (B) Cumulative amounts of sediment sources and sinks averaged over the model domain, where 'dumped' refers to the deep-ocean sink at the right-hand boundary of the domain; (C) Grounding line position from the left-hand boundary of the domain.

under ice sheets, we suggest that this could be the case, due, for instance, to ploughing of plastic sediment by sub-ice bumps (Tulaczyk *et al.*, 2001). Instead of formulating completely different sediment models, here we explore different behaviours by crudely altering sediment model values within the weakly-non-linear framework. Specifically, we increase total sediment transport in Eq. (B3) by an *ad hoc* enhancement factor E. Further work

with alternate rheologies and advection processes is needed to confirm that such modifications are reasonable, and to guide future choices in sub-ice sediment models.

Setting the sediment transport factor $E = 10$ produces quite different behaviour (Fig. 2). The sediments deposited in each advance (after the first three) form barriers to subsequent re-advances, so that new sediment wedges pile up in front of

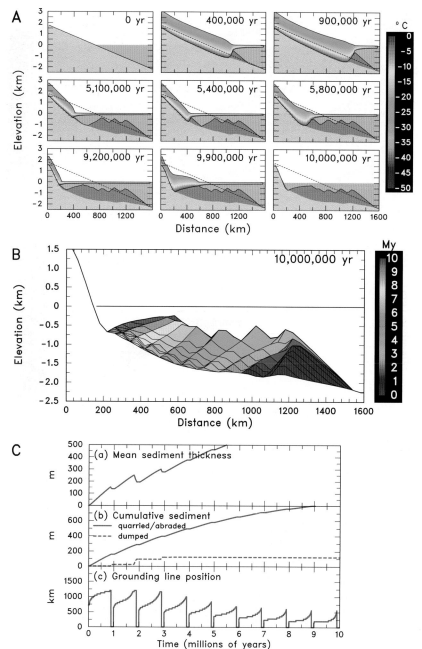

Fig. 2 10 million year simulation with forced retreats every 1 million years. As Fig. 1 except with sediment transport enhancement factor $E = 10$ (see text).

the earlier deposits but do not override them. As shown by the stratal patterns in Fig. 2B, a regular series of retrograding sediment units is formed, one per million years between the imposed retreats. The structure is arguably more realistic than the single laminated triangle in Fig. 1B, but still too regular.

More complex behaviour is shown in Fig. 3, with transport factor $E = 10$ *and* sediment viscos-

ity μ_o reduced to 1×10^9 Pa s. The lower viscosity produces a more slippery bed, thinner ice and slightly less sediment transport than in Fig. 2. The first five re-advances (Fig. 3C) override the previous sediment, forming an early distal laminated wedge as in Fig. 1. The next readvance also overrides this wedge, and transports some of the earlier sediment to the top of the shelf break where it slumps to the deep ocean, forming an erosional

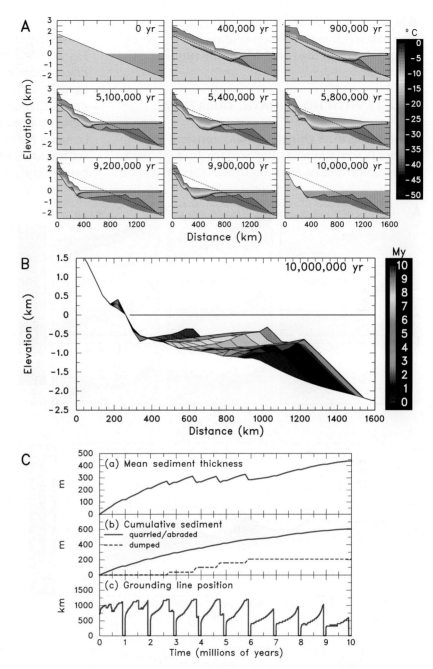

Fig. 3 10 million year simulation with forced retreats every 1 million years. As Fig. 1 except with sediment transport enhancement factor $E = 10$ and sediment viscosity decreased by $\times 0.1$ (to 1×10^9 Pa s).

unconformity at ~5–6 My. After 6 My, some readvances do not override the previous deposits, and an irregular retrograding sequence is formed on the inner shelf until the end of the run. By the last million years, bedrock erosion on the inner shelf (not replaced by sediment) has increased grounding line depths sufficiently to inhibit ice advance, and the last 'proximal' sediment wedge is deposited much closer to the land.

Another effect of the increased basal slipperiness in Fig. 3 is that grounded ice is thinner, especially on the steep inland bedrock slopes, so that basal temperatures are able to reach the freezing point in places. This allows sporadic higher frequency fluctuations due to the MacAyeal-Payne thermal mechanism (basal regions alternately freezing, inhibiting sliding, thickening, then warming and sliding again; MacAyeal, 1993). The resulting breaks

of ice surface slope are evident in Fig. 3A, and propagate downstream to the grounding line triggering
minor retreats in the first 2 My of the run (Fig. 3C).

RESULTS WITH NO CLIMATIC VARIATIONS

If the prescribed climate is held constant (i.e. no
temporal variations in the surface budget parameterization and no forced retreats, otherwise same

as in Fig. 3), *unforced* internal oscillations of the
ice-sediment-bedrock system occur, as shown in
Fig. 4. They are mainly due to sudden groundin-
line retreats triggered by the following mechanism
(deduced from animations). When an advancing
grounding-line sediment wedge reaches and over-
tops a topographic peak in the previous deposits
(usually the outermost 'shelf break'), the maximum
sediment slope of 0.3° is exceeded, and the wedge
begins to slump down the pre-existing outer slope.

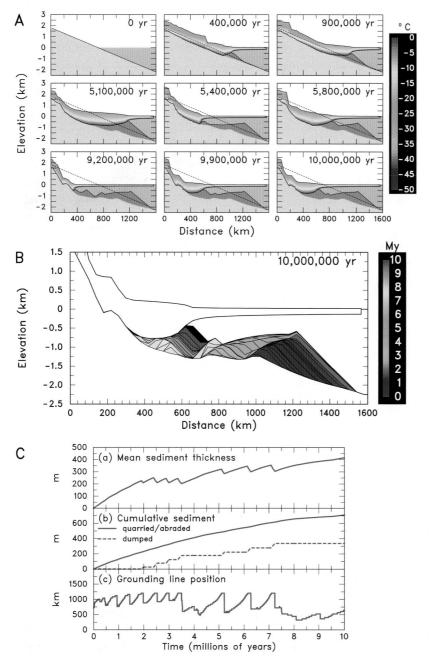

Fig. 4 10 million year simulation with
constant climatic forcing, otherwise same
as in Fig. 3.

The slumping exceeds the supply of deforming sediment transported under the ice, and the net loss of sediment causes a deepening of the grounding line (from ~500 to ~600 m), followed by rapid grounding-line retreat back down the inner slope of the peak. The largest such retreats occur at ~3.5, 5.2, 6.3 and 7.2 My in Fig 4C (panel (c)), and the associated slumping to the deep-ocean sink at the right-hand edge of the domain produces the steps in the 'dumped' curve in panel (b). This same mechanism also occurs just before some of the forced retreats in Fig. 3C.

Although all of the retreats in Fig. 4 are associated with the slumping mechanism, many of them are hastened or triggered by freeze-thaw fluctuations. As described above, higher-frequency basal freeze-thaw fluctuations propagate downstream to the grounding line and cause small transient variations in ice thickness there. Although these fluctuations are not required for the major retreats, they do induce many minor retreats that would not otherwise occur, especially in the first 3 My and after 8 My, and they affect the whole character of the time series in Fig. 4C. If the freeze-thaw switch is turned off (by allowing basal sliding and sediment deformation regardless of temperature), the resulting grounding-line time series is much more regular and smoothly oscillating (not shown).

COMPARISON OF MODEL STRATA WITH OBSERVATIONS

The panels in Figs. 1B–4B above show isochrons of the time of original sediment deposition, which represent 'strata' created by repeated cycles of deposition and erosion. They are plotted by tracking times of original deposition on an Eulerian grid with the same horizontal spacing as the model, and with fine (1-m) regular vertical spacing that shifts up and down with the bedrock (i.e. with $z = 0$ always at the bedrock-sediment interface). As the sediment-top surface moves upwards due to deposition (or downwards due to erosion) through a grid point, the value there is set to the current time (or is reset to null for no sediment). The resulting plots can be compared to observed strata and long-term history inferred from seismic profiles, in the spirit of ten Brink *et al.*'s (1995) studies.

One would not expect much agreement between these simple tests and Cenozoic Antarctic shelf deposits (such as generalized profiles in Cooper *et al.*, 1993). Nevertheless, the more complex stratal patterns in Figs. 3B and 4B do exhibit some features reminiscent of observed seismic profiles. During the first ~3 million years a prograding sequence of shelf breaks is produced in the distal sediments, as wedges are deposited in successive ice advances. This corresponds with prograding shelf breaks noted in the older sections of many seismic profiles by Barker (1995, his Fig. 8). There are some erosional unconformities in the middle parts of the runs, with later proximal sediment packages resting above them. The early-distal/late-proximal depositional sequences in the figures are suggestive of Barker's later aggrading/retrograding shelf breaks. Compared to the initial bedrock profile, there is a trend towards a reverse bathymetric slope with shallower depths on the outer shelf and greater depths on the inner shelf, as more inner-shelf bedrock is eroded and transported to the outer regions. This is a unique feature of continental shelves of glaciated land masses (Anderson, 1999) and has also been modelled by ten Brink and Schneider (1995) and ten Brink *et al.* (1995).

To emphasize the preliminary and unconstrained nature of these results, Fig. 5 shows the sediment isochron pattern for a very different 10 million year run with several model features not included above: a pre-existing continental shelf-like profile, prescribed pelagic sediment deposition, and imposed sea-level variations every 100 ka. The dominant sediment source is pelagic, and the earlier deposits are reworked considerably by sub-ice transport as the grounding line advances to the shelf edge. The resulting stratal pattern shows a pronounced unconformity overlain by the later (9 to 10 My) deposits, and bears little relationship to those above.

CONCLUDING REMARKS

The flowline model described here illustrates several potentially important long-term interactions between ice and sediment in the marine marginal environment. These include thinning and flattening of ice profiles by sediment deformation, shallowing of water depths by sediment accumulation on the continental shelf that allows

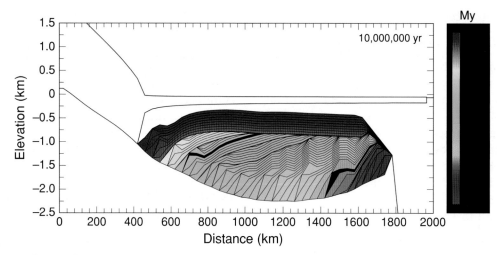

Fig. 5 Isochrons of time of original sediment deposition at the end of a 10 million year run with pre-existing shelf profile, pelagic deposition, 100 kyr sea-level variations, and modified sediment properties.

extensive grounding-line advances and retreats, initiation of drastic retreats by slumping of sediment wedges down pre-existing outer slopes, and stochastic effects of binge-purge fluctuations. A wide variety of sediment distributions and strata is produced on the continental shelf after 10 million years, with some features reminiscent of observed Antarctic margin profiles: notably the trend towards reverse bathymetric slope, and prograding shelf breaks on the outer shelf.

Although the wide variety of patterns produced by uncertain model parameter ranges precludes any definitive conclusions, it suggests what may be possible with more refined models in the future. One tentative conclusion from this study is that within the framework of the weakly non-linear sediment model (Appendix B), relatively featureless sediment distributions are produced (Fig. 1). The richer patterns in Figs. 2 to 5 require modifications that allow different combinations of sediment top velocity and total sediment transport as functions of basal stress, which may be indicative of very different basal sediment mechanisms and rheology (Tulaczyk et al., 2001).

However, it is clear that none of the pictures above bears much overall resemblance to observed profiles, and serious comparisons with Antarctic shelf deposits would be premature. To link Cenozoic climate directly with Antarctic glacimarine deposits, progress will be needed in many modelling areas, including:

1 ascertaining climatic and oceanic forcing of the ice sheet through the Cenozoic on orbital to million year time scales;

2 including forcing factors of eustatic sea level and tectonics;

3 applying 2-D/3-D ice and sediment models to the geographies and histories of individual regions, taking into account geographic distributions of sedimentary sub-basins (De Santis et al., 1999), terrestrial bed properties (Siegert et al., 2005), and overall sediment inventories (Bougamont and Tulaczyk, 2003; Taylor et al., 2004; Jamieson et al., 2005); the single ~1000 km-long 'conveyor belt' for sediment in the flowline model above ignores any lateral sources or sinks;

4 constraining sediment rheologies and transport mechanisms (Tulaczyk et al., 2001) and hydrology (MacAyeal, 1992; Breemer et al., 2002; Flowers and Clarke, 2002);

5 including smaller-scale grounding-line wedge/ bank processes (Powell and Alley, 1997).

ACKNOWLEDGEMENTS

We thank A. Hubbard and an anonymous reviewer for careful reviews that significantly helped the manuscript. This research was funded by the US National Science Foundation under awards ATM-0513402/0513421, ANT-0342484 and ANT-0424589.

REFERENCES

Alley, R.B. (1989) Water pressure coupling of sliding and bed deformation: II. Velocity-depth profiles. *J. Glaciol.*, **35**, 119–129.

Alley, R.B. (1991) Sedimentary processes may cause fluctuations of tidewater glaciers. *Ann. Glaciol.*, **15**, 119–124.

Alley, R.B. and Whillans, I.M. (1984) Response of the East Antarctic Ice Sheet to sea-level rise. *J. Geophys. Res.*, **89**, 6487–6493.

Anderson, J.B. (1999) *Antarctic Marine Geology*. Cambridge University Press, 289 pp.

Anderson, J.B. and Shipp, S.S. (2001) Evolution of the West Antarctic Ice Sheet. In: *The West Antarctic Ice Sheet: Behavior and Environment* (Eds R.B. Alley and R.A. Bindschadler), *Antarctic Research Series Vol. 77*, American Geophysical Union, Washington, D.C., 45–57.

Barker, P.F. (1995) The proximal marine sediment record of Antarctic climate since the late Miocene. In: *Geology and Seismic Stratigraphy of the Antarctic Margin* (Eds A.K. Cooper, P.F. Barker, G. Brancolini), *Antarctic Research Series Vol. 68*, American Geophysical Union, Washington D.C., 25–57.

Blatter, H. (1995) Velocity and stress fields in grounded glaciers: a simple algorithm for including deviatoric stress gradients. *J. Glaciol.*, **42**, 333–344.

Bougamont, M. and Tuylaczyk, S. (2003) Glacial erosion beneath ice streams and ice-stream tributaries: constraints on temporal and spatial distribution of erosion from numerical simulations of a West Antarctic ice stream. *Boreas*, **32**, 178–190.

Boulton, G.S. (1996) Theory of glacial erosion, transport and deposition as a consequence of subglacial sediment deformation. *J. Glaciol.*, **42**, 43–62.

Boulton, G.S. and Hindmarsh, R.C.A. (1987) Sediment deformation beneath glaciers: Rheology and geological consequences. *J. Geophys. Res.*, **92**, 9050–9082.

Brancolini, G., Cooper, A.K. and Coren, F. (1995) Seismic facies and glacial history in the western Ross Sea (Antarctica). In: *Geology and Seismic Stratigraphy of the Antarctic Margin* (Eds A.K. Cooper, P.F. Barker, G. Brancolini), *Antarctic Research Series Vol. 68*, American Geophysical Union, Washington D.C., 209–233.

Breemer, C.W., Clark, P.U. and Haggerty, R. (2002) Modeling the subglacial hydrology of the late Pleistocene Lake Michigan lobe, Laurentide ice sheet. *Geol. Soc. Am. Bull.*, **114**, 665–674.

Clark, P.U. and Pollard, D. (1998) Origin of the mid-Pleistocene transition by ice-sheet erosion of regolith. *Paleoceanography*, **13**, 1–9.

Cooper, A.K., Eittreim, S., ten Brink, U. and Zayatz, I. (1993) Cenozoic glacial sequences of the Antarctic continental margin as recorders of Antarctic ice sheet fluctuations. In: *The Antarctic Paleoenvironment: A Perspective on Global Change, Part Two* (Eds J.P. Kennett, D.A. Warnke), *Antarctic Research Series Vol. 60*, American Geophysical Union, Washington D.C., 75–89.

Dahlgren, K.I.T., Vorren, T.O. and Laberg, J.S. (2002) The role of grounding-line sediment supply in ice-sheet advances and growth on continental shelves: an example from the mid-Norwegian sector of the Fennoscandian ice sheet during the Saalian and Weichselian. *Quatern. Int.*, **95–96**, 25–33.

De Santis, L., Prato, S., Brancoloni, G., Lovo, M. and Torelli, L. (1999) The Eastern Ross Sea continental shelf during the Cenozoic: implications for the West Antarctic ice sheet development. *Global Planet. Change*, **23**, 173–196.

Dowdeswell, J.A. and Siegert, M.J. (1999) Ice-sheet numerical modeling and marine geophysical measurements of glacier-derived sedimentation on the Eurasian Arctic continental margins. *Geol. Soc. Am. Bull.*, **111**, 1080–1097.

Dupont, T.K. and Alley, R.B. (2005) Assessment of the importance of ice-shelf buttressing to ice-sheet flow. *Geophys. Res. Lett.*, **32**, L04503, doi: 10.1029/2004GL022024.

Escutia, C., De Santis, L., Donda, F., Dunbar, R.B., Cooper, A.K., Brancolini, G. and Eittreim, S.L. (2005) Cenozoic ice sheet history from East Antarctic Wilkes Land continental margin sediments. *Global Planet. Change*, **45**, 51–81.

Flowers, G.E. and Clarke, G.K.C. (2002) A multi-component coupled model of glacier hydrology. 1. Theory and synthetic examples. *J. Geophys. Res.*, **107** (B11), 2287, doi: 10.1029/2001JB001122.

Hallet, B. (1996) Glacial quarrying: a simple theoretical model. *Ann. Glaciol.*, **22**, 1–8.

Hambrey, M.J. and Barrett, P.J. (1993) Cenozoic sedimentary and climatic record, Ross Sea region, Antarctica. In: *The Antarctic Paleoenvironment: A Perspective on Global Change, Part Two* (Eds J.P. Kennett, D.A. Warnke), *Antarctic Research Series Vol. 60*, American Geophysical Union, Washington D.C., 91–124.

Hildes, D.H.D., Clarke, G.K.C., Flowers, G.E. and Marshall, S.J. (2004) Subglacial erosion and englacial sediment transport modelled for North American ice sheets. *Quatern. Sci. Rev.*, **23**, 409–430.

Hindmarsh, R.C.A. (1993) Qualitative dynamics of marine ice sheets. In: *Ice in the Climate System.* (Ed. W.R. Peltier), *NATO Advanced Science Series I*: Global Environmental Change, vol. 12, Springer Verlag, Berlin, 67–99.

Hindmarsh, R.C.A. (1996) Stability of ice rises and uncoupled marine ice sheets. *Ann. Glaciol.*, **23**, 105–115.

Hindmarsh, R.C.A. (2004) A numerical comparison of approximations to the Stokes equations used in ice sheet and glacier modeling. *J. Geophys. Res.*, **109**, F01012, doi: 10.1029/2003JF000065, 1–15.

Hindmarsh, R.C.A. and LeMeur, E. (2001) Dynamical processes involved in the retreat of marine ice sheets. *J. Glaciol.*, **47**, 271–282.

Howell, D. and Siegert, M.J. (2000) Intercomparison of subglacial sediment-deformation models: application to the late Weichselian western Barents margin. *Ann Glaciol.*, **30**, 187–196.

Hubbard, A. (1999) High-resolution modeling of the advance of the Younger Dryas ice sheet and its climate in Scotland. *Quatern. Res.*, **52**, 27–43.

Hubbard, A. (2000) The verification and significance of three approaches to longitudinal stresses in high-resolution models of glacier flow. *Geogr. Ann.*, **62A**, 471–487.

Hubbard, A. (2006) The validation and sensitivity of a model of the Icelandic ice sheet. *Quatern. Sci. Rev.*, in press.

Hulbe, C.L. and MacAyeal, D.R. (1999) A new numerical model of coupled inland ice sheet, ice stream and ice shelf flow and its application to the West Antarctic ice sheet. *J. Geophys. Res.*, **104**, 25349–25366.

Huybrechts, P. (1990) A 3-D model for the Antarctic ice sheet: a sensitivity study on the glacial-interglacial contrast. *Clim. Dyn.*, **5**, 79–92.

Huybrechts, P. (1993) Glaciological modeling of the late Cenozoic East Antarctic ice sheet: stability or dynamism? *Geogr. Ann.*, **75A**, 221–238.

Huybrechts, P. (1994) Formation and disintegration of the Antarctic ice sheet. *Ann. Glaciol.*, **20**, 336–340.

Huybrechts, P. (1998) Report of the Third EISMINT Workshop on Model Intercomparison. European Science Foundation, Strasbourg, 105–120.

Huybrechts, P. (2002) Sea-level changes at the LGM from ice-dynamic reconstructions of the Greenland and Antarctic ice sheets during the glacial cycles. *Quatern. Sci. Rev.*, **21**, 203–231.

Jamieson, S.S.R., Hulton, N.R.J., Sugden, D.E., Payne, A.J. and Taylor, J. (2005) Cenozoic landscape evolution of the Lambert basin, East Antarctica: the relative role of rivers and ice sheets. *Global Planet. Change*, **45**, 35–49.

Jenson, J.W., MacAyeal, D.R., Clark, P.U., Ho, C.L. and Vela, J.C. (1996) Numerical modeling of subglacial sediment deformation: Implications for the behavior of the Lake Michigan lobe, Laurentide Ice Sheet. *J. Geophys. Res.*, **101**, B4, 8717–8728.

Kamb, B. (2001) Basal zone of the West Antarctic ice streams and its role in lubrication of their rapid motion. In: *The West Antarctic Ice Sheet: Behavior and Environment* (Eds R.B. Alley and R.A. Bindschadler),

Antarctic Research Series Vol. 77, American Geophysical Union, Washington D.C., 157–199.

MacAyeal, D.R. (1989) Large-scale ice flow over a viscous basal sediment: theory and application to Ice Stream B, Antarctica. *J. Geophys. Res.*, **94**, B4, 4071–4087.

MacAyeal, D.R. (1992) Irregular oscillations of the West Antarctic ice sheet. *Nature*, **359**, 29–32.

MacAyeal, D.R. (1993) Binge-purge oscillations of the Laurentide ice sheet as a cause of the North Atlantic's Heinrich events. *Paleoceanography*, **8**, 775–784.

MacAyeal, D.R. (1996) EISMINT: Lessons in Ice-Sheet Modeling. Department of Geophysical Sciences, University of Chicago, 428 pp.

MacAyeal, D.R. and Thomas, R.H. (1982) Numerical modeling of ice shelf motion. *Ann. Glaciol.*, **3**, 189–193.

Maqueda, M.A.M., Willmott, A.J., Bamber, J.L. and Darby, M.S. (1998) An investigation of the Small Ice Cap Instability in the Southern Hemisphere with a coupled atmosphere-sea ice-ocean terrestrial ice model. *Clim. Dyn.*, **14**, 329–352.

Marshall, S.J. and Clarke, G.K.C. (1997) A continuum mixture of ice stream thermodynamics in the Laurentide Ice Sheet. 1. Theory. *J. Geophys. Res.*, **102**, B9, 20599–20613.

Marshall, S.J., Bjornsson, H., Flowers, G.E. and Clarke, G.K.C. (2005) Simulation of Vatnajokull ice cap dynamics. *J. Geophys. Res.*, **110**, F03009, doi: 10.1029/2004JF000262.

Miller, M.F. and Mabin, M.C.G. (1998) Antarctic Neogene landscapes – in the refrigerator or in the deep freeze? *GSA Today*, Geol. Soc. Amer., **8**(4), 1–8.

Morland, L.W. (1987) Unconfined ice-shelf flow. In: *Dynamics of the West Antarctic Ice Sheet* (Eds C.J. van der Veen and J. Oerlemans), Springer, New York, 99–116.

Oerlemans, J. (2002) On glacial inception and orography. *Quatern. Int.*, **95–96**, 5–10.

Oerlemans, J. (2004a) Correcting the Cenozoic $\delta^{18}O$ deep-sea temperature record for Antarctic ice volume. *Palaeogeogr. Palaeoclimatol. Palaeoecol.*, **208**, 195–205.

Oerlemans, J. (2004b) Antarctic ice volume and deep-sea temperature during the last 50 Ma: a model study. *Ann. Glaciol.*, **39**, 13–19.

Pagani, M., Zachos, J.C., Freeman, K.H., Tipple, B. and Bohaty, S. (2005) Marked decline in atmospheric carbon dioxide concentrations during the Paleogene. *Science*, **309**, 600–603.

Pattyn, F. (2002) Transient glacier response with a higher-order numerical ice-flow model. *J. Glaciol.*, **48**, 467–477.

Pattyn, F. (2003) A new three-dimensional higher-order thermomechanical ice sheet model: Basic sensitivity, ice stream development, and ice flow across subglacial lakes. *J. Geophys. Res.*, **108**, B8, 2382, doi: 10.1029/2002JB002329.

Payne, A.J., Vieli, A., Shepherd, A.P., Wingham, D.J. and Rignot, E. (2004) Recent dramatic thinning pf largest West Antarctic ice stream triggered by oceans. *Geophys. Res. Lett.*, **31**, L23401, doi: 10.1029/2004GL021284, 1–4.

Pekar, S.F. and DeConto, R. (2006) High-resolution ice-volume estimates for the early Miocene: Evidence for a dynamic ice sheet in Antarctica. *Palaeogeogr. Palaeoclimatol. Palaeoecol.*, **231**, 101–109.

Pekar, S.F., Harwood, D. and DeConto, R. (2006) Resolving a late Oligocene conundrum: deep-sea warming versus Antarctic glaciation. *Palaeogeogr. Palaeoclimatol. Palaeoecol.*, **231**, 29–40.

Pollard, D. and DeConto, R.M. (2003) Antarctic ice and sediment flux in the Oligocene simulated by a climate-ice sheet-sediment model. *Palaeogeogr. Palaeoclimatol. Palaeoecol.*, **198**, 53–67.

Powell, R.D. and Alley, R.B. (1997) Grounding-line systems: processes, glaciological inferences and the stratigraphic record. In: *Geology and Seismic Stratigraphy of the Antarctic Margin, Part 2* (Eds P.F. Barker and A.K. Cooper), *Antarctic Research Series Vol. 71*, American Geophysical Union, Washington D.C., 169–187.

Ritz, C., Fabre, A. and Letreguilly, A. (1997) Sensitivity of a Greenland ice sheet model to ice flow and ablation parameters: consequences for the evolution through the last climatic cycle. *Clim. Dyn.*, **13**, 11–24.

Ritz, C., Rommelaere, V. and Dumas, C. (2001) Modeling the evolution of Antarctic ice sheet over the last 420,000 years: Implications for altitude changes in the Vostok region. *J. Geophys., Res.*, **106**, D23, 31943–31964.

Scherer, R.P. (1991) Quaternary and Tertiary microfossils from beneath Ice Stream B – evidence for a dynamic West Antarctic Ice Sheet history. *Global Planet. Change*, **90**, 395–412.

Siegert, M.J., Taylor, J. and Payne, A.J. (2005) Spectral roughness of subglacial topography and implications for former ice-sheet dynamics in East Antarctica. *Global Planet. Change*, **45**, 249–263.

Taylor, J., Siegert, M.J., Payne, A.J., Hambrey, M.J., O'Brien, P.E., Cooper, A.K. and Leitchenkov G. (2004) Topographic controls on post-Oligocene changes in ice-sheet dynamics, Prydz Bay region, East Antarctica. *Geology*, **32**, 197–200.

ten Brink, U.S. and Schneider, C. (1995) Glacial morphology and depositional sequences of the Antarctic continental shelf. *Geology*, **23**, 580–584.

ten Brink, U.S., Schneider, C. and Johnson, A.H. (1995) Morphology and stratal geometry of the Antarctic continental shelf: insights from models. In: *Geology and Seismic Stratigraphy of the Antarctic Margin* (Eds A.K. Cooper, P.F. Barker, G. Brancolini), *Antarctic*

Research Series Vol. 68, American Geophysical Union, Washington D.C., 1–24.

Tomkin, J.H. and Braun, J. (2002) The influence of Alpine glaciation on the relief of tectonically active mountain belts. *Am. J. Sci.*, **302**, 169–190.

Tulaczyk, S., Kamb, W.B. and Engelhardt, H.F. (2000) Basal mechanics of Ice Stream B, West Antarctica 1. Till mechanics. *J. Geophys. Res.*, **105**, B1, 463–481.

Tulaczyk, S.M., Scherer, R.P. and Clark, C.D. (2001) A ploughing model for the origin of weak tills beneath ice streams: a qualitative treatment. *Quatern. Int.*, **86**, 59–70.

van der Veen, C.J. (1985) Response of a marine ice sheet to changes at the grounding line. *Quatern. Res.*, **24**, 257–267.

Vieli, A. and Payne, A.J. (2005) Assessing the ability of numerical ice sheet models to simulate grounding line migration. *J. Geophys. Res.*, **110**, F01003, doi: 10.1029/2004JF000202, 1–18.

Weertman, J. (1974) Stability of the junction of an ice sheet and an ice shelf. *J. Glaciol.*, **13**, 3–11.

Zachos, J., Pagani, M., Sloan, L. and Thomas, E. (2001) Trends, rhythms, and aberrations in global climate 65 Ma to present. *Science*, **292**, 686–693.

APPENDIX A

COMBINED ICE SHEET-SHELF EQUATIONS

The ice flow equations are solved for ice velocities, given the current state of the ice (ice thickness, temperatures, bed topography, basal properties and sea level). As outlined above, we use a heuristic combination of two standard sets of scaled equations, individually called 'shear' (for internal shear $\partial u/\partial z$) and 'stretching' (for vertical-mean longitudinal $\partial u/\partial x$). They are combined by including shear-softening terms due to each type of flow in the other's equations, and more importantly, by (i) using average horizontal velocity $\bar{u} = \bar{u}_i + u_b$ in the stretching equation, where $\bar{u}_i$ is the vertical mean of the internal shear flow and u_b is the basal velocity, (ii) including basal stress dependent on u_b in the stretching equation, and (iii) reducing the driving stress in the shear equations by the gradient of the longitudinal stress from the stretching equations acting on the column above each level.

Symbols are listed in Table A1. Two horizontal dimensions are included, although the paper reports only on 1-D flowline applications. 'Coupling' terms that are not usually found in purely sheet or purely

Table A1 Ice-flow notation and nominal values

x, y	Orthogonal horizontal coordinates (m)
z	Vertical coordinate, increasing upwards from a flat reference plane (m)
z_b	Elevation of ice base (m)
u, u_i, u_b	Horizontal ice velocities in x direction. u = total, u_i = internal deformation, u_b = basal (m yr^{-1})
v, v_i, v_b	Horizontal ice velocities in y direction. v = total, v_i = internal deformation, v_b = basal (m yr^{-1})
H	Ice thickness (m)
h	Ice surface elevation (m)
h_b	Ice base (bedrock or sediment top) elevation (m)
S	Sea level (m)
ρ	Ice density (910 kg m^{-3})
ρ_w	Water density (1000 kg m^{-3})
g	Gravitational acceleration (9.80616 m s^{-2})
A	Ice rheological coefficient (1×10^{-17} yr^{-1} Pa^{-3} at melt point, with temperature dependence as in Ritz et al., 1997, 2001)
n	Ice rheological exponent (3)
B	Basal sliding coefficient between bedrock and ice (0.5×10^{-11} m yr^{-1} Pa^{-2} at melt point, ramping to 0 at $-10°$C)
m	Basal sliding exponent (2)
k	Floating versus basal sliding flag (0 or 1)
$\dot{\varepsilon}_{ij}$	Strain rate components (yr^{-1})
$\dot{\varepsilon}$	Effective strain rate, 2nd invariant (yr^{-1})
σ_{ij}	Deviatoric stress components (Pa)
σ	Effective stress, 2nd invariant (Pa)
μ	$\frac{1}{2}\,\dot{\varepsilon}^{(1-n)/n}$ (yr$^{2/3}$)
LHS_x, LHS_y	Left-hand sides of Eqs. 2a, 2b (Pa)

shelf formulations are identified by a dashed box. Writing Cartesian horizontal ice velocities as $u(x,y,z)$ and $v(x,y,z)$, define the basal ice velocity $u_b(x,y) = u(x,y,z_b)$, and the internal shearing ice velocity $u_i(x,y,z) = u - u_b$, so that $u_i(x,y,z_b) = 0$. Denoting vertical averages through the ice column by an overbar, then $\bar{u} = u_b + \bar{u}_i$ (and similarly for v_b, v_i and $\bar{v}$). The internal shear equations for $u_i(x,y,z)$ and $v_i(x,y,z)$ are

$$\frac{\partial u_i}{\partial z} = 2A[\sigma_{xz}^2 + \sigma_{yz}^2 + \boxed{\sigma_{xx}^2 + \sigma_{yy}^2 + \sigma_{xy}^2 + \sigma_{xx}\sigma_{yy}}]^{\frac{n-1}{2}}\sigma_{xz}$$
(A1a)

$$\frac{\partial v_i}{\partial z} = 2A[\sigma_{xz}^2 + \sigma_{yz}^2 + \boxed{\sigma_{xx}^2 + \sigma_{yy}^2 + \sigma_{xy}^2 + \sigma_{xx}\sigma_{yy}}]^{\frac{n-1}{2}}\sigma_{yz}$$
(A1b)

and the horizontal stretching equations for $\bar{u}(x, y)$ and $\bar{v}(x, y)$ are

$$\frac{\partial}{\partial x}\left[\frac{2\mu H}{\bar{A}^{1/n}}\left(2\frac{\partial \bar{u}}{\partial x} + \frac{\partial \bar{v}}{\partial y}\right)\right] + \frac{\partial}{\partial y}\left[\frac{\mu H}{\bar{A}^{1/n}}\left(2\frac{\partial \bar{u}}{\partial y} + \frac{\partial \bar{v}}{\partial x}\right)\right]$$
$$= \rho g H \frac{\partial h}{\partial x} + \boxed{\frac{k}{B^{1/m}}|u_b^2 + v_b^2|^{\frac{1-m}{2m}}u_b}$$
(A2a)

$$\frac{\partial}{\partial y}\left[\frac{2\mu H}{\bar{A}^{1/n}}\left(2\frac{\partial \bar{v}}{\partial y} + \frac{\partial \bar{u}}{\partial x}\right)\right] + \frac{\partial}{\partial x}\left[\frac{\mu H}{\bar{A}^{1/n}}\left(2\frac{\partial \bar{u}}{\partial y} + \frac{\partial \bar{v}}{\partial x}\right)\right]$$
$$= \rho g H \frac{\partial h}{\partial y} + \boxed{\frac{k}{B^{1/m}}|u_b^2 + v_b^2|^{\frac{1-m}{2m}}v_b}$$
(A2a)

Equations (A2a,b) and their horizontal boundary conditions for unconfined ice shelves are derived for instance in Morland (1982) and MacAyeal (1996). In the zero-order shallow ice approximation, the vertical shear stress (σ_{xz}, σ_{yz}) in Eqs. (A1a,b) would be balanced only by the hydrostatic driving force $-\rho g(h - z)(\partial h/\partial x, \partial h/\partial y)$ acting on the ice column above level z. Here, horizontal stretching forces are included in this force balance (Hubbard, 1999, 2006; Marshall et al., 2005), so that

$$\sigma_{xz} = -\left(\rho g H \frac{\partial h}{\partial x} - \boxed{LHS_x}\right)\left(\frac{h - z}{H}\right),$$

$$\sigma_{yz} = -\left(\rho g H \frac{\partial h}{\partial y} - \boxed{LHS_y}\right)\left(\frac{h - z}{H}\right),$$
(A3)

where LHS_x and LHS_y are the left-hand sides of (A2a) and (A2b) respectively. Because horizontal stretching forces are taken to be vertically uniform and the terms in (A2) are forces on the whole ice thickness, their effect on the ice column above level z is scaled by $(h - z)/H$ in (A3).

Inclusion of the strain softening terms in (A1) and (A2) due to each other's flow requires manipulation of the constitutive relation for ice rheology. In (A2a,b),

$$\mu = \frac{1}{2}(\dot{\varepsilon}^2)^{\frac{1-n}{2n}} \tag{A4}$$

and $\bar{A} = \frac{1}{H}\int A dz$ is the vertical mean of the Arrhenius temperature-dependent coefficient in the constitutive relation

$$\dot{\varepsilon}_{ij} = A(T)(\sigma^2)^{\frac{n-1}{2}}\sigma_{ij} \quad \text{or}$$

equivalently $\dot{\varepsilon}_{ij} = (A(T))^{\frac{1}{n}}(\dot{\varepsilon}^2)^{\frac{n-1}{2n}}\sigma_{ij}$,

where $\dot{\varepsilon}_{ij}$ are strain rates, σ_{ij} are deviatoric stresses, and $\dot{\varepsilon}$ and σ are the second invariants of their respective tensors. The latter are defined by $\dot{\varepsilon}^2 \equiv \sum_{ij}\frac{1}{2}\dot{\varepsilon}_{ij}\dot{\varepsilon}_{ij}$ and $\sigma^2 \equiv \sum_{ij}\frac{1}{2}\sigma_{ij}\sigma_{ij}$. The relationship

$$\dot{\varepsilon}^2 \approx \left(\frac{\partial \bar{u}}{\partial x}\right)^2 + \left(\frac{\partial \bar{v}}{\partial y}\right)^2 + \frac{\partial \bar{u}}{\partial x}\frac{\partial \bar{v}}{\partial y} + \frac{1}{4}\left(\frac{\partial \bar{u}}{\partial y} + \frac{\partial \bar{v}}{\partial x}\right)^2$$
$$+ \left[\frac{1}{4}\left(\overline{\frac{\partial u_i}{\partial z}}\right)^2 + \frac{1}{4}\left(\overline{\frac{\partial v_i}{\partial z}}\right)^2\right] \tag{A5}$$

is used to set μ in (A2a,b), and follows using

$$\dot{\varepsilon}^2 = \dot{\varepsilon}_{xx}^2 + \dot{\varepsilon}_{yy}^2 + \dot{\varepsilon}_{xx}\dot{\varepsilon}_{yy} + \dot{\varepsilon}_{xy}^2 + \dot{\varepsilon}_{xz}^2 + \dot{\varepsilon}_{yz}^2,$$

$$\dot{\varepsilon}_{xx} + \dot{\varepsilon}_{yy} + \dot{\varepsilon}_{zz} = 0, \quad \text{and}$$

$$\dot{\varepsilon}_{xx} = \frac{\partial \bar{u}}{\partial x}, \dot{\varepsilon}_{yy} = \frac{\partial \bar{v}}{\partial y}, \dot{\varepsilon}_{xy} = \frac{1}{2}\left(\frac{\partial \bar{u}}{\partial y} + \frac{\partial \bar{v}}{\partial x}\right),$$

$$\dot{\varepsilon}_{xz} \approx \frac{1}{2}\overline{\frac{\partial u_i}{\partial z}}, \dot{\varepsilon}_{yz} \approx \frac{1}{2}\overline{\frac{\partial v_i}{\partial z}}.$$

The corresponding expression for σ^2 is used in (A1a,b), and the purely horizontal components are obtained in our numerical procedure from

$$\sigma_{xx}^2 + \sigma_{yy}^2 + \sigma_{xy}^2 + \sigma_{xx}\sigma_{yy} =$$
$$\left(\frac{2\mu}{\bar{A}^{1/n}}\right)^2\left[\left(\frac{\partial \bar{u}}{\partial x}\right)^2 + \left(\frac{\partial \bar{v}}{\partial y}\right)^2 + \frac{\partial \bar{u}}{\partial x}\frac{\partial \bar{v}}{\partial y} + \frac{1}{4}\left(\frac{\partial \bar{u}}{\partial x} + \frac{\partial \bar{v}}{\partial y}\right)^2\right] \tag{A6}$$

The basal sliding relation used on the right-hand sides of (A2a) and (A2b) for grounded ice is $\tilde{u}_b = B|\tau_b|^{m-1}\tilde{\tau}_b$, or equivalently $\tilde{\tau}_b = B^{-\frac{1}{m}}|u_b|^{\frac{1-m}{m}}\tilde{u}_b$. Where ice is grounded, i.e. where $\rho_w(S - h_b) < \rho H$ or the ocean

has no access (held back by intervening thicker ice or higher land), then $k = 1$ in the sliding terms, and the ice surface elevation $h = H + h_b$. Where ice is floating, i.e. $\rho_w(S - h_b) > \rho H$ and the ocean has access, then $k = 0$ and $h = S + H(1 - \rho/\rho_w)$.

At each timestep an iteration is performed between the two sets of equations (A1a,b) and (A2a,b). First, the elliptical system (A2) is solved for $\bar{u}$ and $\bar{v}$ by a sparse matrix algorithm, using the previous iteration's $\bar{u}_i$ and $\bar{v}_i$ where needed (to obtain μ via (A4) and (A5), and to relate u_b and v_b to $\bar{u}$ and $\bar{v}$). A standard sub-iteration is used in this step to account for the non-linear dependence of viscosity μ on $\bar{u}$ and $\bar{v}$ in (A4–5). Then (A1a,b) are solved to obtain the internal deformation flow $u_i(z)$ and $v_i(z)$ for each ice column, using LHS_x and LHS_y from the previous solution of (A2) to modify the driving stresses $\rho g H \partial h/\partial x$ and $\rho g H \partial h/\partial y$ in (A3), and using horizontal gradients of the previous $\bar{u}$ and $\bar{v}$ for the strain-softening terms in (A6). This iteration converges naturally to the appropriate scaling depending on the magnitude of the basal drag coefficient. Usually the flow is either almost all internal shear and basal drag balancing the driving stress, with negligible stretching, or is almost all longitudinal stretching balancing the driving stress, with small or no basal drag and negligible internal shear. For a narrow range of small basal drag coefficients, significant amounts of both flow types co-exist. The combination is heuristic because neither set of equations is accurate in the transition region where both $\partial u/\partial x$ and $\partial u/\partial z$ are non-negligible.

APPENDIX B

SEDIMENT MODEL EQUATIONS

The model of sediment deformation under ice follows Jenson *et al.* (1996), assuming a weakly non-linear till rheology (Boulton and Hindmarsh, 1987; Alley, 1989; Boulton, 1996). The same basic model has been applied to North American sediment distributions in the Quaternary (Clark and Pollard, 1998) and to Cenozoic Antarctic terrestrial sediments (Pollard and DeConto, 2003). When basal temperatures are at the ice pressure-melting point, the sediment is assumed saturated, and the effective stress (ice load minus water pressure) is assumed

Table B1 Sediment model notation and nominal values

z	Vertical coordinate, increasing downwards (m)
z_d	Depth below which $u_s = 0$ (m)
u_s	Horizontal sediment velocity (m s^{-1})
h_s	Sediment thickness (m)
f_b	Fractional area of exposed bedrock (0 to 1)
τ_b	Basal ice shear stress (Pa)
μ_0	Newtonian reference sediment viscosity (1×10^{10} Pa s)
D_0	Newtonian reference deformation rate (7.9×10^{-7} s^{-1})
p	Sediment rheologic exponent (1.25)
c	Sediment cohesion (0 Pa)
ϕ	Sediment angle of internal friction (22°)
E	Sediment transport enhancement factor (1)
S_{erode}	Quarrying or abrasion rate of exposed bedrock (m s^{-1})
ρ_s	Sediment density (2390 kg m^{-3})
ρ_b	Bedrock density (3370 kg m^{-3})
ρ_w	Liquid water density (1000 kg m^{-3})
g	Gravitational acceleration (9.80616 m s^{-2})

to be essentially zero at the ice-sediment interface and increases downward due to the buoyant weight of the sediment. The basal shear stress τ_b applied by the ice is assumed to be constant with depth in the sediment. Under these conditions the sediment stress-strain relationship is (Jenson *et al.*, 1996):

$$\tau_b = c + (\rho_s - \rho_w)gz \tan \phi + (2D_0)^{\frac{p-1}{p}} \mu_0 \left(-\frac{\partial u_s}{\partial z}\right)^{\frac{1}{p}} \quad (B1)$$

where z increases downwards from 0 at the ice-sediment interface (see Table B1). As discussed below, the sediment is assumed deep enough to always accomodate the deforming profile, so the bottom sediment-bedrock boundary has no effect. Neglecting sediment cohesion ($c = 0$), Eq. (B1) is integrated between the ice-sediment interface ($z = 0$) and the depth below which u_s is zero [$z_d = \tau_b/(\rho_s - \rho_w) g \tan\phi$], yielding the sediment top velocity

$$u_s(0) = \frac{\tau_b^{p+1}}{(p+1)(\rho_s - \rho_w) g \tan \phi \, (2D_0)^{p-1}\mu_0^p} \quad (B2)$$

This velocity is imparted to the ice and produces additional ice transport in the ice-sheet advection equation (and is equal to u_b in the ice flow model assuming no sliding at the sediment–ice interface). The total horizontal transport of sediment over the deforming column is

$$\int_0^{z_d} u_s dz = \frac{E\tau_b^{p+2}}{(p+2)(p+1)[(\rho_s - \rho_w) g \tan \phi]^2 (2D_0)^{p-1}\mu_0^p} \quad (B3)$$

E is an *ad hoc* transport enhancement factor (see text). This expression is used for sediment transport in the sediment mass continuity equation

$$\frac{\partial h_s}{\partial t} = -\frac{\partial}{\partial x}\left[(1-f_b)\int_0^{z_d} u_s dz\right] + f_b \frac{\rho_b}{\rho_s}S_{erode} \quad (B4)$$

where S_{erode} is the local rate of quarrying or abrasion by basal ice acting on exposed bedrock (Hallet, 1996). It is taken to be proportional to the work done by basal stress, i.e. $S_{erode} = 0.6 \times 10^{-9} \, \tau_b \, u_b$ where the units of τ_b are Pascals. For grid-mean sediment thicknesses h_s less than 0.5 m, the fractional area of exposed bedrock f_b is parameterized as $1 - (h_s/0.5)$, and the patches of sediment are assumed to remain deep enough to accomodate the deforming profile (i.e. deeper than z_d). The basal ice velocity u_b ($= u_s(0)$) is assumed to be uniform within a grid cell, and separate basal shear stresses are calculated over sediment and exposed bedrock patches, which are weighted by $1-f_b$ and f_b respectively to obtain the grid-mean τ_b. For $h_s \geq 0.5$ m, $f_b = 0$.

No sediment deformation, basal sliding or bedrock erosion occurs when the base is below the ice pressure-melting point, when presumably there is insufficient liquid water to support most of the ice load and prevent sediment compaction. For the marine applications of this paper, a maximum slope limit of 0.3° is imposed for submerged ice-free sediment surfaces, crudely representing slumping down the continental slope. Also, any sediment transported or slumped to the right-hand boundary of the domain is immediately removed, representing an infinite sink of sediment to the continental rise.

A brief review on modelling sediment erosion, transport and deposition by former large ice sheets

MARTIN J. SIEGERT*

*School of GeoSciences, Grant Institute, University of Edinburgh, West Mains Road, Edinburgh EH9 3JW, UK
(e-mail: m.j.siegert@ed.ac.uk)

ABSTRACT

Although numerical ice sheet models have been used often to predict the size and dynamics of former ice sheets, few exercises have utilised the geological record to fully constrain model output. Geological evidence can be used to build detailed hypotheses regarding ice sheet and climate history, which can be tested by computer models provided ice flow is coupled to sediment erosion, transport and deposition. Here three examples of how geological data have been used in conjunction with coupled ice-sheet/sediment modelling to comprehend large-scale glacial history are discussed. The first describes how numerical reconstructions of the late glacial Eurasian Ice Sheet have benefited from marine geophysical surveys quantifying sediment fans along the former ice margin. The second reviews how models have been used to determine the likely accumulations of sediments from the initiation and growth of the Antarctic Ice Sheet. The third discusses developments in modelling techniques that allow detailed predictions of sediment deposits and their evolution through time. In addition, a note is provided on how modelling can be used to provide process information concerning former ice sheet instabilities, which can lead to marine sedimentary records such as Heinrich layers.

Keywords Ice sheet, numerical modelling, sediment, subglacial, erosion.

INTRODUCTION

Glaciers and ice sheets have long been recognised as important agents of erosion and deposition. Much of our understanding of ice sheet behaviour comes from the study of glacial geology (e.g. Hambrey, 1994). While quantification of glacial geologic processes has been considered for some time (Drewry, 1986), the ability to fully comprehend large-scale glacial sedimentary systems and, from this, glacial history was restricted until the advent of numerical modelling techniques (and recent advances in computer technology that make modelling of large systems possible).

Ice sheet modelling has progressed over the last decade into a sub-discipline of glaciology with considerable cross-disciplinary importance (e.g. with contributions to the IPCC reports on climate change and to integrated models of the Earth system). During the 1990s, the European Science Foundation (ESF) European Ice Sheet Modelling Initiative (EISMINT) programme allowed intercomparison between ice sheet models and set up a series of 'benchmarks' to which models can be tested (Huybrechts et al., 1996). The outcome of EISMINT is an appropriate level of regulation in ice sheet modelling activity. This service to the scientific community has been developed further with the generation of freely available ice-sheet modelling software, such as GLIMMER (GENIE Land Ice Model with Multiply Enabled Regions, http://glimmer.forge.nesc.ac.uk).

Many ice sheet models incorporate calculations of basal temperature and water production. Such models, when coupled with models of sediment erosion and transfer beneath ice sheets, are capable of testing glaciological hypotheses regarding ice sheet history, derived from geological data. In this review some of the recent advances in our knowledge of ice sheet dynamics and histories, made possible through coupled ice-sheet/sediment models, are highlighted. The aim of the paper is to collate and compare previous research and to identify ways forward for future activities.

BACKGROUND TO SEDIMENT MODELLING IN ICE SHEETS

Ice sheets are capable of eroding, transporting and depositing huge quantities of sediment. For example, marine geological evidence shows that the Bear Island Fan, on the continental shelf break west of the Barents Sea, received material at a rate of over 1 m per 1000 years during the last glaciation as a consequence of sediment delivery by the Eurasian Ice Sheet (Laberg & Vorren, 1996). The mechanism by which the material was transported is likely to be through the deformation (and therefore motion) of the glacier bed as other processes, such as entrainment within basal ice, are difficult to reconcile with the volumes and rates measured. The process of subglacial bed deformation as a means by which ice sheets may flow has been studied for around thirty years, since this 'paradigm shift' was first detailed by Boulton and Jones (1979), and later quantified by Boulton & Hindmarsh (1987). The notion that large ice sheets can grow on sedimentary material which, if water saturated, has little strength caused many glaciologists to consider the very stability of such ice masses (e.g. MacAyeal, 1992; Clark, 1994). These studies made the link between glacial process and the geological record critical to evaluating quantitative ice sheet histories. The following is a simple method, used in several ice sheet models (and as a basis for more sophisticated approaches), by which ice sheet motion is linked with basal sediment transport.

Most ice sheet models are centred on the continuity equation for ice (Mahaffy, 1976), where the time-dependent change in ice thickness is associated with the specific net mass budget:

$$\frac{dH}{dt} = b_s(x, t) - \nabla.F(u, H) \qquad (1)$$

where $F(u,H)$ is the net flux of ice ($m^2\,yr^{-1}$) (the flux of ice being the product of ice velocity, u, and ice thickness, H). The depth-averaged ice velocity, u ($m\,s^{-1}$), is calculated by the sum of depth-averaged internal ice deformation and basal motion (sliding and/or bed deformation). The specific mass budget term b_s, is usually a function of a number of processes including ice sheet surface mass balance (accumulation and ablation) and, where necessary, iceberg calving and ice shelf basal melting. The continuity equation can be applied to a grid of cells (in the case of finite difference schemes) or a net of nodes (in the case of finite elements). In either case, the equation is used to calculate the net change in ice thickness about an ice sheet through understanding ice flow between cells or nodes.

Alley (1990), in one of the first modelling studies of ice-sediment interaction and from which many subsequent models have been based, described the velocity due to the deformation of water-saturated basal sediments, u_b ($m\,s^{-1}$), as:

$$u_b = h_b K_b \frac{(\tau_b - \tau^*)}{N^2} \qquad (2)$$

where K_b is a till deformation softness, h_b is the deforming till thickness (m), and N is the effective pressure (Pa). The till yield strength, τ^*, is:

$$\tau^* = N\tan(\phi) + C \qquad (3)$$

where C is a till cohesion coefficient, and $\tan(\phi)$ is a dimensionless glacier-bed friction parameter. Equation 3 assumes that subglacial sediment behaves as a plastic substance (Boulton & Hindmarsh (1987) refer to it as a nonlinear Bingham fluid). Field observations of sediment deformation have reinforced the opinion that basal sediments deform plastically (e.g. Clarke, 2005), but incorporating such flow effectively into ice flow models is problematic, due to the nonlinearity between stress and strain rates. Choosing an appropriate rheology for subglacial sediments is, clearly, critical to coupled ice-sheet/sediment models, however, and Bougamont & Tulaczyk (2003) and Bougamont et al. (2003) have produced ice-sheet models in which a plastic rheology is used. Such modelling may well mark a way forward for future work.

Sediment can be eroded as well as transported, and Alley (1990) considered this to be controlled by:

$$t = K_t \frac{u_b}{h_b} N \qquad (4)$$

where t is the thickness of till produced in a year and K_t an abrasion softness (3×10^{-9} m Pa^{-1}). One key element to the calculation is the effective pressure, which can be assumed as the difference between ice overburden pressure and the basal water pressure (i.e. as water pressures increase, the effective pressure decreases, increasing sediment deformation).

The strength of subglacial deforming sediment needs to be known for two reasons. First, it dictates the rate of flow of material and, second, it controls the depth to which deformation can occur. This information is needed to understand the transport of sediment. Provided the sediment has at least some integral strength, its deformation will lead to a flux of material in the direction of ice flow. There are two ways in which this flux can be calculated. The simplest is to assume a Eulerian system (Press et al., 1989), in which sediment flow can be calculated by a continuity equation applied to the same grid as used to determine ice flow. Such a system is easy to comprehend and simple to implement (e.g. Dowdeswell & Siegert, 1999). A more sophisticated system involves a Lagrangian approach (Press et al., 1989), where the ice flow model is used to establish sediment flow vectors, which are then used independently to the ice flow model to predict sediment movement. This technique has an advantage of ensuring mass conservation and avoiding numerical diffusion (e.g. Hagdorn & Boulton, 2004), and can be used to track individual packets of material through time and space.

Ice sheets can also erode and transport sediment by entrainment of material into their basal layers. The processes controlling sediment entrainment are complex, however, and not easily quantified. Consequently, very few numerical modelling studies have attempted to build a comprehensive understanding of glacial sedimentary processes (e.g. Tully, 1995; Hildes et al., 2004). One way forward on this issue has been established by Christoffersen et al. (2006), who have shown how ice accretion at Ice Stream C, West Antarctica, is a dominant mechanism by which sediment is incorporated into the ice sheet and, subsequent, transported.

MODELLING SUBGLACIAL SEDIMENTS

Coupled ice-sheet/sediment models have been used to help understand the connection between ice sheet history and the geological record in several regions. Three examples of how ice-sediment models have been used to study former ice sheets are described below: the first investigation provides an assessment of how the geological record can be used to constrain ice sheet history (Dowdeswell & Siegert, 1999); the second shows how modelling can be used to determine where sediment build up may be expected (Pollard & DeConto, 2003); and the third demonstrates a way in which the practice can be developed in future (Hildes et al., 2004).

Eurasian Ice Sheet

Since the Late Cenozoic (2.5 million years ago), a series of submarine sedimentary fans has developed along the continental margins of the Norwegian-Greenland Sea (Dowdeswell et al., 1996; Vorren et al., 1998). The fans are distinct depocentres, being located at the mouths of bathymetric troughs and separated from each other by essentially sediment-free zones (Fig. 1). They are also very large; the Bear Island Fan (the largest) being in excess of 280,000 km^2 in area. The major source of sediments to the fans is from large continental ice sheets that exist during periods of glaciation. Marine geophysics has demonstrated that sediment is delivered to the tops of the fans, from where they are distributed across the wider fan surface by gravity-driven sliding and slumping processes (e.g. Dowdeswell et al., 1996). The rate of sedimentation is restricted to periods when the ice sheet terminates at the shelf break (about 10% of each 100,000 year glacial cycle). Rates of sediment supply are thought to be very high during these active phases (Laberg & Vorren, 1996, estimate over 100 cm per 1000 years across the entire Bear Island Fan). There is an order of magnitude less sedimentation when the ice sheets are either restricted to the continental shelf interior or, as is the case during interglacials such as now, absent.

Dowdeswell & Siegert (1999) adapted Alley's (1990) simple ice-sheet/sediment model to predict ice sheet history by 'matching' model results to geophysical data from the continental margin fans on the eastern side of the Norwegian-Greenland Sea. In this case, the 'match' between model results and observations was undertaken qualitatively. Model output was generated and compared against known records of ice sheet size (such as moraine

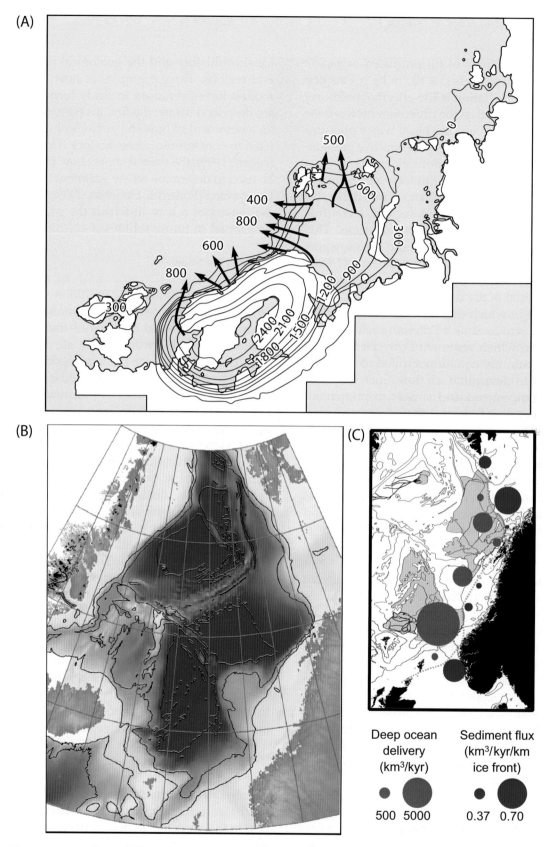

Fig. 1 Measurement and modelling of sediment supply to the western Eurasian margin. (A) The Eurasian Ice Sheet at the LGM. Contours are ice thickness (in m) and arrows denote flow directions, with ice margin velocities noted (in metres per year). Taken from Siegert *et al.* (1999). (B) The positions of major bathymetric trough mouth fans as measured by side-scan sonar, chirp sonar and seismic methods. (C) Rates of sediment supply as calculated by numerical modelling and as measured from geophysical data. The location of (D) the Bear Island Trough Mouth Fan and (E) the North Sea Fan are provided on page 57, in relation to enhanced ice flow as calculated by the ice sheet model illustrated in (A).

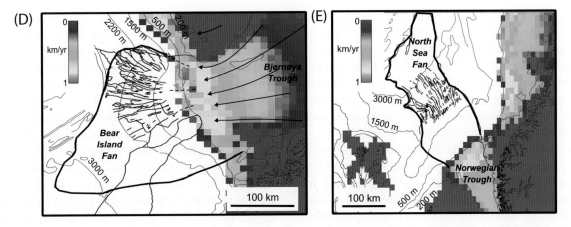

Fig. 1 (*cont'd*)

positions and uplift data). Once the results showed some agreement with these records (gauged through a sensitivity analysis of climate input parameters) a comparison between output and independent geological data (e.g. sediment fans) could then be made. The model calculated the spatial distribution of subglacial sediment over the Eurasian High Arctic, the volume of sediment which accumulates at the shelf edge at the mouths of bathymetric troughs, and the time-dependent variations in the rate of sediment supply to the continental shelf and margin.

The modelling revealed that the most recent glaciation of the Eurasian High Arctic occurred after 28,000 years ago, and that ice streams within bathymetric troughs were active by around 25,000 years ago. The maximum ice thickness over the Barents Sea exceeded 1400 m at around the LGM. At this time, ice extended to the shelf break along both the western Barents Sea margin and the Arctic Ocean margin north of the Barents and Kara seas. Ice streams draining the Barents and Kara seas were present within most major bathymetric troughs during full-glacial conditions. A sedimentation rate of 2–4 cm yr^{-1} was predicted along the mouth of the Bear Island Trough between 27,000 and 14,000 years ago (Fig. 1). This is equivalent to 0.07 to 0.13 cm yr^{-1} averaged over the entire fan. Similarly, high delivery rates of 2–6 cm yr^{-1} of glacial sediments (equivalent to 0.2–0.6 cm yr^{-1} averaged over the fan) were predicted between 27,000 and 12,000 years ago at the mouth of the Storfjorden Trough south of Spitsbergen. The modelled volumes of sediment that accumulate at the continental margin of

the Bear Island and Storfjorden troughs (4,600 km^3 and 900 km^3) are similar to the volumes of Late Weichselian sediment measured over the respective fans using seismic methods (4,200 km^3 and 700 km^3) (Fig. 1). The model also predicted that major glacierfed fan systems would have built up on the northern, Arctic Ocean margin of the Barents and Kara seas, particularly on the continental slope adjacent to the St. Anna and Franz-Victoria troughs.

Ice-sheet decay affected the marine portions of the ice sheet after 15,000 years ago, leaving a northern ice mass between Svalbard and Franz Josef Land which decayed after 13,000 years ago. The transition from a continental-wide ice sheet to one restricted to the northern Barents Sea resulted in a series of sea-floor sediment deposits, that have been measured by marine geophysical surveys around the northwestern shallows.

The modelling demonstrated that ice streams in the Eurasian Arctic were influenced by basal motion (in this case a deforming sediment layer, but it is acknowledged that sliding may also be important to ice stream development), and that the process of sediment deformation led to the delivery of huge quantities of material to the shelf break (Fig. 1). Given the similarity between the Eurasian Ice Sheet and the West Antarctic Ice Sheet, and that deforming sediment is thought to control ice stream dynamics, one might also expect large volumes of sediment to be sent to ice stream mouths and, when the ice sheet expanded during full glacials, to the shelf break. Such analogy may be important when attempting to understand glacial history from the marine geological record in the Ross Sea, such as in the forthcoming Antarctic drilling programme (ANDRILL, http://andrill.org/). Clearly, ice-sheet/sediment modelling, of the type undertaken in the Eurasian Arctic, can provide valuable insights into marine ice sheet dynamics responsible for the accumulation of continental shelf and shelf break sedimentary records.

In recent years, Hagdorn & Boulton (2004) have developed a semi-Lagrangian model in which sediment erosion, transport and deposition in Scandinavia can be tracked over consecutive glacial episodes. The modelling is particularly powerful as sediments from individual glaciations can be monitored through initial transportation and deposition, to reworking and redistribution. The results from such work can be used to understand the glacial history within complex glacial geologic sequences involving sedimentary material from multiple origins.

Antarctic Ice Sheet

The formation of the Antarctic Ice Sheet represents a key episode in the climate history of the Earth. DeConto & Pollard (2003) used a numerical ice-flow model, semi-coupled with a GCM climate model, to demonstrate that the onset of the ice sheet, known to be around 34 million years ago, was an inevitable consequence of atmospheric CO_2 lowering. This result is in contrast to the traditional view of ice sheet initiation in conjunction with the opening of the Drake Passage, so isolating the Antarctic continent climatically with the development of the circumpolar current. While this opening provided a 'trigger' for ice sheet build-up, according the model subsequent CO_2 decline meant that Antarctic Ice Sheet formation was inevitable. Modelling the early glacial history of Antarctica was continued by Pollard & DeConto (2003), through the incorporation of an Alley (1990) type sediment model to their existing ice-climate model, so allowing the erosion, transport and deposition of glacigenic material to be predicted. As the infant Antarctic Ice Sheet would have had an ablating margin, their model also included algorithms to describe fluvial transport and deposition to the continent margin. The results of the model reveal how the presence of deforming basal sediments cause the glacial-interglacial signal of ice sheet initiation,

growth and retreat, to be amplified due to the reduction in basal drag and consequent increases in ice flow velocities (so resulting in more rapid and complete decay). The sediment component of the model allowed a determination of where to expect major deposits of material (Fig. 2), which might be useful in planning future data acquisition programmes. Most material collects at just a few locations, at the mouths of topographic and bathymetric troughs. Consequently, the development of large fans, similar to those observed and modelled in the Eurasian Arctic, should be expected in Antarctica (Fig. 2). Given the period of Antarctic glaciation, and the rates of sedimentation predicted (in some cases over 10 m per 1000 years) such depocentres may be considerable in size.

One place where Pollard and DeConto predict significant build-up of material is around Prydz Bay (offshore of the Lambert-Amery system). To investigate the sedimentological and ice sheet history of this region further, Taylor et al. (2004) used an ice-sheet model to evaluate the glaciological and topographical requirements for the formation of this sedimentary fan. The Lambert-Amery system is the largest drainage pathway of the East Antarctic Ice Sheet, from which a large amount of onshore and offshore geological evidence is available. The records show that an ice stream was established in Prydz Bay during late Miocene–early Pliocene time, in association with the major offshore depocenter. The ice sheet subsequently withdrew to its present position, leaving the Amery Ice Shelf to cover much of the trough. Taylor et al. (2004) revealed, using numerical modelling, that bed morphology change was probably responsible for driving changes in both ice-sheet extent and dynamics in the Lambert-Amery system at Prydz Bay. Changes in bathymetry, caused by sediment erosion is required for shelf-edge glaciation and correlates well with the Prydz Channel fan sedimentation history inferred by geophysical methods. This association suggests a feedback between sediment erosion and glaciation, whereby the current graben is cut to encourage an ice stream responsible for the point-sourced fan development.

Laurentide Ice Sheet

The Late Quaternary ice sheets of North America have been investigated for many decades because

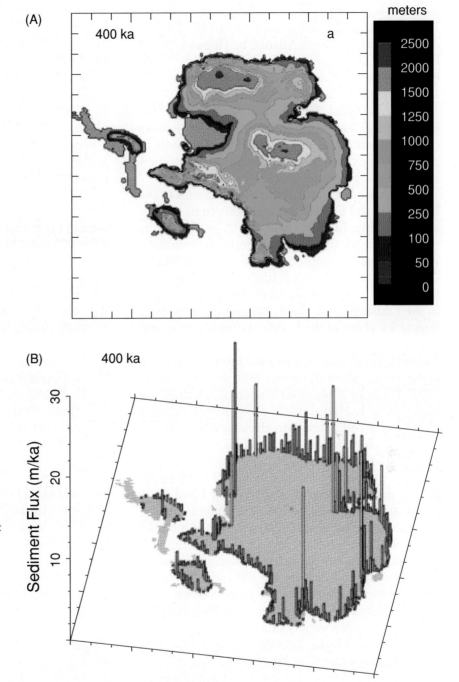

Fig. 2 Modelling sediment transport and deposition by the Antarctic Ice Sheet during a 400,000 year period of its Eocene-Oligocene phase of orbitally-forced growth and decay cycles (~34 million years ago). (A) Average ice thickness of, and (B) sediment deposited by, the ice sheet over the period of the model run. Taken from Pollard & DeConto (2003).

of the well preserved glacial landforms and proglacial lake sequences that now litter the land surface. A comprehensive analysis of these geological data allowed Dyke & Prest (1987) to form a conceptual model of Late Quaternary ice sheet history, involving maps of ice sheet extent through time, and an appreciation of ice flow direction. Subsequent analysis of satellite imagery of glacial landforms has added to the identification of former flow directions and their development in time (Boulton & Clark, 1990; Stokes & Clark, 2001). The acknowledgement that much of these former ice sheets lay over unconsolidated sediment (Alley, 1991) led to the finding that much of the southern margin of the ice sheet was subject to unstable flow (e.g. Clark, 1994). In addition, discrete layers of ice rafted debris discovered in the North Atlantic by Heinrich (1988) have been interpreted by many scientists to be evidence for periodic fast flow of the Laurentide ice sheet in the Hudson Bay region (Bond *et al.*, 1992; Alley & MacAyeal, 1994; Dowdeswell *et al.*, 1995). Each of these insights to the history and dynamics of the Laurentide ice sheet, gained from analysis of glacial geology, has been supplemented by numerical modelling investigations.

Much has been learnt about the dynamics of the Laurentide and Cordilleran ice sheets from numerical modelling. Marshall *et al.* (2002) produced the most comprehensive series of model experiments, which allowed an appreciation of ice sheet volume through time (with a conclusion that as much a 80 m worth of global sea-level change was stored in the Laurentide ice sheet, making it the world's largest ice sheet at the LGM), the production and distribution of meltwater (supraglacial, englacial and subglacial) and the potential influence of basal water on ice sheet dynamics (Marshall & Clarke, 1999). These ice sheet modelling studies formed the background to Hildes *et al.*'s (2004) highly detailed numerical investigation into glacial sediment transport and distribution in glacial North America. Rather than simply relating the erosion of sedimentary material to basal motion, and the transfer of this material through a deforming basal layer (as in most previous models of sediment transfer in ice sheets), Hildes *et al.* (2004) designed a model to account for a full variety of glacial sedimentary processes. They included erosion of sediment by following and developing the model of Tully (1995), which accounts for the process of ice sheet abrasion (englacial basal clasts scratching the subglacial bed) over a variety of lithologies and by the excavation of blocks through the quantification of crack propagation in basal rock. Supraglacial sedimentary processes were ignored in the model, but this is justifiable given the fact that large ice sheets have very little exposures of rock above their surface from which material may fall. As in Marshall *et al.* (2002) the model included a thermodynamic ice flow component, which allowed coupling between the ice sheet and hydrology (including sheet basal flow and groundwater movement). An important input to the model was a depiction of the surface geology (both hard rock

and loose material), from which the model was able to predict the development of sediment sources and sinks, and their effect on ice dynamics (Fig. 3). While the model of Hildes *et al.* (2004) is certainly the most comprehensive attempt to quantify glacial erosion, transport and deposition, the results failed to correspond well with the known distribution of glacial material. Nonetheless, the model represents a benchmark in attempts to fully quantify glacial sedimentary processes, and is a new base from which further studies can be developed.

MODELLING THE SEDIMENTOLOGICAL CONSEQUENCES OF UNSTABLE ICE FLOW

Marine sedimentological investigations have revealed a series of ice rafted debris deposits across the North Atlantic corresponding to the 7000 year periodic production of huge volumes of icebergs, named Heinrich layers (Heinrich, 1988; Bond *et al.*, 1992). Heinrich layers are thought to have formed by the periodic unstable flow of the Laurentide Ice Sheet during the last glaciation (MacAyeal, 1993), as follows. From a relatively stable configuration, the ice sheet slowly builds up over an essentially frozen base. The ice sheet becomes larger and subglacial temperatures rise as a consequence of extra insulation from the cold air on the ice surface and extra heating induced by the deformation of ice. Eventually the basal temperatures reach the pressure melting. As the ice base becomes wet, subglacial sliding and the deformation of subglacial sediments can occur, which for the Laurentide Ice Sheet means enhanced rates of ice flow to the margin. Much of the ice is drained through a single outlet, the Hudson Strait, which becomes the focal point for iceberg release to the North Atlantic. The drainage of so much ice depletes the reserves in the parent ice sheet and it becomes thinner. As it does so, heat is lost from the ice sheet base and, eventually, the ice becomes refrozen to the bed, so reducing the flux of ice to the margin. Once enhanced ice velocities have been curtailed ice-sheet re-growth occurs.

Payne (1995) and Payne & Donglemans (1997) showed, using numerical modelling, how the build-up and decay of large ice sheets has a strong control on the subglacial thermal regime. Under relatively stable external forcing conditions, an ice

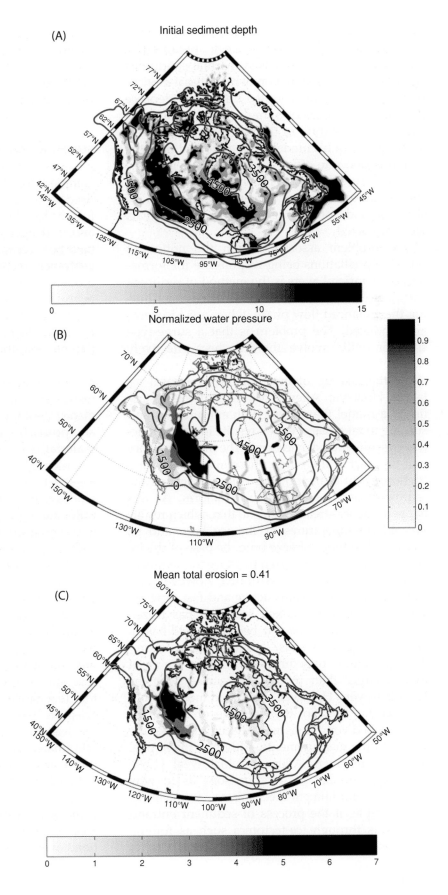

Fig. 3 Numerical modelling of glacial sediment erosion in North America between 60,000 and 10,000 years ago. (A) Sediment distribution prior to ice sheet growth (in metres). (B) Distribution of basal water at the LGM (the figure denotes normalized water pressure). (C) Total amount of sediment erosion after glaciation (in metres). The approximate position of modelled ice thickness contours (given in metres) are superimposed on each panel. Adapted from Hildes *et al.* (2004).

sheet may oscillate between periods of ice growth when the ice is cold-based, and ice decay caused by the attainment of warm-based thermal conditions and the consequent initiation of rapid basal motion. By utilising a thermo-coupled time-dependent ice sheet model, Payne (1995) highlighted the importance of basal sliding to the oscillatory behaviour of ice sheets. In his model, Payne showed how the accumulation rate of ice governed the periodicity of ice sheet oscillations. For low accumulation rates, such as those experienced over the Laurentide ice sheet, periods of between 6000 and 7000 years were predicted.

One problem associated with the notion of unstable oscillations being responsible for formation of Heinrich Layers is the method of sediment entrainment of material within the ice sheet during the enhanced flow phase, when most icebergs are produced. The problem is that a surge-type situation will involve the ice sheet effectively decoupled from the glacier bed, which may not allow the take-up of material into the basal ice layers. However, Alley & MacAyeal (1994) established a model that allows both entrainment and surging to take place, which results in the production of sediment-laden icebergs. The model assumes that the start of the purge phase involves enough frictional heat to melt the ice base, lubricating the sediments and causing them to deform. As this happens the ice stream begins to thin, which results in colder temperatures at the ice base and so water and sediment freeze onto the base of the ice stream. Beneath the refrozen basal ice the sediment is still warm and water-saturated, however, and so the ice stream remains active and fast flowing. When all the water-soaked sediment has frozen onto the ice stream base it becomes 'cold-based' and the purge phase ends. This soft-bed model produces a volume of sediment within the ice sheet that is larger than required to account for the Heinrich layers. However, a hard-bed model would produce far less. So, the presence of both hard and soft bed conditions would seem to make the model fit the measurements. Although this model remains unproven, Alley and MacAyeal (1994) suggest a number of good reasons as to why alternative entrainment mechanisms are less likely. For example, if the process of sediment entrainment was through ice-tectonics such as folding and cavitation, a rate of uptake of material far in

excess of that calculated and measured in modern ice sheets and glaciers would be required to form the Heinrich Layers. Further, a pressure induced basal freezing process results in two orders of magnitude less sediment than is required.

SUMMARY AND FUTURE DEVELOPMENTS

Coupled ice-sheet/sediment models have been used to understand the glacial history of several former ice sheets, both during the last glaciation and in deeper time. The results of these exercises have been compared, with differing success, with the geological record of past glaciation. The simplest models, in which subglacial sediment transfer is related to basal motion, applied to marine settings, predict the build-up of large volumes of material in front of former ice streams. On the margin of the Barents Sea, there is good geological evidence in support of such deposition. When the sophistication of the model increases, and is applied to a terrestrial setting, the success in matching the known glacial geology is reduced. The problem of such an approach may be that the processes by which material is entrained and moved by ice sheets is complex and not easily transferred to traditional ice sheet models. Nevertheless, as pointed out by Hildes et al. (2004), the objective of comparing computer model results with the sedimentary record is essential for 'advancing Quaternary science'.

One area in which this advance may take place involves investigations into glacial sedimentary landform development. By comprehending glaciological requirements for the formation and preservation of known landforms, information on past ice sheets, at a local scale, could be formed. Much work has been undertaken on the identification and measurement of glacial landforms, and qualitative assessment of their implications for ice sheet history (e.g. Stokes & Clark, 2001). Furthermore, some attempts have been made to model the small-scale process responsible for landform development (e.g. Hindmarsh, 1999). Future work may involve using this relatively small-scale evidence, in combination with other data, to form detailed appreciation of large-scale glacial history.

In addition, models capable of a time-dependent analysis of sediment erosion, transport and deposition, over more than one glacial cycle, provide a valuable means by which geological features, formed and influenced by more than one glacial event, can be assessed. The work of Magnus Hagdorn in the development of such models has been particular important on this issue (e.g. Boulton *et al.*, 2003).

Several new scientific programmes are configured to assist the development of these models and to maximise their potential in understanding ice-sheet processes. For example, the Antarctic Climate Evolution (ACE, www.ace.scar.org) programme of the Scientific Committee on Antarctic Research (SCAR) aims to use the predictive ability of numerical models to test hypotheses developed from the geological record. In doing so, ACE will link the numerical modelling community with geologists. Clearly model development is required in conjunction with data collection. The next few years, which will see the International Polar Year and well-established scientific programmes (such as the Antarctic drilling programme, ANDRILL, and the integrated ocean drilling programme, IODP) is likely to witness this dual phase advance. Finally, as a follow up to EISMINT, a new 'ice sheet model intercomparison project' (ISMIP) has been established. One of ISMIP's three themes concerns the incorporation of sediments models, and the role of sediment on unstable ice flow. Results from this project are expected over the next few years.

ACKNOWLEDGEMENTS

Dr. Chris Stokes and Dr. Poul Christoffersen are thanked for their helpful and constructive recommendations.

REFERENCES

Alley, R.B. (1990) Multiple steady states in ice-water-till systems. *Ann. Glaciol.*, **14**, 1–5.

Alley, R.B. (1991) Deforming bed origin for southern Laurentide till sheets? *J. Glaciol.*, **37**, 67–76.

Alley, R.B. and MacAyeal, D. (1994) Ice-rafted debris associated with binge/purge oscillations of the Laurentide Ice Sheet. *Palaeoceanography*, **9**, 503–511.

Bond, G., Heinrich, H., Broecker, W., Labeyrie, L., McManus, J., Andrews, J., Huon, S., Jantschik, R., Clasen, S., Simet, C., Tedesco, K., Klas, M., Bonani, G. and Ivy, S. (1992) Evidence for massive discharges of icebergs into the North Atlantic ocean during the last glacial period. *Nature*, **360**, 245–249.

Bougamont, M. and Tulaczyk, S. (2003) Glacial erosion beneath ice streams and ice-stream tributaries: constraints on temporal and spatial distribution of erosion from numerical simulations of a West Antarctic Ice Stream. *Boreas*, **32**, 178–190.

Bougamont, M., Tulaczyk, S. and Joughin, I. (2003) Response of subglacial sediments to basal freeze-on: II. Application in numerical modeling of the recent stoppage of Ice Stream C, West Antarctica. *J. Geophys. Res.*, **108**, 2223, doi: 10.1029/2002JB001936.

Boulton, G.S. and Hindmarsh, R.C.A. (1987) Sediment deformation beneath glaciers: rheology and sedimentological consequences. *J. Geophys. Res.*, **92**, 9059–9082.

Boulton, G.S. and Clark, C.D. (1990) A highly mobile Laurentide Ice Sheet revealed by satellite images of glacial lineations. *Nature*, **346**, 813–817.

Boulton, G.S. and Jones, A.S. (1979) Stability of temperate ice caps and ice sheets resting on beds with deformable sediment. *J. Glaciol.*, **24**, 29–42.

Boulton, G.S., Hagdorn, M. and Hulton, N.R.J. (2003) Streaming flow in an ice sheet through a glacial cycle. *Ann. Glaciol.*, **36**, 117–128.

Christoffersen, P., Tulaczyk, S., Carsey, F. and Behar, A. (2006) A quantitative framework for interpretation of basal ice facies formed by ice accretion over subglacial sediment. *J. Geophys. Res.*, F01017, doi: 10.1029/2005JF000363.

Clark, P.U. (1994) Unstable behaviour of the Laurentide ice sheet over deforming sediment and its implications for climate change. *Quatern. Res.*, **41**, 19–25.

Clarke, G.K.C. (2005) Subglacial processes. *Annu. Rev. Earth Planet. Sci.*, **37**, 247–276.

DeConto, R.M. and Pollard, D. (2003) Rapid Cenozoic glaciation of Antarctica triggered by declining atmospheric CO_2. *Nature*, **421**, 245–249.

Dowdeswell, J.A. and Siegert, M.J. (1999) Ice-sheet numerical modeling and marine geophysical measurements of glacier-derived sedimentation on the Eurasian Arctic continental margins. *Geol. Soc. Am. Bull.*, **111**, 1080–1097.

Dowdeswell, J.A., Kenyon, N.H., Elverhøi, A., Laberg, J.S., Hollender, F.-J., Mienert, J. and Siegert, M.J. (1996) Large-scale sedimentation on the glacier-influenced Polar North Atlantic margins: long-range side-scan sonar evidence. *Geophys. Res. Lett.*, **23**, 3535–3538.

Dowdeswell, J.A., Maslin, M.A., Andrews, J.T. and McCave, I.N. (1995) Iceberg production, debris rafting and extent and thickness of Heinrich layers (H-1, H-2) in North Atlantic sediments. *Geology*, **23**, 301–304.

Drewry, D.J. (1986) Glacial Geologic Processes. Edward Arnold (Publishers) Ltd., London.

Dyke, A.S. and Prest, V.K. (1987) The late Wisconsin and Holocene history of the Laurentide ice sheet. *Géog. Phys. Quatern.*, **41**, 237–263.

Hagdorn, M. and Boulton, G.S. (2004) Simulating subglacial sediment transport using a semi-lagrangian method. *Geophys. Res. Abstracts*, **6**, 05153.

Hambrey, M.J. (1994) *Glacial Environments*. UCL Press, London. 296 pp.

Heinrich, H. (1988) Origin and consequences of cyclic ice rafting in the Northeast Atlantic Ocean during the past 130,000 years. *Quatern. Res.*, **29**, 143–152.

Hildes, D.H.D., Clarke, G.K.C., Flowers, G.E. and Marshall, S.J. (2004) Subglacial erosion and englacial sediment transport modelled for North American ice sheets. *Quatern. Sci. Rev.*, **23**, 409–430.

Hindmarsh, R.C.A. (1999) Coupled ice-till dynamics and the seeding of drumlins and bedrock forms. *Ann. Glaciol.*, **28**, 221–30.

Huybrechts, P., and Payne, A. and the EISMINT Intercomparison Group (1996) The EISMINT benchmarks for testing ice-sheet models. *Ann. Glaciol.*, **23**, 1–12.

Laberg, J.S. and Vorren, T.O. (1996) The Middle and Late Pleistocene evolution of the Bear Island Trough Mouth Fan. *Global Planet. Change*, **12**, 309–330.

MacAyeal, D.R. (1992) Irregular oscillations of the West Antarctic ice sheet. *Nature*, **359**, 29–32.

MacAyeal, D.R. (1993) Binge-purge oscillations of the Laurentide Ice Sheet as a cause of the North Atlantc's Heinrich events. *Palaeoceanography*, **8**, 775–784.

Mahaffy, M.W. (1976) A Three Dimensional Numerical Model of Ice Sheets: Tests on the Barnes Ice Cap, Northwest Territories. *J. Geophys. Res.*, **81**, 1059–1066.

Marshall, S.J. and Clarke, G.K.C. (1999) Modeling north American freshwater runoff through the last glacial cycle. *Quatern. Res.*, **52**, 300–315.

Marshall, S.J., James, T.S. and Clarke, G.K.C. (2002) North American Ice Sheet reconstructions at the Last Glacial maximum. *Quatern. Sci. Rev.*, **21**, 175–192.

Payne, A.J. (1995) Limit cycles in the basal thermal regime of ice sheets. *J. Geophys. Res.*, **100**, 4249–4263.

Payne, A.J. and Dongelmans, P.W. (1997) Self organisation in the thermomechanical flow of ice sheets. *J. Geophys. Res.*, **102**, 12219–12234.

Pollard, D. and DeConto, R.M. (2003) Antarctic ice and sediment flux in the Oligocene simulated by a climate-ice sheet-sediment model. *Palaeogeogr. Palaeoclimatol. Palaeoecol.*, **198**, 53–67.

Press, W.H., Flannery, B.P., Teukolsky, S.A. and Vetterling, W.T. (1989) Numerical Recipes (The art of Scientific Computing). Cambridge University Press, 702 pp.

Siegert, M.J., Dowdeswell, J.A. and Melles, M. (1999) Late Weichselian glaciation of the Eurasian High Arctic. *Quatern. Res.*, **52**, 273–285.

Stokes, C.R. and Clark, C.D. (2001) Palaeo-ice streams. *Quatern. Sci. Rev.*, **20**, 1437–1457.

Taylor, J., Siegert, M.J., Payne, A.J., Hambrey, M.J., O'Brien, P.E., Leitchenkov, G. and Cooper, A.K. (2004) Late Miocene/early Pliocene changes in sedimentation paths, Prydz Bay, Antarctica: changes in ice-sheet dynamics? Geology, **32**, 197–200.

Tully, M.J.C. (1995) Numerical modeling of erosion and deposition beneath Quaternary ice sheets. Ph.D. thesis, University of Cambridge.

Vorren, T.O., Laberg, J.S., Blaumme, F., Dowdeswell, J.A., Kenyon, N.H., Mienert, J., Rumohr, J. and Werner, F. (1998) The Norwegian-Greenland Sea continental margins: Morphology and late Quaternary sedimentary processes and environment. *Quatern. Sci. Rev.*, **17**, 273–302.

Part 3

Quaternary glacial systems

Stratified, cross-bedded and faulted glaciofluvial sediments of Late Devensian age, Banc-y-Warren, Cardigan, SW Wales. The faulting is believed to have resulted from the melting of a glacier that constrained the deposit. (Photograph by M.J. Hambrey)

Glaciomarine sediment drifts from Gerlache Strait, Antarctic Peninsula

VERONICA WILLMOTT, EUGENE DOMACK*, LAURENCE PADMAN†
and MIQUEL CANALS‡

*Hamilton College, Dept. of Geosciences, 198 College Hill Rd, Clinton, NY, USA (e-mail: edomack@hamilton.edu)
†Earth & Space Research, 3350 SW Cascade Ave, Corvallis, OR, USA
‡GRC Geociencies Marines, Dept. d'Estratigrafia, P. i Geociencies Marines, Universitat de Barcelona, Spain

ABSTRACT

Four sediment cores were collected along the Gerlache Strait, Western Antarctic Peninsula, over two extremely thick, sedimentary accumulations: the Andvord and the Schollaert drifts. The four cores present a sediment facies succession from couplets of laminated diatom ooze and sandy-mud alternating with massive, bioturbated, terrigenous silt/clay unit to a predominantly massive, bioturbated, silty-clay unit. Seismic profiles across the Schollaert Drift show an internal structure defined by wavy stratified reflectors onlapping the northern slope of the strait. We suggest that the formation of the Schollaert Drift was strongly related to the persistence of estuarine conditions during the Middle Holocene along the Gerlache Strait, arising from blockage of the southern entrance by increased glaciation. Such a blockage would have resulted in a retention of fresh water input from melting ice margins and melting sea ice, increased surface temperatures, and reduced exchange with waters originating from the Antarctic Circumpolar Current. In that context, the increase in suspended sediment would have been redistributed by tidal currents leading to enhanced deposition of sediment across the Schollaert Drift.

Keywords Holocene cores, Gerlache Strait, Antarctic Peninsula, sediment drifts.

INTRODUCTION

Present-day glacial-marine sedimentation mechanisms are the main terrigenous sediment suppliers (Anderson et al., 1980; Domack et al., 1994; Isla et al., 2001; McGinnis et al., 1997). During the austral summer, biogenic material input increases due to the high primary production (Gilbert et al., 2003; Nelson, 1988; Palanques et al., 2002; Wefer et al., 1988; Yoon et al., 1994). The fine material settled at depths shallower than 250 m is intensely scoured by currents; thus, in these areas, mainly coarse ice-rafted debris and volcanic clasts remain in the seabed. Sediment accumulation rates have been identified via 210 Pb and 14C radiometric analysis. These rates ranged between 0.02 and 0.5 cm/yr and between 0.01 and 0.3 cm/yr along the Bransfield and Gerlache Straits, respectively (Harden et al., 1992; Isla et al., 2002; Masque et al., 2002; Nelson, 1988).

Sediment drifts are accumulations of hemipelagic sediment contained within a depositional geometry showing evidence of remobilization of foci of accumulation and/or erosion by currents. Based upon models for deep sea deposits, sediment drifts have been related to the interaction of periodic sediment supply from expanded glaciation, contour currents, thermohaline circulation, and dilute sediment gravity flows (Faugeres et al., 1999; Rebesco and Stow, 2002). Although sediment drifts have long been recognized as significant large-scale features of the continental slopes along and near glaciated continental margins, the extent and palaeoenvironmental importance of similar accumulations on the continental shelf of Antarctica remain poorly defined.

Drift deposits were first recognized from the over-deepened inner shelf and fjords of Antarctica in 1998 (Harris et al., 1999) and have since provided important regional palaeoenvironmental archives. Within these shelf drifts, depositional processes and mechanisms (other than ice rafting) differ significantly from their deep-sea counterparts, which include bottom-water formation (Harris and Beaman, 2003) and decelerating coastal/tidal flows (Backman and Domack, 2003; Camerlenghi *et al.*, 2001; Harris and O'Brien, 1998). The sediment architecture has been defined by geophysical surveys that delineate the extent and multi-sourced origin for many of these deposits (Backman and Domack, 2003; Harris and Beaman, 2003; Harris *et al.*, 1999; McMullen *et al.*, 2006). These accumulations of pelagic, hemipelagic, and ice-rafted material are much closer to their glacial sources than are deep water sediment drifts, implying that they may provide superior records of glacial history such as the glacier fluctuations during the Holocene (Cook *et al.*, 2005; Hansom and Flint, 1989); however, their sedimentological character has yet to be investigated in detail.

Leventer *et al.* (Leventer *et al.*, 2002) and Maddison *et al.* (2005) examined the detailed biotic and laminated character of a unique deglacial facies from shelf drifts, which are recognized by rhythmic, laminated couplets of diatom ooze and sandy mud. This distinctive, varved association has not yet been recognized in any modern analogue from the Antarctic or elsewhere. The development of laminations in both siliciclastic and biosiliceous facies in the Antarctic has been related to both the interaction of tidal and estuarine processes in the Antarctic (Domack *et al.*, 2003) and the damping effect of semi-permanent sea ice or ice shelves in glacio-marine settings in general (Dowdeswell *et al.*, 2000) and in the Antarctic (Domack, 1990; Evans *et al.*, 2005). This latter mechanism relates permanent floating ice cover to limitations in infaunal benthic habitation, ice rafting and/or oxygenation, thus favouring the development of laminated facies.

The intent of this paper is to explore the millenial scale variation in sediment lithofacies that is so prominently recorded in the laminated to structureless (bioturbated) siliceous muds of the Gerlache Strait. Traditional models for hemipelagic laminated facies in the marine realm (Pike and Kemp, 1996) call upon bottom water anoxia and the consequent exclusion of infaunal organisms to preserve annual and/or subannual laminations (marine varves). Today, bottom water anoxia within shelf basins of the Antarctic margin rarely takes place because annual sea ice formation densifies surface water which sinks and ventilates basin waters (Jacobs, 1989). Such winter water displaces local bottom water but also replenishes bottom water oxygen levels, thus maintaining conditions conducive for bioturbation (Pike *et al.*, 2001). Hence the preservation of extensive and continuous intervals of well laminated sediment in the Middle Holocene presents a problem for interpreting Antarctic glacial marine palaeoenvironments. We use the physical characteristics of the sediment to develop a model of enhanced estuarine and restricted circulation which, in combination with regional climates, must have allowed annual laminations to be preserved. The spatial and temporal variation in the transition out of the laminated unit into bioturbated unit of the late Holocene must reflect local (sub-basin to sub-basin) changes towards more oxygenated conditions.

MATERIALS AND METHODS

The data in this paper are from sediment cores, oceanographic and geophysical data collected on cruises NBP 99–03, NBP 01–07 and GEBRAP'96 in the western margin of the Antarctic Peninsula. Four jumbo piston cores (JPC) were selected along the length of the Gerlache Strait, between 64° 47′ S and 64° 18′ S (Fig. 1 and Table 1). The jumbo piston cores were shipped to the Antarctic Research Facility, Florida, USA, were they were opened, described, sampled, core-logged and x-rayed. The core chronologies were based on AMS 14C-dated bivalves ((Domack *et al.*, 2003) and Table 2).

Age calibration was performed following methods in Stuiver *et al.* (1988) and Domack *et al.* (1999; 2001) with the software CALIB 5.0 (Stuiver *et al.* 2005).

Digital images, P-wave velocity, Gamma-ray attenuation and magnetic susceptibility (MS) records at 1 cm resolution were obtained with a GEOTEK multi-sensor core logger. Grain-size analysis was performed every 5 cm using a Malvern MasterSizer E® at Hamilton College, NY, in all cores except in JPC-30 where a Coulter® LS 100 was used at the University of Barcelona. The clay and fine to medium silt percentages were chosen as being indicative of terrigenous meltwater plumes and biogenic productivity, respectively (Warner and Domack, 2002). Gravel abundance was determined

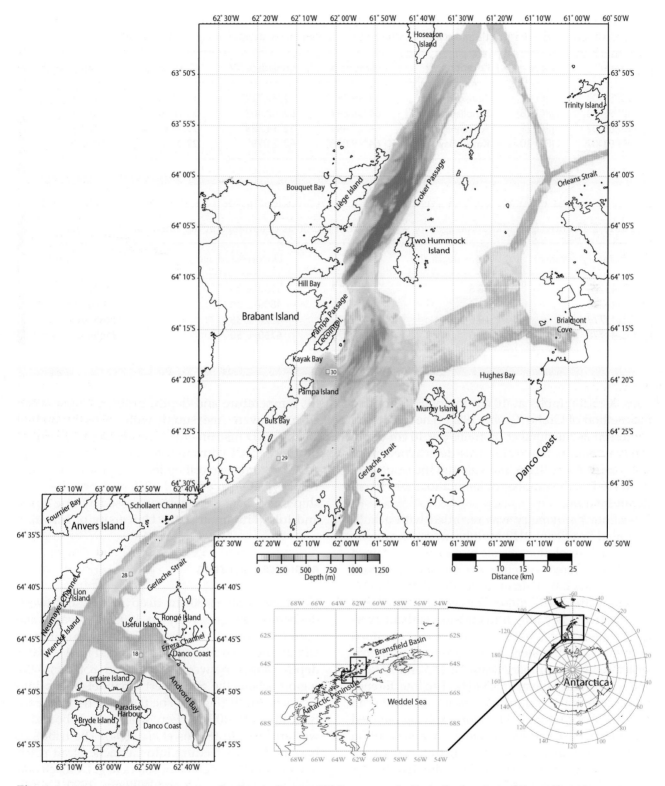

Fig. 1 Location and bathymetry map for the study area. White squares indicate the location of the sediment cores JPC-18, JPC-28, JPC-29 and JPC-30.

Table 1 Location, core length and water depth of the sediment cores studied along the Gerlache Strait

Cruise	Core	Location	Latitude S	Longitude W	Core length	Water depth
NBP99-03	JPC 18	Andvord Drift	64°46.366′	62°49.722′	19.90 m	426 m
NBP99-03	JPC 18	Schollaert Drift	64°38.647′	62°52.407′	20.10 m	673 m
NBP01-07	JPC 29	Avicenna Bay	64°27.499′	62°14.584′	22.25 m	720 m
NBP01-07	JPC 30	Kayak Bay	64°19.058′	62°2.099′	22.26 m	1010 m

Table 2 Uncorrected radiocarbon ages for sediment core JPC-30 as determined at the University of Arizona

Laboratory number	Core depth (m)	$\delta^{13}C$	Uncorrected age yr BP	Carbon source
OS-43774	8.8	−9.17	3960 ± 35	Mollusc
OS-43775	12.6	−0.06	4800 ± 30	Mollusc
OS-43777	19.6	−5.92	6520 ± 35	Mollusc
OS-43778	20.6	−6.72	6760 ± 50	Mollusc

from X-radiographs at 10 cm intervals following the method of Grobe (Grobe, 1987). The results are expressed as the number of grains per 10 cm depth interval. Lamination index was determined from X-radiographs at 10 cm intervals and the results are expressed as the number of laminations per 10 cm depth interval.

Seafloor bathymetry was recorded with a hull-mounted Sea Beam 2112 multibeam sonar system.

The high-resolution seismic profiles were obtained with two different systems; one, obtained on board the Nathaniel B. Palmer used the Bathy-2000W system, with CHIRP pulse. The acoustic signal was a multi-frequency ping transmission with an average frequency of 3.5 kHz. The other was obtained with the TOPAS system, a topographic-parametric system with two primary beams at 21.5 and 18 kHz, and a secondary beam at 3.5 kHz. The interference of the secondary and primary beams generates an acoustic wave which is ideal for resolving detailed sedimentary structures, leading to a resolution greater than 1 m and a typical penetration between 50 and 200 ms in deep sea, unconsolidated marine muds (Canals *et al.*, 1997).

Water mass characteristics were obtained at four stations across the Gerlache Strait (including over the Schollaert sediment drift) during the cruise NBP 0107, using a SeaBird 911 and a CTD (conductiv-ity, temperature and depth) profiler. Upper-ocean currents were measured with a hull-mounted 150 kHz RD Instruments acoustic Doppler current profiler (ADCP) during the NBP 01–07 cruise. The ADCP records two orthogonal components of water velocity between ~50 and ~350 m depth along the ship track. Data were post-processed by Eric Firing and Jules Hummon at the University of Hawaii.

REGIONAL SETTING

Gerlache Strait is 300 m deep in the southwest, deepening progressively to 1200 m towards the northeast end of the strait and between Brabant Island and Two Hummock Island (Fig. 1). Water depth ranges between 600 m and 850 m in the region where the Schollaert Channel opens into Gerlache Strait (Fig. 2). The flanks of the strait are typically 300–400 m deep with a fault-bounded western margin (Evans *et al.*, 2004). The tributary bays and fjords feeding into Gerlache Strait reach maximum water depths of 500 m (e.g. Domack and Ishman, 1993; Domack and Williams, 1990; Griffith and Anderson, 1989).

A persistent surface current along the Gerlache Strait (nutrient-rich 10–30 cm s^{-1}) (Fig. 3), transports warm, Upper Circumpolar Deep Water (UCDW)

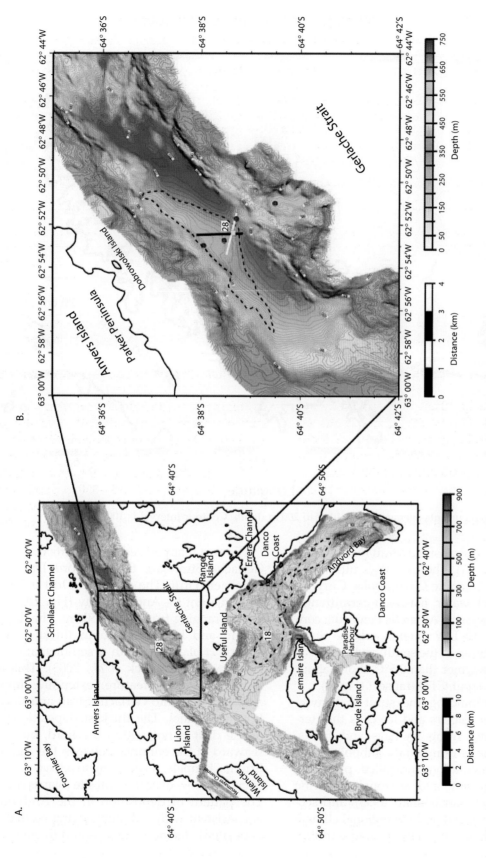

Fig. 2 (A) Bathymetric map of the southern Gerlache Strait (contour interval, 20 m). Yellow squares indicate the location of JPC-18 and JPC-28. Dashed line outlines the outer limit of the Andvord Drift (Harris *et al*., 1999). (B) Detailed bathymetric map around the location of JPC-28 (yellow square). Contour interval is 10 m. Dashed line indicates the minimum extension of the Schollaert Drift. The white and black lines indicate the Chirp profile (Fig. 6A) and the Topas profile (Fig. 6B), respectively. Red dots indicate the positions of CTD casts #21, 22, 23 and 24 (from south to north).

Average ADCP currents 40m-200m

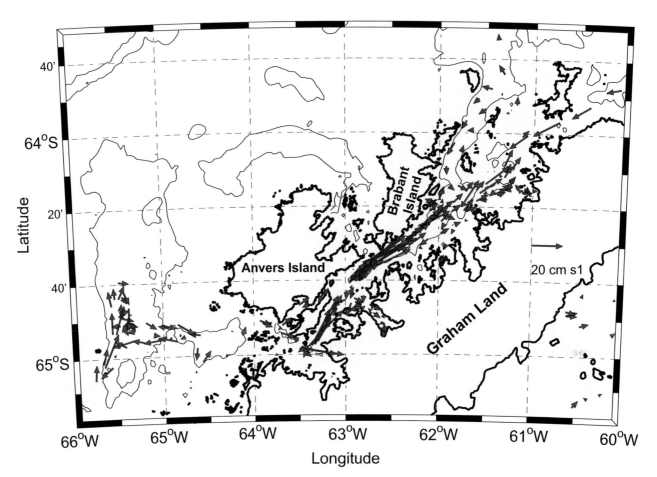

Fig. 3 ADCP measured currents averaged from 40 m to 200 m depth. Contour interval is 1000 m.

northeastward to the Bransfield Strait (Niiler *et al.*, 1990; Zhou *et al.*, 2002). The UCDW is ultimately derived from the Antarctic Circumpolar Current, whose upper level water masses intermittently intrude across the western Antarctic Peninsula continental shelf. The UCDW enters Gerlache Strait through Bismarck Strait in the south, and is seen as subsurface water warmer than 0.5°C above about 200 m in the cross-strait CTD transect (Fig. 4).

The terrain surrounding the strait is heavily glaciated by valley glaciers and ice caps that terminate as tidewater fronts in bays and fjords. The climate regime is dominantly subpolar with contrasts in sedimentation controlled primarily by basin physiography and seasonal estuarine circulation (Domack and Ishman, 1993), and secondarily by sea ice extent and the regional climate gradient from north (subpolar) to south (polar) (Smith *et al.*, 1999).

The two mapped sediment drifts occupying the Gerlache Strait are the Andvord Drift and the Schollaert Drift. The Andvord Drift is located in, and adjacent to, Andvord Bay (Fig. 2A), at water depths ranging from 300 to 500 m. The drift covers 45 km² and has a maximum thickness of 40 m (Harris *et al.*, 1999) with a sedimentation rate of 2.4–5 mm/a (Domack *et al.*, 2003). This deposit is a multi-lobate feature, extending into outer Andvord Bay, into Gerlache Strait and toward the Errera Channel. The thickest depositional centre of the drift is not within the fjord, but rather is seaward of the entrance between the Aguirre and Errera Channels, responding to a local deceleration of current flow over the drift site (Harris *et al.*, 1999). The pattern of deposition in the Andvord Drift is consistent with sediments being eroded (or not deposited) at shallow depths and transported into deeper waters.

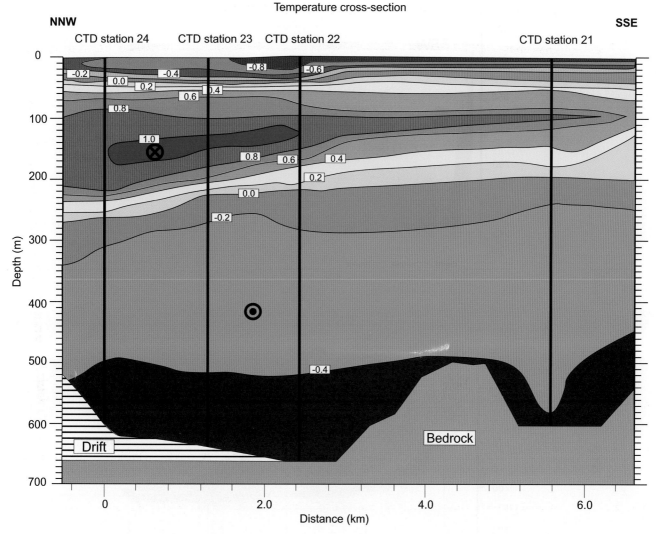

Fig. 4 Temperature cross section from four CTD casts obtained during the cruise NBP 01-07 across the Gerlache Strait at the Schollaert Drift. Circles with inner cross and dot indicate, respectively, northward and southward current along the strait. Location of CTD casts are shown in Fig. 2B.

The Schollaert Drift is located within the Gerlache Strait, between Anvers Island and Danco Coast, and comprises in seismically stratified 70 m-thick deposit that lies outside and mostly south of the Schollaert Channel (Canals *et al.*, 1998). The thickest depositional centre lies in the middle of the Gerlache Strait. The outer limits of the Schollaert drift are not well constrained due to a lack of a complete seismic survey in the area, but its major axis is at least 6 km long (Fig. 2).

The internal structure of the drift is defined by wavy stratified reflectors onlapping the northern slope of the strait. The southern margin contains evidence of semitransparent to chaotic lenses representing debris flow deposits originated

from slumping on the steep side wall of Gerlache Strait (Fig. 5). There is also some evidence of internal gas formation which interrupts reflector continuity.

TIDE MODEL

The ocean current, including the contribution from tides, is the most important factor to consider in sediment drift formation; however, few data other than the NBP 0107 ADCP records are available from the Gerlache Strait. To mitigate this deficiency, we employ a numerical tidal model. Modern tidal conditions were evaluated using the Antarctic

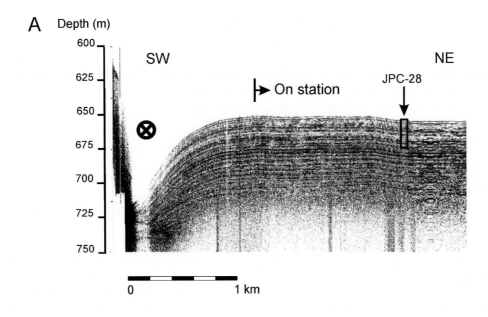

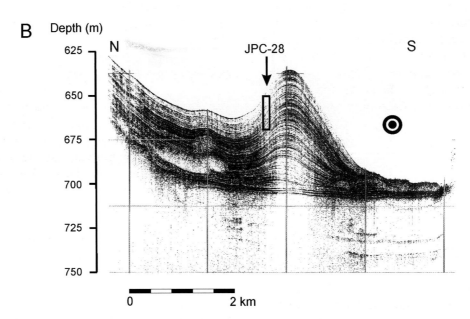

Fig. 5 High resolution seismic profiles obtained across the Schollaert Drift (location is shown in Figure 2B). The projected location of JPC-28 over the profiles is indicated. Circles with inner dot and cross indicate near-bottom current direction as in Fig. 4. (A) Chirp profile obtained during NBP 0107. (B) Topas profile obtained during cruise Gebrap'96.

Peninsula model version 04.01 (AntPen04.01). This model is constructed on a $1/30° \times 1/60°$ (~2 km) grid, and calculates the tides as the solution to the shallow-water wave equations (Robertson *et al.*, 1998) forced at the open boundaries by the circum-Antarctic CATS02.01 tide model (Padman *et al.*, 2002). While AntPen04.01 is a high-resolution model relative to other available Antarctic and global models covering this region, it still fails to resolve tides in many of the smaller channels in the complex topography along the western side of the Antarctic Peninsula.

The model allows us to determine the contribution of tides to the current velocities measured with the hull-mounted ADCP, thus revealing currents due to other causes including wind forcing. By making minor changes to the model configuration, we can also assess whether past changes in basin structure might have led to first-order alterations in sedimentation patterns. Tidal conditions for the ice retreat stage in Gerlache Strait at around 7000 years B.P. were evaluated by rerunning AntPen04.01 with a solid wall across the southern end of Gerlache Strait between Anvers Island and Danco Coast, south of Andvord Bay (hence closing the Neumayer Channel and the strait between Wiencke Island and Danco Coast). We consider this scenario to be realistic since the two connecting channels are very narrow and shallow and thus could easily be choked off by advancing glaciers, and remain choked off for some time as the Strait deglaciated. Our simple approach to modelling of tides for distinct glacial stages assumes that the effect of the modified topography does not extend to the open boundaries of the model, which is confirmed by looking at the details of specific tidal constituents in the two model runs. Furthermore, no other modifications were made to the topography to account for other variations in ice shelf extent elsewhere, or sea level variability. We also assume that global tides for this time are not significantly different from modern tides.

RESULTS

Core descriptions

Andvord Drift (JPC-18)

The 20 m long JPC 18 was recovered from outer Andvord Bay at 428 m water depth in a sediment drift deposit (Harris *et al.*, 1999) (Fig. 2). The core stratigraphy (Fig. 6) is characterized by thicker bioturbated intervals (high MS sections) and thinner laminated beds that consist of diatom ooze with minimal silt and clay (low MS sections). The top interval of JPC-18 is composed of 7 meters of bioturbated silty-clay with high MS. Below this interval the core is characterized by an inter-laminated diatom ooze and terrigenous silt/clay from 7 m to 20 m core depth (Carlson, 2001). These laminated horizons alternate at a decimetre scale with bioturbated and disrupted laminations, which are indicative of slower rates of deposition and biological disturbance of primary physical structures (Domack *et al.*, 2003). This transition in sediment-

ary structures is paralleled by changes in particle size-distribution and ice-rafted debris concentration. The age model (Fig. 6B) for this core was based on 8 Accelerator Mass Spectrometric (AMS) radiocarbon dates based on in-situ pelecypod shells (Domack *et al.*, 2003).

Schollaert Drift (JPC-28)

The 20 m long JPC-28 was taken at a water depth of 672 m to sample the deepest portion of the Schollaert Drift (Fig. 2). The sediment in JPC-28 (Fig. 6A) is characterized from core top to ~11 m by laminated diatom muds with intermittent bioturbation below which are rhythmically laminated diatom muds and oozes, with very few instances of bioturbation (Kirkwood *et al.*, 2004; Wright, 2000). The age model for this core was constructed on 10 AMS radiocarbon dates based on in-situ pelecypod shells (Domack *et al.*, 2003; Kirkwood *et al.*, 2004).

Avicenna Bay (JPC-29)

The 22 m long JPC 29 was obtained in Gerlache Strait at 720 m depth (Fig. 1) at 36.5 km distance from JPC-28. This core consists predominantly of bioturbated diatom mud alternating with disrupted clayey silt laminations (Fig. 6). Well-defined thick laminations occur from ~10 m to core bottom. A subtle change in sediment colour occurs between 18.6 m and 20 m core depth, where laminations change from an alternation between grey and olive grey to dark grey and black.

Kayak Bay (JPC-30)

The sediment core JPC-30 recovered 22 m of sediment from the Gerlache Strait at 1010 m depth (Fig. 1), 18.5 km from JPC-28 and ~8.5 km from Kayak Bay. JPC-30 consists predominantly of bioturbated diatom mud, with silt/clay and ooze interbedded laminations, which are more frequent and more marked from ~10 m to the core bottom (Fig. 6A). Those clayey silts consist of alternating light and dark lamination, each millimetres to centimetres thick. The limits between different facies are transitional and no sharp contacts were observed.

The age model was constructed with four AMS radiocarbon dates based on *in situ* shells (Table 2).

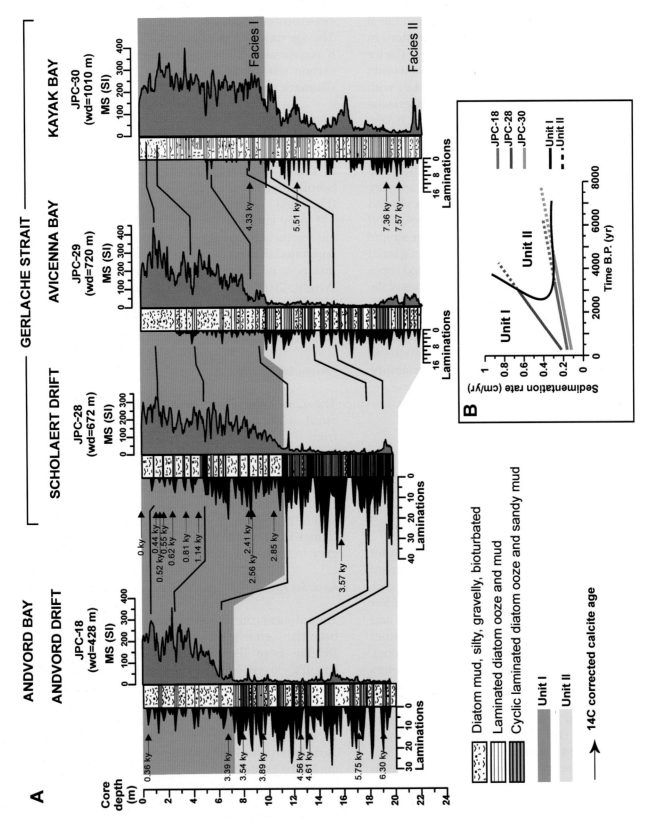

Fig. 6 (A) Correlation panel for sediment cores JPC-18, JPC-28, JPC-29 and JPC-30. Corrected radiocarbon ages and tie lines are shown. See Table 2 for JPC-30, and Domack *et al.* (2003) for JPC-18 and JPC-28 uncorrected radiocarbon ages. The lamination index and the magnetic susceptibility (MS) parameter are also shown. (B) Sedimentation rate of JPC-18, JPC-28 and JPC-30.

The core top was correlated with the jumbo trigger core and a Kasten core obtained in the same location to identify the water-sediment interface. Age–depth correlations show that the core covers the last 7700 years during which the sedimentation rate averaged ~3 mm/a.

Sediment facies and sedimentary units

The composition of the four cores along the strait is dominated by three sediment facies: homogeneous bioturbated diatom silty mud, laminated diatom ooze and sandy mud. The combination of

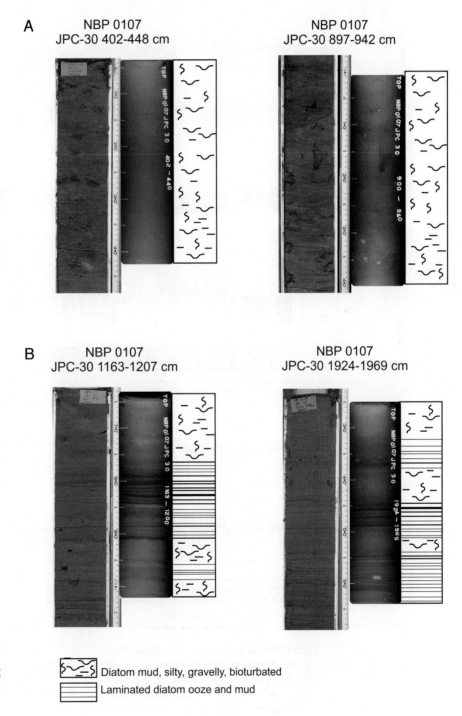

Fig. 7 Pictures, X-radiographs and schematic lithological log from selected sections in JPC-30 showing the two units described in the text. (A) Unit I. (B) Unit II.

Diatom mud, silty, gravelly, bioturbated
Laminated diatom ooze and mud

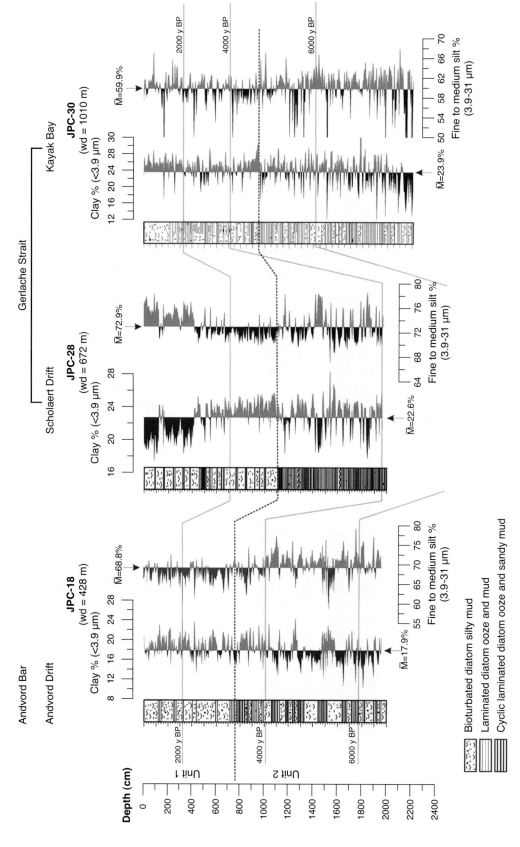

Fig. 8 Correlation panel between the three AMS radiocarbon dated cores in the study area. For each core graphic log, IRD content (in number of grains per 10 cm interval), fine to medium silt content (in volume %) and clay content (in volume %) are displayed. Clay and fine to medium silt contents over each core average are represented in gray, while those under the average are in black with respect to the median. Grain size parameters for JPC-18 and JPC-28 modified from Domack *et al.* (2003). Some time lines are shown in gray. The black dashed line marks the boundary between the sedimentary Units I and II.

these three sediment facies in different proportions lead to two main sedimentary units which are widely recognized in nearby areas (Domack *et al.*, 2003; Kirby *et al.*, 1998; Leventer *et al.*, 2002; Maddison *et al.*, 2005).

Unit I consists of a predominantly massive, bioturbated, silty-clay (Fig. 7A) rarely interrupted by laminated, mud-bearing diatom oozes. Although silt is the dominant size fraction, clay percent tends to reach values over the median, while fine to medium silt percent reaches minimum values (Fig. 8). Ice-rafted debris (IRD) content is variable (Fig. 8) and sand-size particles are present throughout the sequence in low amounts (<10%). This unit extends from 6 to 10 m below the sediment surface. Magnetic susceptibility in this unit ranges from 50 to 400 SI.

Unit II consists of couplets of laminated diatom ooze and sandy-mud alternating with massive, bioturbated, terrigenous silt/clay (Fig. 7B). The occurrence with respect to depth of massive (bioturbated) and laminated sediments is shown in the lithological column (Fig. 6A). The percentage of clay tends to be lower while silt increases considerably. Ice-rafted debris and sand-size particles are present throughout the sequence in low amounts (<10%) (Fig. 8). Magnetic susceptibility in this unit ranges from 0 to 100 SI units.

Age of deposits and sedimentation rates

The age model for cores JPC-18, JPC-28 and JPC-30 was constructed using the corrected and calibrated ages. A second order polynomial equation was used to relate time with core depth using least squares regression. Sedimentation rate positively increases with time, reaching at least an average of 0.8 cm/yr in the case of JPC-28. Sedimentation rate is always higher in the laminated interval (Unit II). The shift from Unit II to Unit I occurs at sedimentation rate values lower than 0.3 cm/yr in JPC-18 and JPC-30, while the same shift in JPC-28 takes place at sedimentation rates below 0.65 cm/yr.

Correlation

The correlation between adjacent core sites was achieved using a combination of physical properties, particle size, ice-rafted debris, sedimentary structures and radiocarbon data, allowing the con-

struction of a composite stratigraphic correlation over 70 km distance (Fig. 6A).

Currents

The tidal component of the velocities from the underway ADCP during cruise NBP 0107 (Fig. 3) is consistent with predicted currents from our tide model. Roughly half of the near-surface current in Fig. 3 can be ascribed to tides: the remainder is associated with the persistent, wind-forced northward drift through Gerlache Strait. We anticipate that tides comprise a much greater fraction of total currents in the deep water of the Strait, but lack the data at this time to confirm this. Modelled modern tidal current speeds display the well-known fortnightly spring/neap variability in strength (Fig. 9A), which arises from the superposition of the lunar and solar gravitational potentials and changes of the declination of the moon's orbit relative to the equator. Semi-annual periodicity in currents also occurs (Fig. 9B), arising from the difference between tidal forcing at equinoxes and solstices.

Blocking the southern entrance to the Gerlache Strait, representing glacial conditions near 7000 years B.P., significantly changes tidal current speeds (Fig. 9). The blockage forces the tidal energy flux, which propagates southward into Gerlache Strait along the west coast of the Antarctic Peninsula, into the Schollaert Channel. Tidal currents throughout much of the Gerlache Strait are affected, with stronger currents in the centre (where the Schollaert Drift lies) and weaker currents in Andvord Bay and Croker Passage (Fig. 9). There is almost no tidal flow just north of the model wall.

Beyond modifying tidal flow, blocking the southern entrance to Gerlache Strait will also influence the net wind-forced ocean and sea-ice transport throughout the Strait. We anticipate that the blockage would reduce the supply of nutrient-rich UCDW water to the region (the relatively warm subsurface layer seen in CTD profiles (Fig. 4) in the modern Strait.

DISCUSSION

Harris *et al.* (1999) proposed a model in which the formation of Andvord Drift was the response of decelerating currents flowing out from Andvord Bay into the Gerlache Strait. This deceleration allowed

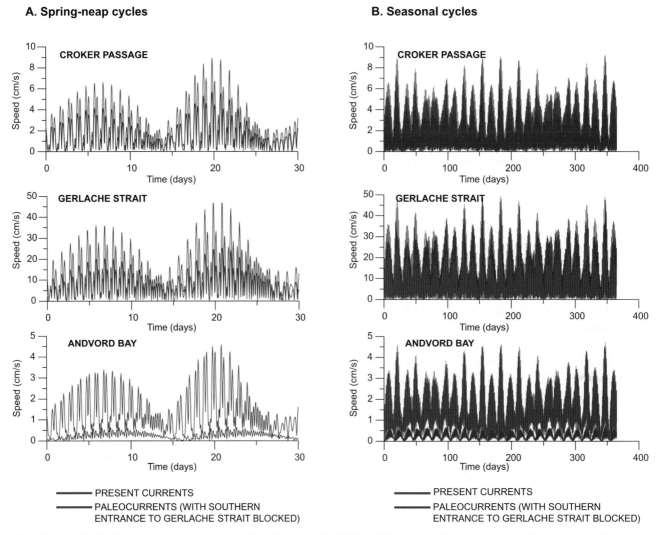

Fig. 9 Modelled tidal current speeds for modern (blue) and Middle Holocene (red) glaciation conditions (i.e. with open or ice-blocked southern entrance to the Gerlache Strait). (A) Spring/neap (fortnightly) variability. (B) Seasonal variability, showing semi-annual cycle superimposed on spring/neap cycles. See Fig. 1 for geographic reference.

the suspended sediment to accumulate on the sea floor. In this paper, we suggest that the formation of the Schollaert Drift was strongly influenced by the presence of estuarine conditions in the Gerlache Strait area during the Middle Holocene, and the role of mean and tidal currents in distributing the sediments along the Gerlache Strait.

Although the outer limits of the sediment drift are not well determined, the coherency and integrity of sediment and acoustic facies along the Gerlache Strait support the continuity (or at least the appearance of similar extremely thick contourite-drifts) to the north of the strait, at least to Kayak Bay in Brabant Island, where the sediment cores JPC-29 and JPC-30 are located.

Variability of oceanic conditions

Blocking the southernmost part of the Gerlache Strait creates a more estuarine circulation regime, by reducing net ocean and sea-ice transport through the Strait and enhancing the relative importance of tidal currents. During the melt season, glacial meltwater and meltwater from sea ice would be retained more effectively in the inner reaches of the fjord complex, where both mean and tidal flows are expected to be reduced. Thus, surface salinities would be decreased and suspended sediment concentration increased, a fundamental change that could take place independently of a climate shift toward greater melting and runoff. In addition, a comparison of the modelled tidal flows near the Schollaert drift for modern and glaciated conditions indicates a much larger range in energy between neap (minima in current velocities) and spring (maxima) tides for the glaciated state; see the Gerlache Strait time series in Fig. 9A. Wider ranges of spring/neap variability in currents would have enhanced the overall efficiency of deposition, and higher sedimentation rates across the Schollaert drift would be expected when the southern entrance to the Strait was blocked.

Origin of sedimentary units

Sediments from Gerlache Strait show the existence of two sedimentary units originated by different depositional processes. Unit I, massive bioturbated silt/clay, corresponds to the sediment settling through melt water plumes, vertical flux of siliceous organisms, ice rafting, and reworked by bottom currents and strong bioturbation activity. Unit II corresponds to the alternation of sediment settling from meltwater plumes and the vertical flux of siliceous organisms. The preservation of a well defined layered pattern in Unit II is associated with the absence of bioturbation. Reduced bioturbation in contourites can be related to high-energy conditions at the sea bottom (Yoon and Chough, 1993), high sedimentation rate (Laberg and Vorren, 2004) or periodic stagnation of near-bottom waters, generating anoxic conditions at the sea bottom (Sivkov *et al.*, 2002). Although tidal currents of up to ~50 cm s^{-1} are inferred to have occurred through sections of the Gerlache Strait (Fig. 9) when the southernmost end of the strait was blocked with ice, this velocity is not extremely high for such environments. At present sediment accumulation rates identified by 210Pb and 14C radiometric analysis ranges from 0.01 to 0.3 cm/yr over the Gerlache Strait, and has been constant in the last 100 yr (Isla *et al.*, 2002; Masque *et al.*, 2002); however, during the sedimentation of Unit II sedimentation rates ranged from 0.25 to 1 cm/yr (Fig. 6B). Bulk density obtained with GRAPE (gamma-ray attenuation porosity evaluation) every 1 cm shows that variations in density are minor (ranging from 1.15 to 1.6 g/cm^3). In that context, mass flux variations are overwhelmingly related to changes in sedimentation rate, and not to variations in sediment density. More likely, the primary cause of formation and preservation of these inter-layered

sediments is greater water stratification under enhanced estuarine conditions in combination with greater sedimentation rates.

The drop in magnetic susceptibility is paralleled by an increase of the laminated index, as shown in Fig. 6. The magnetic susceptibility parameter cumulates all iron mineral concentration with pronounced emphasis on ferromagnetic species. This drop in magnetic susceptibility can be related to dilution of magnetic minerals, different source of terrigenous input, or diagenesis of the magnetic minerals. A more detailed study of the sediment geochemistry in JPC-30 reveals a combination of the two latter factors as the cause of this drop in magnetic susceptibility (Willmott *et al.*, 2006).

The change between laminated diatom ooze and sandy mud (facies II) and bioturbated silt/clay (Unit I) is not synchronous in distinct locations within the Gerlache Strait and other locations such as Palmer Deep (Leventer *et al.*, 2002; Leventer *et al.*, 1996), indicating that the presence of well-defined laminations of diatom ooze and terrigenous silt/clay is due mostly to local rather than to regional factors.

In Unit II the diatom ooze and sandy mud laminations alternate with massive, bioturbated, terrigenous silt/clay (Fig. 7B). This alternation probably reflects the inter-annual variability of the three main processes affecting the WAIS climate: the El Niño Southern Oscillation (ENSO) (Stammerjohn and Smith, 1996; Turner, 2004), the Antarctic Oscillation (AO) also referred to as Southern Annular Mode (SAM) (Thompson and Solomon, 2002) and the position of the axis of the circumpolar low-pressure trough or Atmospheric Convergence Line (ACL) (Enomoto and Ohmura, 1990).

We consider that the combination of enhanced estuarine conditions associated with increased glaciation during the Middle Holocene, together with the differences between the spring and neap tidal conditions in the Gerlache Strait during that epoch, was the triggering mechanism of the formation of laminated sediments in this environment. In spring, the melting of sea ice would cause intense blooms that would lead to a spring biogenic deposition (Leventer, 1991; Maddison *et al.*, 2005). At the beginning of the summer, the sand silt and clay transported by ice, are suspended in the water column following ice melt and deposited in the sediment (Domack *et al.*, 2005; Leventer *et al.*, 2002). During the summer the water masses of the Gerlache Strait would be more stratified, as a result of the strong estuarine conditions produced by large fresh water input from melting ice margins, melting sea ice, warmer surface temperatures, and no exchange with the nutrient-rich UCDW water originating from the Antarctic Circumpolar Current. The resulting open water conditions throughout the strait and the lower ice melting rate would have reduced the formation of estuarine conditions, initiating the definitive shift from laminated diatoms ooze couplets (Unit II) to the massive, bioturbated, silty clay (Unit I).

CONCLUSIONS

The analysis of sedimentary and acoustic facies support a contourite origin for the sediment accumulations in Gerlache Strait. The stratigraphy of the sediment drift reveals the succession of two sedimentary facies. Unit I consists of a predominantly massive, bioturbated, silty-clay; while Unit II consists of couplets of laminated diatom ooze and sandy-mud alternating with massive, bioturbated, terrigenous silt/clay. The substitution of Unit II by Unit I appears to be a response to a change in the estuarine conditions that may have persisted during the Middle Holocene in the Gerlache Strait area. The fluctuating estuarine conditions may have acted as a controlling mechanism for the formation of laminated diatoms ooze couplets vs massive, bioturbated silt/clay, controlling the stratification of water masses and the volume of suspended sediment. The difference between spring (maxima in current velocities) to neap (minima) in tidal flow velocity would have led to enhanced deposition of sediment across the Schollaert Drift when the southern entrance to the Strait was blocked by ice.

ACKNOWLEDGEMENTS

We cordially thank the crew of R/V Nathaniel B. Palmer, technicians and scientists who participated in the NBP 9903 and NBP 0107 cruises where the sediment cores were acquired, and people from the Antarctic Research Facility (Florida, USA) for their help in sampling, description and core logging of the sediment cores. R. Gilbert, M. Rebesco, and the Editor, N. Glasser, provided many valuable comments and suggestions on the original manuscript. This work was supported by National Science Foundation Grants OPP-9615053 (ED) and OPP-0338101 (LP). We also acknowledge support from the Spanish 'Programa Nacional de Investigación', project REN2000–0896/ANT (COHIMAR). GRC Geociències Marines of the University of Barcelona, Spain, is supported by 'Generalitat de Catalunya' autonomous government through its excellency research groups program (ref. 2005 SGR-00152). This is ESR contribution number 53.

REFERENCES

Anderson, J.B., Kurtz, D.D., Domack, E.W. and Balshaw, K.M. (1980) Glacial and glacial marine sediments of the Antarctic continental shelf. J. Geol., 88: 399–414.

Backman, E. and Domack, E.W. (2003) Depositional Architecture and Seafloor Mapping of the Vega Drift, Erebus and Terror Gulf, Antarctic Peninsula. In: Fall Meet. Suppl. (Ed. E.T. AGU), 84 (46), pp. Abstract PP31B-0256. Eos Trans. AGU.

Camerlenghi, A., Domack, E.W., Rebesco, M., Gilbert, R., Ishman, S.E., Leventer, A., Brachfeld, S.A. and Drake, A. (2001) Glacial morphology and post-glacial contourites in northern Prince Gustav Channel (NW Weddell Sea, Antarctica). Mar. Geophys. Res., 22: 417–443.

Canals, M. and GEBRAP Team. (1997) Evolución geológica del Margen Pacífico de la Antártida Occidental: Expansión del fondo Oceánico y Tectónica de Placas, Universidad de Barcelona, Barcelona.

Canals, M., Estrada, F., Urgeles, R. and GEBRAP Team. (1998) Very high-resolution seismic definition of glacial and postglacial sediment bodies in the continental shelves of the northern Trinity Peninsula region, Antarctica. Ann. Glaciol., 27: 260–264.

Carlson, D. (2001) High resolution study of the Andvord Drift, Western Antarctic Peninsula, Antarctica. Unpublished MS thesis, Hamilton College, Clinton, NY, 41 pp.

Cook, A.J., Fox, A.J., Vaughan, D.G. and Ferrigno, J.G. (2005) Retreating glacier fronts on the Antarctic Peninsula over the past half-century. Science, 308: 541–544.

Domack, E.W. (1990) Laminated terrigenous sediments from the Antarctic Peninsula: the role of subglacial and marine processes. In: Glacimarine environments: Processes and sediments (Eds J.A. Dowdeswell and J.D. Scourse), 53, pp. 91–103. Geol. Soc. Spec. Pub., London.

Domack, E.W., Duran, D., Leventer, A., Ishman, S.E., Doane, S., McCallum, S., Amblas, D., Ring, J., Gilbert, I.M. and Prentice, M.L. (2005) Stability of the Larsen B ice shelf on the Antarctic Peninsula during the Holocene epoch. Nature, 436: 681–685.

Domack, E.W., Foss, D.J.P., Syvitski, J.P.M. and McClennan, C.E. (1994) Transport of suspended particulate matter in an Antarctic fjord. *Mar. Geol.*, **121**: 161–170.

Domack, E.W. and Ishman, S.E. (1993) Oceanographic and physiographic controls on modern sedimentation within Antarctic fjords. *Geol. Soc. Am. Bull.*, **105**: 1175–1189.

Domack, E.W., Leventer, A., Root, S., Ring, J., Williams, E., Carlson, D., Hirshorn, E., Wright, W., Gilbert, R. and Burr, G. (2003) Marine sedimentary record of natural environmental variability and recent warming in the Antarctic Peninsula. In: *Antarctic Peninsula Climate Variability* (Eds E.W. Domack *et al.*), *Antarctic Research Series*, **79**, pp. 205–224. American Geophysical Union, Washington, D.C.

Domack, E.W. and Williams, C.R. (1990) Fine structure and suspended sediment transport in three Antarctic fjords. *Contrib. Ant. Res.*, **50**: 71–89.

Dowdeswell, J.A., Whittington, R.J., Jennings, A.E., Andrews, J.T., Mackensen, A. and Marienfeld, P. (2000) An origin for laminated glacimarine sediments through sea-ice build-up and suppressed iceberg rafting. *Sedimentology*, **47**: 557–576.

Enomoto, H. and Ohmura, A. (1990) The influences of atmospheric half-yearly cycle on the sea ice in the Antarctic. *J. Geophys. Res.*, **95**: 9497–9511.

Evans, J., Dowdeswell, J.A. and Ó Cofaigh, C. (2004) Late Quaternary submarine bedforms and ice-sheet flow in Gerlache Strait and on the adjacent continental shelf, Antarctic Peninsula. *J. Quat. Sci.*, **19**: 397–407.

Evans, J., Pudsey, C.J., OCofaigh, C., Morris, P. and Domack, E. (2005) Late Quaternary glacial history, flow dynamics and sedimentation along the eastern margin of the Antarctic Peninsula Ice Sheet. *Quat. Sci. Rev.*, **24**: 741–774.

Faugeres, J.C., Stow, D.A.V., Imbert, P. and Viana, A. (1999) Seismic features diagnostic of contourite drifts. *Mar. Geol.*, **162**: 1–38.

Gilbert, R., Chong, A., Dunbar, R. and Domack, E.W. (2003) Sediment trap records of glacimarine sedimentation at Muller Ice Shelf, Lallemand Fjord, Antarctic Peninsula. *Arctic Alpine Res.*, **35**: 24–33.

Griffith, T.W. and Anderson, J.B. (1989) Climatic control of sedimentation in bays and fjords of the northern Antarctic Peninsula. *Mar. Geol.*, **85**: 181–204.

Grobe, H. (1987) A simple method for the determination of ice-rafted debris in sediment cores. *Polarforschung*, **57**: 123–126.

Hansom, J.D. and Flint, C.P. (1989) Holocene ice fluctuations of Brabant Island, Antarctic Peninsula. *Antarct. Sci.*, **1**: 165–166.

Harden, S.L., DeMaster, D.J. and Nittrouer, C.A. (1992) Developing sediment geochronologies for high-latitude continental shelf deposits: A radiochemical approach. *Mar. Geol.*, **103**: 69–97.

Harris, P.T. and Beaman, R.J. (2003) Processes controlling the formation of the Mertz Drift, George Vth continental shelf, East Antarctica: evidence from 3.5 Hz sub-bottom profiling and sediment cores. *Deep Sea Res. II*, **50**: 1463–1480.

Harris, P.T., Domack, E.W., Manley, P., Gilbert, I.M. and Leventer, A. (1999) Andvord drift: a new type of inner shelf, glacial marine deposystem from the Antarctic Peninsula. *Geology*, **27**: 683–686.

Harris, P.T. and O'Brien, P.E. (1998) Bottom currents, sedimentation and ice-sheet retreat facies successions on the Mac Robertson shelf, East Antarctica. *Mar. Geol.*, **151**: 47–72.

Isla, E., Masqué, P., Palanques, A., Sanchez-Cabeza, J.A., Bruach, J.M., Guillén, J. and Puig, P. (2002) Sediment accumulation rates and carbon burial in the bottom sediment in a high-productivity area: Gerlache Strait (Antarctica). *Deep Sea Research II*, **49**: 3275–3287.

Isla, E., Palanques, A., Alva, V., Puig, P. and Guillen, J. (2001) Fluxes and composition of settling particles during summer in an Antarctic shallow bay of Livingston Island, South Shetlands. *Polar Biol.*, **24**: 670–676.

Jacobs, S.S. (1989) Marine controls on modern sedimentation on the Antarctic continental shelf. *Mar. Geol.*, **85**: 121–153.

Kirby, M.E., Domack, E.W. and McClennen, C.E. (1998) Magnetic stratigraphy and sedimentology of Holocene glacial marine deposits in the Palmer Deep, Bellingshausen Sea, Antarctica: implications for climate change? *Mar. Geol.*, **152**: 247–259.

Kirkwood, G., Domack, E.W. and Brachfeld, S. (2004) Solar versus tidal forcing of centennial to decadal scale variability in marine sedimentary records from the Western Antarctic Peninsula (Abstract). In: *AGU Fall Meet. Suppl., PP51E-1360.* AGU, San Francisco.

Laberg, J.S. and Vorren, T. (2004) Weichselian and Holocene growth of the northern high latitude Lofoten Contourite Drift on the continental slope of Norway. *Sed. Geol.*, **164**: 1–17.

Leventer, A. (1991) Sediment trap diatom assemblages from the northern Antarctic Peninsula region. *Deep Sea Research Part A*, **38**: 1127–1143.

Leventer, A., Domack, E.W., Barkoukis, A., McAndrews, B. and Murray, J.W. (2002) Laminations from the Palmer Deep: A diatom-based interpretation. *Paleoceanography*, **17**: 10.1029/2001PA000624.

Leventer, A., Domack, E.W., Ishman, S.E., Brachfeld, S.A., McClennen, C.E. and Manley, P. (1996) Productivity cycles of 200–300 years in the Antarctic Peninsula Region: Understanding linkages among the sun, atmosphere, oceans, sea ice and biota. *Geol. Soc. Am. Bul.*, **108**: 1626–1644.

Maddison, E., J., Pike, J., Leventer, A. and Domack, E.W. (2005) Deglacial seasonal and sub-seasonal diatom record from Palmer Deep, Antarctica. *J. Quat. Sci.*, **20**: 435–4446.

Masqué, P., Isla, E., Sanchez-Cabeza, J.A., Palanques, A., Bruach, J.M., Puig, P. and Guillen, J. (2002) Sediment accumulation rates and carbon fluxes to bottom sediments at the Western Bransfield Strait (Antarctica). *Deep Sea Res. II*, **49**: 921–933.

McGinnis, J.P., Hayes, D.E. and Driscoll, N.W. (1997) Sedimentary processes across the continental rise of the southern Antarctic Peninsula. *Mar. Geol.*, **141**: 91–109.

McMullen, K., Domack, E., Leventer, A., Olson, C., Dunbar, R. and Brachfeld, S. (2006) Glacial morphology and sediment formation in the Mertz Trough, East Antarctica. *Palaeogeogr., Palaeoclimatol., Palaeocol.*, **231**: 169–180.

Nelson, D.M. (1988) Biogenic silica and carbon accumulaion in the Bransfield Strait, Antarctica, Unpublished MSc thesis, North Carolina State University, USA.

Niiler, P., Illeman, J. and Hu, J.H. (1990) RACER: Lagrangian and drifter observations of surface circulation in the Gerlache and Bransfield Straits. *Antarct. J. US*, **25**: 134–137.

Padman, L., Fricker, H.A., Coleman, R., Howard, S. and Erofeeva, S. (2002) A new tidal model for the Antarctic ice shelves and seas. *Ann. Glaciol.*, **34**: 247–254.

Palanques, A., Isla, E., Puig, P., Sanchez-Cabeza, J.A. and Masqué, P. (2002) Annual evolution of downward particle fluxes in the Western Bransfield Strait (Antarctica) during the FRUELA project. *Deep Sea Res. II*, **49**: 903–920.

Pike, J. and Kemp, A.E.S. (1996) Records of seasonal flux in Holocene laminated sediments, Gulf of California. In: *Palaeoclimatology and palaeoceanography from laminated sediments* (Ed A.E.S. Kemp), **116**: 157–169. Geological Society Special Publication.

Pike, J., Bernhard, J.M., Moreton, S.G. and Butler, I.B. (2001) Microbioirrigation of marine sediments in dysoxic environments: Implications for early sediment fabric formation and diagenetic processes. *Geology*, **29**: 923–926.

Rebesco, M. and Stow, D.A.V. (2002) Seismic expression of contourites and related deposits. *Mar. Geophys. Res.*, **22**: 303–308.

Robertson, R.A., Padman, L. and Egbert, G.D. (1998) Tides in the Weddell Sea. In: *Ocean, ice, and atmosphere interactions at the Antarctic Continental Margin* (Eds S.S. Jacobs and R.F. Weiss), *Antarctic Research Series*, **75**: 341–369. American Geophysical Union, Washington, D.C.

Sivkov, V., Gorbatskiy, V., Kuleshov, A. and Zhurov, Y. (2002) Muddy contourites in the Baltic Sea: an example of a shallow-water contourite system. In: *Deep-water contourites systems: modern drifts and ancient series, seismic and sedimentary characteristics* (Eds D.A.V. Stow, J.C. Faugeres, J.A. Howe, C.J. Pudsey and A. Viana), **22**, pp. 7–20. The Geological Society of London, London.

Smith, D.A., Hofmann, E., E., Klinck, J.M. and Lascara, C.M. (1999) Hydrography and circulation of the West Antarctic Peninsula Continental Shelf. *Deep. Sea Res. I*, **46**: 925–949.

Stammerjohn, S.E. and Smith, R.C. (1996) Spatial and temporal variability of western Antarctic Peninsula sea ice coverage. In: *Foundations for Ecological Research West of the Antarctic Peninsula* (Eds R.M. Ross, E.E. Hofmann and L.B. Quetin), *Antarctic Research Series*, **70**: 81–104. American Geophysical Union, Washington, D.C.

Stuiver, M., Reimer, P., Bard, B., Beck, J.W., Burr, G.S., Hughen, K.A., Kromer, B., McCormack, G., Van der Plicht, J. and Spurk, M. (1998) INTCAL98 radiocarbon age calibration, 24,000-0 cal. BP. *Radiocarbon*, **40**: 1041–1083.

Stuiver, M., Reimer, P. and Reimer, R.W. (2005) Calib 5.0. (www program and documentation) /http://radiocarbon.pa.qub.ac.uk/calib/S.

Thompson, D.W.J. and Solomon, S. (2002) Interpretation of recent southern hemisphere climate change. *Science*, **296**: 895–899.

Turner, J. (2004) The El Niño Southern Oscillation and Antarctica. *Int. J. Climatol.*, **24**: 1–31.

Warner, N.R. and Domack, E.W. (2002) Millennial- to decadal-scale paleoenvironmental change during the Holocene in the Palmer Deep, Antarctica, as recorded by particle size analysis. *Paleoceanography*, **17**.

Wefer, G., Fischer, G., Fueetterer, D. and Gersonde, R. (1988) Seasonal particle flux in the Bransfield Strait, Antarctica. *Deep Sea Res.*, **35**: 891–898.

Willmott, V., Domack, E.W. and Canals, M. (2006) *Marine productivity and terrigenous supply in Holocene sediments from the Antarctic Peninsula. A geochemical and sedimentological approach.* Abstract: OS-082. Ocean Sciences Meeting. Honolulu, Hawaii.

Wright, W.N. (2000) The Schollaert sediment drift: an ultra high resolution paleoenvironmental archive in the Gerlache Strait, Antarctica. MS, Hamilton College, Clinton, NY, 70 pp.

Yoon, H.I., Han, M.H., Park, B.-K., Oh, J.-K. and Chang, S.-K. (1994) Depositional environment of near-surface sedimentss, King George Basin, Bransfield Strait, Antarctica. *Geo-Mar. Let.*, **14**: 1–9.

Yoon, S.H. and Chough, S.K. (1993) Sedimentary characteristics of Late Pleistocene bottom current deposits, Barents Sea slope off northern Norway. *Sed. Geol.*, **82**: 33–45.

Yoon, H.I., Han, M.H., Park, B.-K., Oh, J.-K. and Chang, S.-K. (1994) Depositional environment of near-surface sediments, King George Basin, Bransfield Strait, Antarctica. *Geo-Mar. Let.*, **14**: 1–9.

Zhou, M., Niiler, P.P. and Hu, J.H. (2002) Surface currents in the Bransfield and Gerlache Straits, Antarctica. *Deep. Sea Res.*, 267–280.

Sedimentary signatures of the Waterloo Moraine, Ontario, Canada

H.A.J. RUSSELL*, D.R. SHARPE* AND A.F. BAJC†

*Geological Survey of Canada, 601 Booth Street, Ottawa, Ontario, KIA 0E8, Canada
†Ontario Geological Survey, 933 Ramsey Lake Road, Sudbury, ON, P3E 6B5, Canada (E-mail: hrussell@nrcan.gc.ca)

ABSTRACT

The Waterloo Moraine is a stratified moraine with an area of ~400 km^2 and thickness of 60 m. Its sedimentary record contains evidence of high-magnitude meltwater discharge and rapid sedimentation. Sedimentary structures and sediment architecture are presented from several sites that record depositional events of subcritical flows, supercritical flows and hydraulic jumps. The sedimentological signature of these events is: (i) a variety of large scale cross-strata, (ii) climbing cross-stratification, (iii) antidune cross-stratification and (iv) steep-walled scours with diffusely graded fills. These depositional signatures can be attributed to two depositional environments. A conduit or esker setting is interpreted for the observed large-scale cross-beds with bimodal and openwork gravel, whereas a subaqueous fan setting is inferred for the remainder of the studied deposits. The subaqueous fan setting is interpreted in terms of the jet-efflux model where rapid streamwise deceleration of the inertia-dominated jet is recorded by antidunes produced by supercritical flow; steep-walled scours and diffusely graded fills formed beneath hydraulic jump processes; and climbing dunes record subcritical flow. The paper demonstrates that stratified deposits of the Waterloo Moraine consist of organized deposits that can be interpreted within the current understanding of flow dynamics and depositional sedimentary facies models. This provides a step toward improved understanding of the spatial heterogeneity of the moraine sediment and development of predictive models for improved understanding of the hydrogeological character of the Waterloo Moraine.

Keywords Glaciofluvial, moraine, esker, subaqueous fan, antidune, hydraulic jump, southern Ontario

INTRODUCTION

Moraines form a prominent element of the southern Ontario landscape that were first mapped by Taylor (1913) in the early 20th century. Subsequent work has refined both the extent and form of these landforms (see Barnett, 1992), and classified them as either till moraines or kame moraines (Chapman and Putnam, 1943; Fig. 1B). Kame moraines have been inferred to be deposited at the margin of inactive ice, whereas till moraines are inferred to be thrust into position by advancing ice and thus mark the extent of ice advance (Chapman and Putnam, 1943). Comparatively few studies of the sedimentology of moraines have been completed in southern Ontario (e.g. Karrow and Paloschi,

1996). Consequently, 60 years after Chapman and Putnam's work, understanding of many moraines in southern Ontario is incomplete. One exception is the Oak Ridges Moraine (ORM), which has been the focus of significant sedimentological and stratigraphic studies (e.g. Barnett et al., 1998; Sharpe et al., 2002).

Inferences about how moraines form can only be made from a combination of landform and detailed sedimentological analysis (e.g. Warren and Ashley, 1994). To reconstruct the origin of moraines in southern Ontario requires improved description of moraine sediment. Previous analyses of bedforms, eskers, and moraines in southern Ontario have provided significant constraints and insights into the processes active during deposition of these

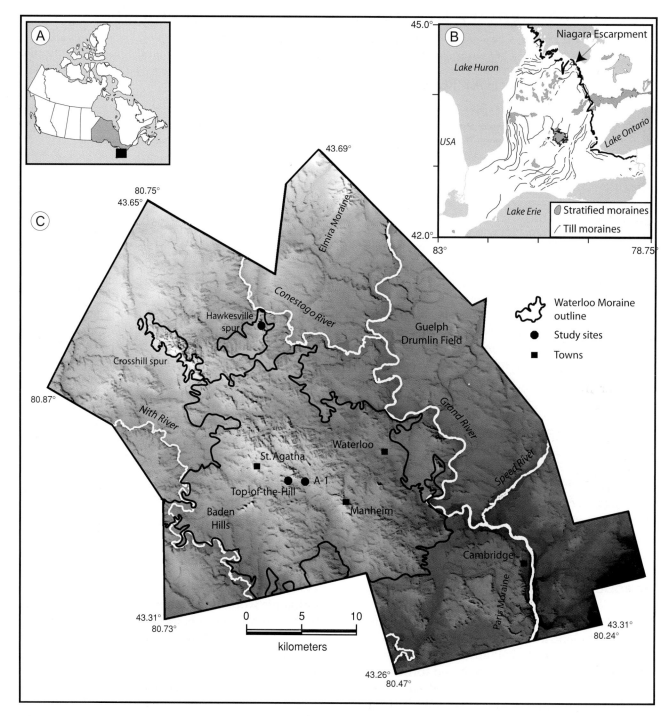

Fig. 1 (A) Location of Ontario (shaded) in Canada and location of southern Ontario (black box). (B) Distribution of stratified and till moraines in southern Ontario. (C) Hill-shaded DEM of the Waterloo Moraine with study sites. The moraine generally has a greater surface roughness than surrounding terrain.

landforms (e.g. Saunderson, 1976; Shaw and Gorrell, 1991; Brennand, 1994; Benvenuti and Martini, 2002; Russell and Arnott, 2003). These processes include fluidal or hyperconcentrated meltwater flows that

may be subcritical or supercritical depending upon flow depth, velocity, and density. Inferences about the depositional flow are possible from the style and texture of sedimentary deposits, stratigraphic architecture, and streamwise sediment facies transitions. For example, Shaw and Gorrell (1991) interpret large, metre-scale foresets to represent deposition of large dunes within tunnel channels from deep, turbulent subcritical meltwater flows. Brennand and Shaw (1996) interpret diffusely graded sandy antidune stratification to represent critical-supercritical flow conditions within an esker. The flow transformation between supercritical and subcritical flows is marked by a hydraulic jump. Using sediment stratification and sedimentary architecture, Russell and Arnott (2003) constrained steep-walled scour deposits in the Oak Ridges Moraine to an origin by hydraulic-jumps. Consequently, it is likely that detailed sedimentological studies of moraines, in this case the Waterloo Moraine, can yield significant insight into depositional controls on moraine formation and hence ice sheet events.

An improved understanding of moraines in southern Ontario is of particular interest due to increased concern over source water protection for potable groundwater supplies (TEC, 2004). Routing of water from the surface to groundwater, and subsequently to discharge points is controlled by moraine topography and three-dimensional sediment facies architecture. Studies that emphasize geological controls on the groundwater system, such as those in the ORM (Howard *et al.*, 1995; Sharpe *et al.*, 2002; Meriano and Eyles, 2003; Holysh *et al.*, 2004), north of Lake Ontario (Fig. 1), are thus required to understand watershed flow systems. Such studies provide an important contribution to understanding areas of watershed recharge and advance understanding for the protection of source water in Ontario watersheds.

Paper objective

Sedimentary deposits of stratified moraines in southern Ontario have commonly been dismissed as chaotic; lacking any stratigraphic and sedimentary structure organization that could provide insights into the origin of the landforms. This paper demonstrates that much of the stratified material of the Waterloo Moraine can be described within the well-developed framework of sedimentary analysis developed during the past forty years (e.g. Simons *et al.*, 1965; Miall, 1996). A number of sedimentary structures and textures are described and are attributed to depositional processes within a flow

dynamics framework. The intent of the paper is to draw attention to the range of processes recorded in the Waterloo Moraine and, consequently, the potential of the moraine to provide insight into meltwater processes in southern Ontario during moraine formation. Additionally, the hydrogeological significance of improved understanding of the moraine sedimentology is indicated.

Geological setting and previous work

Southern Ontario is an area that is covered, almost completely, by surficial sediment (e.g. Barnett, 1992). Bedrock crops out along escarpments, the largest being the Niagara Escarpment, and in river valleys. Surficial deposits extend back to lllinoian age, with Lake Ontario bluffs forming the type section for much of the pre Late-Wisconsinan stratigraphy (Coleman, 1932; Karrow, 1974). A notable part of the Quaternary landscape and deposits are moraines, which form a subparallel network of ridges around the margin of the Niagara cuesta (Fig. 1b). These moraines were first mapped by Taylor and classified as either till moraines or kame moraines (Chapman and Putnam, 1943). Geological knowledge of many of these moraines is limited to surficial geological mapping at a 1:50,000 scale (e.g. Karrow, 1993).

The Waterloo Moraine is a prominent topographic feature with an elevation range of 330 to 400 masl. Local topographic highs can rise above 420 m and have relief of ~50 m, for example the Baden Hills and Crosshill Spur (Figs. 1C, 2). The moraine consists of a main southeast-northwest elongate ridge and a number of associated ridges that are orientated radial to the main moraine (e.g. Crosshill, Hawkesville; Karrow, 1993). The Waterloo Moraine sediment stratigraphically overlies Catfish Creek and lower till (Fig. 3, Bajc and Karrow, 2004). The moraine consists of sand, gravel, and minor amounts of muddy diamicton and associated fine-textured glacilacustrine deposits (Duckworth, 1983; Bajc and Karrow, 2004). Detailed descriptions of the Hawkesville Spur indicate that it is predominantly gravel (Bowes, 1976). Depending upon the stratigraphic position of the diamicton, it has been correlated on the basis of matrix texture, clast lithology and matrix chemistry with upper and middle Maryhill, Tavistock, Mornington and Port Stanley tills (Karrow, 1993).

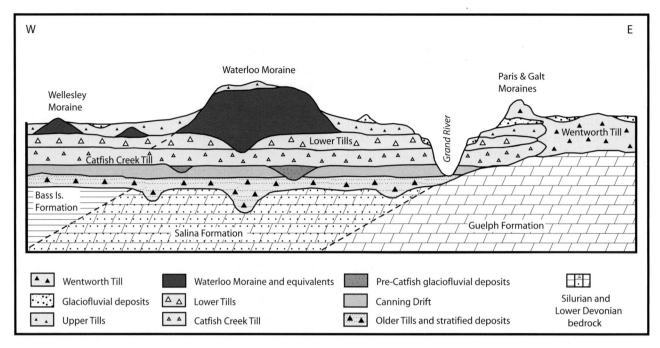

Fig. 2 Conceptual model of the regional stratigraphic context of the Waterloo Moraine (Bajc and Karrow, 2004).

On the basis of terrain analysis and till stratigraphy, the Waterloo Moraine has been interpreted as an interlobate moraine deposited between converging flow of the Georgian Bay – Lake Huron ice lobes from the north and west, Lake Ontario lobe from the east, and the Lake Erie lobe from the south (Karrow, 1973; Chapman and Putnam, 1984). A proposed interlobe re-entrant and meltwater flux in this re-construction was attributed to climatic warming (Karrow, 1974). On the basis of surficial geological mapping and sedimentary facies analysis, the moraine has been interpreted to consist of esker, subaqueous fan and deltaic deposits (Harris, 1970; Bowes, 1976; Karrow and Paloschi, 1996; Bajc and Karrow, 2004).

Study sites

Mapping of the Waterloo Moraine indicates that it is commonly <60 m thick (Fig. 3, Bajc and Karrow, 2004). Analysis of sediment textures indicates that the moraine is coarsest in the southeast, where it is predominantly sand and gravel and that it fines westward. West of St Agatha the moraine is predominantly fine sand, silt, and muddy diamicton (Karrow, 1993). The moraine also fines laterally to the northeast and southwest from a gravel rich

core, beneath the high ground between Manhiem and St. Agatha, to mud. A cored borehole located at the Top-of-the-Hill site encountered bedrock at 91 m depth. This hole was drilled from the pit floor and intercepted sand and gravel to a depth of 60 m, below which 28 m of diamicton overlies Palaeozoic bedrock (Fig. 3). Core recovery for the Waterloo Moraine unit was 88%, with a mud-sand-gravel composition of 17–62–21 %. From this one borehole, 98% of the mud occurs in the basal 10 m of the moraine (Fig. 3). Average bed thickness in the core is 60 and 30 cm respectively for 25 sand and 6 gravel beds, excluding 7 outlier beds 2–8 m thick.

The sedimentary features presented in this paper were observed at three sites, Hawkesville Spur, Top-of-the-Hill and A-1 aggregate pits (Fig. 1). The Hawkesville Spur is located to the north of the main Waterloo Moraine deposit and south of the Elmira Moraine. The Top-of-the-Hill and A1 sites are ~14 km to the southeast of the Hawkesville Spur, in the main body of the Waterloo Moraine

Hawkesville Spur

The Hawkesville Spur occurs in the northern part of the area. It is truncated by the Conestoga River

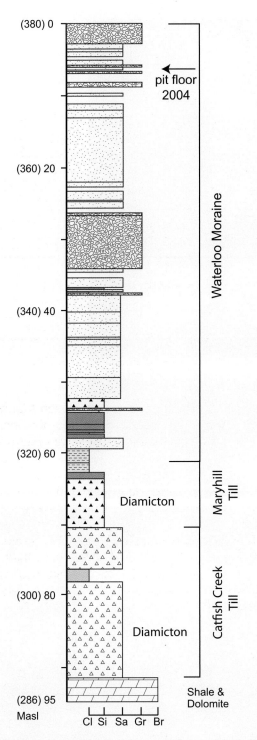

Fig. 3 Log of borehole that penetrated to bedrock at the Top-of-the-Hill site. Note the sand and gravel character of the moraine succession with a basal mud rich interval at 60 m depth. (Borehole from Regional Municipality of Waterloo.)

to the north and extends southward for 3 km as a well-defined ridge. Ridge relief is <50 m, with a maximum elevation of ~400 m, decreasing toward the south. It has a width of <500 m in the north, and broadens southward to 1.2 km (Fig. 4; Bajc and Karrow, 2004). Much of the original ridge morphology has been destroyed by the extensive, long-term aggregate operations. The northern two-thirds of the ridge is mapped as glaciofluvial sand and gravel with flanking glaciofluvial sand that completely crosses the southern, lower relief portion of the ridge (Karrow, 1993). The sedimentary deposits of the ridge are predominantly gravel with large tabular gravel cross-sets exposed in sections up to 25 m high (Fig. 5; Bowes, 1976). Bowes studied three pits in the ridge, examined 23 sections, and presented 10 measured sections. Tabular cross-bedded gravel beds were measured in 8 of the sections, with up to 7 stacked and nested beds recorded in an individual section. Maximum bed-thickness was 9 m and the average thickness of 33 beds was 4 m. The mean palaeoflow direction varied both up-section and along the length of the ridge; however, there was a strong trend toward the south (Bowes, 1976). The deposits fine from cobble gravel in the north to more sand-rich deposits in the south. Boulder sizes of 45 cm were reported from the north with average maximum clast sizes of 19 to 23 cm (Bowes, 1976). At the south end of the landform a coarsening upward succession is present, consisting of ripple and plane laminated silts and silty very fine sand grading up to pebble gravel. The entire sequence is capped by a fine-grained diamicton.

The ridge flanks have been interpreted as ice-contact slopes (Karrow, 1993), and internally drained depressions along the flanks of the ridge are interpreted as kettles (Bowes, 1976, Karrow, 1993). Within the excavated part of the ridge there is a silt filled depression with diamicton at the base. This diamicton unconformably overlies glaciofluvial sand and gravel of the Hawkesville spur. Wood fragments collected from silts immediately overlying the diamicton have a C-14 age of 12, 135 ± 75 yrs BP (OGS 2478). The depression is interpreted to be a lacustrine kettle fill (Bajc and Karrow, 2004). Based on sediment facies and scale of gravel cross-beds, the ridge sediment has been interpreted as gravel bar deposits (Kuehl, 1975) and ice-contact deltaic deposits (Bowes, 1976).

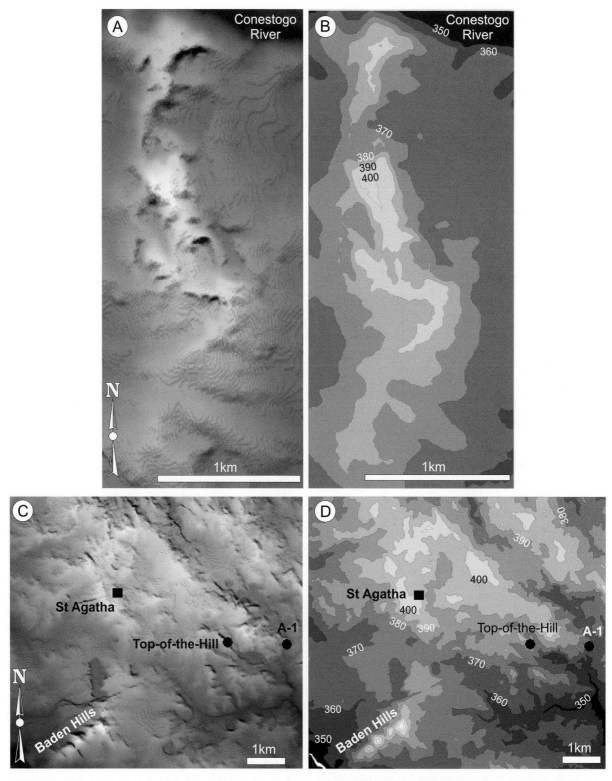

Fig. 4 General landform setting of the study sites. (A) Hill-shaded DEM of the Hawkesville spur. (B) Ten-metre elevation contour map of the Hawkesville spur. Note decreasing elevation and broadening of the ridge form toward the south. (C) Hill-shaded DEM of the central part of the Waterloo Moraine with the Top-of-the-Hill and A-1 study sites indicated. (D) Ten-metre elevation contour map derived from the DEM. Note increasing elevation toward the northwest and location of A-1 site on main southeast-northwest ridge and the Top-of-the-Hill site lateral to the ridge.

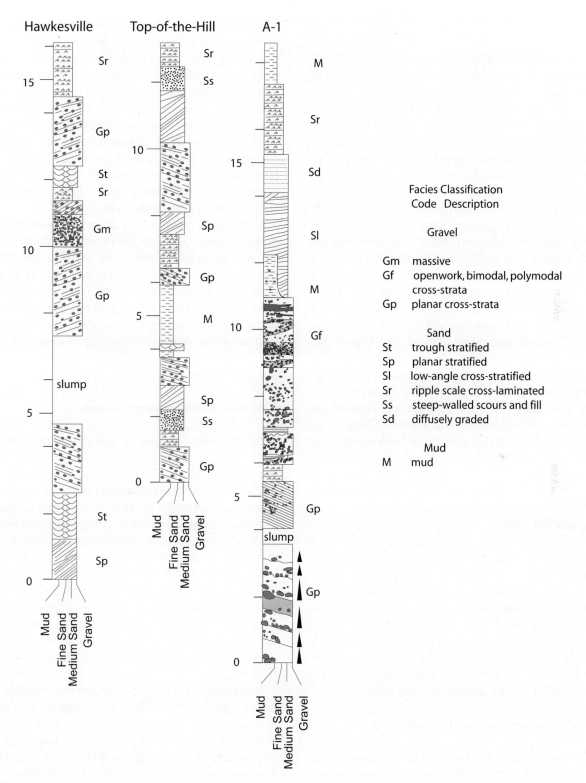

Fig. 5 Composite sedimentological logs for the three study sites. The Hawkesville spur log is derived from data in Bowes (1976). Note the tripartite succession at the Top-of-the-Hill and A-1 site of sand and gravel, mud, and sand and gravel. Facies codes modified from Miall (1996).

Top-of-the-Hill and A-1

The Top-of-the-Hill and A-1 sites are located in the main body of the Waterloo Moraine, 4–5 km north of the Baden Hills, a northeast trending glacio-fluvial ridge with a maximum elevation of 425 m (Fig. 4; Karrow, 1993). The sites are 1.5 km apart, have surface elevations of 390 m, and occur within a relatively large area mapped as ice-contact glacio-fluvial sand (Bajc and Karrow, 2004). One hundred to five hundred metres to the north, the sand is overlain by a muddy diamicton, the Maryhill Till. The topography is generally hummocky to gently rolling (Bajc and Karrow, 2004). The two pits are 15–20 m deep and contain a tripartite succession of gravel, mud and sand (Fig. 5). The A-1 site is the coarser deposit and consists of a lower gravel unit that is being mined, an overlying mud unit, and an upper unit that fines upward from medium sand to laminated silt and massive mud. The Top-of-the-Hill site is a shallower broader excavation with a similar tripartite stratigraphy but overall finer sediment calibre (Fig. 5). Palaeoflow measurements at the two sites place the A-1 site up-flow of the Top-of-the-Hill site (Bajc and Karrow, 2004). The A-1 and Top-of-the-Hill sites have been interpreted to be esker-subaqueous fan deposits (Bajc and Karrow, 2004).

SEDIMENTOLOGY

Data presented in this paper were collected in 2004 during a series of multi-day trips and a week of focused fieldwork. Four notable styles of stratification observed in three active aggregate pits are discussed and attributed to subcritical and supercritical flows; i) medium and large-scale cross-stratification, ii) climbing dune-scale cross-stratification, iii) antidune stratification, and iv) steep-walled scours and associated fills.

Medium- and large-scale cross-stratification

Cross-stratification is observed in sediment from centimetre to metre scale and as non-climbing and climbing forms. Individual cross-strata within beds may consist of a single sediment facies or a number of distinct sediment facies. On the basis of bed thickness, dip of cross-strata, sediment facies

association, and nature of basal contact, four styles of medium- and large-scale cross-strata are reviewed: i) medium-scale, planar-tabular cross-beds; ii) large-scale cross-stratified beds; iii) cross-beds of bimodal and openwork gravel; and iv) large-scale low-angle clinoforms. Descriptions are based on individual examples rather than representing a composite of many beds from different locations, which is common in sediment facies descriptions. Few observations and an absence of physical measurements lessen the extent and reliability to which styles i and ii can be described.

Grading in cross-sets is described following the nomenclature of Allen (1984, in Kleinhans, 2004). Tangential grading refers to grading parallel to the direction of sediment transport. Consequently, tangential grading in cross-strata is normal when the largest particles are deposited toward the top of the cross-strata. Perpendicular grading is upward, normal to the plane of flow. In this case, the grading is normal when the largest particles occur at the base of the unit.

(i) Medium-scale planar-tabular cross-strata

Medium-scale, tabular, cross-strata are common in the Hawkesville Spur, a northwest extension of the moraine (Figs. 1, 4, 5). Tabular beds of 1–2 m thickness have been observed with near to angle-of-repose, normally graded cross-strata (Fig. 6A). The base of the cross-set is abrupt, sub-horizontal, with infrequent cobble-sized clasts. Individual cross-strata are ~0.4 m thick and consist of: i) normally graded polymodal gravel, ii) bimodal gravel and iii) openwork gravel. The polymodal and bimodal gravel are predominantly clast supported. The apparent palaeoflow direction is southeast, parallel to the landform orientation. Similar cross-beds are observed at the top of the large-scale cross-strata (Fig. 6C).

Interpretation: On the basis of the tabular geometry, planar bedding, cross-strata dip-angle, and sediment facies this bedding style is interpreted as the product of bedload transport and deposition within dunes (Harms *et al.*, 1975). The scale of bedding described here, along with the stratigraphic setting is similar to the smaller tabular gravel cross-beds described by Bowes (1976) in a more extensive study of the Hawkesville Spur.

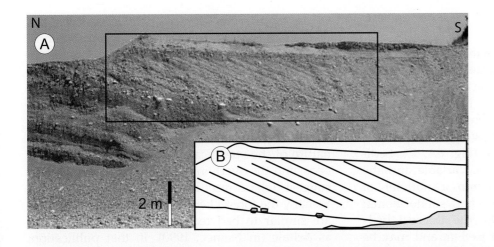

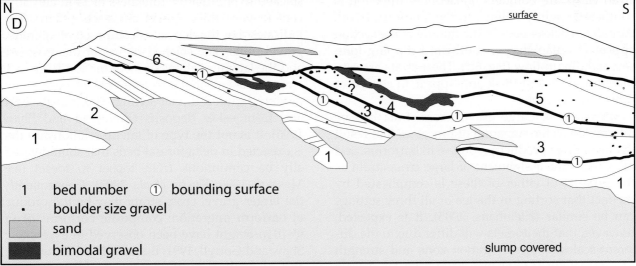

Fig. 6 Cross-stratified beds present in the Hawkesville spur. (A) Tabular ~2.5 m thick, normally graded cross-beds. (B) Line drawing of foreset strata in (A). (C) Succession of 10–15 m high stacked, large-scale cross-stratified beds (2, 3, 5, 6). Bed 4 consists of medium-scale backset beds. Face is ~20 m high and ~70 m long. Note bed 6 is similar to cross-strata in (A). (D) Line drawing from photo mosaic presented in (C). (Palaeoflow in all images is from left to right.)

(ii) Large-scale cross-strata

In contrast to the simple planar-tabular geometry described above, style ii consists of a succession of cross-stratified beds that have an estimated, stacked thickness of 10–15 m (Fig. 6C). No measurements of this face were completed due to access restrictions. Consequently, the following description is based on photos of this face (Fig. 6A, C). Sediment facies and dip angle are variable and the succession can be divided into a number of beds separated by high-angle and subhorizontal disconformities (Fig. 6C, D). The thickest beds have an estimated height of 6–8 m, and may be significantly thicker depending upon continuity beneath the extensive slump along the basal half of the face. Within beds the bedding includes subhorizontal (e.g. bed 1), steeply dipping (e.g. beds 2, 6, 7), and backset beds (e.g. bed 4). In at least one bed (bed 2) the cross-strata exhibit a down-dip transition from steeply dipping to subhorizontal. Sediment facies consist of sand, sandy gravel, polymodal framework and matrix supported gravel, bimodal framework gravel, and minor openwork gravel. These sediment facies are preferentially concentrated in various beds. For example, bed 2 is predominantly sandy gravel. The largest oversized clasts are boulders. Palaeoflow direction is southeast, slightly oblique to the landform trend. Apparent backset beds in the middle of the section appear to onlap steeper southward dipping foresets and dip up-flow (Fig. 6C). These strata are estimated to be 1–2 m thick and appear to consist of polymodal gravel.

Interpretation: Large gravel cross-strata can be interpreted as the signature of processes that result in one of three products, i) deltas, ii) barforms, and iii) dunes. The assignment of large cross-strata as the product of either of these is complicated by the fact that sorting in the lee of all three settings can be similar (Kleinhans, 2005). It is expected, however, that the toesets will differ due to the different scales of the separation zone and strength of return flow (Kleinhans, 2005). The large-scale cross-strata of the Hawkesville Spur have previously been interpreted as the product of barform migration (Kuehl, 1975 in Bowes, 1976) and delta progradation (Bowes, 1976; Karrow, 1993). Traditionally it has been assumed that dunes could not form in gravel sediment (e.g. Carling, 1999), con-

sequently interpretation of gravel cross-strata has been biased to barform and delta interpretations. It is apparent, however, that there is no reason to believe that dunes are restricted from forming in gravel (Carling, 1999; Lunt et al., 2004). Thus, the following interpretation considers evidence supporting each of the three possibilities.

The large scale of the cross-strata, the composite nature of the stratification, the onlapping backset beds, and the overlying planar-tabular cross-set all suggest delta foreset bedding (e.g. Nemec, 1990). Visually the foreset bedding and faint backset beds are similar to bed features described by Colella as deltaic (in Nemec, 1990). In that publication, the delta foreset beds with chute-fill backsets are interpreted to be deposited beneath a granular hydraulic-jump (e.g. Nemec, 1990). At odds with a deltaic interpretation is the lateral relationship with stacked beds (see below, Bowes, 1976) and the absence of texturally distinct toeset beds. Texturally distinct toesets in a gravel dominated glaciogenic delta is a feature clearly detailed by Kostic et al. (2005).

Observations of this study and a re-assessment of Bowes (1976) sections suggest that an alternative to the delta interpretation is probable. Bowes describes 1–3 m thick tabular gravel cross-sets stacked to cumulative thickness of 14 m and adjacent to very thick gravel cross-sets (~8 m). This indicates that the deposit is composed of aggraded gravel units. The stacked nature of the cross-sets and the lateral relations of medium and large scale cross-strata are similar to relationships observed in modern braided fluvial settings (Lunt et al., 2004) and from esker deposits (Brennand and Shaw, 1996). It is not the type of lateral relationship that is expected in delta foreset beds that should generally be continuous from topset to toeset (e.g. Massari, 1996; Kostic et al., 2005). Consequently, the larger gravel cross-strata may be the product of barform migration; composite barforms of up to 10 m height have been observed in eskers (e.g. Shaw and Gorrell, 1991; Brennand and Shaw, 1996;) and ice-walled conduit deposits (Russell et al., 2003). In the model of Brennand and Shaw (1996), large foreset beds progressively backstep over the stacked, medium-scale foreset beds. No such relationship is observed directly in the Hawkesville Spur deposits; however, this could be the result of inadequate lateral continuity of the exposure.

Nevertheless, the foreset dip, bed texture and grading, and toeset texture are consistent with a barform foreset interpretation.

Where both the morphology (landform) and stratal architecture are observed, relatively confident interpretations of a dune versus delta or bar origin for large cross-strata is possible (Shaw and Gorrell, 1991). Depositional facies described by Shaw and Gorrell (1991) have similarities to the Hawkesville Spur deposits; however, the Hawkesville Spur cross-strata are better graded, show more gradational facies changes, and have less evidence of reactivation surfaces. Large variations in sandy dune-scale foreset dimensions are known to be the product of trough scour (Leclair, 2002). Similar processes may explain the variation in foreset dimension in the Hawkesville Spur. It is generally believed, however, that bed armouring would preclude or reduce the extent of trough scour in a gravel system (Leclair, personal commun. 2006).

(iii) Medium-scale cross-sets of bimodal and openwork gravel

A hierarchal series of bedding surfaces are identified in the lower gravel unit of the tripartite stratigraphic succession at the A-1 site (Fig. 5). In the lower gravel unit, a stacked succession of cross-sets is bounded by subhorizontal erosional surfaces (Fig. 7A). The highest, <1.5 m high bed occur immediately above the slump and are overlain by more sand rich and finer gravel beds. A mud rich unit that locally contains gravel overlies the lower gravel unit disconformably. This mud unit is disconformably overlain by cross-stratified sand and local gravel horizons (Fig. 7A).

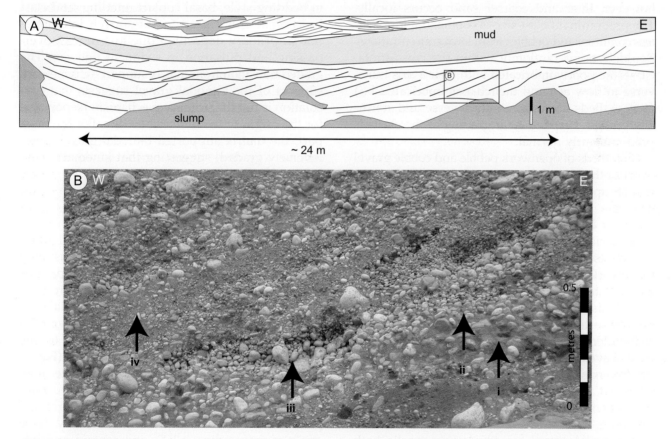

Fig. 7 Gravel, medium-scale lateral-accretion cross-stratification at the A-1 site. (A) Line drawing of the stratigraphic architecture and bedding relationships in sand and gravel. Note middle mud package separates basal gravel rich unit from overlying sand rich succession (see Fig. 5). (B) Details of bedding style and composition of many of the 1–2 m high cross-stratified bedsets at the A-1 site. Note presence of (i) bimodal matrix supported sand gravel, (ii) framework bimodal gravel, (iii) openwork gravel, and (iv) polymodal gravel. (Palaeoflow is from left to right.)

Cross-strata <1.5 m high at the base of the section dip at up to 20° to the west for most of the section. In the western part of the section the beds dip at a similar angle toward the east, forming a large trough cross-set (Fig. 7A). The bed is composed of three distinct cross-strata facies, a) bimodal gravel, B) openwork gravel, C) polymodal gravel. Openwork gravel systematically overlies bimodal gravel and is, in turn, overlain by polymodal gravel. These three facies compose variable proportions of individual cross-strata (Fig. 7B). Individual cross-beds are <0.75 m thick, and pinch and swell laterally. Bimodal, sand-cobble gravel occurs at the base of cross-strata and can form the lower three quarters of the cross-strata. This unit is matrix-supported, with a few cobbles or boulders 'floating' in the sand matrix. It becomes better sorted upward and with increasing gravel content is framework supported. The matrix grain-size is commonly medium sand, however, finer and coarser sand occurs locally. The predominant clast size is cobble; boulders are less common, and pebbles form local concentrations. Boulders occur randomly within the cross-strata. Clasts are poorly imbricated with their a-axis transverse to flow a(t) and their b-axis imbricated b(i) to flow. Beds are tangentially reverse and normally graded. Perpendicular grading of the clasts is predominantly normal.

Cross-beds of openwork pebble and cobble gravel overlies the bimodal gravel with both gradational and abrupt basal contacts. Openwork gravel cross-beds are generally thinner, have better sorting, and have less extensive lateral continuity than the other facies. Thickness can be limited to only a few clasts up to a maximum thickness of 0.5 m. Openwork cross-beds can pinch and swell laterally as the extent of matrix fill changes. The thinnest cross-beds are commonly discontinuous, have the poorest sorting, and are generally formed of pebble to boulder sized clasts. In places openwork gravel clasts have a coating of mud. Tangential sorting is predominantly reverse; however, normal sorting is observed. Perpendicular sorting is predominantly normal with minor occurrences of reverse grading observed.

Polymodal gravel cross-beds can overlie both of the previous units with either gradational or abrupt lower contacts. The gravel can be well-sorted pebble to poorly sorted pebble and cobble. Beds may be >1 m thick, but most are <0.5 m thick.

Interpretation: Style (iii) cross-beds of bimodal and openwork gravel is interpreted on the basis of facies transitions and geometry as barform foresets. The multimodal sediment facies of the foresets and stacked nature of cross-strata all suggest a compound barform. The geometry, bedding succession and facies associations of barforms are well documented from fluvial settings (Miall, 1996; Allen, 1983a) and in glaciofluvial tunnel channel and esker deposits (Shaw and Gorrell, 1991; Benvenuti and Martini, 2002; Brennand and Shaw, 1996). The origin of openwork and framework gravel is uncertain and include: primary sorting (e.g. Carling, 1990; Shaw and Gorrell, 1991), infiltration; (Kleinhans, 2005), and post-deposition removal by groundwater (Browne, 2002). Similar bimodal-openwork gravel couplets have been described from glaciofluvial deposits of the Rhine (Sigenthaler and Huggenberger, 1993). In this study the variations in bedding style, basal contact, grading, and clast concentration are used to support an interpretation of primary deposition due to sediment overpassing and pulsation in sediment delivery. The graded transition from matrix support to framework support of the bimodal gravel suggests that percolation is not the primary mechanism responsible for the origin of the bimodal texture. For the most part, the matrix supported bimodal gravel is not reversely graded, suggesting that kinematic sieving is of limited importance.

(iv) Large-scale low-angle clinoforms

Low-angle clinoforms from the Top-of-the-Hill pit are the fourth style of cross-stratification illustrated (Fig. 8). The section extends for ~30 m subparallel to palaeoflow and is ~7 m high. Five-metre high clinoforms of sand and gravel are overlain by massive, horizontal and cross-bedded polymodal gravel that is in turn overlain by medium to fine sand. The proximal-clinoform dip is ~10° (Fig. 8) and is sub-horizontal distally (westward). The clinoforms consist of medium-scale cross-bedded sand and sandy gravel. Individual beds are stratified parallel to clinoform dip, or alternatively, have angle-of-repose dips (>26°), and contain multiple divergent and truncated dipping beds. The dune-scale cross-strata down climb locally within the clinoform sets. Distally the gravel beds fine and diminish in extent. Gravel is predominantly matrix-

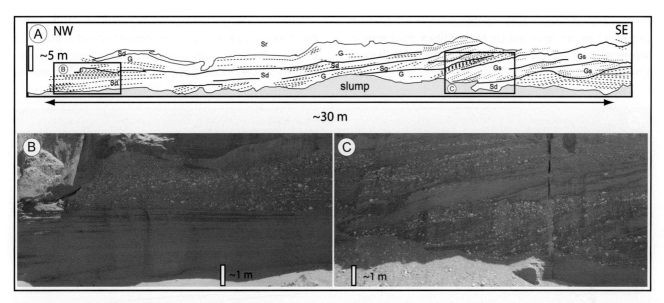

Fig. 8 Large-scale low-angle clinoforms from the Top-of-the-Hill site. (A) Line drawing illustrates the lateral continuity of low-angle clinoforms and intervening dune-scale cross-strata. Sediment calibre fines from proximal (right) to distal with a commensurate change in bedding style from cross-stratified gravel to plane bedded and low-angle cross-stratified medium sand. Sediment facies codes are: G gravel, Gs sandy gravel, Sd diffusely graded sand, Sr small-scale cross-laminated sand. (B) Close up of distal plane-bedded medium sand. (C) Proximal medium-scale cross-bedded sand and gravel. Location of (B) and (C) are indicated by rectangle and respective letters. (Palaeoflow is from left to right.)

supported pebble and cobble with weak clast imbrication. Boulder size clasts are rare. Framework cobble gravel is present as thin <0.3 m thick beds, contains clast clusters, and has weak imbrication. Where observed, clast imbrication is predominantly a (t), b (i). Sand units contain angle of repose, low-angle cross-bedding, and diffusely graded structures in medium sand.

Interpretation: On the basis of the clinoform scale, dip angle and lateral continuity, the low-angle clinoforms, style (iv), is interpreted to be a barform. The down climbing nature of the gravel cross-beds, and uniformity of dip direction indicate that there was strong flow attachment along the bar foresets. The distal and stratigraphically upward fining of the beds suggests a possible mouth-bar form deposited under conditions of waning flow (e.g. Wright, 1977). The style of the barform bedding supports an interpretation for a friction- dominated, sediment-laden jet discharge as is common to a number of settings, including: fluvial channels (Allen, 1983a; Haszeldine, 1983), crevasse-splays and tidal deltas (Wright, 1977), and subaqueous fans (Russell and Arnott, 2003). The clinoforms in the Waterloo Moraine are more diverse than the

examples cited, consisting of at least three sediment facies with a variety of bedding styles: i) steep, cross-stratified sand and gravel, ii) low-angle, cross-stratified sand, and iii) diffusely graded sand. This suggests that, in the Waterloo Moraine example, the flow regime was more variable than in the examples cited above.

Medium-scale climbing cross-stratification

Climbing cross-stratification is observed at a number of different scales and textures ranging from fine sand to pebble and cobble gravel throughout the Waterloo Moraine; however, coarser climbing bedforms were only observed in the Top-of-the-Hill, and A-1 sites (Fig. 1). The most common scale of climbing cross-strata is ripple-scale, cross-laminated fine sand. This small-scale, climbing cross-stratification is virtually ubiquitous throughout sandy Waterloo Moraine deposits. The cross-sets can be either stoss-erosional or stoss-depositional with variable angles of climb. The cross-strata can form tabular sets overlying a broad range of sediment textures. Much less common are 0.25 to 2 m thick climbing cross-strata of sand, gravelly sand,

and gravel. Cross-strata of this scale and texture occur in bed-sets that are up to 4 m thick, and within steep-walled scours associated with diffusely graded sand.

In a rare example of climbing dunes, a transition from stoss-erosional, sandy, cross-stratification to large-scale, stoss-depositional sandy gravel cross-stratification occurs (Fig. 9). The underlying stoss-

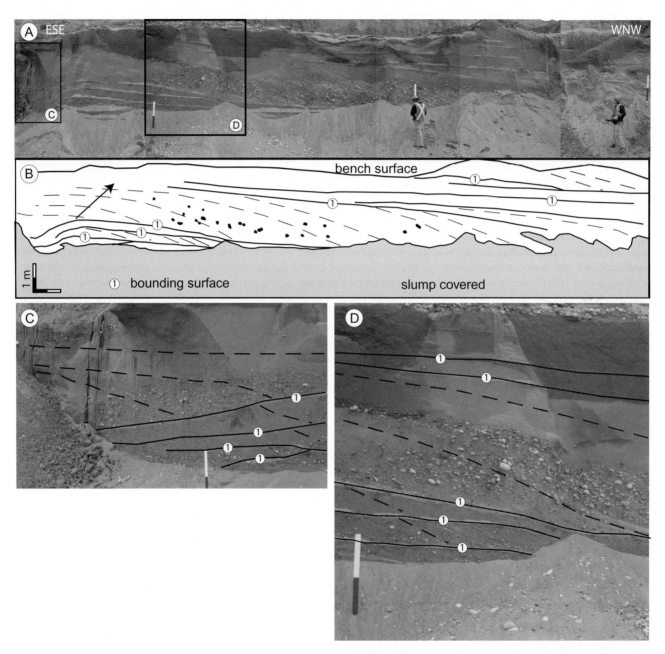

Fig. 9 Climbing dune-scale cross-stratification from the Top-of-the-Hill site. (A) Photomosaic and (B) line drawing approximately parallel to flow direction. Flow was from left to right. Note the sand dominated stoss-beds compared to the gravel concentrated on the foreset beds. Angle of climb of the cross-strata is ~40°, indicated by arrow in (B). (C) Photo of proximal part of climbing cross-strata showing subhorizontal stoss beds, and underlying truncated stoss-erosional climbing beds. (D) Close-up of gravel rich foresets and underlying stoss-erosional beds. Dashed lines illustrate the approximate trend of stratification. Location of (C) and (D) are indicated by rectangle and letters in (A). Scale in photos is a metre stick with 0.5 m increments.

erosional sets are gravelly coarse sand, 20 to 40 cm thick. These cross-sets are overlain by a 1 m thick set of stoss-depositional climbing cross-beds of coarse sand and gravel dipping toward the south-west, with foreset beds dipping at 12–26°. Toward the top of the cross-strata they are sand-rich with a shallow dip (<15°), and in the lower-part they are more gravel rich and steeper dipping (>20°). The angle of climb is ~40°. The stoss beds are predominantly coarse sand with minor pebbles. The brink-point or transition from the stoss side to foreset bed is poorly defined due to the very faint bedding. The cross-strata display well-developed, tangential, normal grading with rare reverse grading. Gravel concentration does not always reach a peak at the base of the cross-strata, some cross-strata are predominantly sand at the toeset. The largest cobble clasts are ~18 by 6 cm, however, most of the gravel is pebble calibre.

Interpretation: Climbing cross-stratification is a characteristic of flows experiencing rapid loss of transport capacity and consequently deposition from both bedload and suspended load (e.g. Ashley *et al.*, 1982; Hiscott, 1994). The stratification here is very similar to cross-stratification attributed to humpback dunes (Allen, 1983b). Allen attributed the gravel sand sorting along the foresets to an overpassing mechanism. A similar interpretation is likely here, the gravel travelled preferentially on the sand bed until it entered the zone of flow separation on the foreset, and a zone of low shear stress. In this case, the transport of gravel down the foreset by secondary processes such as avalanching or grain flows was limited by the rapid rate of bed-aggradation.

Some of the best known examples of climbing, ripple-scale, cross-stratification are described from glaciolacustrine deposits (e.g. Jopling and Walker, 1968). Description of climbing dune-scale cross-stratification is less common, particularly dune-scale, stoss-depositional cross-stratification; however, in eskers and subaqueous fan deposits it is not uncommon (e.g. Gorrell and Shaw, 1991; Brennand and Shaw, 1996). Climbing dune-scale cross-stratification is less common as dunes have strong flow separation and associated flow reattachment that produces extensive downflow erosion. Additionally, under equilibrium conditions dunes scale to the flow depth. Consequently, in settings with limited flow depth, local bed aggradation reduces

flow depth and increases flow velocities, and in turn, the climbing bedforms are eroded. Flow depth or thickness is not similarly constrained in sub-aqueous fans where the flow occurs as a density underflow and bed aggradation does not necessarily reduce the flow depth.

Antidune stratification

Two distinct styles of stratification observed in the Top-of-the-Hill site are discussed: i) sand-dominated, low-angle cross-stratification, ii) low-angle, concave-up erosional surfaces with a single clast thickness cobble layer, and diverse association of cross-stratification. The first style consists of low-angle cross-stratification with inclined strata in both upflow and downflow directions, divergent strata, and erosional truncations. The cross-strata increase in dip downward along the bed up to 18°. Cross-strata exhibit both downlapping and offlapping relationships with the basal contact. Locally, pebbles and cobbles are concentrated in the lower part of backset beds (Fig. 10A). Individual cross-strata become thinner up dip. Stacked beds are up to 4 metres thick and continue laterally for tens of metres.

The second style of cross stratification occurs in coarse sand, gravelly sand and gravel with multiple scales of scour surfaces and cross-stratification. This pit face is located lateral to extensive steep-walled scours documented in the next section of the paper. The largest scours may have an undulating and broad asymmetric scallop shape, extending >8 m longitudinally, and with relief of <0.40 m (Fig. 10B). Smaller less continuous scours commonly lack a gravel concentration. The stacked succession of scours forms an overlapping network of asymmetric concave-up surfaces. The largest scour surfaces commonly have a discontinuous single layer of cobble clasts that may have clast clusters and an a(t) b(i) imbrication. Where pebbles predominate the clast layer may be up to 10 cm thick. Gravel along largest scour surfaces does not exhibit any association with gravel in overlying cross-stratified units, where the gravel is commonly of finer calibre. In addition, the largest scour surfaces are overlain by fill with predominantly low-angle, undulating cross-stratification. Smaller scours commonly have pebble clasts that can be associated with the toe-set of cross-bedding in the scour fill.

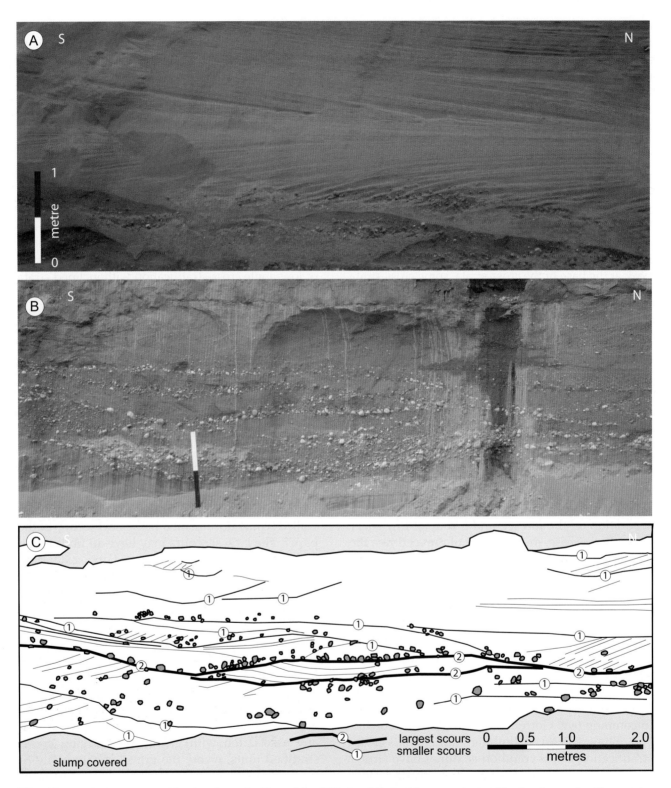

Fig. 10 Antidune cross-stratification from the Top-of-the-Hill site. (A) Antidune cross-stratification in sand with gravel concentration on lower backset. (B) Antidune scour and cross-stratification in sandy gravel. Largest scour surfaces are inferred to be the product of antidune erosion. (C) Line drawing of (B) highlighting the hierarchy bedding surfaces (indicated by numbers). Note single grain armour of imbricate clasts along scour surface. Palaeoflow was from left to right in (A) and was away from the viewer in (B).

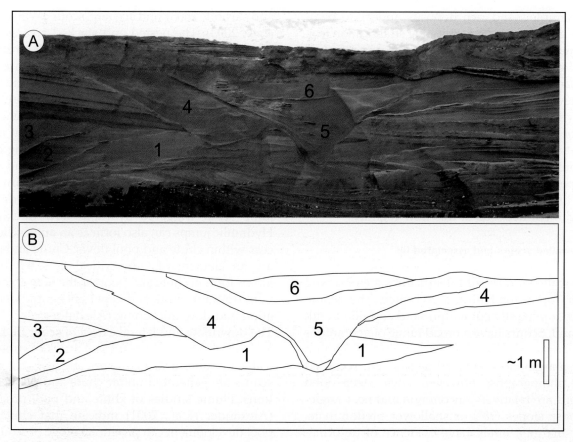

Fig. 11 Stacked and nested scour fills from the Top-of-the-Hill site. Orientation of the face is unknown. (A) Photograph of stacked and nested scours filled with diffusely graded sand eroded into cross-laminated fine sand and cross-bedded medium sand. (B) Trace of the stacked and nested nature of scours. Numbers refer only to the sequence of the stacked scours. Orientation of this section is unknown. Palaeoflow was perpendicular to the page, sense up.

The dip angle of near-planar cross-beds are <20°. Additional stratification in the succession includes diffusely graded coarse sand, trough and tabular cross-stratified medium sand with isolated pebbles, and horizontal laminated medium sand.

Interpretation: On the basis of bedding style, cross-strata dip-angle, and direction of dip the low-angle stratification is interpreted to be the product of antidunes or in-phase waves. Antidunes form when the water surface or a rheological surface within the flow and bed surface are in-phase. Antidune deposits consist of a variety of sediment facies and low-angle cross-stratification (Alexander *et al.*, 2001). Under depositional conditions gravel concentration on the backset-bed has been documented, and attributed to decrease in flow velocity and transport competence. Erosional antidune structures consist of low-amplitude undulating surfaces that truncate underlying stratification (Brennand and

Shaw, 1996). In the example presented (Fig. 10B), the scour surfaces form part of a complex succession of various bedding styles and sediment facies. The interpretation of antidune erosional surfaces is based on the lateral continuity of the scour, and the association of a thin, commonly single clast horizon directly on the scour surface with clast clusters. Furthermore, there appears to be no connection between the clast horizon and the clast distribution in the overlying cross-stratified gravelly sand. The size of the largest clasts distributed along the scour surface also shows no similarity. The majority of literature on antidunes describes sand or gravel structures from shallow water flows (e.g. Shaw and Kellerhals, 1977; Alexander and Fielding, 1997; Blair, 2000). It is apparent from the literature that the style of antidune cross-stratification is variable and is controlled, in part, not only by flow dynamics (depth, velocity) but also sediment calibre.

For example, sand rich deposits appear to commonly generate low-angle dipping stratification (e.g. Alexander *et al.*, 2001), whereas gravel rich sediment forms graded openwork to matrix rich couplets and clast clusters (e.g. Blair, 2000). Admittedly, the antidune interpretation for the broad scours is speculative; however, within the stratigraphic context, and for the reason outlined above, it is reasonable. As noted by Carling and Shvidchenko (2002), the interpretation of deposits and structures produced by supercritical flows and hydraulic jumps continue to be limited by the paucity of experimental data on sediment deposits of this flow regime.

Steep-walled scours and associated fill

Steep-walled scour and fill structures are observed in the Top-of-the-Hill and A-1 sites. The scours are predominantly cut within sand and filled with fine sand. Scours have a broad range of geometries from steep-walled to shallow margins. Maximum scour margin slopes of ~50 degrees are recorded from photographs; however, such steep-sided margins are relatively uncommon and near angle-of-repose slopes (28°) or shallower predominate. Scour margins rarely have evidence of deformation within the flanking sediment, and faulting or slumping of the margin is extremely rare. Scours may be up to 3 m deep and 20 m wide. Cross-section geometry is variable, with both broad, circular transverse profiles and narrower V-notched cross-sections. Nested scours occur where the fill of one scour contains additional scour and fill features. Scours can be stacked to form sets 5–6 m thick. Where scours are underlain by gravel they are generally shallower with little incision into the substrate. Internal fill of the scours is predominantly diffusely graded fine to medium sand. In rare cases fills contain climbing small-scale cross-lamination, dune-scale cross-stratification that can be climbing, and plane-bed sand. In some sections, multiple stacked or nested scours have a complex fill succession of all or many of these structures (Fig. 12). Occasionally, silty sand intraclasts up to 50 cm in length were observed. Gravel is a rare component of the scour fills.

Interpretation: Steep-walled scours and associated fills have been interpreted as the product of slumping (e.g. Cheel and Rust, 1982), erosion beneath a hydraulic jump (Gorrell and Shaw, 1991; Russell

and Arnott, 2003) and iceberg scour (Winsemann *et al.*, 2003). The nested and stacked arrangement of scours, the diverse suite of sediment structures within the fill, and the lack of deformation associated with scour margins, indicate erosion and rapid fill beneath turbulent flows. This interpretation is further supported by the lack of evidence for water escape structures within the fills that might be expected in slump deposits experiencing retrogressive failure and liquefaction. Gorrell and Shaw (1991) and Russell and Arnott (2003) both invoked a single erosional mechanism involving hydraulic-jump scour triggered by rapid flow thickening. Hydraulic jumps can also form as an autocyclic process within chute and pool flows. Chute and pools are an element of supercritical flow. Hydraulic jumps form in chute and pools as the flow enters the pool from the chute and experience rapid deceleration and flow thickening (Alexander *et al.*, 2001). The downflow and lateral extent of scour fill facies, and variability of scour fill structures of supercritical and subcritical bedforms suggest that some scours are generated under chute and pool conditions. Flume studies of chute and pool processes (Alexander *et al.*, 2001) indicate that chute and pool development can produce a range of scour and fills that are similar to elements observed here.

DISCUSSION

Depositional environments of the Waterloo Moraine

The Waterloo Moraine has been interpreted previously to be an ice-marginal glaciofluvial – glaciolacustrine depositional system composed of either deltaic outwash (Harris, 1970) or esker-subaqueous fan deposits (Bajc and Karrow, 2004). The bedforms and sediment facies presented in this paper are not individually diagnostic of any single environment. Interpretation of depositional settings commonly requires integration of individual sediment facies with observations from the broader site setting and landform analysis. Fortunately, detailed surficial sediment mapping completed by previous authors (Karrow, 1993) allows the integration of isolated observations into a broader landform context that permits improved depositional inferences. Key observations presented in this paper that support the esker – subaqueous fan depositional

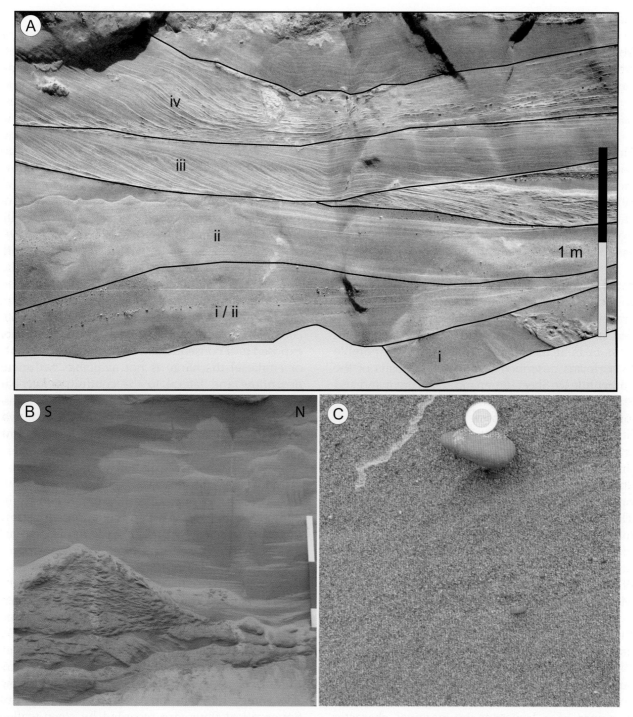

Fig. 12 Scour-fill characteristics from the Top-of-the-Hill site. (A) Stacked scours with various styles of fill, (i) diffusely graded fill, (ii) diffusely graded antidune stratification, (iii) dune-scale scour-fill cross-stratification, (iv) climbing dune-scale cross-stratified sand. Scale bar is one metre long. (B) Steep-walled scour eroded into small-scale cross-laminated fine sand and filled with diffusely graded fine sand, stick is 1.10 m long. (C) Rare 'floating' oversized clast in a diffusely graded scour-fill, coin diameter is 2.7 cm.

model include: i) scale of cross beds, ii) evidence of supercritical flow deposits, iii) transformation from supercritical to subcritical flow, iv) climbing cross-strata, and v) inter-stratification with glacio-lacustrine mud.

Conduit – esker setting

The Hawkesville Spur and A-1 sites (Fig. 1) occur within landforms that have been interpreted previously to have an ice-supported origin (Bowes, 1976; Karrow, 1996; Bajc and Karrow, 2004). At Hawkesville, large-scale cross-strata indicate flow from the north-northwest toward the south-southeast. The palaeoflow direction, and coarse sediment at the Hawkesville site record subcritical flow and bedload transport within either confined or open conduits. The presence of diamicton overlying the gravel deposits supports an interpretation of deposition in an ice-roofed conduit setting. The scale of cross-beds at the Hawkesville site is comparable to barforms described from esker deposits of the Laurentide Ice Sheet (Brennand, 1994; Brennand and Shaw, 1996) and ice-walled conduits of Icelandic jökulhlaup events (Russell et al., 2001). The well documented development and evolution of the ice-walled conduit during the 1996 Icelandic jökulhlaup provides a modern analogue for the Hawkesville Spur. A major difference, however, is the likely presence of ice cover of the Hawkesville Spur. The diamicton provides additional support for the interpretation that the large-scale planar cross-beds of the Hawkesville Spur represent barform foresets rather than deltaic foresets. A similar reinterpretation of medium to large-scale foresets in the Oak Ridges Moraine was suggested by Brennand (1994).

The depositional setting and interpretation of the cross-stratification at the A-1 site is less clear as the deposit is within a much larger part of the Waterloo Moraine. Surficial geological mapping suggests an ice-contact position was located to the south-southwest and the latest stage of sedimentation was of a lacustrine ice-marginal character (Bajc and Karrow, 2004). The gravel described here form part of a unit at ~10 m depth within the moraine stratigraphy. The style of foresets, and texture suggest deposition by bedload processes of a glaciofluvial system flowing from the southeast to northwest. More specifically, the deposits are interpreted as conduit barforms of either a closed, or open ice-walled conduit.

Subaqueous fan setting

Previous work in the area interpreted the Top-of-the-Hill deposits to be of subaqueous fan origin (Karrow, 1996; Bajc and Karrow, 2004). The mud horizon within the tripartite stratigraphic succession at the site is interpreted as a lacustrine basin event (Fig. 5). Climbing dunes, antidunes, and scours with diffusely graded fills beneath the mud unit, record sedimentation and erosion by a flow experiencing rapid transformation as happens at a zone of flow expansion. This is characteristic of subaqueous fan settings (e.g. Rust and Romanelli, 1975; Gorrell and Shaw, 1991; Brennand, 1994; Brennand and Shaw, 1996; Russell and Arnott, 2003). The conduit mouth-bar clinoforms are similar to low-angle delta clinoforms and are interpreted to record rapid deposition and progradation at an ice-marginal conduit mouth (Russell and Arnott, 2003). The upflow continuation of the foresets is not exposed and consequently evidence for subglacial or englacial discharge is not available. Sediment downflow and lateral to the conduit mouth-bar clinoforms indicate rapid bed aggradation and transformation from supercritical to subcritical flows. Climbing bedforms record a rapid loss of flow capacity and simultaneous traction and suspension sedimentation. Rapid bed aggradation is also indicated by antidune cross-stratification and scour fills. These types of stratification are rare in environments with limited accommodation space, low depositional rates and bed reworking. In contrast to glacigenic subaqueous fans, few depositional settings preserve a spectrum of supercritical flow deposits, and consequently the understanding of the fluid dynamics and depositional record of supercritical bedforms are poorly developed compared to that of subcritical bedforms (e.g. Bridge, 2003).

The suite of bedforms attributed to the subaqueous fan setting in this paper provides a continuum of forms deposited under a regime of diminishing flow energy and transport competence, and capacity. Supercritical flows are recorded by steep-walled scours, and antidune stratification in diffusely graded beds interpreted as the record of chute and pool regime (Simons et al., 1965; Alexander et al., 2001). For some scours, erosion beneath a hydraulic jump at a transition of rapid flow thickening may be an alternative interpretation (Gorrell and Shaw, 1991; Russell and Arnott, 2003). The climbing, dune-scale cross-beds record a rapid loss of transport capacity by a flow overloaded with suspension sediment. In this case, the gravel content of the cross-stratification indicates continued bedload transport, whereas the climbing dune-form indicates a contemporary phase of suspension sedimentation.

Waterloo – Oak Ridges Moraine comparison

Stratified moraines in southern Ontario have a record of anomalously thick deposits of sand and gravel. The sedimentary architecture of these deposits is poorly understood (Cowan *et al.*, 1978) and consequently, comparison of the Waterloo Moraine is only possible with the Oak Ridges Moraine (ORM) (Barnett *et al.*, 1998). The ORM is the most extensively studied landform in southern Ontario, and along with the Brampton and Guelph eskers (Saunderson, 1976; Benvenuti and Martini, 2002), and eskers north of Kingston (Brennand, 1994) is the only one whose sedimentology has been described in bed-scale detail (Barnett *et al.*, 1998; Sharpe *et al.*, 2002; Russell and Arnott, 2003; Russell *et al.*, 2003).

Similar bedform styles and scour scales are documented from both moraines. The largest cross-beds of the Hawkesville Spur are comparable in scale to large cross-beds described from channel fills north of the ORM (Shaw and Gorrell, 1991), from buried tunnel-channel fills beneath the moraine (Sharpe *et al.*, 2003), and from the Brighton area (Sharpe *et al.*, 2003). Bimodal cross-bedding has received less prominence in ORM descriptions but are known from the eastern ORM near Brighton and from tunnel-channel fill sediment north of the moraine (Shaw and Gorrell, 1991; Sharpe *et al.*, 2003). Climbing bedforms, antidunes, and scour fills are all well documented in the ORM and occur in similar subaqueous fan settings (Russell and Arnott, 2003). The similarity of bedding styles between the two moraines provides a preliminary indication that similar meltwater processes are responsible for the origin of both landforms. It requires additional palaeohydraulic work on the Waterloo Moraine to better constrain the nature of the meltwater flux.

Groundwater implications

The hydrogeological characteristics of aquifers are closely linked to sediment texture and depositional processes and environments (e.g. Fogg, 1998). Based on the limited data presented here, a number of considerations on the scale, probable distribution, thickness, and spatial variability of aquifer properties of the Waterloo Moraine can be made. Data have been presented from three sites, Hawkesville Spur, A-1, and Top-of-the-Hill, which illustrate contrasting depositional controls on the spatial arrangement and scales of structures in the moraine that are closely related with the principal ice-conduit or subaqueous-fan environments.

The two sites (A-1 and Top-of-the-Hill) in the southeast part of the Waterloo Moraine have palaeo-flow orientations that indicate a proximal to distal transition from the southeast to northwest with a corresponding decrease in sediment calibre and bed thickness. Transverse to the palaeoflow there is a distinct shift in depositional settings from conduit to subaqueous fan. For example the Top-of-the-Hill site is located lateral to the main ridge axis that the A-1 site is located in. Consequently, the most prolific aquifer targets are likely associated with ridge areas in the southeast part of the moraine. This relationship is supported by the location of the Manheim well field, a major municipal well field (Duckworth, 1983), near the crest of the moraine ridge in the southeast part of the landform (Fig. 1).

A number of hydrogeological characteristics can be associated with the conduit and subaqueous fan depositional environments. Conduit deposits (sites A-1 and Hawkesville) are commonly an order of magnitude coarser than the subaqueous fan sediment. Gravel within the conduit setting is cobble calibre with boulders being common, and are more frequently framework in character. Within foresets, particularly in the lower half of the foreset, openwork gravel is common. In contrast, framework and bimodal gravel have not been documented from the fan deposits of the Waterloo Moraine. Consequently, conduit deposits are much more likely to have high hydraulic conductivities that will form preferential flow paths. Openwork gravel beds are the most extensively developed example of high conductivity beds. Similar facies documented in Germany have measured hydraulic-conductivities of 10^{-1} to 10^{-2} m s^{-1} (Kostic *et al.*, 2005). The openwork gravel occurs in a similar facies association overlying bimodal sand-cobble gravel in both settings. In the Waterloo Moraine, the openwork gravel forms beds up to 0.75 m thick in cross-sets 1–1.5 m thick and longitudinal cross-set continuity of tens of metres. The larger channel scale that hosts these cross-sets remains to be assessed but is greater than 1 km longitudinally and 300 m in width.

Subaqueous-fan deposits are finer grained and are locally characterized by steep-walled scours that form nested and stacked arrangements that truncate underlying, more tabular strata (Fig. 11). In this setting the upper-medium sand calibre of scour fills may provide good vertical connectivity through silt and fine-sand rich beds that are less permeable. Subaqueous fans are also characterized by extensive downflow extension with strong longitudinal sorting. The proximal subaqueous fan, from which examples are illustrated in this paper, form only a small fraction of the total fan area. Improved depositional models may allow prediction of aquifer properties in those areas. The remainder of the fan setting is characterized by fine sand of intermediate hydraulic conductivity that provides extensive reservoirs for the high-yield proximal sediment.

CONCLUSION

The Waterloo Moraine with an area of ~400 km^2 is one of a number of stratified moraines in southern Ontario. These moraines form important recharge areas in many of the principal watersheds of southern Ontario (e.g. Howard *et al.*, 1995). Consequently, improved understanding of their sedimentology and their role in groundwater recharge is warranted. This improved understanding is particularly important in light of present source water initiatives within Ontario that emphasizes well field protection using time-of-travel modelling estimates (e.g. TEC, 2004).

The recognition of high-energy deposits further refines the existing depositional model for the Waterloo Moraine (Bajc and Karrow, 2004). The subaqueous fan deposits have elements that are similar to published models for the ORM (Russell and Arnott, 2003). This emerging model of the Waterloo Moraine suggests that the ORM is not unique, but that high-energy, voluminous sedimentation events were perhaps a common element of stratified moraine formation in southern Ontario. Further, previous recognition of the Waterloo Moraine as an esker – subaqueous fan system is supported and expanded upon. This work indicates that sedimentological analysis of stratified moraines can yield improved understanding of the deglacial events of the Laurentide Ice Sheet in southern Ontario.

This initial assessment of sedimentary structures and facies of the Waterloo Moraine indicate that additional sedimentological study can answer a number of questions on aquifer architecture, heterogeneity, and character key to improved hydrogeological modelling of the Waterloo Moraine aquifers (e.g. Poeter and McKenna, 1999). Improved input of this type of data will be critical to development of robust groundwater models required by the provincial source water protection legislation (TEC, 2004).

ACKNOWLEDGEMENTS

The authors are grateful to the respective pit operators for granting access to their pits. T. Barry provided assistance with the graphics. Reviews at the GSC by R. Knight and D. Cummings greatly improved the clarity of the manuscript. Critical reviews by the journal reviewers T. Brennand and an anonymous reviewer contributed to the authors making key improvements to the manuscript. Funding for this work was provided through the Ground Water Program of the Geological Survey of Canada. This is contribution 2005701 of the Earth Science Sector, Natural Resources Canada.

REFERENCES

Alexander, J., Bridge, J.S., Cheel, R.J. and Leclair, S.F. (2001) Bedforms and associated sedimentary structures formed under supercritical water flows over aggrading sand beds. *Sedimentology*, **48**, 133–152.

Alexander, J. and Fielding, C. (1997) Gravel antidunes in the tropical Burdekin River, Queensland, Australia. *Sedimentology*, **44**, 327–337.

Allen, J.R.L. (1983a) Studies in fluviatile sedimentation: bars, bar-complexes and sandstone sheets (low-sinuosity braided streams) in the brownstones (L. Devonian), Welsh Borders. *Sediment. Geol.*, **33**, 237–293.

Allen, J.R.L. (1983b) Gravel overpassing on humpback bars supplied with mixed sediment: examples from the Lower Old Red Sandstone, southern Britain. *Sedimentology*, **30**, 285–294.

Ashley, G.M., Southard, J.B. and Boothroyd, J.C. (1982) Deposition of climbing-ripple beds; a flume simulation. *Sedimentology*, **29**, 67–79.

Bajc, A.F. and Karrow, P.F. (2004) 3-dimensional mapping of Quaternary deposits in the Regional Municipality of Waterloo, southwestern Ontario. Geological Association of Canada and Mineralogical Association of

Canada, Joint Annual Meeting, 2004, Fieldtrip FT-7, St Catharines, Ontario, 72 pp.

Barnett, P.J. (1992). Quaternary geology of Ontario. In: *Geology of Ontario* (Eds P.C. Thurston, H.R. Williams, R.H. Sutcliffe and G.M. Stott), pp. 1011–1088. Ontario Geological Survey, Special Volume 4, Part 2, Toronto.

Barnett, P.J., Sharpe, D.R., Russell, H.A.J., Brennand, T.A., Gorrell, G., Kenny, F. and Pugin, A. (1998) On the origins of the Oak Ridges Moraine. *Can. Jour. Earth Sci.*, **35**, 1152–1167.

Benvenuti, M. and Martini, I.P. (2002) Analysis of terrestrial hyperconcentrated flows and their deposits. In: *Flood and Megaflood Processes and Deposits: Recent and Ancient Examples* (Eds I.P. Martini, V.R. Baker and G. Garzonm), pp. 167–193. Special Publication Number 32 of the International Association of Sedimentologists.

Blair, T.C. (2000) Sedimentology and progressive tectonic unconformities of the sheetflood-dominated Hell's Gate alluvial fan, Death Valley, California. *Sediment. Geol.*, **132**, 233–262.

Bowes, E. (1976) *Sedimentology of the glaciofluvial deposits of Woolwich and Pilkington Townships*: Unpublished MS thesis, University of Wilfred Laurier, Waterloo, 216 pp.

Brennand, T.A. (1994) Macroforms, large bedforms and rhythmic sedimentary sequences in subglacial eskers, south-central Ontario: implications for esker genesis and meltwater regime. *Sediment. Geol.*, **91**, 9–55.

Brennand, T.A. and Shaw, J. (1996) The Harricana glaciofluvial complex, Abitibi region, Quebec; its genesis and implications for meltwater regime and ice-sheet dynamics. *Sediment. Geol.*, **102**, 221–262.

Bridge, J.S. (2003) *Rivers and Floodplains: Forms, Processes and Sedimentary Record*. Blackwell Science, Oxford, 491 pp.

Browne, G.H. (2002) A large-scale flood event in 1994 from the mid-Canterbury Plains, New Zealand, and implications for ancient fluvial deposits. In: *Flood and Megaflood Processes and Deposits: Recent and Ancient Examples* (Eds I.P. Martini, V.R. Baker and G. Garzon), pp. 99–112. International Association of Sedimentologists, Special Publication Number 32.

Carling, P.A. (1990) Particle over-passing on depth-limited gravel bars. *Sedimentology*, **37**, 345–355.

Carling, P.A. (1999) Subaqueous gravel dunes. *J. Sedim. Res.*, **69**, 534–545.

Carling, P.A. and Shvidchenko, A.B. (2002) A consideration of the dune:antidune transition in fine gravel. *Sedimentology*, **49**, 1269–1282.

Chapman, L.J. and Putnam, D.F. (1943) The moraines of southern Ontario. *Trans. Roy. Soc. Canada*, **37**, 33–41.

Chapman, L.J. and Putnam, D.F. (1984) *The Physiography of Southern Ontario*. Ontario Geological Survey, *Special Volume 2*. 270 pp.

Cheel, R.J. and Rust, B.R. (1982) Coarse grained facies of glacio-marine deposits near Ottawa, Canada. In: *Research in Glacial, Glacio-fluvial, and Glacio-lacustrine Systems* (Eds A.R. Davidson, W. Nickling and B.D. Fahey), *Proceedings – Guelph Symposium on Geomorphology*, **6**, pp. 279–295. University of Guelph, Guelph, Ontario Canada.

Coleman, A.P. (1932) The Pleistocene of the Toronto region. Ontario Department of Mines, Map no. 42g, scale 1:63,360, Toronto.

Cowan, W.R., Sharpe, D.R., Feenstra, B.H. and Gwyn, Q.H.J. (1978) Glacial geology of the Toronto-Owen Sound area. In: *'78: Field Excursion Guidebook*, pp. 1–16. Geological Association of Canada.

Duckworth, P.B. (1983) Artificial recharge potential of aquifers in the Region of Waterloo Ontario, Canada. In: *International Conference on Groundwater and Man*, **3**, pp. 375–384.

Gorrell, G. and Shaw, J. (1991) Deposition in an esker, bead and fan complex, Lanark, Ontario, Canada. *Sediment. Geol.*, **72**, 285–314.

Harms, J.C., Southard, J.B., Spearing, D.R. and Walker, R.G. (1975) *Depositional Environments as Interpreted From Primary Sedimentary Structures and Stratification Sequences*. SEPM Lecture notes for Short Course No. 2, 161 pp.

Harris, S.A. (1970) The Waterloo kame-moraine, Ontario, and its relationship to the Wisconsin advances of the Erie and Simcoe lobes. *Zeitschrift für Geomorphologie*, **33**, 487–509.

Haszeldine, R.S. (1983) Descending tabular cross-bed sets and bounding surfaces from a fluvial channel in the Upper Carboniferous coal field of north-east England. In: *Modern and Ancient Fluvial Systems* (Eds J.D. Collinson and J. Lewin), **6**, pp. 449–456. Special Publication of the International Association of Sedimentologists.

Hiscott, R.N. (1994) Loss of capacity, not competence, as the fundamental process governing deposition from turbidity currents. *J. Sed. Res. A*, **64**, 209–214.

Holysh, S., Davies, S.D. and Goodyear, D. (2004) 28. Project Unit 04–028. An investigation into buried valley aquifer systems in the Lake Simcoe Area. Ontario Geological Survey, Open File Report 6145, 58–1 to 58–6 pp.

Howard, K.W.F., Eyles, N., Smart, P.J., Boyce, J.I., Gerber, R.E., Salvatori, S.L. and Doughty, M. (1995) The Oak Ridges Moraine of southern Ontario: a groundwater resource at risk. *Geoscience Canada*, **22**, 101–120.

Jopling, A.V. and Walker, R.G. (1968) Morphology and origin of ripple-drift cross-lamination, with examples from the Pleistocene of Massachusetts. *J. Sed. Petrol.*, **38**, 971–984.

Karrow, P.F. (1973) The Waterloo kame-moraine, a discussion. *Zeitschrift fur Geomorphologie*, **17**, 126–133.

Karrow, P.F. (1974) Till stratigraphy in parts of south-western Ontario. *Geol. Soc. Am. Bull.*, **85**, 761–768.

Karrow, P.F. (1993) Quaternary geology Stratford – Conestogo area. Ontario Geological Survey, Report 283, 104 pp.

Karrow, P.F. and Paloschi, G.V.R. (1996) The Waterloo kame moraine revisited: new light on the origin of some Great Lake region interlobate moraines. *Z. Geomorph. N F*, **40**, 305–315.

Kleinhans, M.G. (2004) Sorting in grain flows at the lee side of dunes. *Earth Sci. Rev.*, **65**, 75–102.

Kleinhans, M.G. (2005) Grain-size sorting in grain flows at the lee-side of deltas. *Sedimentology*, **52**, 291–311.

Kostic, B., Becht, A. and Aigner, T. (2005) 3-D sedimentary architecture of a Quaternary gravel delta (SW-Germany): Implications for hydrostratigraphy. *Sediment. Geol.*, **181**, 143–171.

Kuehl, G.A. (1975) *Sedimentology of the Hakesville moraine*: Unpublished B.S. thesis, University of Waterloo, Waterloo, Canada, 56 pp.

Leclair, S.F. (2002) Preservation of cross-strata due to the migration of subaqueous dunes: an experimental investigation. *Sedimentology*, **49**, 1157–1180.

Lunt, I.A., Bridge, J.S. and Tye, R.S. (2004) A quantitative, three-dimensional depositional model of gravelly braided rivers. *Sedimentology*, **51**, 377–414.

Massari, F. (1996) Upper-flow-regime stratification types on steep-face, coarse grained, gilbert-type progradational wedges (Pleistocene, southern Italy). *J. Sed. Res.*, **66**, 364–375.

Miall, A.D. (1996) *The Geology of Fluvial Deposits: Sedimentary Facies, Basin Analysis, and Petroleum Geology*. Springer, Berlin; New York, xvi, 582pp.

Meriano, M. and Eyles, N. (2003) Groundwater flow through Pleistocene glacial deposits in the rapidly urbanizing Rouge River-Highland Creek watershed, City of Scarborough, southern Ontario, Canada. *Hydrogeol. J.*, **11**, 288–303.

Nemec, W. (1990) Aspects of sediment movement on steep delta slopes. In: *Coarse-grained deltas*. (Eds A. Colella and D.B. Prior), *Special Publication of the International Association of Sedimentologists*, **10**, pp. 29–73. Blackwell, Oxford, International.

Poeter, E.P. and McKenna, S.A. (Eds) (1999) *Combining geological information and inverse parameter estimation*. (Eds G.S. Fraser and J.M. Davis), *Hydrogeology Models of Sedimentary Aquifers*. SEPM Concepts in Hydrogeology and Environmental Geology No. 1, 171–188pp.

Russell, H.A.J. and Arnott, R.W.C. (2003) Hydraulic Jump and Hyperconcentrated Flow Deposits of a Glacigenic Subaqueous Fan: Oak Ridges Moraine, Southern Ontario. *J. Sed. Res.*, **73**, 887–905.

Russell, H.A.J., Arnott, R.W.C. and Sharpe, D.R. (2003) Evidence for rapid sedimentation in a tunnel channel, Oak Ridges Moraine, southern Ontario, Canada. *Sediment. Geol.*, **160**, 33–55.

Rust, B.R. and Romanelli, R. (1975) Late Quaternary subaqueous outwash deposits near Ottawa, Canada. In: *Glaciofluvial and Glaciolacustrine Sedimentation* (Eds A.V. Jopling and B.C. McDonald), *SEPM Spec. Publ.*, **23**, pp. 177–192.

Saunderson, H.C. (1976) Paleocurrent analysis of large-scale cross-stratification in the Brampton Esker, Ontario. *J. Sed. Petrol.*, **46**, 761–769.

Sharpe, D.R., Hinton, M.J., Russell, H.A.J. and Desbarats, A.J. (2002) The need for basin analysis in regional hydrogeological studies: Oak Ridges Moraine, Southern Ontario. *Geoscience Canada*, **29**, 3–20.

Sharpe, D.R., Pugin, A., Pullan, S.E. and Gorrell, G. (2003) Application of seismic stratigraphy and sedimentology to regional investigations: an example from Oak Ridges Moraine, southern Ontario, Canada. *Can. Geotechnical J.*, **40**, 711–730.

Shaw, J. and Gorrell, G. (1991) Subglacially formed dunes with bimodal and graded gravel in the Trenton drumlin field, Ontario. *Géographie Physique Quaternaire*, **45**, 21–34.

Shaw, J. and Kellerhals, R. (1977) Paleohydraulic interpretation of antidune bedforms with applications to antidunes in gravel. *J. Sed. Petrol.*, **47**, 257–266.

Sigenthaler, C. and Huggenberger, P. (1993) Evidence of dominant pool preservation in Rhine gravel. In: *Braided rivers* (Eds J.L. Best and C.S. Bristow), pp. 291–304. Geol. Soc. Lond.; Special Publication, 75.

Simons, D.B., Richardson, E.V. and Nordin, C.F. (1965) Sedimentary structures generated by flow in alluvial channels. In: *Primary Sedimentary Structures and Their Hydrodynamic Interpretation* (Ed. G.V. Middleton), *SEPM Spec. Publ.*, **12**, pp. 34–52.

Taylor, F.B. (1913) The moraine systems of southwestern Ontario. *Roy. Can. Inst. Trans.*, **10**, 57–79.

TEC (2004) Watershed-Based Source Protection Planning: Science-based decision-making for protecting Ontario's drinking water resources: A Threats Assessment Framework. Technical Experts Committee (TEC), Report to the Minister of the Environment.

Warren, W.P. and Ashley, G.M. (1994) Origins of the ice-contact stratified ridges (eskers) of Ireland. *J. Sed. Res.*, **A64**, 433–449.

Winsemann, J., Asprion, U., Meyer, T., Schultz, H. and Victor, P. (2003) Evidence of iceberg-ploughing in a subaqueous ice-contact fan, glacial Lake Rinteln, NW Germany. *Boreas*, **32**, 386–398.

Wright, L.D. (1977) Sediment transport and deposition at river mouths: A synthesis. *Geol. Soc. Am. Bull.*, **88**, 857–868.

Estimating episodic permafrost development in northern Germany during the Pleistocene

GEORG DELISLE*, STEFAN GRASSMANN*, BERNHARD CRAMER*,
JÜRGEN MESSNER† and JUTTA WINSEMANN‡

*Bundesanstalt für Geowissenschaften und Rohstoffe (BGR), Stilleweg 2, 30655 Hannover, Germany (e-mail: G.Delisle@bgr.de)
†Landesamt für Bergbau, Energie und Geologie (LBEG), Stilleweg 2, 30655 Hannover, Germany
‡Institut für Geologies, Leibniz Universität, Hannover, Callinstraße 30, 30167 Hannover

ABSTRACT

Climate variations in central Europe during the Weichselian can be retraced with reasonably good confidence on the basis of proxy data such as botanical macrofossils and pollen content in Weichselian sediments, for which good age control is available. The availability of proxy data from pre-Eemian subaerial deposits tends to be too erratic in space and time to enable us to reconstruct from them, with confidence, a continuous record of northern Germany's regional climate for the whole Pleistocene, in particular prior to the times of major glacial advances from the north. The likely duration and maximum depth of the occurrence of permafrost in northern Germany throughout the Pleistocene, however, can be estimated. This assumes that the Pleistocene climate trend archived in oxygen isotope records from marine sites (such as e.g. ODP-sites 659 and 677) can be considered in broad terms to reflect the global climatic variations. Here we discuss the approach of using a previously presented reconstruction of the mean annual ground temperatures (MAGT) of the last 120 ky for northern Germany as a calibration tool to reconstruct, from marine proxy data, MAGT-estimates for northern Germany during the Pleistocene. These MAGT-data are the basis of calculations to estimate permafrost depth fluctuations in northern Germany. Calculations for the growth and decay of permafrost are presented for a vertical sediment column and a 75 km long profile modelled after a seismic line that crosses in an east-west direction several salt domes in the subsurface of the German North Sea sector. The latter model intends to demonstrate the interplay between variations in terrestrial heat flow caused by the presence of salt structures and changing climatic conditions over time on the development and decay of permafrost. Depending on the applied climate curve, the maximum vertical extent of permafrost during the Pleistocene is estimated to be c. 130 m (MAGT$_{ODP677}$-reconstruction) or 170 m (MAGT$_{ODP659}$-reconstruction). We favour the MAGT$_{ODP659}$-reconstruction, since its results correlate well with available geological evidence for cold stages with indications of permafrost development during the Pleistocene.

Keywords northern Germany, permafrost, oxygen isotope records, ODP-659, ODP-677, heat flow, climate.

INTRODUCTION

The presence of permafrost in northern Germany during various glacial stages of the Pleistocene is beyond doubt. Evidence of relict ice-wedge structures within Elsterian- Saalian- and Weichselian-stage deposits has been repeatedly observed in excavations of the lignite mining districts of East Germany (e.g. Eissmann, 1997). Figure 1 shows a vegetation pattern observable today, which mimics the pattern of relict ice-wedge polygons near Wolfsburg (Lower Saxony). The vegetation pattern was formed when open space, produced by the melt-out of former ice wedges was filled with

Fig. 1 Relict ice-wedges mimicked by vegetation pattern today near Wolfsburg/northern Germany. Finer soil material collected in former ice-wedges results in denser vegetation along the outline of the former ice wedges (photo courtesy of J. Merkt).

finer grained material in comparison with the surrounding material. The resulting soil contrast favoured new growth of denser vegetation along the outline of the former ice wedges.

Such relict permafrost features point to palaeo-climatic conditions in northern Germany, which are typical for Svalbard or Central Siberia today. However, it appears that no direct indication of the former vertical extent of permafrost during cold stages has been imprinted into the sub-surface sedimentary record. It seems that the only viable approach available to us, to derive realistic estimates of the probable former depth extent of permafrost, is to estimate this parameter with numerical simulation of the freezing process in soils as a function of the presumed climatic changes in the past. Climate is the controlling factor of soil surface temperatures and, if a region remains exposed to mean annual sub-zero temperatures, serves as the 'engine' triggering permafrost growth and decay. Recent advances in climate research, primarily based on marine proxy records, have established a clear understanding of the sequence of glacial and interglacial stages globally (e.g. Thiedemann et al., 1989; Shackleton et al., 1990; Thiedemann et al., 1994). On land, disregarding the special case of proxies from ice sheets, no comparable climate archives are available. Details of climatic conditions on land can only be reconstructed on a regional scale for some segments of the geological past, wherever favourable fossil and pollen records are available in the sedimentary record.

PAST CLIMATE IN NORTHERN GERMANY

Climatic variations in central Europe during the Weichselian can be retraced with reasonable confidence on the basis of proxy data such as botanical macrofossils and pollen content in Weichselian sediments, for which good age control is available. Such a reconstruction has been presented by Caspers and Freund (1997) for north-central Europe on the basis of the palaeo-botanic and palaeo-faunistic investigations of lake deposits, drill cores and sediment exposures. Their findings were transformed into a 'first attempt'-reconstruction of the mean annual ground temperatures (MAGT) during the Weichselian, which also served as the basis for an attempt to assess the depth extent of permafrost in northern Germany during that time (Delisle et al., 2003). The reader is referred to this paper for a discussion of the ambiguities inherent in this type of approach.

The availability of proxy data from pre-Eemian sub-aerial deposits, however, tends to be too sparse to enable us to reconstruct with confidence a continuous record of the regional climate for the entire Pleistocene, in particular prior to the times of the major advances of the Fennoscandian ice-sheets from the north. This paper explains our approach chosen to develop, on the basis of two available marine proxy records (ODP-sites 659 and 677), MAGT-curves for northern Germany throughout the Pleistocene period.

CONTROLLING PARAMETERS OF PERMAFROST DEVELOPMENT

Several factors control soil surface temperatures. Air temperatures at the air/soil-interface, the type of vegetation, snow cover, surface slope and material properties of the surface material (albedo properties) are the controlling factors. Permafrost typically includes an annual 'active layer' on top, which is the surface layer exposed to the annual freeze/thaw process. Our numerical model does not deal with such short term effects. Rather it calculates the depth extent of former permafrost. At the top of the permafrost, a MAGT-curve is presumed, based on the soil temperature at 10 m depth, i.e. immediately below the annual thermal disturbance caused by the seasons. MAGT is in a complex way interconnected with the mean annual air temperature (MAAT). A thorough discussion of this interrelation can be found in Brown and Péwé (1973), which shows that the southern boundary of permafrost occurrences roughly coincides with the −1°C mean annual air isotherm of Canada. They also note that a temperature difference of +3.5°C between MAGT and MAAT is common in Canada in terrain with a MAAT of −4°C. The MAGT-approximation, as discussed in Delisle *et al.* (2003), is in close agreement with these findings. Therefore, we consider this MAGT correction to be realistic in general, but recognize at the same time that under specific land-surface conditions vastly different relationships might occur.

In addition, various soil parameters influence permafrost growth and decay. Of major importance is the overall content of fluid in the soil, which at the freezing point will freeze (melt) after the latent heat content has been released (taken up). The amount of latent heat released or being taken up is a key component that controls the rate of permafrost growth or decay. The freezing/melting point of ice in permafrost is also dependent on soil type. The depression of the freezing point in particular in fine-grained soils is a well known effect (see e.g. Tsytovich, 1957; Washburn, 1979). However, for most clayey to sandy soils, more than 85% to 97% of the pore water is frozen when the soil temperature falls below −0.6°C. To account for this effect, we have chosen in our calculations as an approximation a mean value of −0.6°C as the effective freezing point, where all latent heat has been released (upon freezing) or will start to be taken up (upon melting). An extensive summary of the dependence of the depression of the freezing point on soil type is given in Figure 4.1 of Washburn (1979). Finally, the thermal conductivity of soils and the heat-flow density from the Earth's interior are additional controlling factors.

CONSTRUCTING THE MAGT RECORD

We use two types of data-sets to derive estimates of the former MAGT for northern Germany: (i) the previously presented MAGT-curve for northern Germany (Figure 2; Delisle *et al.*, 2003), and (ii) the climatic variations for the Pleistocene as derived from benthic foraminifera collected from cores of

the boreholes ODP-659 and ODP-677 (Figs 3, 4). ODP-659 was drilled offshore from West Africa (Thiedemann *et al.*, 1989), and ODP-677 in the eastern Pacific (Shackleton *et al.*, 1990; Shackleton, 1996). Both foraminiferal records represent the climatic trend at their drilling sites during the Pleistocene. For comparison, we have used both records to translate these into MAGT-temperatures for northern Germany. We have fixed the recorded maximum $\delta^{18}O$ values during the peak of the Weichselian with the minimum value of the previously reconstructed MAGT curve for northern Germany. Likewise, the minimum values of the $\delta^{18}O$ reached during the Eemian stage with the maximum Eemian ground temperature of the MAGT-record are also fixed. Using these two end points, a linear relationship between $\delta^{18}O$-values and MAGT-values is developed. The resulting 'reconstructions' of the MAGT in northern Germany during the Pleistocene are shown in Figures 3 and 4. The previously published MAGT-record for Eemian times to the present has been inserted into the overall reconstructions. Details of the climate variability of the Holocene have not been included in this curve. Instead, a constant ground-surface temperature of 9°C is assumed.

The amplitudes and time extent of Holocene climatic optima and minima are yet not fully understood and are controversially discussed in the literature (see e.g. Mann *et al.*, 1999; Moberg *et al.*, 2005). Attempts to invert borehole temperatures to identify climatic signals are typically successful in the identification of the climatic signals of the 'Little Ice Age', but not so effective for earlier climatic changes.

RECONSTRUCTION OF PERMAFROST IN NORTHERN GERMANY

As a first step, sediment columns with an upper surface exposed to the MAGT-curves derived from ODP-659 and ODP-677 are considered (Figs 3, 4). The average porosity of the sediments is taken as 20%, the average thermal conductivity as 2.0 W m^{-1} K^{-1}. In the case of MAGT$_{ODP677}$, only the time span of the last 1.15 million years is considered, since the MAGT$_{ODP677}$-reconstruction for the Pleistocene (Fig. 4) does not suggest episodes of prolonged sub-zero temperatures at earlier

G. Delisle et al.

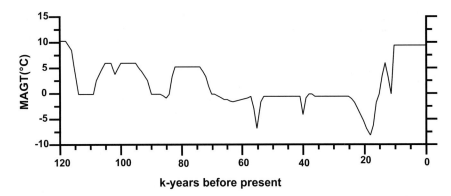

Fig. 2 Reconstruction of mean annual ground temperature (MAGT) according to Delisle *et al.*, 2003).

times. However, a time span of 2 million years is considered in the case of MAGT$_{ODP659}$, since this reconstruction shows prolonged episodes of sub-zero temperatures throughout the whole time period.

In addition to the thermal surface conditions, permafrost aggradation and decay is also controlled at the base by the heat flow from the interior of the Earth. The climatic cooling trend throughout the Quaternary has resulted in a transient subsurface temperature field in the upper crust. Detailed thermal measurements in super-deep boreholes, such as the KTB-drilling site in southeast Germany, have provided evidence for a 'Quaternary climate signal' in the crust down to depths of about 4 km (Clauser *et al.*, 1997; Kohl, 1998). The same studies show that, as a consequence, present-day measured heat-flow density values tend to be too low in the depth range of 0 m–1500 m. We have adopted at the base of our numerical model, the lowermost grid points represent a depth of 11,200 m below surface, a heat-flow density of 75 mW m^{-2}. As a consequence of the climatic deterioration contained in both of our MAGT-curves, a 'Quaternary climatic signal' is automatically produced in the subsurface, and the calculated heat-flow density values for the uppermost 1 km of the crust are reduced to values around 50 to 60 mW m^{-2}. This is in agreement with measured values for northern Germany, in particular for the German North Sea sector (Haenel, 1988). We use a numerical model described earlier (Delisle, 1998a) for the calculation of permafrost aggradation and decay. In particular, the model includes the effect of release or uptake of latent heat during freezing or melting of permafrost, but which otherwise limits itself to a purely heat-conductive approach of the problem. The two

presented models calculate the time-dependent vertical extent of the permafrost zone during each cold stage (Figs 3C, 4C). It should be noted that the base of the permafrost zone is not a sharply defined boundary in reality since, as discussed above, partial melting will occur in the boundary zone wherever near-zero ground temperatures are approached. In the context of the numerical model presented, the boundary defines the level, where release (freezing) or uptake (melting) of latent heat has been completed.

Both MAGT-reconstructions for northern Germany predict permafrost in northern Germany during the glacial/interglacial cycles in Central Europe during the last 1 million years. The permafrost reconstruction based on the MAGT$_{ODP659}$-curve predicts a maximum permafrost depth of about 170 m during the late Pleistocene (Fig. 3C), while MAGT$_{ODP677}$ predicts only about 130 m (Fig. 4C). The MAGT$_{ODP677}$ reconstruction suggests that climatic conditions were too warm for the formation of permafrost during the period of 2–1.2 million years ago. However, the MAGT$_{ODP659}$ reconstruction hind casts prolonged sub-zero ground temperature episodes and phases of permafrost development during the early Pleistocene. This is in good agreement with geological evidence for permafrost development in the Menapian, Eburonian and Tiglian stages, as has been reported by various workers. A summary of the available evidence was presented by Vandenberghe (2001). It appears reasonable that the MAGT$_{ODP659}$ reconstruction offers a superior result in comparison to MAGT$_{ODP677}$. The climate in Europe is strongly influenced by the circulation pattern of the Atlantic water masses, which should be reflected by the oxygen isotope record from the ODP-site 659. The timing of climatic

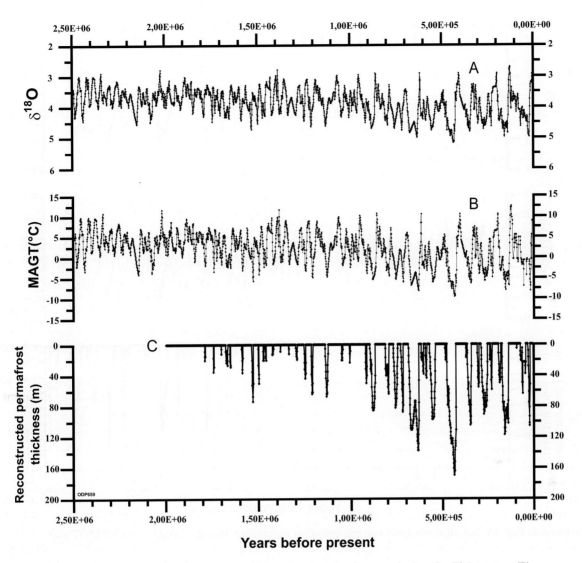

Fig. 3 (A) δ^{18}O-record from the ODP 659 drill hole reflects the climatic changes during the Pleistocene. (B) Reconstruction of the MAGT for northern Germany based on a comparison with the record from ODP 659. The previously published MAGT-record from Eemian times to the present (Fig. 2) has been inserted into the overall reconstruction. (C) Calculated vertical extent of permafrost during the Pleistocene based on the MAGT$_{ODP659}$ curve.

fluctuations recorded by the Pacific-based ODP-site 677 is comparable to the ODP-659-record. However, the amplitudes of the climatic fluctuations recorded by the Atlantic-based ODP-659 are more pronounced in comparison with ODP-677.

The model calculates, in addition, the variation of heat-flow density at various depths in response to changes in MAGT's induced by the climatic forcing. A decrease in MAGT increases the temperature gradient in the vertical direction, since the temperature adjustment at depth is slowed down by the low thermal conductivity of sediments. In consequence, the near-surface heat-flow density is increased. The opposite is true if the MAGT's warm up.

An intriguing result of the models is the apparent strong modulation of the heat-flow density in the uppermost 1 km of the sediment column throughout time. Strong positive heat-flow excursions up to +220 mW m^{-2} are observed at times of rapid cooling and the opposite (up to –175 mW m^{-2}) during episodes of climatic rebound (Fig. 5). A detailed examination of the situation for the last 25,000 years demonstrates this strong dependence.

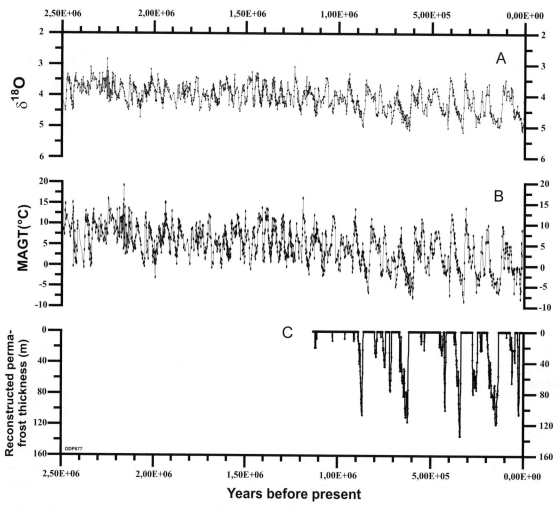

Fig. 4 Same as Fig. 3A, but all reconstructions are based on the oxygen isotope record of the ODP677-site.

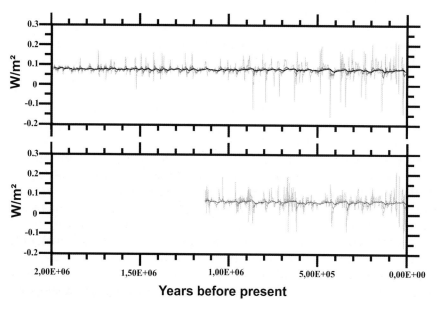

Fig. 5 Calculated variations of heat-flow density induced by climatic change based on the $MAGT_{ODP659}$ curve (top) and the $MAGT_{ODP677}$ (bottom).

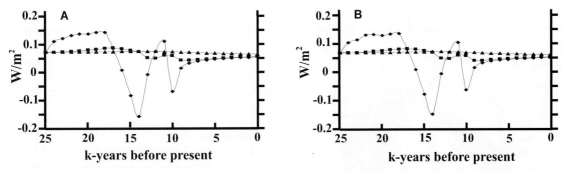

Fig. 6 Heat-flow density distribution near the ground surface (diamonds), at 250 m depth (triangles) and at 1 km depth (boxes) based on the sediment column model for (a) $MAGT_{ODP659}$ curve, and (b) the $MAGT_{ODP677}$ curve. The calculated heat-flow value of c. 50 mW m^{-2} near the ground surface compares with model results.

Figure 6 show the calculated heat-flow densities, q, at the air/ground interface at 250 m and 1 km depth. Within this period, q varies for both models between −163 and +143 mW m^{-2} near the surface and between 41 and 86 mW m^{-2} at 1 km depth. Clearly, the initial background heat flow of 75 mW m^{-2} is perturbed most of the time. Today, in consequence of the series of glacial stages in the past, the heat-flow density near the surface according to both models is near 53 mW m^{-2} and at 63 mW m^{-2} at 1 km depth, which is in agreement with field data (Haenel, 1988). Additional effects such as convective heat transport by ground-water are likely to modify the subsurface temperature and the heat-flow density distribution on a local, sometimes regional scale. Ground-water flow under ice sheets and in the foreland of ice sheets may have a substantial impact on lateral heat transfer (Boulton *et al.*, 1995). A special study of the groundwater regime over the Gorleben salt dome in northern Germany, a proposed site for a nuclear waste repository, has revealed evidence of short-term enhanced ground-water flow under permafrost in the foreland of glaciers (Boulton *et al.*, 2001).

MODELLING OF EPISODIC PERMAFROST ALONG A 75 KM LONG E-W PROFILE ACROSS A SHELF REGION, GERMAN NORTH SEA SECTOR

The presence of salt domes in the subsurface leads to a redistribution of heat flow (Delisle, 1998b). Higher thermal conductivity of rock salt results in enhanced heat flow through the salt structures at the expense of the surrounding country rock.

Enhanced heat flow over salt domes impedes the development of permafrost. Reduced heat flow in the adjacent country rock favours deeper penetration of the lower permafrost boundary. To illustrate this relationship, we have chosen a 75 km long east-west profile across the shelf area of the German North Sea Sector, which crosses four salt diapirs (Fig. 7). Using the same reconstructions of the Pleistocene MAGT for northern Germany as above, we calculated the episodic permafrost development for these profiles. However, the thermal effects of short term glaciation of this area during the Elsterian- and Saalian-glacial periods were specifically excluded, since such a treatment would have required a coupled glacier/climate model, which is beyond the scope of this paper. Based on a geological interpretation for the vertical section of this profile developed from seismic data (Baldschuhn *et al.*, 2001), we have subdivided the subsurface into a multitude of elements, to which we have attributed the pertinent thermal conductivity value based on sediment type present (Table 1). Thermal conductivity values were taken from Zoth and Haenel, 1988.

The resulting images of permafrost distribution for both MAGT-reconstructions are shown in Figures 8a and 8b. Both results appear to reflect well our perception of the sequence of cold stages in northern Germany for the last 1 million years. The Weichselian, Saalian and Elsterian stages are clearly represented as well as the Cromer and Bavel stages. The maximum attained permafrost thickness is the same as in the previously discussed sediment column. The presence of salt domes tends to reduce the maximum depth of overlying

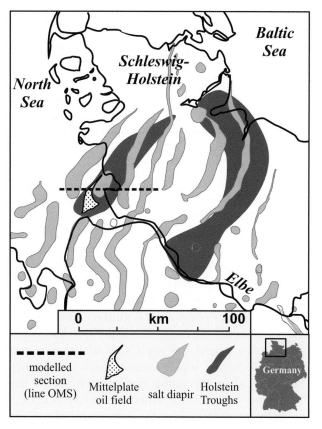

Table 1 Thermal conductivity values for the numerical model

Sediment[1]	Thermal conductivity value in $W\,m^{-1}\,K^{-1}$
Clay	1.7
Sand	2.2
Limestone	3.1
Rock salt	$3.6394-0.007*T(°C)$

[1] The chosen value for rock salt is temperature-dependent.

Fig. 7 Location of east-west profile across the shelf area of the German sector of the North Sea for which we present estimates of the vertical extent of permafrost during cold stages of the Pleistocene Epoch.

current or palaeo-saltwater occurrences to our model predictions should be kept in mind.

CONCLUSIONS

permafrost by as much as 40 m (Figs 8A, B). On the basis of the $MAGT_{ODP677}$ curve, again little to no permafrost is expected for earlier Pleistocene Periods. However, the $MAGT_{ODP659}$ curve results in additional prolonged permafrost episodes in the period between 2 and 1 million years ago, in agreement with the available geologic evidence (Vandenberghe, 2001).

Salty groundwater is common in the vicinity of salt domes. The typical depth extent of saltwater deposits depends on a number of factors and cannot be easily generalized for the sedimentary basins in northern Germany in space and time. Depending on their depth and the salt content, which results in a depression of the freezing point, they may have further impeded permafrost aggradation in the past, particularly at depth and on a local scale. Nevertheless, fresh-water occurrences today frequently extend below the 150 m level, and probably have done so in the past. The limitation induced by

Both MAGT-reconstructions for northern Germany result in climate curves, from which reconstructions of the vertical extent of permafrost in northern Germany during cold stages can be derived. These curves are in broad agreement with earlier published concepts for the glacial/interglacial cycles of Central Europe during the last 1 million years. $MAGT_{ODP659}$ predicts a maximum permafrost depth of about 170 m and $MAGT_{ODP677}$ a value of about 130 m during short time periods during the late Pleistocene. The $MAGT_{ODP677}$ reconstruction suggests that climatic conditions were too warm for the formation of permafrost for the period of 2–1.2 million years ago. However, the $MAGT_{ODP659}$ reconstruction hindcasts for the whole Pleistocene prolonged sub-zero ground temperature episodes and phases of permafrost development, which are in good temporal agreement with dated permafrost phases based on geological work. Thus, the $MAGT_{ODP659}$ reconstruction represents the more realistic approximation, which appears reasonable considering the significantly shorter lateral distance of the ODP-site 659 to Germany in comparison with the ODP-site 677. The increased heat flow over salt domes, present in large numbers in northern Germany, impedes the downward aggradation of permafrost. Our profile suggests reduction of the maximum vertical extent of permafrost over salt domes by up to 40 m in comparison with the neighbouring host rock. The repeated phases of drastic climatic changes in Central Europe and the resulting 'Quaternary climate signal' in the upper crust are reproduced by our model. In tune with climatic changes, heat-flow density measurable near the surface fluctuated significantly in the past. Our model confirms earlier predictions that the climatic warming during the Holocene Epoch resulted in a reduced heat-flow density at the surface in comparison with the heat-flow density at depth.

OUTLOOK

Advances of the Scandinavian ice sheets during Pleistocene times greatly influenced the near-surface

geology of Schleswig-Holstein. Glacial processes related to the advances of ice sheets include additional loading, changes of the sediment-surface temperature, convective heat transport induced by outflow of subglacial melt-water and the formation of permafrost in the foreland. For example, on the Norwegian continental margin Johansen *et al.* (1996) and Solheim *et al.* (1996) showed that these processes have a significant effect on the development of the subsurface temperature and pressure fields, and furthermore alter the hydrodynamic conditions in the sedimentary basins. We are now in the process of investigating how glacial processes have acted on the physicochemical habitat and the petroleum system of the Northwest German Basin in Schleswig-Holstein. Despite the fact that glacial

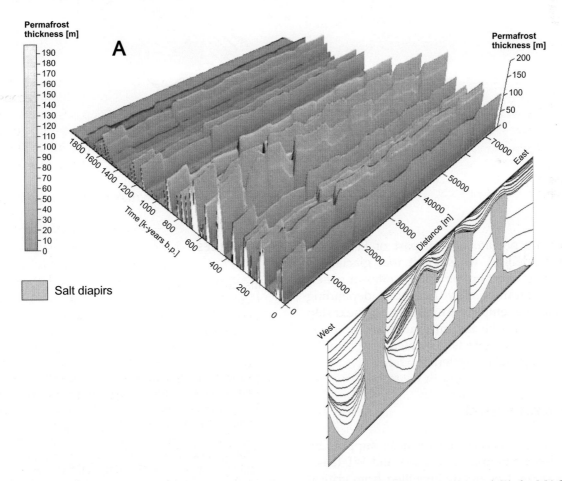

Fig. 8 Calculated variations in depth extent of permafrost based on (A) the MAGT$_{ODP659}$ curve, and (B) the MAGT$_{ODP677}$ curve. The right-hand axes represent a 75 km long profile crossing several salt structures (see Fig. 6). The axis on the left denotes time. Visible in both figures are reductions of permafrost thickness by up to 40 m at positions where the profile crosses four salt domes. Only the model based on the MAGT$_{ODP659}$ curve (A) reproduces permafrost episodes during the early Pleistocene in agreement with field evidence.

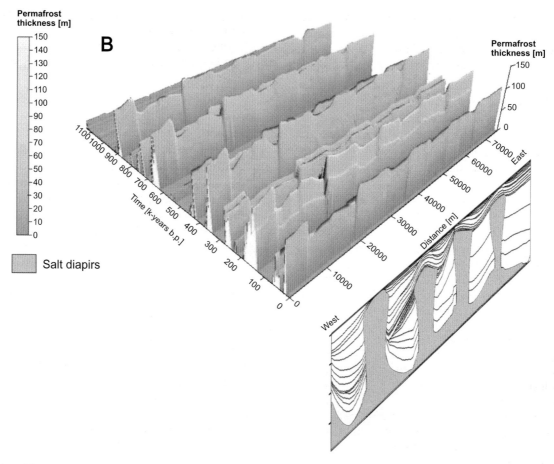

Permafrost thickness [m]

B

Salt diapirs

Fig. 8 (*cont'd*)

processes are not an integral part of Petroleum Systems Modelling, we suggest that glaciations have a strong effect on the timing, rates of generation and migration of hydrocarbons, depending on the absolute temperature history. The next step of this work will be the integration of the baseline data of glacial effects presented here into the current 2D petroleum system model.

ACKNOWLEDGEMENTS

We gratefully acknowledge financial support by DFG under the grants CR 139 1–1 and WI 1844/ 4–1. The paper has greatly benefited from critical reviews of an earlier version by two anonymous reviewers. The assistance of A. Bartels in preparing the figures is gratefully acknowledged.

REFERENCES

Baldschuhn, R., Binot, F. Fleig, S. and Kockel, F. (2001) Geotektonischer Atlas von Nordwest-Deutschland und dem deutschen Nordsee-Sektor. *Geol. Jb.*, Reihe A, **153**, 95 pp.

Brown, R.J.E. and Péwé, T.L. (1973) Distribution of permafrost in North America and its relationship to the environment: a review, 1963–1973. In: *North American contribution – Permafrost – Second Int. Conf.*, 13–28.7.1973, Yakutsk, U.S.S.R., National Academy of Sciences, Washington D.C., 770 pp.

Boulton, G., Caban, P.E. and van Gijssel, K. (1995) Groundwater flow beneath ice sheets: part I – large scale patterns. *Quat. Sci. Rev.*, **14**, 545–562.

Boulton, G., Gustafson, G., Schelkes, K., Casanova, J. and Moren, L. (2001) Palaeohydrogeology and geoforecasting for performance assessment in geosphere repositories for radioactive waste disposal (Pagepa).

Report: EU 19784 – Nuclear science and technology series, Office for Official Publications of the European Communities, Luxemburg, 147 pp.

Caspers, G. and Freund, H. (1997) Die Vegetation und Klimaentwicklung des Weichsel-Früh- und –Hochglazials im nördlichen Mitteleuropa. In: Freund and Caspers (Eds) Vegetation und Paläoklima der Weichsel-Kaltzeit im nördlichen Mitteleuropa – Ergebnisse paläobotanischer, – faunistischer und geologischer Untersuchungen. *Schriftenreihe der Deutschen Geologischen Gesellschaft,* Heft **4**, 201–249.

Clauser, Ch., Giese, P., Huenges, E., Kohl, T., Lehmann, H., Rybach, L., Safanda, J., Wilhelm, H., Windloff, K. and Zoth, G. (1997) The thermal regime of the crystalline continental crust: Implications from KTB. *J. Geophys. Res.,* 102, B8, 18417–18441.

Delisle, G. (1998a) Numerical simulation of permafrost growth and decay. – *J. Quatern. Sci.,* **13**(4), 325–333.

Delisle, G. (1998b) The evolution of the natural temperature field of a salt dome in geological time. *Journal of Seismic Exploration,* **7**, 251–264.

Delisle, G., Caspers, G. and Freund, H. (2003) Permafrost in north-central Europe during the Weichselian: how deep? – *Proc. of 8th International Conference on Permafrost,* Zurich, 187–191.

Eissmann, L. (1997) Das quartäre Eiszeitalter in Sachsen und Nordostthüringen. *Altenburger Naturwissenschaftliche Naturforschungen,* **8**, 1–98.

Haenel, R. (1988) Atlas of geothermal resources in the European Community, Austria and Switzerland. *Report EUR/Commission of the European Communities,* 11026, Schaefer Verlag, Hannover, 74 pp.

Johansen, H., Fjeldskaar, W. and Mykkeltveit, J. (1996) The influence of glaciation on the basin temperature regime. *Global Planet. Change,* **12**, 437–448.

Kohl, T. (1998) Palaeoclimatic temperature signals – can they be washed out? *Tectonophysics,* **291**, 225–234.

Mann, M.E., Bradley, R.S. and Hughes, M.K. (1999) Northern hemisphere temperatures during the past millennium: Inferences, uncertainties, and limitations. *Geophys. Res. Lett.,* **26**, 759–762.

Moberg, A., Sonechkin, D.M., Holmgren, K., Datsenko, N.M. and Karlen, W. (2005) Highly variable Northern Hemisphere temperatures reconstructed from low- and high-resolution proxy data. *Nature,* **433**, 613–617.

Shackleton, N. (1996) Timescale Calibration, ODP 677. *IGBP PAGES/World Data Center-A for Paleoclimatology Data Contribution Series # 96–018.* NOAA/NGDC Paleoclimatology Program, Boulder CO, USA.

Shackleton, N.J., Berger, A. and Peltier, W.R. (1990) An alternative astronomical calibration of the lower Pleistocene timescale based on ODP Site 677. *Trans. Roy. Soc. Edinb. Earth Sci.,* **81**, 251–261.

Solheim, A., Riis, F., Elverhoi, A., Faleide, J.I., Jensen, L.N. and Cloetingh, S. (1996) Impact of glaciations on basin evolution; data and models from the Norwegian margin and adjacent areas – Introduction and summary. *Global Planet. Change,* **12**, 1–9.

Thiedemann, R., Sarntheim, M. and Stein, R. (1989) Climatic changes in the western Sahara: Aeolomarine records of the last 8 million years. *Proc. ODP Sci. Results,* **108**, 241–278.

Thiedemann, R., Sarntheim, M. and Shackleton, N.J. (1994) Astronomic timescale for the Pliocene Atlantic $\delta^{18}O$ and dust flux records of Ocean Drilling Program site 659. *Paleoceanography,* **9**, 4, 619–638.

Tsytovich, N.A. (1957) The mechanics of frozen ground. *McGraw-Hill Book Comp.* 426 pp.

Vandenberghe, J. (2001) Permafrost during the Pleistocene in northwest and central Europe. In: *Permafrost Response on Economic Development, Environmental Security and Natural Resources* (Eds Paepe, R. and Melnikov, V.), pp. 185–194, Kluver Academic Publishers, Dordrecht, Netherlands.

Washburn, A.L. (1979) *Geocryology: a survey of periglacial processes and environments,* Edward Arnold Publishers, London, UK, 406 pp.

Zoth, G. and Haenel, R. (1988) Thermal conductivity. In: *Handbook of Terrestrial Heat Flow Density Determination* (Eds Haenel, R., Rybach, L. and Stegena, L.), pp. 449–466, Kluver Academic Publishers, Dordrecht, Netherlands.

Lake-level control on ice-margin subaqueous fans, glacial Lake Rinteln, Northwest Germany

JUTTA WINSEMANN, ULRICH ASPRION *and* THOMAS MEYER

Institut für Geologie, Leibniz Universität Hannover, Callinstraße 30, D-30167 Hannover, Germany (e-mail: winsemann@geowi.uni-hannover.de)

ABSTRACT

During the maximum advance of the Early Saalian Scandinavian Ice Sheet ice-dammed lakes developed within valleys of the Northwest German Mountain ranges. The blocking of the River Weser valley led to the formation of glacial Lake Rinteln where most of the sediment was deposited by melt-water. Two subaqueous fan complexes have been identified, which formed lakeward of retreating ice-lobes during an overall lake-level rise. The sediment transport on the fans has been dominated by sustained gravity-flows, mainly cohesionless debris flows or high- and low-density turbidity currents, reflecting discharge of semi-continuous meltwater flows. Intercalations of surge-type high- and low-density turbidites increase towards the mid- and lower-fan slopes, indicating more ice-distal and periodic deposition. Individual fan bodies commonly have a coarse-grained proximal core of steeply dipping gravel, overlain by gently to steeply dipping mid- to outer fan deposits. During glacier retreat commonly fine-grained sediments, rich in ice-rafted debris, were deposited on the ice-distal and ice-proximal slopes of the abandoned fans. Climbing-ripple cross-laminated sand may onlap coarse-grained upper fan gravel and in some cases overtop the older fan deposits. Phases of glacier still-stands are characterised by fan systems that display an upward flattening of fan clinoforms and minor vertical facies changes. The position of ice marginal fans was controlled by the combination of bedrock topography and water depth. At the eastern lake margin topographic highs served as pinning points for the retreating glacier and facilitated ice margin stabilisation. A strong lake-level fall probably triggered a major drainage event that tapped previously unconnected reservoirs of englacial and subglacial meltwater.

Keywords Subaqueous ice-contact fans, facies architecture, depositional processes, melt-water outburst, Pleistocene, Germany.

INTRODUCTION

Ice-contact deposits provide important palaeo-environmental information because they are a record of the physical environment immediately before and during deglaciation. The understanding of the specific depositional processes, the resultant facies, and their spatial distribution within glacially influenced subaqueous environments has been developed from various glaciolacustrine settings (e.g. Clemmensen & Houmark-Nielsen, 1981; Teller & Clayton, 1983; Thomas, 1984; Smith & Ashley, 1985; Eyles & Clark, 1988; Fyfe, 1990; Martini, 1990; Mastalerz, 1990; Sharpe & Cowan, 1990; Gorrell & Shaw, 1991; Martini & Brookfield, 1995; Ashley, 1995; Teller, 1995; Brookfield & Martini, 1999; Bennett *et al.*, 2002; Knudsen & Marren, 2002; Lajeunesse & Michel, 2002; Richards, 2002; Russell & Arnott, 2003; Johnsen & Brennand, 2006). Depositional processes and the resulting facies architecture are highly variable because the margins of ice-contact lakes continually change as glaciers advance or retreat and as ice melts and sediment is released. Glacier-fed lakes exist both during the advancing and retreating phases of a glacier but, during ice advance, proglacial deposits are commonly over-ridden and incorporated into the base of the ice. Consequently most glacial lake records are of deglaciation (e.g. Ashley, 1995). The spatial distribution of sediments in glacier-fed lakes is a function of the mechanisms of sediment distribution and lake level. Where glaciers terminate on land the sediment injection point in the basin is directly related to water level. Unblocking of outlets may cause catastrophic drainage and almost instantaneous and very large irregular drops in lake level (e.g. Ashley, 1995; Brookfield & Martini, 1999).

Rising lake levels often correspond with glacial advances into a basin, as the glacier isostatically depresses the land and blocks drainage outlets. This causes a shift in the locus of proglacial sedimentation to deeper waters. During high water levels, the glacier injection point may be underwater at the base of the slope. During low water levels, the injection point of the glacier may be on land. Since the lake level is controlled by the position of drainage outlets, water levels may change dramatically, abruptly and independently of the glacial input point. The basin can empty or fill with water rapidly, with only slight changes in glacier position. In this case accomodation space has no relation to the glacier injection point. (Martini & Brookfield, 1995; Brookfield & Martini, 1999). Where glaciers terminate underwater, sediment input points can fluctuate independently of lake level. Sediment may be supplied by subglacial, englacial or supraglacial meltwater flows at the ice front and by rainout from a floating glacier. Keeping the lake level at highstand, input points of the glacier system vary according to the position and nature of the ice front. As a glacier retreats out of the basin, a series of retreating fining-upwards subaqueous fans form and depending on the regional topography, the deposits consist of a series of laterally stacked bodies which stratigraphically onlap and become younger in the direction of ice retreat. When the glacier readvances into the basin it may deposit a coarsening-upwards subaqueous outwash section capped by an erosion surface. Furthermore local fluctuating ice fronts through ice-front calving may produce successive subaqueous fans, overlain by thin mud drapes (e.g. Ashley, 1995; Brookfield & Martini, 1999).

Ice cliffs or floating ice tongues form where ice terminates at the grounding line or zone at which ice entering a water body comes afloat. Ice cliffs develop when lake water is relatively deep adjacent to the ice. Frequently, calving icebergs maintain the cliff face and thus ice-cliffted margins are likely to occur only on active glacier termini where ice is

replenished. As the ratio of melting to calving increases a more parabolic surface profile is produced. Under these later circumstances a cliff will remain grounded unless the conditions required for floating are met (Powell & Domack, 1995). Ice-ramps occur only at lake margins of stagnating glaciers where water partially covers melting ice (Ashley, 1995).

The relationship between calving rates and water depth means that glaciers terminating in deep water are potentially unstable and vulnerable to catastrophic retreat by rapid calving. Ice margin stability is therefore favoured by the presence of pinning points or constrictions in the enclosing lake basin. Such constrictions in the valley occupied by a glacier reduce losses by calving and if the calving rate is less than the ice flux, the glacier will thicken and stabilize. However, if the glacier pulls back from a pinning point into deeper water, the calving rate may exceed the ice flux, and rapid retreat will result (e.g. Benn & Evans, 1998).

Glaciers occupying a widening basin should be relatively stable, because calving rates will increase if the glacier advances and decreases if the glacier retreats. On the other hand, glaciers occupying narrowing basins tend to be unstable. An advance of the glacier will reduce the calving rate, encouraging further advance, whereas retreat will increase the calving rate, leading to accelerating deglaciation result (e.g. Benn & Evans, 1998). Bottom topography has a similar influence on ice margin stability (Thomas, 1979). Where the bed slopes away from the ice margin, a small increase in ice thickness will result in the advance of the grounding line into deeper water where calving rates are higher. On the other hand, where the bed slopes towards the ice-margin, a small increase in ice thickness will initiate unstable advance into shallowing water, and a decrease in ice thickness will cause accelerating retreat into deeper water (Benn & Evans, 1998).

Sedimentation at the grounding line zone varies with glacial thermal regime. Under polar conditions sediment supply is low because glacial erosion is less effective and meltwater is scarce or absent; groundling line sediment is derived only from the base of the ice. The groundling line of temperate glaciers is the one where the largest volume of sediment is deposited and large quantities of fluvial bedload and suspended load can be transported and deposited by jets (Powell & Domack, 1995). In glaciolacustrine environments, sediment-laden meltwater is generally denser than the surrounding lake water, and will tend to produce underflows. Deposition on grounding line fans is therefore likely to be dominated by mass flows, with comparatively minor inputs from high-level suspended sediment (Benn & Evans, 1998). If the ice terminus remains stable for a long period of time, a grounding line fan may aggrade to lake level and form an ice-contact delta (Lønne, 1995). Alternatively, if the grounding line migrates rapidly, sheet-like deposits form. Larger conduits systems may be destabilised, if the glacier advances into the lake basin and water depth is sufficiently high and may be converted into a linked cavity system (Fowler, 1987; Fyfe, 1990). Smaller conduits or cavities are more unstable and have smaller effluxes, which more easily mix with lake water, so constraining the distance of sediment dispersal. Therefore smaller fans will be deposited which tend to overlap to form sediment aprons. The areal extent of these smaller conduits or cavity system depends on the size of the zone in which lake water pressure influences conduit water pressures, and is determined by lake water depth, the subglacial discharge, ice surface profiles and bedrock topography (Fyfe, 1990; Sharpe & Cowan, 1990).

The study reported here focuses on subaqueous fan systems occurring on the northern and eastern margin of glacial Lake Rinteln. Depositional processes, principal architectural elements and fan stacking patterns will be discussed and related to lake-level changes, topography and glacier termini dynamics.

STUDY AREA AND PREVIOUS RESEACH

The study area is located south of the North German Lowlands, mainly built-up by Mesozoic sedimentary rocks and characterized by several mountain ridges up to 400 m high and broad valleys. The blocking of the River Weser Valley by the Middle Pleistocene (Older Saalian) Scandinavian Ice Sheet led to the formation of glacial Lake Rinteln (e.g. Spethmann, 1908; Thome, 1983; 2001; Klostermann, 1992; Winsemann *et al.*, 2004). Two ice-lobes advanced into the Weser Valley, blocking the east and northward drainage. Field data indicate a maximum ice thickness of *c.* 200 m for these ice

lobes; the terminal ice thickness of the northern Ice Sheet has been approximately 200–350 m (e.g. Skupin *et al.*, 2003). The ice advance from the western lake margin successively closed lake outlets leading to an overall lake-level rise (Winsemann *et al.*, 2004). These overspills are located at the southwestern lake margin, lie at altitudes between *c.* 89–215 m a.s.l. and decrease in altitude towards the west (Figs 1 and 2).

The ice marginal depositional systems are characterized by coarse-grained deltas and subaqueous fans (Fig. 2) mainly deposited from high-energy meltwater flows (Winsemann *et al.*, 2004; Hornung *et al.*, 2007). Well logs and several clay pits record the widespread occurrence of up to 20 m thick fine-grained lake bottom sediments ('Hauptbeckenton'), overlying older Quaternary fluvial deposits of the River Weser or bedrocks in the basin centre (*c.* 55–75 m a.s.l.). At the basin margins fine-grained glaciolacustrine deposits occur also at higher topographic levels up to 130 m a.s.l. (e.g. Rausch, 1975; Deutloff *et al.*, 1982; Rohde, 1994; Könemann, 1995; Wellmann, 1998). The delta complexes are thought to reflect a relatively stable position of the ice-margin in front of the mountain ridge. Subaqueous fans reflect more unstable ice-fronts of smaller ice-lobes that advanced into the lake basins and were subject to periodic calving and short-term oscillations (Winsemann *et al.*, 2004).

Most previous workers have assumed a subaerial origin of these ice-marginal deposits and an end moraine, kame, or glaciofluvial formation has been assumed (Grupe, 1925; 1930; Naumann, 1922; Naumann, 1927; Naumann & Burre, 1927; Stach, 1930; Lüttig, 1954; 1960; Miotke, 1971; Seraphin, 1972; 1973; Rausch, 1975; Könemann, 1995; Wellmann, 1998). However, the deposits were rarely described and former interpretations have been exclusively based on the geomorphology and petrography of ice-marginal deposits.

Re-examination of outcrops indicate a subaqueous origin of these ice-marginal deposits (Winsemann *et al.*, 2003; 2004; Hornung *et al.*, 2007). Several sediment characteristics indicative of subaqueous fan versus subaerial deposition were recorded (e.g. Shaw, 1975; Rust, 1977; Cheel & Rust, 1982; Nemec & Steel, 1984; Thomas, 1984; Powell, 1990; Sharpe & Cowan, 1990; Ashley, 1995; Lønne, 1995; Plink-Björklund & Ronnert, 1999; Sohn, 2000; Krzyszkowski, 2002; Krzyszkowski & Zielínski,

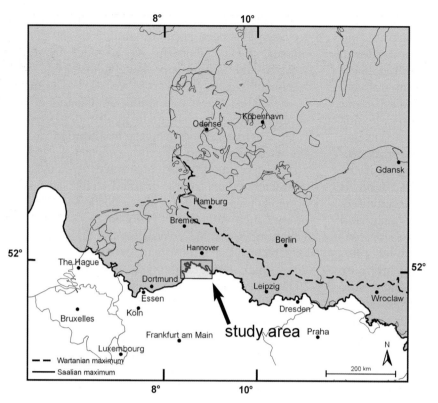

Fig. 1 Maximum extent of the Middle Pleistocene Ice Sheets in Central Europe. Shown are the extent of the Early Saalian 'Drenthe Haupt-Vorstoß' and the extent of the Late Saalian 'Wartanian' Ice Sheet (modified from Ehlers & Gibbard, 2004).

2002; Pisarska-Jamroży, 2006). Data critical to this re-interpretation include the recognition of subaqueous jet-efflux deposits, turbidites, thick ripple-drift units, ice-rafted debris and the occurrence of an iceberg scour on top of coarse-grained ice-marginal deposits.

The stratigraphic record indicates a retreat of active ice, which occurred by calving. Ice-margin fluctuations during retreat are recorded from glaciotectonic deformation of overridden previously deposited lake sediments (e.g. Lüttig, 1960; Wellmann, 1998; Winsemann *et al.*, 2004). Evidence for rapid calving is given by the abundant occurrence of dropstones within the lake bottom sediments (Rausch, 1975).

The duration of the ice-dammed lake can only be estimated because varve deposits of the basin centre have been partly eroded by the River Weser. At the eastern basin margin 2.5 m thick varves have been deposited in about 50 years (Kulle, 1985). Consequently, the duration of glacial Lake Rinteln was probably very short and has been a few hundred years or less. At the initial stage, glacial Lake Rinteln had its level at an altitude of c. 55 m a.s.l. The lake level then rose by as much as 100 m to a highstand of c. 160–180 m a.s.l. The large-scale

stacking pattern of the Emme delta complex indicates an overall lake-level rise, interrupted only by minor short-term lake-level falls. A very strong lake-level fall occurred from a lake-level of c. 160–170 m, which led to the formation of an incised valley in the Emme delta and a subaerial exposure of the Porta subaqueous fan complex (Winsemann & Asprion, 2001; Winsemann *et al.*, 2004).

METHODS

A total of 15 exposures were examined in order to document the regional pattern and character of subaqueous fan sediments. The outcrops were studied by lateral mapping and vertical measured sections across two- and three-dimensional exposures. The sections were measured at the scale of individual beds, noting grain size, bed thickness, bed contacts, and bed geometry, internal sedimentary structures and palaeocurrent directions. Landform genesis and deglacial palaeogeography are inferred from landform-sediment relationships (e.g. Warren & Ashley, 1994).

The field study was supported by the use of ground-penetrating radar (GPR). The GPR device

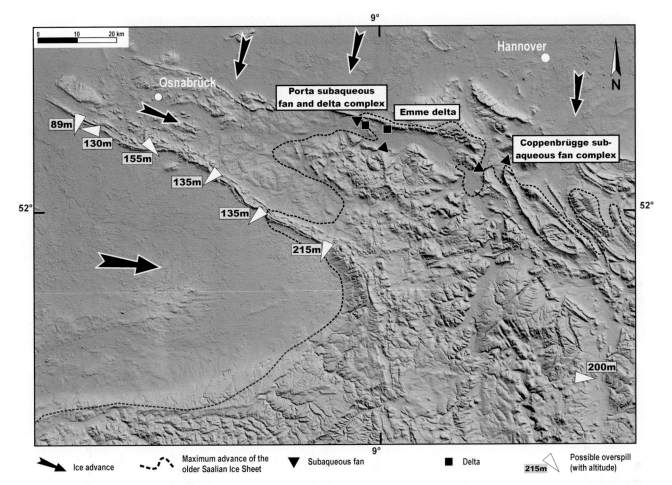

Fig. 2 Hill-shaded relief model showing the maximum extent of the older Saalian Ice Sheet and associated ice-marginal depositional systems. Possible overspills of the Lake Basin were located on the southwestern and southeastern margin. Data compiled from Thome (2001), Winsemann *et al.* (2004) and Hornung *et al.* (2007).

used was a GSSI SIR-10 together with a 100 MHz bistatic antenna. Processing was limited to a minimum to avoid information loss and artefacts. Therefore all lines have undergone bandpass filtering for noise reduction only. No migration was applied due to velocity uncertainties and limited data quality enhancements.

Sedimentary sections that illustrate the most characteristic sedimentary facies and stacking pattern are presented in this article. Additional sedimentary logs are presented in Winsemann *et al.* (2003; 2004), Hornung *et al.* (2007).

SUBAQUEOUS FAN SEDIMENTATION

Two subaqueous fan systems were identified in the study area, overlying fine-grained lake-bottom

sediments and patchy occurrences of till (Figs 2–15). The Coppenbrügge fan complex is located at the eastern lake margin and consists of several small fan bodies, deposited on a hummocky low-angle basin slope (Figs 2 and 4). In contrast, the Porta subaqueous fan complex consists of larger fan bodies, deposited on a nearly flat lake bottom surface in front of the Porta Westfalica pass (Figs 2 and 13). Measured palaeoflow directions and clast composition indicate that meltwater flows were the main source of sediment (e.g. Wellmann, 1998; Winsemann *et al.*, 2004).

Sedimentary facies and facies associations

Twelve facies types were defined on the basis of grain size, bed thickness, bed contacts, and sedimentary structures (Table 1). The terminology for

Table 1 Classification of sedimentary facies

Facies	Facies description	Bed contacts	Bed thickness	Interpretation
F1: Massive conglomerates	Massive clast- or matrix-supported gravel. Gravel is mainly pebble- to cobble-sized; rare occurrence of boulders. The matrix (5–60 vol. %) consists of fine- to medium-grained sand. Larger clasts can be oriented parallel to dip and show a steeply imbricate clast fabric a(p) a(i) with the a-axes dipping in upslope direction.	Sharp	5–50 cm	Massive clast- or matrix-supported gravel with sharp bed contacts and a sandy matrix indicates deposition from non-cohesive debris flows (Shanmugam, 2000); the steep-clast fabric indicates laminar shear during or immediately after the flow's stop (Nemec, 1990).
F2: Inversely graded gravel	Cobble- to pebble-sized clast-supported gravel with inverse distribution grading. The matrix (5–25 vol. %) consists of fine- to coarse-grained sand.	Sharp	17–25 cm	Sharp bed contacts and the occurrence of inverse distribution grading indicate deposition from non-turbulent debris-flows by freezing. Sediment is supported by matrix strength, dispersive pressure and buoyant lift (Johannson & Stow, 1995; Shanmugam, 1996; 2000).
F3: Normally graded gravel	Clast-supported pebble- to boulder-sized gravel with normal distribution or coarse-tail grading. The matrix (5–30 vol. %) consists of fine- to coarse-grained sand.	Sharp or erosive	10–60 cm	Normal distribution grading and erosive bed contacts indicate continuous deposition from suspension of waning high-density turbidity currents (R3, cf. Lowe, 1982; Kneller, 1995).
F4: Massive pebbly sand	The sand is fine- to coarse-grained and makes up 60–95 vol. %. Clasts are commonly pebble to cobble-sized; rare occurrence of boulders.	Sharp or erosive	2–30 cm	Deposition from sandy debris flows by freezing (Shanmugam, 1996; 2000) or high-density turbidity currents by gradual aggradation (S3, cf. Lowe, 1982; Kneller, 1995).
F5: Inversely graded pebbly sand	Pebbly sand with inverse distribution grading. The sand is fine- to coarse-grained and makes up 60–95 vol. %. Clasts are commonly pebble to cobble-sized; rare occurrence of boulders.	Sharp	10–25 cm	Pebbly sand with non-erosive basis and inverse distribution grading indicates deposition from non-turbulent sandy debris-flows. Sediment is supported by matrix strength, dispersive pressure, and buoyant lift (Johannson & Stow, 1995; Mulder & Cochonat, 1996; Mohrig et al., 1998; Shanmugam, 1996; 2000).
F6: Normally graded pebbly sand	Pebbly sand with normal distribution or coarse-tail grading. The sand is fine- to coarse-grained and clasts pebble- to cobble sized (2–30 vol. %).	Sharp or erosive	5–30 cm	Erosive bed contacts and normal grading indicate deposition from suspension of waning high-density turbidity currents (S3, cf. Lowe, 1982; Kneller, 1995).
F7: Diffusely stratified pebbly sand and gravel	Diffusely stratified pebbly sand and gravel. Individual layers are commonly 1–5 cm thick and show erosive or gradational contacts. The pebble or gravel layers are mainly matrix-supported with a fine- to coarse-grained sandy matrix.	Erosive	20–45 cm	Diffuse stratification in gravelly beds indicates traction deposition from turbulent flows (Kneller, 1995; Lowe, 1982). Thick beds with no vertical trends suggest that the flow conditions were constant during a longer time period (Kneller & Branney, 1995; Kneller, 1995). Gravelly intervals or clast trains probably indicate periods of higher shear velocity within a fluctuating quasi-steady high-density turbidity current (Plink-Björklund & Ronnert, 1999).

Facies	Description		Interpretation
F8: Planar-parallel stratified pebbly sand	Planar-parallel stratified pebbly sand. The sand is fine- to coarse-grained and clasts commonly pebble- to cobble sized (2–35 vol. %). Individual layers are 0.3–3 cm thick. Some beds display a fining-upward.	Erosive or sharp　10–75 cm	Planar-parallel stratification in sandy beds, which fine upwards indicates traction deposition from turbulent waning high-density flows (S1 cf. Lowe, 1982; Kneller, 1995).
F9: Planar or trough cross-stratified sand and gravel	Fine- to coarse-grained sand, pebbly sand and gravel with medium- and large-scale planar and trough cross-stratification. Gravel is pebble- to boulder sized, clast-supported and matrix-poor to openwork. The matrix consists of coarse sand and granules and commonly increases upwards. Local occurrences of small lenses (50 cm wide, 30 cm thick) of matrix-poor (<10 vol. %), pebble- to cobble-sized gravel at the base or within a cross-set. Foreset beds show angular to tangential basal contacts. Troughs are 20–800 cm wide and 4–150 cm deep.	Sharp or erosive　10–300 cm	Medium-scale planar and trough cross-stratified sand, pebbly sand and gravel is interpreted to represent 2-D and 3-D dunes. This deposition requires turbulent flows that are sustained at a relatively constant discharge for longer periods (Kneller & Branney, 1995; Plink-Björklund & Ronnert, 1999; Mulder & Alexander, 2001). Large-scale well graded cross-stratified gravel is interpreted to has been deposited from gravelly turbulent flows in the leeward flow separation eddy of preformed deep scours (Allen, 1982; Carling & Glaister, 1987). Open-work fabric indicates intensive outwash of sandy grain-fraction during initial peak flow (Nemec et al., 1999). Isolated lenses of coarse gravel indicate intense slipface avalanching combined with discrete collapses.
F10: Ripple cross-laminated sand	Fine- to coarse-grained ripple cross-laminated sand. Some beds show a thin basal unit with planar-parallel lamination or scattered pebbles. Ripples are planar or trough cross-laminated and beds commonly show a fining-upward where a lamination with eroded ripple stoss sides passes upwards into lamination with preserved stoss sides and into draping lamination. Some beds display a coarsening-upward or a lower coarsening and upper fining interval.	Sharp or erosive　10–60 cm	Thick fining-upward beds with climbing-ripple cross-lamination indicate deposition from waning, sustained low-density turbidity flows (Mulder & Alexander, 2001). Coarsening-upward beds indicate waxing flows commonly produced during the waning stage of longer-lived flows, e.g. generated by flood events (Kneller, 1995). The scattered pebbles are interpreted as coeval debris fall from the steep upper fan slope (Nemec et al., 1999).
F11: Normally graded sand to clay beds	Individual beds consist of intervals of normally graded or massive coarse-grained sand or pebbly sand, that fines upwards into planar-parallel laminated and ripple-cross laminated medium- to fine-grained sand and silt, laminated silt, and finally into laminated or massive mud or clay. Beds are most commonly 'incomplete' and contain both 'top-absent' or 'base-absent' successions.	Sharp or erosive　3–85 cm	Normal grading and fining of individual beds reflect deposition from waning surge-type low-density turbidity currents (Ta-d, cf. Bouma, 1962; Lowe, 1982; Kneller, 1995).
F12: Deformed beds	Deformed beds include folds and deformed or undeformed enclosures of other lithofacies.	Erosive　10–300 cm	Deformation structures and bed geometry suggest formation by sliding and slumping, probably triggered by scouring or channel formation and channel undercutting (Nemec et al., 1999; Plink-Björklund & Ronnert, 1999; Russell & Arnott, 2003).

gravel characteristics is after Walker (1975). The fabric notation uses symbols a and b for the clast long axes, with indices (t) and (p) denoting axis orientation transverse or parallel to flow direction, and index (i) denoting axis imbrication. The notation of turbidites, Tabcd, refers to Bouma divisions (cf. Bouma 1962), S1–3; R1-R3 to Lowe divisions (cf. Lowe, 1982).

Facies types can be grouped into 7 major facies associations, which are characterized by distinct depositional processes, grain size, bed geometries and channel aspect ratios (Table 2). Deposits of the upper fan (FA1) are distinctly coarse grained with relatively few sand or silt beds (Fig. 3A). Beds mainly consist of massive clast-supported pebble- to cobble-sized gravel with a fine- to coarse-grained sand matrix, which may contain scattered clasts of diamicton. Towards the distal upper fan slope (FA2) more intercalations of graded or stratified gravel, pebbly sand and sand beds can be observed.

Table 2 Classification of facies associations

Facies association	Sedimentary facies	Gravel/sand/mud ratio	Geometry	Depositional processes
Upper fan				
FA1	F1, F2, F3, F4, F6, F9, F11	90:10:0	Low- to high-angle bedding (3°–34°)	Cohesionless debris flows, tractional deposition and suspension fallout from sustained, highly concentrated turbulent flows.
FA2	F1, F2, F3, F4, F5, F6, F8, F9, F10, F11	40:60:0	Low- to high-angle bedding (5°–25°)	Cohesionless debris flows, tractional deposition and suspension fallout from sustained and surge-type turbulent density flows.
Mid fan				
FA3	F1, F3, F4, F6, F9, F10, F11,	30:70:5	Low- to high-angle bedding (6°–15°)	Tractional deposition and suspension fallout from sustained and surge-type turbulent density flows. Less commonly cohesionless debris flows.
FA4	F1, F3, F4, F5, F6, F8, F9, F11	20:75:5	Low- to high-angle bedding (3°–25°)	Mainly tractional deposition and suspension fallout from surge-type turbulent density flows. Less commonly sustained turbulent flows or cohesionless debris flows.
FA5	F1, F9, F11	5:90:5	High-angle bedding (14°–29°)	Mainly tractional deposition from sustained turbulent density flows. Less commonly surge-type turbulent flows and cohesionless debris flows.
FA6	F1, F9, F10, F11	2:93:5	Low-angle bedding (3°–9°)	Suspension fallout and tractional deposition from low-density surge-type and sustained turbulent flows. Rare cohesionless debris flows.
Lower fan				
FA7	F11	1:90:10	Low-angle bedding (2°–5°) or draping lamination	Suspension fallout and tractional deposition from surge-type low-density turbidity currents.

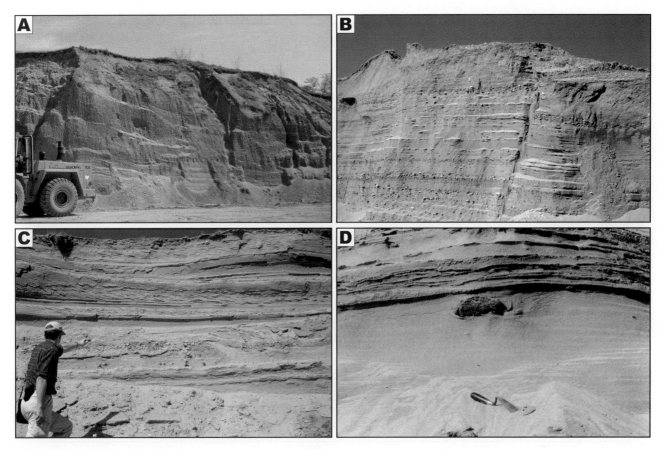

Fig. 3 Examples of lithofacies and facies associations. (A) Massive, matrix-supported gravel (F1) of the proximal upper-fan (FA1), Steinbrink open pit. (B) Proximal mid-fan facies (FA4), consisting of diffusely and planar-parallel stratified pebbly sand and sand (F7, F8, F11), Heerburg open pit. (C) Ball-and pillow structures in distal mid-fan deposits (FA6), Steinbrink open pit. (D) Thick diffusely stratified sand bed (F7), overlain by thin- to medium-bedded turbidites displaying Bouma Ta-c divisions (F11). Note dropstone within the diffusely stratified sand, distal mid-fan facies (FA6), Pampel open pit. Trowel for scale is 30 cm.

Subordinate thin beds of massive silt or silty sand occur. Deposits of the proximal mid-fan slope (FA3) consist of massive, planar-parallel stratified, planar or trough cross-stratified pebbly sand and climbing-ripple cross-laminated sand alternating with channelized massive or normally graded gravel and pebbly sand. Downslope these deposits pass into alternations of planar parallel-stratified, planar or trough cross-stratified and ripple cross-laminated fine- to coarse-grained sand, alternating with massive, inversely graded, normally graded, diffusely stratified or planar parallel-stratified pebbly sand (Figs 3B and 3D). Less commonly massive or normally graded gravel beds can be observed (FA4, FA5). Towards the distal mid-fan and outer fan slope (FA6, FA7) multiple stacked climbing-ripple-cross-laminated sand units (Fig. 3C) or alternations of fine-grained sand, silt, and mud

occur, in which individual beds fine upwards. Scattered pebbles can frequently be observed and are mainly concentrated in mud layers.

Depositional processes

The variety of sedimentary facies identified in Table 1 reflects a wide range of depositional processes and high rates of sedimentation. The sediment transport on the proximal upper fan slope was dominated by sustained gravity-flows, mainly cohesionless debris flows, reflecting discharge of semi-continuous meltwater flows (Nemec *et al.*, 1999) or resedimentation of subaqueous outwash material (e.g. Lønne, 1995). Subordinately intercalated normally graded gravel and massive pebbly sand, deposited from surge-type high-density turbidity flows probably indicate slope

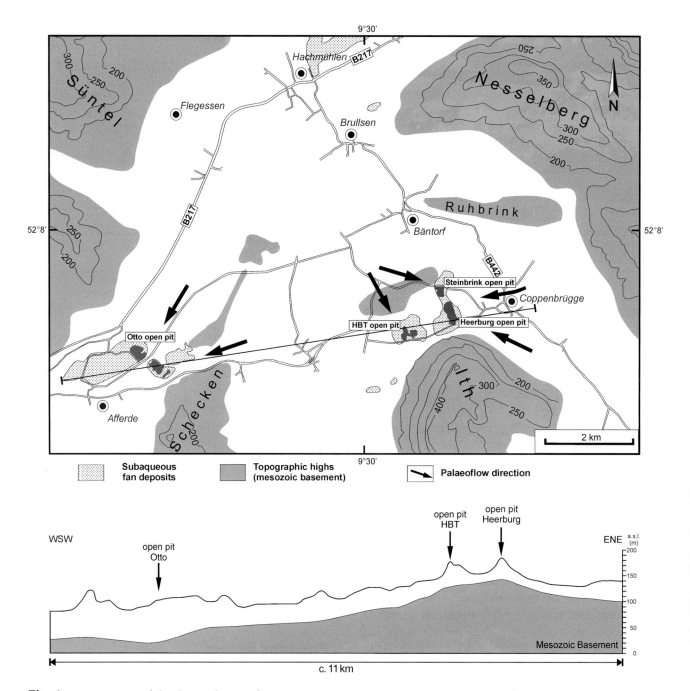

Fig. 4 Location map of the Coppenbrügge fan systems.

instability events triggered by the shear of over-riding flows or by gravity (Nemec *et al.*, 1999; Plink-Björklund & Ronnert, 1999). The intercalation of stratified gravel and pebbly sand increases towards the distal upper fan zone, indicating a change in flow regime and tractional deposition from sustained and surge-type high- and low-density turbidity flows. Steady currents require

rather strong deceleration, or considerable increase in concentration to initiate deposition (Kneller, 1995). Therefore deposition was probably triggered by deceleration in the slope break region (Lowe, 1988; Allen, 1991) causing vertical flow divergence (flow-thickening, *sensu* Kneller, 1995), rapid flow deceleration and deposition. The finer-grained sandy material moved further downslope where it

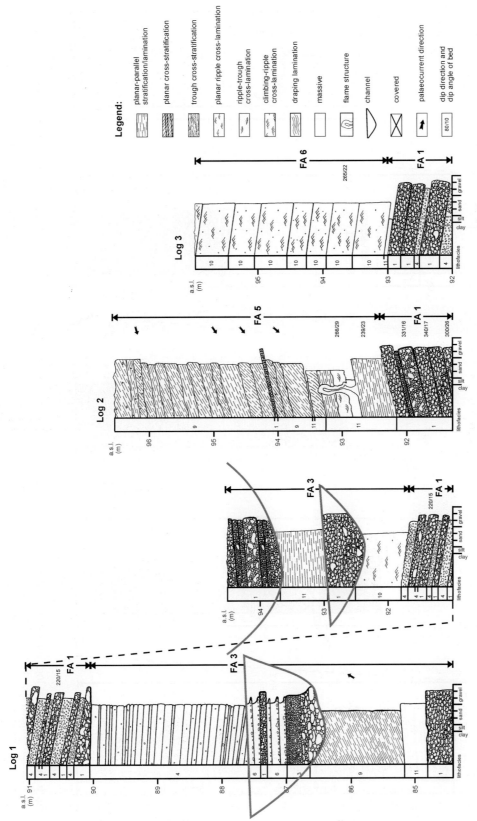

Fig. 5 Sedimentological log measured in the Otto open pit. The logs show facies types, facies associations (FA), palaeocurrent directions (arrows) and directions of bed dip. For location see Figs 4 and 7.

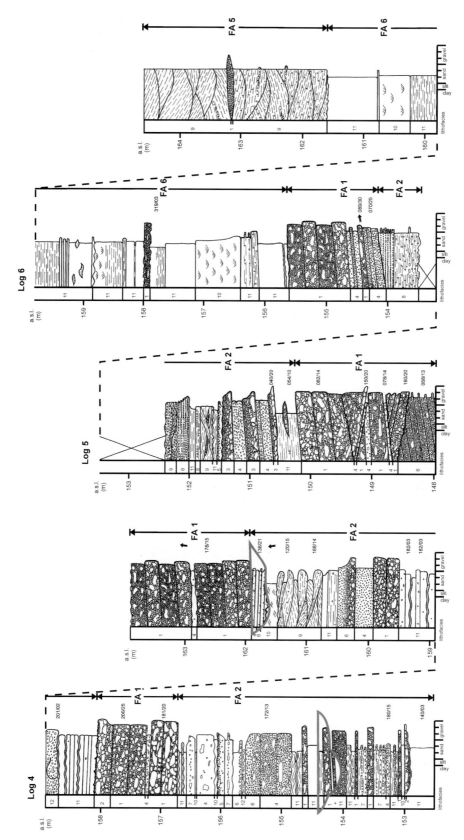

Fig. 6 Sedimentological logs measured in the HBT open-pit. The logs show facies types, facies associations (FA), palaeocurrent directions (arrows) and directions of bed dip. For location see Figs 4 and 7; for key see Fig. 5.

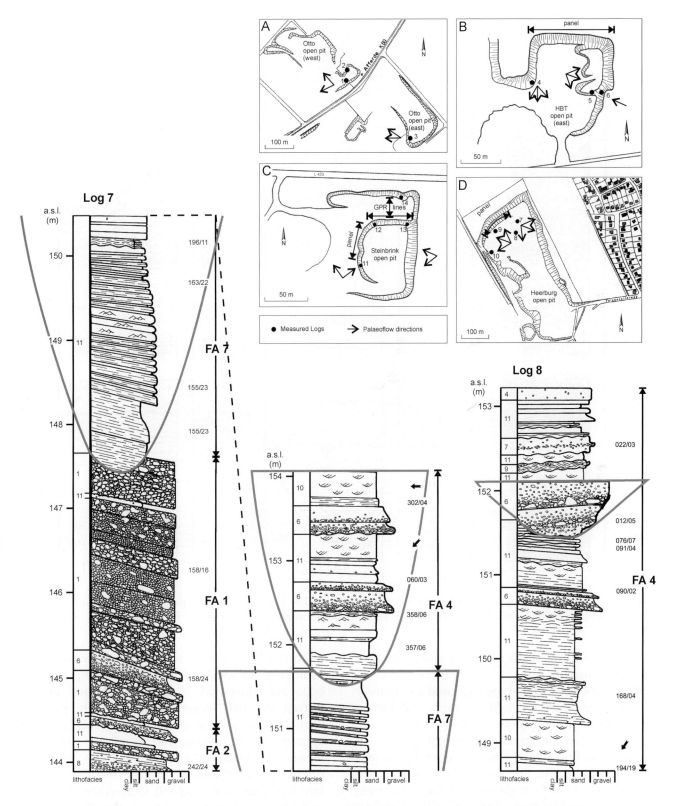

Fig. 7 Sedimentological logs measured in the Heerburg open pit. The logs show facies types, facies associations (FA), palaeocurrent directions (arrows) and directions of bed dip. For location see Figs 4 and 7; for key see Fig. 5. Detail maps of the open pits show locations of measured logs, panels and georadar lines.

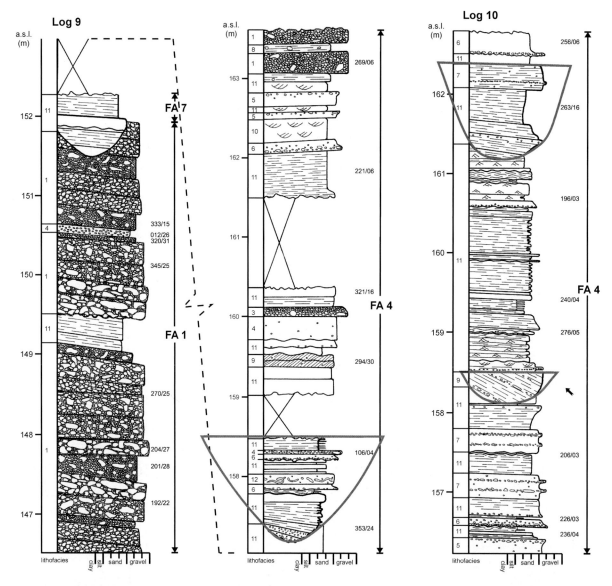

Fig. 8 Sedimentological logs measured in the Heerburg open pit. The logs show facies types, facies associations (FA), palaeocurrent directions (arrows) and directions of bed dip. For location see Figs. 4 and 7; for key see Fig. 5.

was deposited from both sustained and surge-type turbidity currents.

Intercalations of surge-type high- and low-density turbidites increase towards the mid- and lower-fan slope, indicating more ice-distal and periodic deposition. On the upper- and proximal mid-fan slope, these flows were confined in channels and chutes, before spreading out at the mid-fan slope to form low- to high-angle foreset beds with large-scale cross-stratification, planar-parallel stratification or ripple cross-lamination. High discharge phases are indicated by the formation of 2D- and 3D-dunes due to quasi-steady effluxes and

an increased sediment supply (Nemec *et al.*, 1999; Mulder & Alexander, 2001). Thick fining-upward beds are attributed to individual flood events during the melt season or alternatively, may be attributed to an autocyclic process of fan-head aggradation and erosion, or upper fan-slope collapses. The observed scattered pebbles are interpreted as coeval debris fall from the steep upper fan slope (Nemec *et al.*, 1999). During low discharge periods transport processes on the slope where dominated by volumetric relatively small, sandy turbulent flows and debris-flows due to slope-instability. Scours filled with deformed strata or

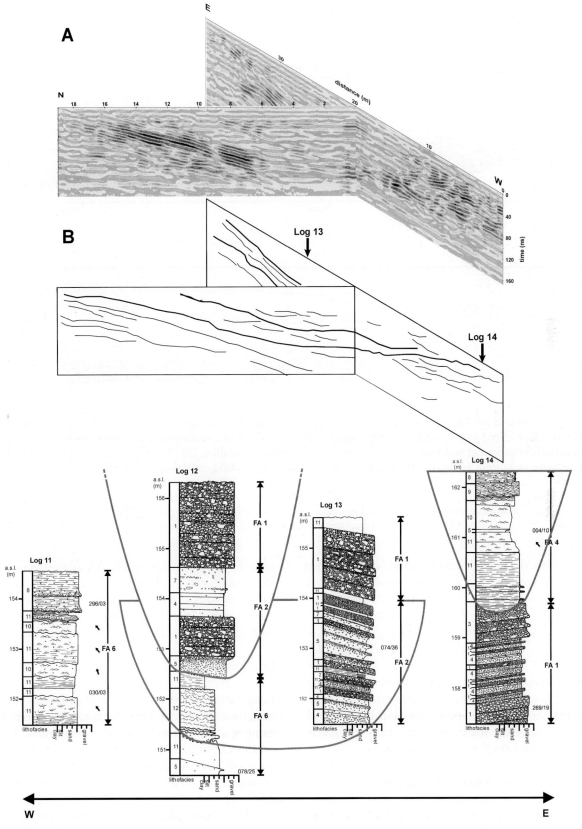

Fig. 9 GPR fence diagram and sedimentological logs measured in the Steinbrink open pit. The logs show facies types, facies associations (FA), palaeocurrent directions (arrows) and directions of bed dip. The radar lines show the southward dip of the erosional structure, observed in log 12 and 13 and interpreted to represent a major slump scar. This erosional structure shallows towards the north and is filled with eastward dipping strata. For location see Figs. 4 and 7; for key see Fig. 5.

A

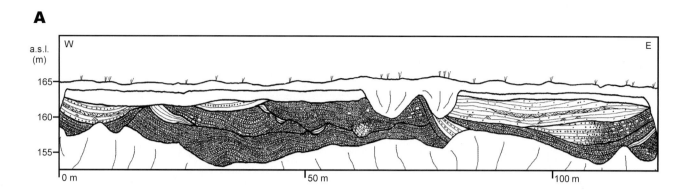

B

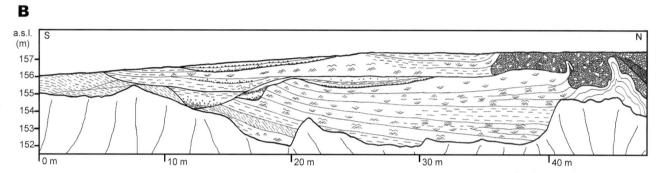

Fig. 10 (A) Sketch of an outcrop section exposed in the HBT open pit, showing upper-fan deposits, unconformably overlain by channelized mid-fan deposits. Beds steeply dip (14°–35°) into southwesterly to southeasterly directions. The sketch is parallel to foreset bedding, not showing the depositional inclination. (B) Sketch of an outcrop section exposed in the western Steinbrink open pit, showing distal mid-fan deposits (FA6) gently dipping (3–9°) into northeasterly directions. In the upper part of the sections small-scale chutes can be observed, 5–20 m wide and 0.6–2 m deep and filled with massive, normally graded or graded-stratified pebbly sand and sand. At the northern part of the outcrop the western flank of a large erosional structure can be observed, filled with upper-fan gravel. The margin of this erosional structure has been deformed by gravitational faulting and dewatering. For location see Figs. 4 and 7; for key see Fig. 5.

massive or diffusely graded sand and pebbly sand record rapid cut-and-fill processes on the fan slope, probably associated with discharge peaks or hydraulic jumps at the mouth of larger chutes or at a break in the slope gradient (e.g. Postma & Cruickshank, 1988; Powell, 1990; Prior & Bornhold, 1990; Gorell & Shaw, 1991; Nemec *et al.*, 1999; Russell & Arnott, 2003). Towards the lower-fan slope high suspension fallout rates promoted the formation of climbing-ripple cross-lamination and graded or massive sand, silt and mud interbeds with glacial debris dumped by icebergs.

The Coppenbrügge subaqueous fan complex

The Coppenbrügge fan complex is located at the eastern margin of glacial Lake Rinteln and consists of various small sediment bodies. Fan sediments are exposed in several gravel pits at an altitude

of *c.* 85–160 m (Figs 4–11). Measured palaeoflow directions are highly variable. In the western-most sections flows have been mainly towards the southwest. Upsection flows are towards south-easterly and westerly directions (Figs 4 and 7).

Facies architecture and fan stacking patterns

Four fan systems could be recognized, characterized by vertically or laterally stacked moderately to steeply dipping fan bodies (Fig. 12). Individual fan bodies commonly have a coarse-grained proximal core of steeply dipping upper fan gravel, disconformably overlain by sandy outer- to mid fan deposits (Fig. 11).

The stratigraphically lowest fan system (Fan I) is exposed in the open-pit Otto (Figs 4, 5 and 12). These deposits form part of an up to 60 m thick sedimentary succession, overlying glaciolacustrine mud.

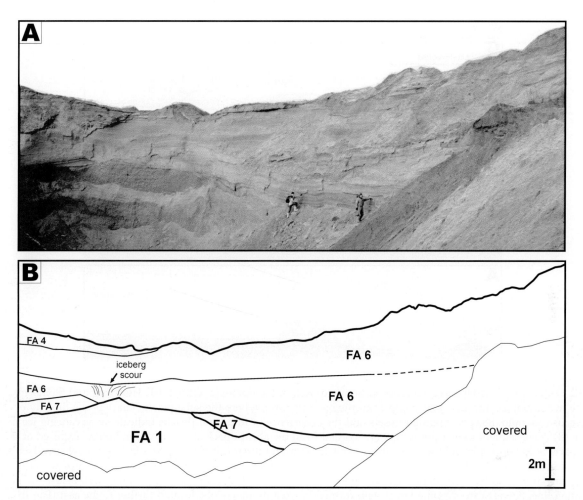

Fig. 11 (A) Geometrical relationship between two ice-contact fans, interpreted to have been formed during ice retreat and a subsequent ice stillstand (open-pit Heerburg). (B) Fine-grained silt-mud-alternations with dropstones (FA7) and climbing-ripple cross-laminated sand beds (FA6) unconformably overly upper-fan deposits (FA1). The former position of the ice-front is indicated by the boundary between basinward-dipping clinoforms (FA1) and landward-dipping to subhorizontal clinoforms (FA7; FA6). Note relic of an iceberg scour on top of the eroded fan body. Palaeoflow directions are up-slope towards westerly directions. For location see Figs 4 and 7.

The exposed sedimentary sequence is deformed, displaying folds and thrusts, dipping eastward. Fan II overlies a till and is exposed at an altitude of *c.* 144–155 a.s.l. in the open-pit HBT and Heerburg (Figs 4, 6, 7, 10B and 12B). Fan II dips steeply east and south, and has been deposited on the lee side of a northwest to southeast-trending basement high, consisting of Lower Jurassic rocks (Fig. 12B). Fan II is disconformably overlain by southwest to northwest-dipping mid- to upper-fan deposits (fan III). The proximal upper fan deposits of Fan III are deeply truncated and unconformably overlain by westward-dipping outer- to upper-fan deposits (Fig. 11). In the open-pit Steinbrink Fan III deposits

are cut by a *c.* 50 m wide and 7 m deep northwest to southeast-trending U-shaped erosional structure, laterally filled with coarse-grained gravel and pebbly sand (Figs 9 and 10B). Foresets are disconformably overlain by westward-dipping upper fan gravel of a new prograding fan system (Fan IV). In the open-pit Heerburg the proximal upper fan deposits of Fan III are unconformably overlain by west-dipping outer- and mid-fan deposits, which onlap the eastward-dipping erosion surface (Fig. 11) and are increasingly preserved and exposed at the lower fan's back-slope (Fig. 11). Flow directions, obtained from climbing-ripple sequences indicate a transport direction towards the west, climbing up the truncated

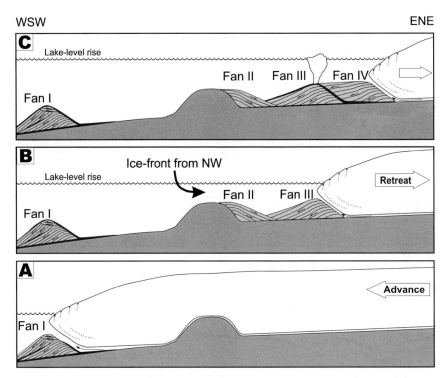

Fig. 12 Depositional model based on cross sections exposed in the open-pits Otto, HBT, Heerburg and Steinbrink. (A) The maximum ice-advance is recorded from the open-pit Otto, where the subaqueous fan deposits directly overly glaciolacustrine mud. Steeply dipping foresets and the occurrence of folds and thrusts indicate an advancing ice margin (Fan I). (B) A subsequent ice-margin retreat is recorded from the abrupt back-stepping of fan bodies, exposed at the lower parts of the open-pits HBT and Heerburg. This ice-margin retreat is attributed to a fast lake-level rise. Two fan systems formed that differ in palaeoflow directions, indicating two major meltwater effluxes. Fan II dips towards easterly directions and has been deposited on the lee side of a NW–SE trending basement high, well below the glacier's grounding line. A westward-dipping fan system formed in front of an ice-lobe located further to the east (Fan III), onlapping Fan II. (C) A subsequent ice-margin retreat led to the abandonment of Fan II and Fan III. Gently landward-dipping to subhorizontal clinoforms, onlapping Fan III indicate the deposition of a new fan system (Fan IV) in front of a stabilized ice-margin.

foreset slope. On top of this sequence a prominent iceberg scour mark can be observed, overlain by coarsening-upwards mid-fan deposits (Fig. 11).

Interpretation

Fan I is related to a major glacier stillstand. The steep dip (20–30°) and the observed folds and thrusts indicate a subsequent glacier re-advance (Fig. 12A). Ice-front advance often provides an asymmetric sediment-profile parallel to the direction of sediment transport, reflecting erosion on the ice-proximal side of the fan and deposition on the ice-distal side (e.g. Boulton, 1986; Alley, 1991; Lønne, 1995; Lønne & Syvitsky, 1997; Lønne; 2001). A subsequent fan that was abandoned and a shift of depocentres towards more up-slope positions is recorded from

the abrupt back-stepping of fan bodies exposed in the open-pits HBT and Heerburg. This ice-margin retreat is attributed to a rapid lake-level rise (Fig. 12B). The glacier became grounded at a basement high, serving as a pinning point. The deposition of fan II then occurred below the level of the glacier's underwater grounding line, down-lapping the pre-existing basement slope. The upper-fan deposits, exposed in the open-pit HBT are exceptionally rich in Jurassic claystone clasts, which have been derived from the adjacent basement high.

Fan II is disconformably overlain by southwest to northwest-dipping mid- to upper-fan deposits, indicating the deposition of a new fan system, prograding from the east (Fan III, Fig. 12B). The large erosional feature exposed in the open pit Steinrink

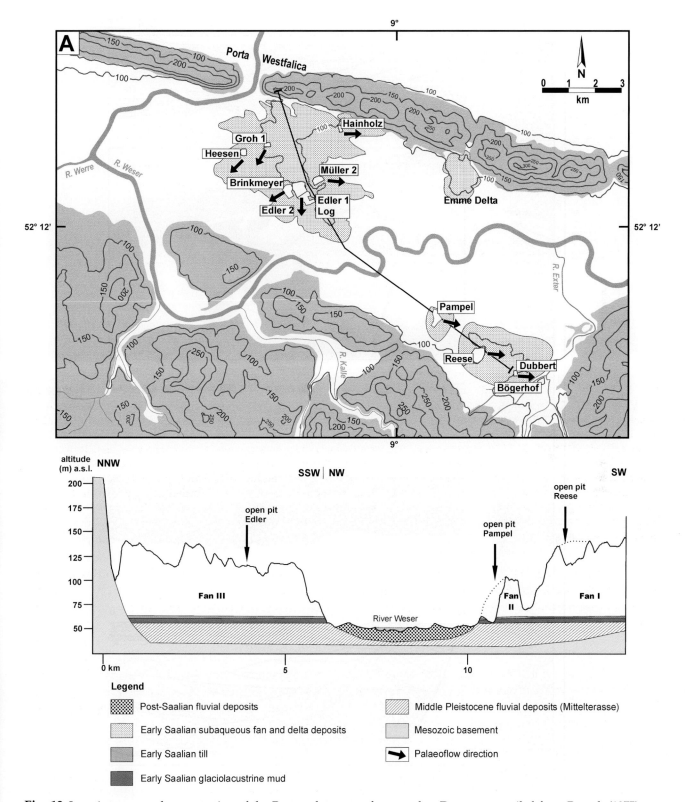

Fig. 13 Location map and cross-section of the Porta subaqueous fan complex. Data are compiled from Rausch (1975), Rohde (1994), Wellmann (1998), Könemann (1995) and Hornung *et al.* (2007).

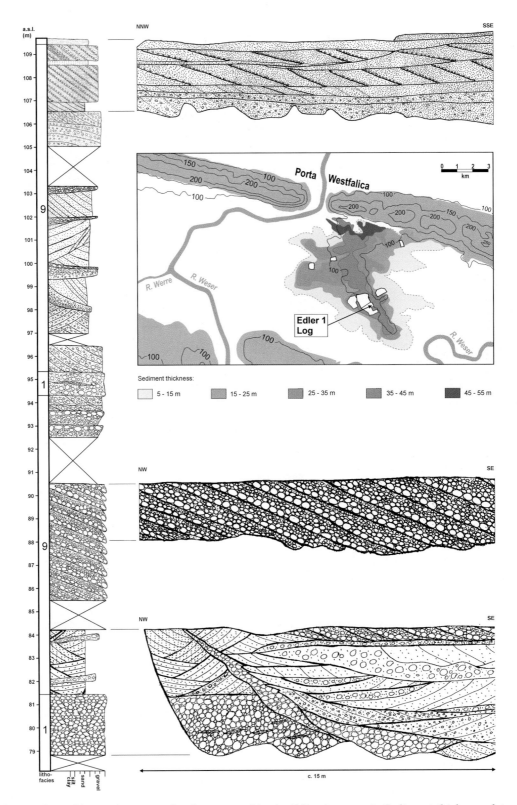

Fig. 14 Sedimentological log and outcrop sketch measured in the Edler 1 open pit. Sediment thickness data are complied from Könemann (1995).

WSW   ENE

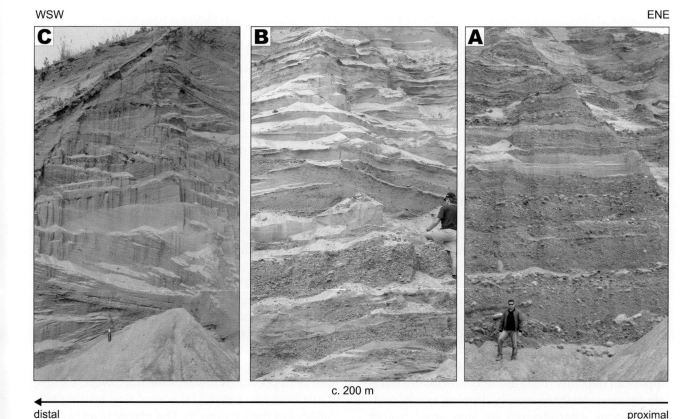

c. 200 m

distal   proximal

Fig. 15 Facies associations of the Porta subaqueous fan complex. Photographs show a downflow succession of upper- to mid-fan sediments that is interpreted to has been deposited under hydraulic-jump condition during flow expansion at the mouth of a subglacial conduit. (A) Highly scoured massive gravel, overlain by cross-stratified gravel and planar-parallel stratified pebbly sand, Brinkmeyer open pit. (B) Highly scoured cross-stratified gravel, pebbly sand and sand, Brinkmeyer open pit (Photo by B. Garlt). (C) Large-scale trough-cross stratified sand, Edler 2 open pit. Trowel for scale is 30 cm.

(Fig. 9) shallows towards the northwest and is interpreted to represent a major slump scar (e.g. Postma, 1984b), probably indicating a southeast-ward-directed slope failure of the ice-proximal back-slope. Both foresets and the slump-scar-fill are disconformably overlain by westward-dipping upper-fan gravel, interpreted to represent proximal deposits of a new prograding fan system (Fan IV). In the open pit Heerburg proximal upper-fan deposits of Fan III are unconformably overlain by westward-dipping outer- to mid-fan deposits, which onlap the eastward dipping erosion surface (Fig. 11). Flow directions, obtained from climbing-ripple sequences indicate a transport direction towards the west, climbing up the truncated foreset slope, indicating a glacier retreat and deposition on the fan's ice-proximal backslope slope (cf. Lønne, 1995; 2001). Ice-rafted debris and the occurrence of

a prominent ice scour mark cut into outer- and mid-fan deposits demonstrate a rapid retreat of the glacier and an abrupt cut-off of the sediment flux to the fan. These deposits of the early abandonment phase have been partly removed by later erosion. The overall coarsening upwards of the uppermost section indicates the progradation of a new fan system (Fan IV) from the east (Fig. 12C). Gently landward dipping to subhorizontal clinoforms and little facies variations point to the formation of an aggrading wedge, related to a glacier stillstand (e.g. Lønne & Syvitsky, 1997).

The Porta subaqueous fan and delta complex

The Porta subaqueous fan complex is located at the northern margin of glacial Lake Rinteln, south of the Porta Westfalica pass. It overlies glaciolacustrine mud and patchy occurrences of till (Könemann 1995, Wellmann, 1998, Hornung *et al.*, 2007) and is exposed in various gravel pits at an altitude of *c.* 70–130 m. Palaeoflow directions are mainly towards southwesterly and southeasterly directions (Figs 2 and 13).

Facies architecture and fan stacking patterns

Three fan systems have been recognized (Fig. 13), characterized by vertically and laterally stacked moderately to steeply dipping fan bodies. One characteristic section is presented, which is exposed in the northernmost subaqueous fan system (fan III). Additional logs and a detailed interpretation of sedimentary facies are presented in Hornung *et al.* (2007).

The stratigraphically lowest fan is exposed in the open-pits Dubbbert and Reese south of the River Weser (Fig. 13). These deposits form part of an up to 60 m thick sedimentary succession, overlying glaciolacustrine mud. Coarse-grained upper-fan deposits (FA1) are unconformably overlain by moderately to steeply dipping sandy mid-fan deposits, characterized by large-scale trough cross-stratified or graded-stratified sand and gravel (FA3-FA5). The exposed sedimentary sequence is partly deformed, displaying thrusts, dipping towards the northwest (Wellmann, 1998).

Deposits of fan II are exposed in the open pit Pampel and consist of upper fan deposits, unconformably overlain by moderately to steeply dipping mid-fan deposits, commonly displaying normally graded, planar-parallel or large-scale trough cross-stratified sand (FA5-FA6; Fig. 3D). Fan III is exposed in several gravel pits north of the River Weser. This northernmost fan system has a typical fan-shaped geometry and is unconformably overlain by glaciofluvial braid-plain deposits (Winsemann & Asprion, 2001).

The greatest thickness of fan deposits is recorded from a central, *c.* 1 km wide, northwest to southeast-trending zone (Fig. 14). Deposits, exposed in this central zone (open pit Brinkmeyer, Edler 1 and Edler 2) differ from fan deposits exposed in the south or in adjacent open pits, where deposits of debris flows and turbidity currents form aggradational successions of upward-steepening foreset beds inclined at 6° to 26°. In contrast, deposits of the central fan zone are characterized by a flat to low-angle, distally steepening geometry (3°–10°) and consist of highly scoured, massive, normally graded, planar-parallel or cross-stratified gravel and pebbly sand (Figs 14 and 15). The exposed succession shows an overall fining-upward trend and is truncated by an irregular erosion surface, exposed at an altitude of *c.* 100–105 m a.s.l. and overlain by subhorizontal sandy braid plain and shallow-water mouth bar deposits.

Interpretation

Fan I is related to a relative stable ice-front, indicated by thick vertically aggraded fan deposits. The observed deformation probably records minor seasonal oscillations of the ice-margin (Alley, 1991; Lønne, 1995; Lønne, 2001). A subsequent ice-margin retreat is inferred from the deposition of fan II, exposed in the open-pit Pampel. Truncated upper fan deposits are unconformably overlain by moderately to steeply dipping, fining-upward sandy mid-fan deposits. The fan abandoned and shift of depocentres towards the northwest (Fig. 13) is attributed to a rapid lake-level rise, leading to a collapse of the ice margin. The glacial front became re-stabilized in front of the Porta Westfalica pass, serving as a pinning point. After this ice-margin retreat fan III evolved in front of the Porta Westfalica pass. Highly scoured gravelly and sandy deposits from the central fan fan zone (Figs 14 and 15) indicate sedimentation from coarse-grained hyper-concentrated turbulent flows (e.g. Mulder & Alexander 2001), typical for the proximal region of ice-contact fans and distal intra-conduit deposits (Gorell & Shaw 1991, Russell & Arnott 2003; Hornung *et al.*, 2007). Parts of the downflow succession are in accordance with facies assemblages previously described for deposition under hydraulic-jump condition during flow expansion (Gorrell & Shaw, 1991; Russell & Arnott, 2003; Hornung *et al.*, 2007). Hornung *et al.* (in press) interpreted these deeply scoured gravelly sediments to have been deposited from a friction-dominated plane-wall jet at the mouth of a subglacial meltwater tunnel. Large till clasts in coarse-grained unsorted massive gravel point to the erosion of subglacial till and rapid sedimentation from hyper-concentrated flows at the conduit mouth. Deposits of the jet's zone of flow establishment consist of highly scoured, coarse-grained, clast- to matrix-supported gravel (Figs 14 and 15A). Deposits of the jet's proximal zone of flow transition are characterized multiple scoured, normally graded beds and cross-stratified clast-supported gravel, passing downflow into scoured planar parallel-stratified and cross-stratified pebbly sand and sand (Figs 15A and B). Deposits of the jet's proximal zone of established flow consist of trough-cross-stratified sand and pebbly sand interpreted as 3-D dunes formed by a steady, thick, turbulent current (Fig. 15C).

The lithofacies association is similar to those described by Bornhold & Prior (1990) and Russell & Arnott (2003) as being of jökulhlaup origin. A drop in lake level may therefore have triggered a catastrophic drainage event that tapped previously unconnected reservoirs of englacial and subglacial meltwater (e.g. Sharpe & Cowan, 1990; Mokhtari Fard *et al.*, 1997). Subsequently these deposits must have been draped by fan sediments and overlain by glaciofluvial braid plain deposits (e.g. Sharpe, 1988 Warren & Ashley, 1994).

DISCUSSION

The extent, morphology, and sedimentary facies of subaqueous fan deposits indicate deposition into a lake at the margin of a lobate, grounded ice sheet (e.g. Ashley *et al.*, 1991). The Coppenbrügge subaqueous fan complex consists of relatively small-scale fan bodies, deposited on a hummocky low-angle basin slope. The sedimentary facies demonstrate that the fan deposits accumulated from an easterly and northerly direction as several small subaqueous fans, indicating small conduits with minor effluxes which more easily mix with lake water, so constraining the distance of sediment dispersal (e.g. Fyfe, 1990; Powell, 1990). Deposits are texturally mature and therefore mainly represent resedimented outwash material (e.g. Lønne, 1995). Probably coarse bedload was deposited close to the tunnel mouth as the efflux jet decelerated, building up steep, unstable slopes at the glacier margin. Slope failure and renewed sediment discharge from the tunnel mouth fed mass flows that transported sediments radially away from the margin (e.g. Lonne, 1995). As flows travelled over the fan surface, turbulent mixing with the overlying water resulted in their progressive dilution and the transformation into turbidity currents.

After a phase of maximum ice-advance, accompanied by the deposition of steep ice-contact subaqueous fan deposits and deformation of fan deposits, a rapid back-stepping of fan bodies towards up-slope positions occurred. Fans comprise fining-upward packages of gravel, sand and mud and rhythmically laminated fine-grained sediments partly drape the fans. Climbing-ripple cross-laminated sand may onlap coarse-grained upper fan gravel and in some cases overtop the older fan deposits (Fig. 11). This retrogradational trend is attributed to a retreating ice-margin associated with a rapid lake-level rise in the range of 30–40 m, which is thought to has occurred from a supposed 135 m lake-level. Bedrock highs acted as pinning points for the retreating glacier and, after the re-establishment of subglacial drainage systems, ice-marginal sediment accumulated from restricted point sources, giving rise to small isolated subaqueous fans (Figs 4 and 12). The lack of subaerial topset facies demonstrates that the retreat was fast and fans did not reach the contemporary water–level (e.g. Lønne, 1995).

The Porta subaqueous fan complex consists of 3 fan systems, deposited on a flat lake-bottom surface. Similar to the eastern basin margin, a phase of maximum ice advance was followed by an ice-margin retreat, indicated by fan abandonment and a shift of depocentres towards the northwest. The flat-bottom topography and deep water may have encouraged rapid ice wastage. The ice margin destabilized and rapidly retreated towards the northwest and became re-stabilized near to the Porta Westfalica pass (Fig. 13). The large size of the northern Porta fan system (fan III) is attributed to the position in front of this pass, where a stable meltwater tunnel facilitated the construction of a larger subaqueous fan. The frequent occurrence of tractive structures in gravelly and sandy fan deposits indicates sustained and high-energy flows associated with high discharges (e.g. Powell, 1990; Lønne, 1995; Cutler *et al.*, 2002). Hornung *et al.* (2007) interpreted deposits of the central zone as subaqueous products of a supercritical plane-wall jet with hydraulic jump. The direct link of the sublacustrine depositional system with an ice front has been inferred on the basis of the sedimentary facies, which is consistent with the notion of a subaqueous effluent jet (e.g. Gorell & Shaw, 1991; Mokhtari Fard *et al.*, 1997; Russell &

Arnott, 2003) and the regional extent of the Saalian ice front. The dimension of jet efflux deposits is larger than that of previously described examples from the Laurentide Ice Sheet (e.g. Gorrell & Shaw, 1991; Russell & Arnott, 2003) and therefore might indicate a catastrophic drainage event. The level of glacial lake Rinteln must have partly controlled the potentiometric surface by acting as a base level for water stored in and under the Saalian Ice Sheet. High levels of glacial Lake Rinteln caused a relatively high potentiometric surface in the ice sheet. A rapid lake level fall would have produced steeper hydraulic gradients near the ice margin (e.g. Waite, 1985; Gustavson & Boothroyed, 1987; Sharpe & Cowan, 1990). A drop in lake level may therefore have triggered a catastrophic drainage event that tapped previously unconnected reservoirs of englacial and subglacial meltwater (e.g. Sharpe & Cowan, 1990; Mokhtari Fard *et al.*, 1997; Fisher *et al.*, 2002). Such powerful subglacial flows erode sediment from the ice and ground surface, transporting vast quantities of sediment to the ice margin. A strong lake-level fall is evidenced by the widespread truncation of the subaqueous fan deposits and the superposition of glaciofluvial braid plain deposits. This prominent incision occurred down from the supposed 160 m a.s.l. highstand, corresponding well with the formation of an incised valley in the Emme delta (Winsemann *et al.*, 2004).

Therefore, rapid lowering of the lake is thought to have triggered widespread outbursts of subglacial meltwater and the deposition of jet efflux deposits either in front of a subglacial conduit. This strong lake-level fall led to a subaerial exposure of the fan-system and the subsequent deposition of glaciofluvial braid plain sediments.

CONCLUSIONS

The geometry and sedimentary facies of subaqueous fan deposits indicate deposition into a lake at the margin of a multi-lobate, grounded ice sheet. As deglaciation continued, the ice lobes receded and a widespread lake formed in the gap between them. An overall lake-level rise led to the deposition of various fan systems, which rapidly backstepped towards the lake margins. The position of ice-marginal fans was controlled by the combination of bedrock topography and water depth. At the eastern lake margin bedrock highs served as pinning points, whereas a flat-bottom topography at the northwestern lake margin probably caused a more rapid ice wastage, because the ice terminated in deeper water and the calving rate may have exceeded the ice flux, resulting in rapid retreat.

Individual fan bodies commonly have a coarse-grained proximal core of steeply dipping gravel, overlain by gently to steeply dipping mid- to outer fan deposits. Subaqueous fans resulting from ice-frontal advance are characterized by steeply dipping progradational foresets with basinward progradation or vertically stacking. During glacier retreat commonly fine-grained sediments, rich in ice-rafted debris, were deposited on the ice-distal and ice-proximal slopes of the abandoned fans. Phases of glacier still-stands are characterised by fan systems with an upward flattening of fan clinoforms and minor vertical facies changes.

A strong lake-level fall probably triggered a major drainage event that tapped previously unconnected reservoirs of englacial and subglacial meltwater. This meltwater outburst led to the deposition of jet-efflux deposits at the mouth of a subglacial conduit. The Porta fan complex subsequently emerged and became overlain by braid-plain and shallow-water mouth bar deposits.

ACKNOWLEDGEMENTS

We thank H.-B. Deters, M. Bussmann, K. Bussmann, O. Meinsen, T. Fisher, P. Rohde and H. Russell for discussion; V. Arrué, B. Garlt, T. Fandré and A. Freund for technical assistance, help with field work and drafting. The comments of reviewers, J. Knight and J. Piotrowski, and guest editor P. Christoffersen are greatly appreciated.

REFERENCES

Allen, J.R.L. (1982) Sedimentary Structures: Their Character and Physical Basis. *Dev. Sedimentol.*, **30**, 1–679.

Allen, J.R.L. (1991) The Bouma division A and the possible duration of turbidity currents. *J. Sedim. Petrol.*, **61**, 291–295.

Alley, R.B. (1991) Sedimentary processes may cause fluctuations of tidewater glaciers. *Ann. Glaciol.*, **15**, 119–124.

Ashley, G.M. (1995) Glaciolacustrine Environments. In: *Modern Glacial Environments* (Ed. J. Menzies), pp. 417–444. Butterworth-Heinemann, Oxford.

Ashley, G.M., Boothroyed, J.C. and Borns, H.W., Jr. (1991) Sedimentology of late Pleistocene (Laurentide) deglacial-phase deposits, eastern Maine; an example of a temperate marine grounded ice sheet margin. In: *Glacial Marine Sedimentation, Paleoclimatic Significance* (Eds J.B. and G.M. Ashley), Geol. Soc. Am. Spec. Pap., **261**, 107–125.

Benn, D.I. and Evans, D.J.A. (1998) *Glaciers & Glaciation*. Arnold, London, 734 pp.

Benn, D.I. (1996) Subglacial and subaqueous processes near a glacier grounding line: sedimentological evidence from a former ice-dammed lake, Achnasheen Scotland. *Boreas*, **25**, 23–36.

Bennett, M.R., Huddart, D. and Thomas, G. (2002) Facies architecture within a regional glaciolacustrine basin: Copper River, Alaska. *Quatern. Sci. Rev.*, **21**, 2237–2279.

Bornhold, B.D. and Prior, D.B. (1990) Morphology and sedimentary processes on the subaqueous Noeick River delta, British Columbia, Canada. In: *Coarse-Grained Deltas* (Eds A. Colella and B.D. Prior), *Int. Assoc. Sedimentol. Spec. Publ.*, **10**, 169–181.

Boulton, G.S. (1986) Push moraines and glacier-contact fans in marine and terrestrial environments. *Sedimentology*, **33**, 667–698.

Boulton, G.S. (1990) Sedimentary and sea level changes during glacial cycles and their control on glacimarine facies architecture. In: *Glacimarine Environments: Processes and Sediments* (Eds J.A. Dowdeswell and J.D. Scourse), *J. Geol. Soc. London Spec. Publ.*, **53**, 15–52.

Bouma, A. (1962) *Sedimentology of some Flysch Deposits. A Graphic Approach to Facies Interpretation*. Elsevier Publications, Amsterdam, 168 pp.

Brennand, T.A. (1994) Macroforms, large bedforms and rhythmic sedimentary sequences in subglacial eskers, south-central Ontario: implications for esker genesis and meltwater regime. *Sed. Geol.*, **91**, 9–55.

Brookfield, M.W. and Martini, I.P. (1999) Facies architecture and sequence stratigraphy in glacially influenced basins: basic problems and water-level/glacier input-point controls (with an example from the Quaternary of Ontario, Canada). *Sed. Geol.*, **123**, 183–197.

Carling, P.A. and Glaister, M.S. (1987) Rapid deposition of sand and gravel mixtures downstream of a negative step: the role of matrix infilling and particle over-passing in the process of bar-front accretion. *J. Geol. Soc. London*, **144**, 543–551.

Cheel, R.J. and Rust, B.R. (1982) Coarse grained facies of glaciomarine deposits near Ottawa, Canada. In: *Research in Glaciofluvial and Glaciolacustrine Systems* (Eds R. Davidson-Arnott, W. Nickling and B.D. Fahey), pp. 279–295. Geo Books, Norwich.

Chough, S.K. and Hwang, I.G. (1997) The Duksung fan delta, SE Korea: growth of delta lobes on a Gilbert-type topset in response to relative sea-level rise. *J. Sed. Res.*, **67**, 725–739.

Clemmensen, L.B. and Houmark-Nielsen, M. (1981) Sedimentary features of a Weichselian glaciolacustrine delta. *Boreas*, **10**, 229–245.

Cutler, P.M., Colgan, P.M. and Mickelson, D.M. (2002) Sedimentologic evidence for outburst floods from the Laurentide Ice Sheet margin in Wisconsin, USA: implications for tunnel-channel formation. *Quatern. Int.*, **90**, 23–40.

Deutloff, O., Kühn-Velten, H., Michel, G. and Skupin, K. (1982). Erläuterungen zu Blatt C3918 Minden, Geol. Kt. Nordrh.-Westf. 1:100 000, Krefeld, 80 pp.

Ehlers, J. and Gibbard, P. (2004) *Quaternary Glaciations. Extent and Chronology Part I: Europe*. Elsevier, Amsterdam, 488 pp.

Eyles, N. and Clark, B.M. (1988) Storm-influenced deltas and ice scouring in a late Pleistocene glacial lake. *Geol. Soc. Am. Bull.*, **100**, 793–809.

Fisher, T.G., Clague, J.J. and Teller, J.T., (2002) The role of outburst floods and glacial meltwater in subglacial and proglacial landform genesis. *Quatern. Int.*, **90**, 1–4.

Fowler, A.C. (1987) Sliding with cavity formation. *J. Glaciol.*, **33**, 255–267.

Fyfe, G. (1990) The effects of water depth on ice-proximal glaciolacustrine sedimentation: Salpausselkä I, southern Finland. *Boreas*, **19**, 147–164.

Gorell, G. and Shaw, J. (1991) Deposition in an esker, bead and fan complex, Lanark, Ontario, Canada. *Sed. Geol.*, **72**, 285–314.

Grupe, O. (1925) Führer zu den Exkursionen der DGG vor und nach der Hauptversammlung in Münster 1925. *Schr. Ges. Förd. Westf. Wilhelms-Universität Münster*, **7**, 1–107.

Grupe, O. (1930) Die Kamesbildungen des Weserberglandes. *Jb. Preuss. Geol. La.-Anst.*, **51**, 350–370.

Gustavson, T.C. and Boothroyed, J.C. (1987) A depositional model for outwash, sediment sources, and hydrologic characteristics, Malaspina Glacier, Alaska: A modern analog of the southeastern margin of the Laurentide Ice Sheet. *Geol. Soc. Am. Bull.*, **99**, 292–302.

Hornung, J.J., Asprion, U. and Winsemann, J. (2007) Jet-efflux deposits of a subaqueous ice-contact fan, glacial Lake Rinteln, northwestern Germany. *Sed. Geol*, **193**, 167–192.

Johansson, M. and Stow, D.A.V. (1995) A classification scheme for shale clasts in deep water sandstones. In: *Characterization of deep-marine Clastic Systems* (Eds A.J. Hartley and D.J. Prosser), *J. Geol. Soc. London Spec. Publ.*, **94**, 221–241.

Johnsen, T.F. and Brennand, T.A. (2006) The environment in and around ice-dammed lakes in the moderately high relief setting of the southern Canadian Cordillera. *Boreas*, **35**, 106–125.

Klostermann, J. (1992) *Das Quartär der Niederrheinischen Bucht – Ablagerungen der letzten Eiszeit am Niederrhein*. Geol. La. Nordrh.-Westf., Krefeld, 200 pp.

Kneller, B.C. (1995) Beyond the turbidite paradigm: physical models for deposition of turbidites and their implications for reservoir prediction. In: *Characterization of deep-marine Clastic Systems* Hartley (Eds A.J. Hartley and D.J. Prosser), *J. Geol. Soc. London Spec. Publ.*, **94**, 31–49.

Kneller, B.C. and Branney, M.J. (1995) Sustained high-density turbidity currents and the deposition of thick massive sands. *Sedimentology*, **42**, 607–616.

Knudsen, O. and Marren, P.M. (2002) Sedimentation in a volcanically dammed valley, Brúarjökull, northeast Iceland. *Quatern. Sci. Rev.*, **21**, 1677–1692.

Könemann, P. (1995) Der Sand- und Kiesabbau im Wesertal an der Porta Westfalica – Ökonomische und ökologische Untersuchungen zur Belastung und Inwertsetzung des Naturraumpotentials. *Hannov. Geogr. Arb.*, **50**, 1–216.

Krzyszkowski, D. (2002) Sedimentary successions in ice-marginal fans of the Late Saalian glaciation, southwest Poland. *Sed. Geol.*, **149**, 93–109.

Krzyszkowski, D. and Zieliński, T. (2002) The Pleistocene end moraine fans: controls on their sedimentation and location. *Sed. Geol.*, **149**, 73–92.

Kulle, S. (1985) Drenthe-stadiale Staubecken-Sedimente (Pleistozän) und ihr Lagerungsverband aus zwei Aufschlüssen im Wesertal zwischen Rinteln und Hameln. *Unpubl. Diploma thesis*, Univ. Hannover, 58 pp.

Lajeunesse, P. and Michel, A. (2002) Sedimentology of an ice-contact glacimarine fan complex, Nastapoka Hills, eastern Hudson Bay, northern Québec. *Sed. Geol.*, **152**, 201–220.

Lønne, I. (1995) Sedimentary facies and depositional architecture of ice-contact glaciomarine systems. *Sed. Geol.*, **98**, 13–43.

Lønne, I. (1997) Facies characteristics of a proglacial turbidite sand-lobe at Svalbard. *Sed. Geol.*, **109**, 13–35.

Lønne, I. (2001) Dynamics of marine glacier termini read from moraine architecture. *Geology*, **29**, 199–202.

Lønne, I. and Syvitski, J.P. (1997) Effects of the readvance of an ice margin on the seismic character of the underlying sediment. *Mar. Geol.*, **143**, 81–102.

Lowe, D.R. (1982) Sediment Gravity Flows II: Depositional Models with Special References to the Deposits of High-Density Turbidity Currents. *J. Sed. Petrol.*, **52**, 279–297.

Lowe, D.R. (1988) Suspended-load fallout rate as an independent variable in the analysis of current structures. *Sedimentology*, **35**, 765–776.

Lunkka, J.P. and Gibbard, P. (1996) Ice-marginal sedimentation and its implications for ice-lobe deglaciation patterns in the Baltic region: Pojhankangas, western Finland. *J. Quatern. Sci.*, **11**, 377–388.

Lüttig, G. (1954) Alt- und mittelpleistozäne Eisrandlagen zwischen Harz und Weser. *Geol. Jb.*, **70**, 43–125.

Lüttig, G. (1960) Neue Ergebnisse quartärgeschichtlicher Forschung im Raum Alfeld – Hameln – Elze. *Geol. Jb.*, **77**, 337–390.

Martini, I.P. (1990) Pleistocene glacial fan deltas in southern Ontario, Canada. In: *Coarse-Grained Deltas* (Eds A. Colella and B.D. Prior), *Int. Assoc. Sedimentol. Spec. Publ.*, **10**, 281–295.

Martini, I.P. and Brookfield, M.E. (1995) Sequence analysis of Upper Pleistocene (Wisconsinan) glaciolacustrine deposits of the north-shore bluffs of Lake Ontario, Canada. *J. Sed. Res.*, **B65**, 388–400.

Mastalerz, K. (1990) Diurnally and seasonally controlled sedimentation on a glaciolacustrine foreset slope: an example from the Pleistocene of eastern Poland. In: *Coarse-Grained Deltas* (Eds A. Colella and B.D. Prior), *Int. Assoc. Sedimentol. Spec. Publ.*, **10**, 297–309.

Meier, M.F. and Post, A. (1987) Fast tidewater glaciers. *J. Geophys. Res.*, **92** (B9), 9051–9058.

Miotke, F.-D. (1971) Die Landschaft an der Porta Westfalica. Teil 1: Die Naturlandschaft. *Jb. Geogr. Ges. Hannover 1969*, 1–256.

Mohrig, D., Whipple, K.X., Hondzo, M., Ellis, C. and Parker, G. (1998) Hydroplaning of subaqueous debris flows. *Geol. Soc. Am. Bull.*, **110**, 387–394.

Mokhtari Fard, A., Gruszka, B., Brunnberg, L. and Ringberg, B. (1997) Sedimentology of a glaciofluvial deposit at Ekeby, East Central Sweden. *Quatern. Sci. Rev.*, **16**, 755–765.

Mulder, T. and Alexander, J. (2001) The physical character of subaqueous sedimentary density flows and their deposits. *Sedimentology*, **48**, 269–299.

Mulder, T. and Cochonat, P. (1996) Classification of offshore mass movements. *J. Sed. Petrol.*, **66**, 43–57.

Naumann, E. (1922) *Erläuterungen der geologischen Karte von Preußen, Bl. Rinteln, Lfg. 233*. Berlin, 46 pp.

Naumann, E. (1927) *Erläuterungen zur geologischen Karte von Preußen, Bl. Eldagsen, Lfg. 265, Nr. 2088*. Berlin, 57 pp.

Naumann, E. and Burre O. (1927) *Erläuterungen zur geologischen Karte von Preußen, Bl. Hameln, Lfg. 251, Nr. 2087*. Berlin, 77 pp.

Nemec, W. (1990) Aspects of sediment movement on steep delta slopes. In: *Coarse-Grained Deltas* (Eds A. Colella and B.D. Prior), *Int. Assoc. Sedimentol. Spec. Publ.*, **10**, 29–73.

Nemec, W. and Steel, R.J. (1984) Alluvial and coastal conglomerates: their significant features and some comments on gravelly mass-flow deposits. In: *Sedimentology of Gravels and Conglomerates* (Eds E.H. Koster and R.J. Steel), *Can. Soc. Petrol. Geol. Mem.*, **10**, 1–31.

Nemec, W., Lønne, I. and Blikra, L.H. (1999) The Kregnes moraine in Gaudalen, west-central Norway: Anatomy of a Younger Dryas proglacial delta in a paleofjord basin. *Boreas*, **28**, 454–476.

Pisarska-Jamroży, M. (2006) Transitional deposits between the end moraine and outwash plain in the Pomeranian glaciomarginal zone of NW Poland: a missing component of ice-contact sedimentary models. *Boreas*, **35**, 126–141.

Plink-Björklund, P. and Ronnert, L. (1999) Depositional processes and internal architecture of late Weichselian ice-margin submarine fan and delta settings, Swedish west coast. *Sedimentology*, **46**, 215–234.

Postma, G. (1984a) Mass-flow conglomerates in a submarine canyon: Abrioja fan-delta, Pliocene, southeast Spain. In: *Sedimentology of Gravels and Conglomerates* (Eds E.H. Koster and R.J. Steel), *Can. Soc. Petrol. Geol. Mem.*, **10**, 237–258.

Postma, G. (1984b) Slumps and their deposits in fan delta front and slope. *Geology*, **12**, 27–39.

Postma, G. (1990) Depositional architecture and facies of river and fan deltas: a synthesis. In: *Coarse-Grained Deltas* (Eds A. Colella and B.D. Prior), *Int. Assoc. Sedimentol. Spec. Publ.*, **10**, 13–27.

Postma, G. (1995) Sea-level-related architectural trends in coarse-grained delta complexes. *Sed. Geol.*, **98**, 3–12.

Postma, G. and Cruickshank, C. (1988) Sedimentology of a late Weichselian to Holocene terraced fan delta, Varangerfjord, northern Norway. In: *Fan Deltas: Sedimentology and Tectonic Settings* (Eds W. Nemec and R.J. Steel), pp. 144–157. Blackie, London.

Powell, R.D. (1990) Glacimarine processes at grounding-line fans and their growth to ice-contact deltas. In: *Glacimarine Environments: Processes and Sediments* (Eds J.A. Dowdeswell and J.D. Scourse), *J. Geol. Soc. London Spec. Publ.*, **53**, 53–73.

Powell, R.D. and Domack (1995) Modern Glaciomarine Environments. In: *Modern Glacial Environments* (Ed. J. Menzies), pp. 445–486. Butterworth-Heinemann, Oxford.

Prior, D.B. and Bornhold, B.D. (1990) The underwater development of Holocene fan deltas. In: *Coarse-Grained Deltas* (Eds A. Colella and B.D. Prior), *Int. Assoc. Sedimentol. Spec. Publ.*, **10**, 75–90.

Rausch, M. (1975) Der 'Dropstein-Laminit' von Bögerhof und seine Zuordnung zu den Drenthe-zeitlichen Ablagerungen des Wesertals bei Rinteln. *Mitt. geol. Inst. Univ. Hannover*, **12**, 1–86.

Rausch, M. (1977) Fluß-, Schmelzwasser- und Solifluktionsablagerungen im Terrassengebiet der Leine und der Innerste. *Mitt. Geol. Inst. Univ. Hannover*, **14**, 1–84.

Richards, A.E. (2002) Self-organisation, fractal scaling and cyclicity in Late Midlandian glacio-deltaic sediments associated with glacial Lake Blessington, Co. Wicklow. *Sed. Geol.*, **149**, 127–143.

Rohde, P. (1994) Weser und Leine am Berglandrand zur Ober- und Mittelterrassenzeit. *Eiszeit. Gegenw.*, **44**, 106–113.

Russell, H.A.J. and Arnott, R.W.C. (2003) Hydraulic-jump and hyperconcentrated-flow deposits of a glacigenic subaqueous fan: Oak Ridge Moraine, southern Ontario, Canada. *J. Sed. Res.*, **73**, 887–905.

Seraphim, E.T. (1972) Wege und Halte des saalezeitlichen Inlandeises zwischen Osnig und Weser. *Geol. Jb., Reihe A*, **3**, 1–85.

Seraphim, E.T. (1973) Eine saalezeitliche Mittelmoräne zwischen Teutoburger Wald und Wiehengebirge. *Eiszeit. Gegenw.*, **23/24**, 116–129.

Shanmugam, G. (1996) High-density turbidity currents: Are they sandy debris flows? *J. Sed. Res.*, **66**, 2–10.

Shanmugam, G. (2000) 50 years of the turbidite paradigm (1950s–1990s): Deep-water processes and facies models – a critical perspective. *Mar. Petrol. Geol.*, **17**, 285–342.

Sharpe, D.R. (1988) Glaciomarine fan deposits in the Champlain Sea. In: *The Late Quaternary Development of the Champlain Sea Basin* (Ed. N.R. Gadd), *Geol. Assoc. Can. Spec. Pap.*, **35**, 63–82.

Sharpe, D.R. and Cowan, W.R. (1990) Moraine formation in northwestern Ontario: product of subglacial fluvial and glacilacustrine sedimentation. *Can. J. Earth Sci.*, **27**, 1478–1486.

Skupin, K., Speetzen, E. and Zandstra, J.G. (2003) Die Eiszeit in Nordost-Westfalen und angrenzenden Gebieten Niedersachsens. *Geologischer Dienst Nordrhein-Westfalen*, Krefeld, 96 pp.

Sohn, Y.K., (2000) Depositional processes of submarine debris flows in the Miocene fan deltas, Pohang basin, SE Korea with special reference to flow transformation. *J. Sed. Res.*, **70**, 491–503.

Shaw, J. (1975) Sedimentary successions in Pleistocene ice-marginal lakes. In: *Glaciofluvial and Glaciolacustrine Sedimentation* (Eds A.V. Jopling and B.C. McDonald), *SEPM Spec. Publ.*, **23**, 281–303.

Smith, N.D. and Ashley, G.M. (1985) Proglacial lacustrine environments. In: *Glacial Sedimentary Environments* (Eds G.M. Ashley, J. Shaw and N.D. Smith), *SEPM Short Course*, **16**, 135–215.

Southard, J.B. and Boguchwal, L.A. (1990) Bed configurations in steady unidirectional water flow part 2. Synthesis of flume data. *J. Sed. Petrol.*, **60**, 658–679.

Spethmann, H. (1908) Glaziale Stillstandslagen im Gebiet der mittleren Weser. *Mitt. Geogr. Ges. Lübeck*, **22**, 1–17.

Stach, E. (1930) Die Eisrandbildungen an der Porta Westfalica. *Jb. Preuss. Geol. La.-Anst.*, **51**, 174–187.

Teller, J.T. (1995) History and drainage of large ice-dammed lakes along the Laurentide Ice Sheet. *Ouatern. Int.*, **28**, 83–92.

Teller, J.T. and Clayton, L. (1983) Glacial Lake Agassiz. Geol. Assoc. Can. Spec. Pap., **26**, p. 451.

Thomas, R.H. (1979) Ice shelves: a review. *J. Glaciol.*, **24**, 273–286.

Thomas, G.S.P. (1984) A late Devensian glaciolacustrine fan-delta at Rhosesmor, Clwyd, North Wales. *J. Geol.*, **19**, 125–141.

Thome, K.N. (1983) Gletschererosion und –akkumulation im Münsterland und angrenzenden Gebieten. *Neues Jb. Geol. Paläontol. Abh.*, **166**, 116–138.

Thome, K.N. (2001) Jüngere Erdgeschichte des nördlichen Sauerlandes und des südlichen Münsterlandes im Rahmen von Exkursionen. *Decheniana*, **154**, 182–209.

Waite, R.B., Jr. (1985) Case for periodic, colossal jökulhlaups from Pleistocene glacial Lake Missoula. *Geol Soc. Am. Bull*, **96**, 1271–1286.

Walker, R.G. (1975) Conglomerates: sedimentary structures and facies models. In: *Depositional Environments as Interpreted from Primary Sedimentary Structures and Stratification Sequences* (Eds J.C. Harms, R.G. Walker and D. Spearing), *SEPM Spec. Publ. Short Course Lectures Notes*, **12**, 133–161.

Warren, W.P. and Ashley, G.M. (1994) Origins of the ice-contact stratified ridges (eskers) of Ireland. *J. Sed. Res.*, **A64**, 433–449.

Wellmann, P. (1998) Kies-/Sandkörper im Wesertal zwischen Rinteln und Porta Westfalica. *Mitt. Geol. Inst. Univ. Hannover*, **38**, 203–212.

Winsemann, J. and Asprion, U. (2001) Glazilakustrine Deltas am Südhang des Wesergebirges: Aufbau, Entwicklung und Kontrollfaktoren. *Geol. Beiträge Hannover*, **2**, 139–157.

Winsemann, J., Asprion, U., Meyer, T., Schultz, H. and Victor P. (2003) Evidence of iceberg ploughing in a subaqueous ice-contact fan, glacial Lake Rinteln, Northwest Germany. *Boreas*, **32**, 386–398.

Winsemann, J., Asprion, U. and Meyer, T. (2004) Sequence analysis of early Saalian glacial lake deposits (NW Germany): evidence of rapid local ice margin retreat and related calving processes. *Sed. Geol.*, **165**, 223–251.

Seasonal controls on deposition of Late Devensian Glaciolacustrine Sediments, Central Ireland

CATHY DELANEY

Department of Environmental and Geographical Sciences, Manchester Metropolitan University, John Dalton Extension, Chester St., Manchester M1 5GD, UK (e-mail: c.delaney@mmu.ac.uk)

ABSTRACT

Laminated proglacial glaciolacustrine sediments dating from the Late Devensian (22–10 Ka BP) from central Ireland were examined using a combination of detailed logging and Scanning Electron Microscope (SEM) microfabric analyses. The sediments are rhythmically laminated and consist of coarser, pale silt layers which alternate with darker clay layers containing occasional thin laminae of fine sand and coarse silt. The pale silt layers contain single or multiple normally graded laminae, erosional surfaces and soft sediment deformation structures, indicating deposition from multiple high density underflows. The dark clay layers have sharp upper and lower contacts and an internal fabric consistent with deposition from a combination of flocculation and grain-by-grain deposition. Silt laminae within the clay layers are interpreted as sporadic turbidity underflows. The sediments are interpreted as annually laminated (varved). Varves deposited close to the ice margin showed considerable spatial variation in thickness and adjacent sequences could not be correlated; however sequences of medial varves separated by 500 m were correlatable by thickness. As glaciolacustrine deposits are widespread throughout the Irish midlands, it is likely that a varve chronology could be constructed.

Keywords Varve, microfabric, flocculation, turbidity current, Ireland

INTRODUCTION

Annually laminated lake sediments (termed varves) have been used extensively to establish high resolution chronologies and examine environmental change on a local and regional scale. In proglacial lakes, clastic varves are formed due to seasonally-controlled changes in meltwater discharge: summer layers of sand and silt are formed from deposition from underflows and interflows, while winter layers are formed by deposition of clay after autumn overturning of the thermocline (Ashley, 1995). Varve thickness is controlled by interannual variation in discharge and sediment fluxes. Long-term chronologies are constructed by measuring varve thicknesses at sites across hydrologically linked proglacial lake basins and correlating from site to site to establish the relative ages of varve sequences. These chronologies have been used to reconstruct the timing and dynamics

of deglaciation, to date particular events and to correlate events on land with ice core and marine records (Andrén *et al.*, 1999; Ridge, 2003). Such chronologies have improved understanding of the interactions between changing terrestrial ice volumes, meltwater discharges, ocean circulation and climate change (e.g. Andrén *et al.*, 1999; Lindeberg and Ringberg, 1999; Boulton *et al.*, 2001). However, to date, this method of dating has been of limited use as no chronology has been extended to before 15.4 [14]C Ka BP.

The aim of this paper is to establish whether glaciolacustrine sediments in the Irish Midlands are annually laminated, and whether they can be correlated in order to create a varve chronology. Glaciolacustrine deposits laid down around the margins of the decaying British-Irish Ice Sheet (BIIS) after 20 Ka BP are found across the Irish Midlands, extending northwards along the Shannon basin and southwards as far as the Southern Ireland end

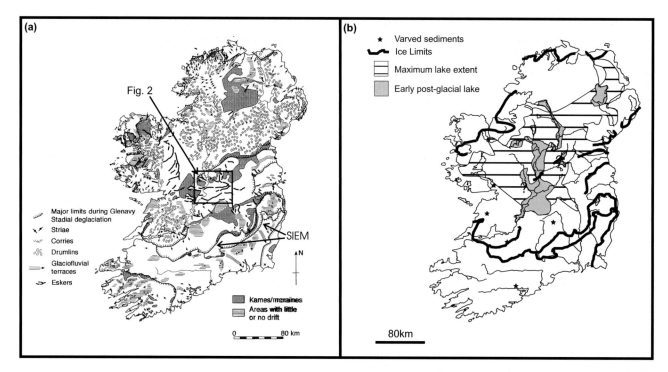

Fig. 1 (a) Quaternary glacial landforms in Ireland, showing positions of eskers, drumlins, ice margins and the location of Fig. 2 (redrawn by P. Coxon, after McCabe, 1987). SIEM = Southern Ireland End Moraine. (b) Possible extent of proglacial lakes in Ireland. Maximum lake extent is after Van der Meer and Warren (1997) and is based on present extent of catchments; early post-glacial lake extent is after Mitchell (1986).

moraine (SIEM) and beyond (Fig. 1; Van der Meer and Warren, 1997; Delaney, 2002). The sediments are rhythmically laminated (Van der Meer and Warren, 1997; Long and O'Riordan, 2001), and are similar in appearance to proven varved sediments found elsewhere. Potentially, they could be used to construct a varve chronology to date the recession of the BIIS from 20–10 Ka BP. During this time the BIIS was characterised by rapidly changing dynamics, including multiple shifts in ice flow directions, the establishment of deforming bed conditions, and episodic retreat and readvance of ice margins (McCabe, 1996; McCabe and Clark, 1998; Clark and Meehan, 2001; Delaney, 2002). These changes have been suggested to correlate with Atlantic-wide changes in thermohaline circulation, meltwater pulses from other ice sheets, abrupt sea-level rise and Heinrich (ice-rafting) events (McCabe and Clark, 1998; Scourse *et al.*, 2000, Clark *et al.*, 2004). However, accurate correlations of the timing of these changes is problematic, as suitable material for dating is hard to find, and the methods used to date these materials are subject to errors of over

1,000 years for this period (Waelbroeck *et al.*, 2001; Bowen *et al.*, 2002). A varve chronology could provide a high-resolution chronology for this period which could be tied to ice and marine records using tephrachronology and isotopic dating.

VARVE FORMATION

Rhythmically laminated, coarse-fine couplets consisting of a lower, silt-dominated and an upper, clay-dominated layer are common in glaciolacustrine sediments, and can form due to a variety of processes. These include: (1) diurnal variation in inflow (e.g. Ringberg, 1984); (2) subseasonal variations due to short-term (hours to days) changes in inflow (e.g. Lambert and Hsü, 1979; Hambley and Lamoureux, 2006); (3) seasonal variations resulting in two separate peaks in sediment input due to late spring snowmelt and summer ice melt (e.g. Smith, 1978); (4) deposition from slump-generated surge currents, usually turbidity currents, from unstable lake margins (Ashley, 1975; Hambley and Lamoureux, 2006); (5) and annual variation in discharge involving summer deposition from under-flows, overflow-interflows and surge currents in summer and suspension settling of fine silts and clays in winter, resulting in the formation of varves (Smith and Ashley, 1985; Ashley, 1995). Annually laminated, or varved, sediments form within lakes deep enough to be thermally stratified, so that during autumn overturning fine particles in the epilimnion are brought towards the bottom. This clears the lake of suspended sediment and deposits a clay drape across the entire lake (Ashley, 1995).

Based on differences in transport and depositional processes, Smith and Ashley (1985) and Ashley (1995) have identified a number of sedimentary criteria for distinguishing true varves from other deposits. These are: (1) a relatively sharp contact between the summer and winter layers which suggests a break in sedimentation, and may represent the autumn overturn; (2) no overall grading of the summer layer, as the layer accumulates from a variety of flows over weeks to months; (3) Fining upwards within the winter layer, reflecting suspension deposition of a limited sediment supply; (4) the presence of trace fossils (Lebensspuren) on bedding planes within the summer layer and on top of the winter layer, indicating periods of non-deposition; (5) winter layer thickness shows little variation from year to year, as it represents approximately the same amount of time, while summer layer thickness can vary considerably, reflecting variation in rapid sedimentation events from year to year.

Scanning Electron Microscope (SEM) analysis of microfabrics within rhythmites can also be used to infer depositional mechanisms. O'Brien and Pietraszek-Mattner (1998) have shown that most of the sediment in glaciolacustrine rhythmites has a microfabric consistent with deposition by floccula-tion, indicating episodic deposition rather than con-tinuous rain-out of dispersed sediment, but that the final millimetre of each rhythmite has a strong preferred orientation, which they interpreted as due to the reorientation of clay grains by bioturbation during periods of non-deposition.

While it is clear that clay deposited in winter is likely to have a distinctive sedimentological signature, it is also possible to form false varves by splitting this winter layer. Turbidity currents which occur within the winter months could easily be mis-

interpreted as summer deposits (Ashley, 1976; Shaw and Archer, 1978; Shaw *et al.*, 1978).

GEOLOGICAL SETTING AND SITE DESCRIPTION

The sediments to be discussed were deposited in a proglacial lake, Glacial Lake Riada, which formed in this basin during the late Midlandian (Weichselian) after 22 Ka BP and covered much of the Midlands during the last glacial termination (Figs 1b, 2A; Van der Meer and Warren, 1997; Delaney, 2002). The area is underlain by Carboniferous Dinantian limestones (Gatley *et al.*, 2005) and forms a lowlying (c. 30–100 m O.D; Fig. 2A) basin which rises slowly westwards, northwards and eastwards. Within the basin pre-Quaternary topography is masked by extensive glacigenic deposits and by Holocene raised bog. Glaciofluvial deposits, including eskers and kames, form the topographic high points over much of the basin (Fig. 2A; Warren and Ashley, 1994; Delaney, 2002).

The existence of Glacial Lake Riada was con-trolled by a combination of topography and the existence of ice lobes to the north, west and south-west (Delaney, 1995). Recession of these lobes is likely to have caused changes in lake extent and water depth; however, temporal variations in lake extent are not yet understood. The extent of the lake immediately prior to the start of deposition at the sites discussed below has been mapped using ice-contact deltas and subaqueous outwash fans; the evidence indicates that water levels were at around 92 m O.D. during this time and the sedi-ments were deposited in water depths of around 30 m (Fig. 2A; Delaney, 1995, 2002).

The two cores described below were taken from underneath reclaimed raised bog at Knocknanool (Kn) and Rooskagh (Ro) townlands, west of Lough Ree in Co. Roscommon, Ireland (Fig. 2b). The core sites are thought to lie a short distance north of the most southerly position reached by a readvance during deglaciation (Fig. 2; Delaney, 2001, 2002), some time between c. 17–11 Ka BP (Knight *et al.*, 2004). This readvance appears to be a relatively local event, since associated margins further east are not associated with readvance (Delaney, 2002; Meehan, 2004). The core sites lie towards the edge of the main glacial lake basin, within a sub-basin infilled with raised bog which

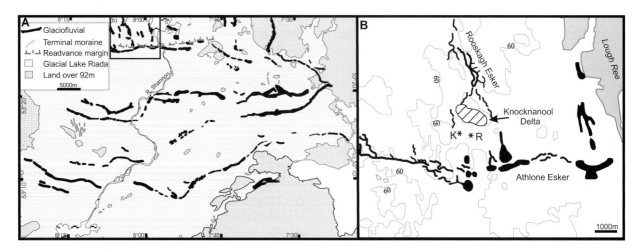

Fig. 2 (A) Map of the central Irish Midlands showing the extent of Glacial Lake Riada and glaciofluvial and ice-marginal landforms. (B) Location of core sites and relationship with adjacent glacial features. K = Knocknanool core; R = Rooskagh core.

has been partly cut for turf and then reclaimed. The basin is separated from other peat-filled basins to the east and south by eskers and other deglacial deposits (Fig. 2). The sides of the basin are delimited by bedrock to the west, by kames and the Rooskagh Esker to the east and north east, and by the Athlone Esker to the south. The core sites lie approximately 1500 m north of the Athlone Esker, 1600 m (Knocknanool) and 1000 m (Rooskagh) west of the Rooskagh Esker, approximately 600 m south of the Knocknanool delta kame and are c. 500 m apart (Fig. 1).

METHODS

Cores were retrieved in 0.5 m long sections using a hand-operated Russian corer to minimise compaction of the sediments and to avoid contamination of the cores. Two cores were collected at each site; the sampling depth in the second borehole was offset by approximately 0.25 m in relation to the first core, so that a complete undisturbed sequence was retrieved. The cores were wrapped in clingfilm and tinfoil to retain moisture and brought to the laboratory for description.

In the lab, each core section was unwrapped and air-dried overnight, so that any laminations would be clearly visible. When dry, the core surface was cleaned off using a sharp knife, an initial examination made, any potential winter layers

marked with pins, and the core section then photographed. Cores were then examined in detail and logged at a sub-millimetre scale using a moving stage microscope with an 8x magnification. Features noted included colour, estimate of particle size, grading, any lamination and soft sediment deformation structures. A preliminary identification of winter clay laminae was made, based on colour, sharpness of upper and lower contacts and grain size estimates. The thickness of individual clay laminae and the total thickness of sediment between consecutive clay laminae was measured to the nearest 0.01 mm. Individual core sections were visually correlated with overlapping sections and laminae numbered accordingly. Where individual laminae appeared in more than one core section, the average thickness was calculated and the range was obtained based on maximum and minimum thicknesses.

Samples for SEM microfabric analysis were taken from the Knocknanool core. Samples were prepared for microfabric analysis following the method of O'Brien and Pietraszek-Mattner (1998), but using a sharp knife instead of a diamond disc to trim blocks to size.

RESULTS

The stratigraphy is shown in Figure 3 and is similar in both cores – c. 0.35 m thick peat overlies c. 2.2 m of calcareous silts and clays (marls) before

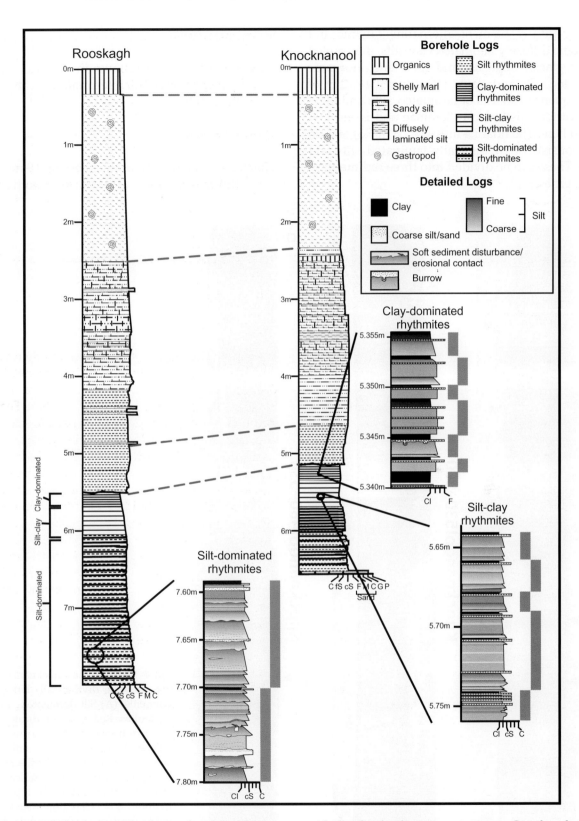

Fig. 3 Stratigraphic logs of Knocknanool and Rooskagh cores, with details of sedimentary structures. Grey bars beside detailed logs indicate thickness of individual couplets.

passing into c. 1.6 m of alternating organic and inorganic diffusely laminated clay-silts, interpreted as Lateglacial deposits, and then into inorganic laminated silts and clays. The upper 1.1–1.2 m of this sequence consists of laminated silts and clay-silts without distinct clay laminae and is not discussed here. The final 1.4 m(Kn) – 2.3 m(Ro) of each core consists of rhythmically laminated silts and clays. These can be divided into three sedimentary facies types.

Sedimentary facies

Sedimentary facies are shown in Figures 3, 4 and 5 and descriptions and interpretations are summarised in Table 1.

Silt-dominated rhythmites

This facies reaches a total thickness of 0.9 m(Kn) – 1.9 m(Ro) and is found at the base of both cores,

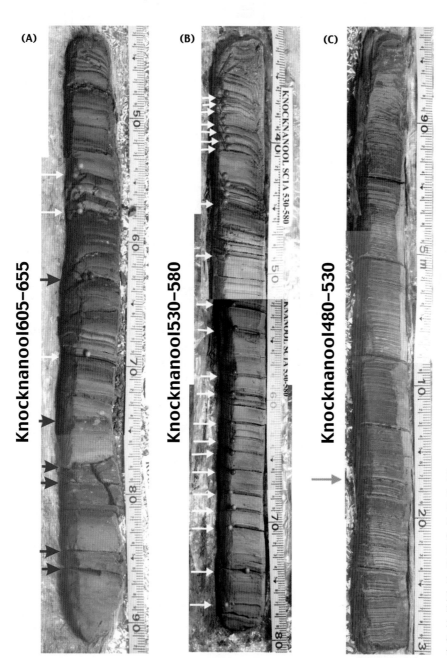

Fig. 4 Facies in the Knocknanool core, showing rhythmically laminated sediments. (A) Silt-dominated rhythmites. Red arrows indicate erosional horizons; white arrows indicate the position of fine units at the top of rhythmites. (B) Silt-clay rhythmites; white arrows indicate position of fine units at the top of rhythmites. (C) Clay-dominated rhythmites (base of core only). The arrow marks the change to rhythmites without clay laminae.

Table 1 Facies description and interpretation

Facies	Coarse unit characteristics	Fine unit characteristics	Interpretation
Silt-dominated rhythmites	Coarse- to fine-grained silt and clayey silt; multiple, usually normally graded laminae; frequent soft sediment deformation structures and erosion surfaces, dropstones, some flocculation	Sharp upper and lower contacts; normally graded silty clay to clay; occasional coarse-grained silt laminae; always underlain by coarse-grained silt; soft sediment deformation structures common on upper contact	Annual varves, summer deposition from turbidity currents and interflows, winter deposition from suspension and occasional underflows
Silt-clay rhythmites	Coarse- to fine-grained silt and clayey silt; 2–5 normally graded laminae with occasional coarse-grained silt laminae; occasional soft sediment deformation structures and burrows; flocculation common	As in silt-dominated rhythmites, but soft sediment deformation structures rare on upper contact, silt laminae less common	Annual varves, as above, but erosional surfaces are uncommon
Clay-dominated rhythmites	Coarse- to fine-grained silt and clayey silt; either 3 laminae coarse-fine-coarse, or a single lamina which may be graded. Rare burrows.	As in silt-dominated rhythmites, but without underlying coarse silt, and silt laminae within clay are 1–2 grains thick	Coarse-fine-coarse sequences with clay cap are interpreted as annual varves; origin of couplets with unlaminated coarse layer is unclear

immediately overlying diamicton. Couplets consist of a lower coarse (silt and some fine sand) unit, which is much thicker (20–120 mm vs. 0.5–8 mm) than the overlying fine unit (composed of silty clay and clay; Figs 3, 4A). The coarse units show no overall grading in particle size, but consist of multiple ungraded or normally graded laminae between 0.5–40 mm thick, with occasional thin, but distinct, laminae of clay-free, coarse silt and sand (Figs 3, 5A). Normally graded laminae are composed of coarse to fine, clayey silt and are initially thin (<1.5 mm), but then thicken towards the centre of each coarse unit; often laminae thin and fine again towards the top of the unit. Contacts between laminae are sharp, planar or irregular, and are frequently marked by soft sediment deformation, consisting of load casts and flame structures (Fig. 5A). Flame structures are 1–3 mm high and may be tilted and oriented in one direction. They are often associated with small rip-up clasts in the overlying coarse lamina, indicating partial erosion of the underlying sediment. Occasional granules and fine pebbles have caused deflection down-

wards of the underlying beds and are interpreted as dropstones.

Thin, pale lamina of coarse silt and sand may be present at the base of the coarse unit, which is usually erosional on the underlying fine unit. The uppermost lamina in the coarse unit is always composed of pale coarse silt and sand, and is commonly between 0.5–1.2 mm thick (Fig. 5B).

The upper fine unit is dark grey in colour, is composed of silty clay and clay and usually exhibits normal grading. Both the lower and upper contacts are sharp (Fig. 6A, B). The upper contact is frequently irregular, is often marked by flame structures c. 1 mm high and occasional rip-up clasts of clay are visible at the base of the overlying coarse unit. The fine unit may also contain thin, nearly white laminae of coarse silt and sand, similar to that seen at the top of the coarse unit (Fig. 5B).

Interpretation: Silt-dominated rhythmites are interpreted as annual varves, as they exhibit most of the characteristics suggested by Smith and Ashley (1985) and Ashley (1995), including: sharp contacts between summer and winter layers

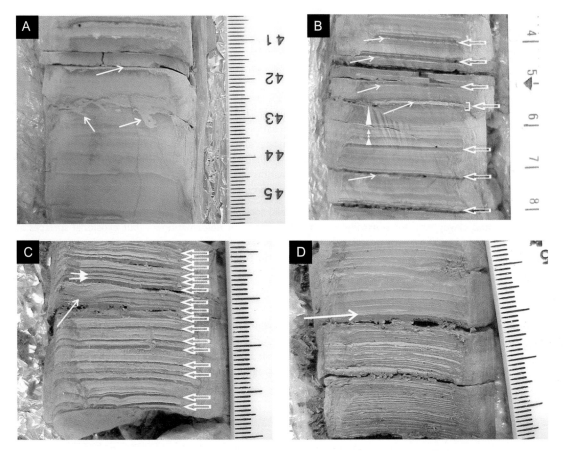

Fig. 5 (A) Close-up of coarse unit in silt-dominated rhythmites showing soft sediment deformation structures and erosional surfaces overlain by paler coarse-grained silt and fine-grained sand. (B) Close-up of silt-clay rhythmites showing multiple normally graded laminae within coarse units and pale laminae of coarse-grained silt and fine-grained sand immediately below fine units. Fine units are marked by outlined arrows, normally graded laminae by triangles and coarse-grained laminae by open arrows. (C) Close-up of transition between silt-clay and clay-dominated rhythmites, showing coarse-fine-coarse grading within coarse units (closed-head arrows) and possible burrow cross-cutting two fine units (open arrow). Fine units indicated by outlined arrows. (D) Close-up of clay-dominated rhythmites overlain by non-varved rhythmites, showing coarse units consisting of single, non-graded laminae alternating with fine units of similar thickness. Boundary between clay-dominated rhythmites and non-varved sediments marked by arrow. Major units on scale bars are centimetres.

consistent with deposition after autumn over-turning; fining upwards in the winter layer; and considerable variation in summer layer thickness reflecting variable rainfall and slumping of sediments, but very little in winter layer thickness, unless silt laminae are present. The sediments resemble descriptions of proximal varves deposited adjacent to the ice margin (Smith, 1978; Ringberg and Erlström, 1999).

Normally graded laminae containing coarser laminae within the summer layers are interpreted as turbidity current underflows. These are related to a variety of causes, including episodic increases in meltwater and sediment influx due to rainfall (Lambert and Hsü, 1979; Blass *et al.*, 2003; Hambley and Lamoureux, 2006) and surge currents generated by slumping along the lake margins (Smith and Ashley, 1985; Hambley and Lamoureux, 2006). Other normally graded units without internal lamination may represent deposition from underflows and interflows caused by meltwater inputs (Hambley and Lamoureux, 2006). The fine-coarse-fine variation in lamina thickness in many summer layers indicates that seasonal variation in meltwater generation was the dominant (but not the only) control on sedimentation here. Occasional thin

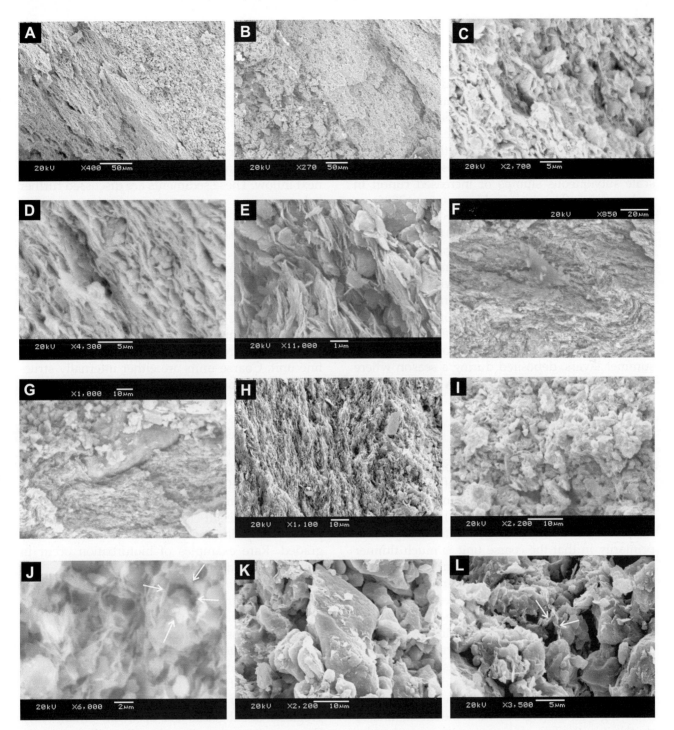

Fig. 6 SEM images of fine units in rhythmically laminated facies. Top of unit is at top right of all photos. (A) Contact between fine clay unit and overlying coarse unit. (B) Contact between coarse unit and overlying fine unit. (C) Microfabric at base of fine unit. Note combination of platy clay grains and prolate silt grains and poorly developed preferential orientation of clay domains. (D) Microfabric at top of fine unit, showing dominantly clay sized material, with strongly developed preferred orientation. Coarse grains are interpreted as dropstones. (E) Close-up of top of fine unit, showing arrangement of grains in domains with face-to-face contacts. (F) Folding within clays at top of a fine unit. (G) Distortion of microfabric at the top of a fine unit due to emplacement of a coarse grain. (H) Wavy distortion of clay microfabric at top of fine unit. Top of unit is to right of photo. (I) Microfabric at base of a normally graded lamina. Some preferred orientation is visible. (J) Close-up of the base of a normally graded lamina in a silt-clay rhythmite, showing multiple edge-to-edge contacts between clay domains. Outline of part of large floc is indicated by arrows. (K) Arrangement of grains in a coarse silt lamina in a clay-dominated rhythmite. No preferred orientation is seen. (L) Coarse unit, clay-dominated rhythmite, showing development of chains of clay domains consistent with flocculation.

laminae of clay-poor, coarse silt and sand may be associated with higher energy underflows where finer suspended sediments bypassed the core sites (Smith, 1981), or they may represent sporadic aeolian deposition within the lake. Similarly, the occurrence of a coarse lamina at the top of each summer layer may represent deposition by small-scale subaqueous slumps or increased runoff in early autumn (Hambley and Lamoureux, 2006), or reflect increased aeolian deposition at this time of year (Lamoureux and Gilbert, 2004). Evidence of deformation and erosion indicates that dewatering and autocompaction of sediments had not yet occurred. Silt laminae within winter layers are also interpreted as turbidity currents, again most likely due to periodic slumping of lake margins during winter low stage (Shaw *et al.*, 1978; Ashley 1995). However, it is possible that some of these are summer layers, deposited during a season where little or no melting occurred, or when meltwater was diverted to another site.

Silt-clay rhythmites

These rhythmites overlie the silt-dominated rhythmites and have a total thickness of c. 0.3 m. They are transitional between silt-dominated and clay-dominated rhythmites (see below), but are more similar to silt-dominated rhythmites. They differ primarily in that the coarse unit is much thinner than in the silt-dominated rhythmites (<30 mm, usually <20 mm thick), and contains 1–5 massive and normally graded laminae between 0.2–10 mm thick (Figs 3B, 5B). Occasional laminae of coarse sand or clayey silt are also present. Possible evidence of bioturbation in the form of burrows is identifiable at the top of normally graded units. These consist of shallow (c. 0.5 mm thick), steep-sided pits cut into the underlying sediment and infilled with coarser overlying material (Fig. 5C). As in the silt-dominated rhythmites, the base of many units is marked by a strongly normally graded lamina which changes colour to dark grey upwards; again the coarsest sediment is at the top of each unit, and consists of a thin lamina of white sand or coarse silt (Fig. 5B). Fine units are similar to those in the silt-dominated rhythmites, and may contain up to 3 laminae of white silt or sand (Fig. 5B).

Interpretation: Silt-clay rhythmites are also interpreted as varves, deposited further away from the ice margin than silt-dominated rhythmites. The number of laminae within each summer layer is much less than in silt-dominated rhythmites, indicating fewer depositional episodes. The absence of erosional surfaces and soft sediment deformation in summer layers indicates that sediment had time to settle and compact prior to the start of the next inflow. These sediments are discussed further below.

Clay-dominated rhythmites

Silt-clay rhythmites fine upwards into clay-dominated rhythmites, which reach a maximum thickness of 0.2 m (Figs 4C, 5C,D). They are much thinner than silt-dominated and silt-clay rhythmites (total couplet thickness 0.3–2.6 mm), and the coarse unit is between 0.5–3 times the thickness of the fine unit. Coarse units are either internally structureless, in which case they consist of coarse silt, or they contain a thin lamina of silty clay within medium silt, usually followed by a thin lamina of coarse silt immediately below the fine unit (Fig. 5C). Structureless coarse units closely resemble the thin laminae of coarse material seen within the fine units of silt-dominated rhythmites. Fine units in the clay-dominated rhythmites are similar to fine units in the silt-clay rhythmites, and are composed of silty clay and clay, which is often normally graded. Rare examples of bioturbation occur in both coarse and fine units.

Interpretation: Clay-dominated rhythmites differ considerably from silt-dominated and silt-clay rhythmites. The very thin coarse layers and the absence of massive or normally graded laminae of clay-silt indicate that sediment influx to the core sites is much lower than before. This may reflect the recession of the ice margin out of the basin, or the trapping of sediment within a newly formed proglacial or subglacial lake upstream (Smith, 1981). Coarse units with a silt – clay-silt – silt pattern are similar to those described by Smith (1978), and are interpreted as summer deposits reflecting two separate inflow maxima, one from late spring/early summer snowpack melting and the second from late summer glacial melting. However, structureless coarse layers are less clearly associated with summer melting, and these laminae closely resemble the silt laminae found in many silt-dominated and silt-clay rhythmite winter layers. It is impossible to say whether these are full summer layers, turbidity current deposits within winter layers, or whether the coarse-fine couplets were formed due to some process other than annual variation in lake stratification and meltwater influx.

Microfabric textures

S.E.M. images of sediments are shown in Figure 6. Fine units show a distinct change upwards. The basal part of the lamina consists of relatively poorly sorted prolate fine silt and platy clay particles (Fig. 6C). Clay particles are arranged in domains (stacks of particles; Bennett *et al.*, 1991) and exhibit dominantly face-to-face contacts consistent with dispersed deposition. However, some edge-to-face and edge-to-edge contacts are observed, and while these can be partly attributed to tilting of the clay domains against larger silt grains, some small flocs are observed (Fig. 6C).

As particle size reduces and sorting improves upwards through the laminae, preferred orientation is increasingly well-developed (Figs 6A,D,E). The top 0.1 mm of most laminae consists almost entirely of clay domains deposited face-to-face and displaying a strong parallel orientation; coarser grains are interpreted as dropstones (Fig. 6G). Where the upper contact exhibits soft sediment deformation structures, or is disturbed by emplacement of coarse grains, the fabric may be folded (Figs 6F,G), or exhibit a wavy structure (Fig. 6H).

Coarse units in silt-dominated and silt-clay rhythmites are largely composed of relatively poorly sorted sediment, consisting of a combination of blocky and prolate single particles, and domains of platy clay particles (Figs 6I,J). Sediment is generally unoriented (Fig. 6I), but poorly developed preferred orientation is seen towards the top of some normally graded units. Sediment is more porous than in fine units, edge-to-edge contacts between clay domains are common, and chains of linked domains occur, indicating flocculation (Fig. 6H,L; O'Brien and Pietraszek-Mattner, 1998). However, clay domains are absent from the very coarse laminae found immediately below and within fine units (Fig. 6K). Coarse units in clay-dominated rhythmites may contain clay domains, in which case flocs are identifiable, or resemble the very coarse laminae found in the fine units of silt-dominated and silt-clay rhythmites.

LAMINAE THICKNESSES

Total thicknesses for all varves are shown in Figures 7A and B, together with variation in thickness across adjacent cores where that information is available. Varve thickness decreases upwards through each core, but with considerable variation from varve to varve. Variation in thickness of couplets also decreases upwards.

Comparisons of total varve thicknesses for each varve type are shown in Figures 7C-F. No correlation in thickness was found between the sites for silt-dominated rhythmites (varves) (Figs 7C,D). However, comparison of silt-clay and the lowermost clay-dominated rhythmites (varves) shows a clear correspondence in total thickness variation up-core between Kn varves 28–72 and Ro varves 30–70 (Figs 7E,F). Four rhythmites present in the Knocknanool core are not seen in the Rooskagh core. The three lowest rhythmites (Kn varves 37, 40, 47) are very thin (<1 mm), consist of a single normally graded silt layer with a clay cap, and lack the coarse silt lamina seen at the base of the winter layer in other varves. Consequently, these rhythmites are considered to be false varves caused by single winter underflows. The winter layer of the fourth rhythmite (Kn varve 67) has evidence of soft sediment deformation and partial erosion along the upper contact; it is overlain by a thin lamina of coarse silt and then by a typical summer layer sequence. In the Rooskagh core the equivalent to Kn varve 66 (Ro varve 65) contains a silt lamina. In this case it appears likely that deposition of coarse silt from an underflow has caused disturbance of the underlying winter layer at Knocknanool, but almost entirely removed the varve at Rooskagh. Comparison between the cores indicates that deposition started at the Rooskagh site between 2–4 years before Knocknanool. No correlation was seen in rhythmite thicknesses for clay-dominated rhythmites.

INTERPRETATION AND DISCUSSION

Proximally deposited sediments within the proglacial lake at Knocknanool exhibit a clear annual control on their deposition, resulting in well-defined couplets. Summer deposits resemble those described elsewhere (e.g. Smith and Ashley, 1985;

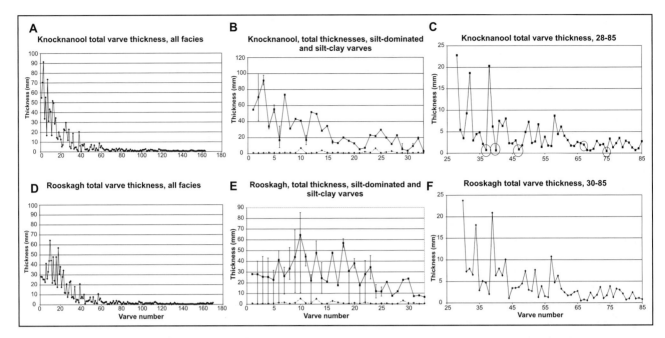

Fig. 7 (A) Total varve thickness, all facies types, Knocknanool core. (B) Total varve thickness, all facies types, Rooskagh core. (C) Varve thicknesses, silt-dominated rhythmites and lower silt-clay rhythmites, Knocknanool core. Error bars indicate variation in thickness for individual varves. (D) Varve thicknesses, silt-dominated rhythmites and lower silt-clay rhythmites, Rooskagh core. Error bars indicate variation in thickness for individual varves. (E) Varve thickness, varves 28–85, Knocknanool core. Circled points are varves with no equivalent in the Rooskagh core. (F) Varve thickness, varves 30–85, Rooskagh core.

Ringberg and Erlström, 1999) and appear to have been deposited dominantly from underflows caused by pulses of meltwater entering the lake basin. Some of this sediment was deposited as flocs, although coarser laminae contain no evidence of flocculation. Fabrics consistent with deposition by flocculation were also identified by Long and O'Riordan (2001) in glaciolacustrine deposits at the southern end of Lough Ree. Flocculation can be a significant process in cold freshwater environments where sediment organic content is low and can be caused either by bioflocculation due to the presence of bacteria (Droppo *et al.*, 1998), or electrochemical flocculation (Woodward *et al.*, 2002). Controls on the proportion of flocculated sediment and size of flocs include: suspended sediment concentration which can be a limiting factor if low (e.g. Malmaeus, 2004); shear stress, as higher shear stresses result in break-up of flocs; and dissolved ion concentration, particularly Ca^{2+} and Mg^{2+} ions (Tsai *et al.*, 1987). As the bedrock throughout the Irish Midlands is predominantly limestone, ion concentrations and suspended sediment concentration are unlikely

to have fallen below the limits needed to allow flocculation, and it appears that coarser, well-sorted laminae are likely to have formed from flows with higher shear values, so that flocs failed to deposit at this point in the lake basin.

Winter laminae also contain evidence of minor flocculation at their base, which is consistent with relatively high suspended sediment concentrations immediately after overturning. Subsequently, the upward-fining of sediment and the apparent absence of flocs is consistent with a reduction in suspended sediment concentration and maximum available particle size as sediment deposition progresses without the input of fresh material. Some caution is necessary in interpreting the development of a preferred orientation at the top of each lamina, as this type of fabric may be formed either by deposition of dispersed (non-flocculated) grains, by break-up of flocs during deposition due to high shear (Bennett *et al.*, 1991; Partheniades, 1991), or by post-depositional alteration of fabric by bioturbation (O'Brien and Pietraszek-Mattner, 1998). However, there is no evidence for high shear during deposition, and the absence of burrows shows that bioturbation was unlikely to have been important, at least in silt-dominated rhythmites.

While finer layers exhibit most of the characteristics of typical winter varve layers, they are thinner than similar examples seen in Scandinavia and New England (O'Brien and Pietraszek-Mattner, 1998; Ringberg and Erlström,1999). This can be at least partly explained for silt-dominated rhythmites by partial erosion of the upper surface of the winter laminae, indicated by the presence of soft sediment deformation structures and rip-up clasts. However, winter layers remain thin as evidence of erosion disappears in silt-clay rhythmites. One possibility is that a relatively high proportion of fine silt and clay was removed from the water column by flocculation during the summer months, so that relatively little material remained to be deposited during winter. Alternatively, the winter freeze-up may have been relatively brief during ice recession in this area, compared to ice-marginal sites in Scandinavia and North America.

The occurrence of erosion surfaces in both summer and winter layers of silt-dominated rhythmites also explains why correlation of these varves is not possible. At the Rooskagh site in particular, comparison of adjacent cores indicates that winter layers present in one core have been completely removed in a second core taken less than 5 m away (e.g. Fig. 7D, varves 2–5). Even where winter layers have been preserved in both cores, there can be considerable variation in both winter layer thickness and total varve thickness (e.g. Fig. 7D, varves 7–10), indicating that further erosion may have occurred, possibly removing other winter layers. Evidence for erosion is more frequently seen in the Rooskagh core, which probably reflects the closer proximity of this site to the point discharge which formed the Knocknanool delta.

Correlation was also not possible between the higher type 3 rhythmites, where coarse units were of similar thickness to fine units. This is probably because summer layers and winter turbidites are indistinguishable in these distal sediments. However, correlation of thicknesses is possible for those silt-clay and clay-dominated varves between 20–2 mm in thickness which contain summer layers consisting of more than one lamina. Within a 30 varve sequence, 3 possible false varves were detected, indicating that the cores are between

1–4 years apart in age. However, if erosional partings and evidence of soft sediment deformation in silt-dominated varves are considered, the error for each core is up to 43 years for Rooskagh and up to 9 years for Knocknanool. Nevertheless, despite the effect of erosion, it is noteworthy that the difference in the total number of rhythmites between sites is just 4 (166 at Knocknanool, 170 at Rooskagh).

CONCLUSIONS

Varved sediments are present in glaciolacustrine sediments underlying Knocknanool Bog, Co. Roscommon. Summer layers are characterised by deposition from underflows and interflows, partly due to flocculation, with thickness controlled by seasonal meltwater generation. Winter layers are deposited mostly as dispersed grains, but with minor flocculation at the base of layers. Occasional underflow deposition also occurs in winter. The varves exhibit proximal to distal fining up-core. Correlation between cores is possible if medial to distal varves between 20–0.5 mm thick are used; local errors may be as high as 43 years per core.

A number of conclusions can be drawn about the use of similar sediments in the Irish Midlands to construct a varve chronology, based both on the work detailed here and previous work in Scandinavia (De Geer 1940). Core sites should be selected with reference to lake bottom topography and sediment thicknesses in order to select sites where the thickest possible sequences with the least likely disturbance are located and sites should also be located away from points where meltwater discharge is likely to have caused repeated scouring of the lake bottom (De Geer, 1940). In order to maximise the chances of locating erosional horizons, multiple cores (at least two, preferably more) should be taken at each core site. Core sites should be as close together as feasible – a maximum of 500 m is suggested – so that visual correlation between cores is possible. Correlation between cores should be based on medial to distal varves only, where thickness variations clearly correspond; the age for each core site can then be calculated by adding on the number of proximal varves below the correlated part of the core. Finally, other dating means, including thermoluminescence dating (e.g. Berger, 1984) and tephrachronologies (where possible) should be used to check varve chronologies, once constructed.

ACKNOWLEDGEMENTS

I thank Roy Gibson and Eamonn Delaney for help with fieldwork. This research was partly funded by the Manchester Geographical Society.

REFERENCES

Andrén, T., Björck, J. and Johnsen, S. (1999) Correlation of Swedish glacial varves with the Greenland (GRIP) oxygen isotope record. *J. Quatern. Sci.*, **14**, 361–371.

Ashley, G.M. (1975) Rhythmic sedimentation in glacial lake Hitchcock, Massachusetts-Connecticut. In: *Glaciofluvial and Glaciolacustrine Sedimemtation* (Eds A.V. Jopling and B.C. McDonald), Society of Economic Paleontologists and Mineralogists Special Publication 23, pp. 304–320.

Ashley, G.M. (1995) Glaciolacustrine environments. In: *Glacial Environments, Volume 1: Modern Glacial Environments: Processes, Dynamics and Sediments* (Ed. J. Menzies), pp. 417–444, Butterworth-Heinemann, Oxford.

Bennett, R.H., O'Brien, N.R. and Hulbert, M.H. (1991) Determinants of clay and shale microfabric signatures: processes and mechanisms. In: *Microstructure of Fine-Grained Sediments* (Eds R.H. Bennett, W.R. Bryant and M.H. Hulbert), pp. 5–32. Springer-Verlag, New York.

Berger, G.W. (1984) Thermoluminescence dating studies of glacial silts from Ontario. *Can. J. Earth Sci.*, 21, 1393–1399.

Blass, A., Anselmetti, F.S. and Ariztegui, D. (2003) 60 years of glaciolacustrine sedimentation in Steinsee (Sustenpass, Switzerland) compared with historic events and instrumental meteorological data. *Eclogae Geol. Helv.*, **96**, Supplement 1, S59–S71.

Boulton, G.S., Dongelmans, P., Punkari, M. and Broadgate, M. (2001) Palaeoglaciology of an ice sheet through a glacial cycle: the European ice sheet through the Weichselian. *Quatern. Sci. Rev.*, **20**, 591–62.

Bowen, D.Q., Phillips, F.M., McCabe, A.M., Knutz, P.C. and Sykes, G.A. (2002) New data for the last glacial maximum in Great Britain and Ireland. *Quatern. Sci. Rev.*, **21**, 89–101.

Clark, C.D. and Meehan, R.T. (2001) Subglacial bedform geomorphology of the Irish Ice Sheet reveals major configuration changes during growth and decay. *J. Quatern. Sci.*, **16**, 483–496.

Clark, P.U., McCabe., A.M., Mix, A.C. and Weaver, A.J. (2004) Rapid rise of sea level 19,000 years ago and its global implications. *Science*, **304**, 1141–1144.

De Geer, G. (1940) *Geochronologia Suecica Principles.* Kungliga Svenska Vetenskapliga Akademica, Handlingen, 3:18:6, 367 pp.

Delaney, C. (1995) *Sedimentology of Late Devensian Deglacial Deposits in the Lough Ree Area, Central Ireland.* Unpublished Ph.D. thesis, Trinity College, Dublin University.

Delaney, C. (2001) Morphology and sedimentology of the Rooskagh Esker, Co. Roscommon. *Irish J. Earth Sci.,* **19**, 5–22.

Delaney, C. (2002) Sedimentology of a glaciofluvial landsystem, Lough Ree area, Central Ireland: implications for ice margin characteristics during Devensian deglaciation. *Sed. Geol.*, **149**, 111–126.

Droppo, I.G., Jeffries, D., Jaskot, C. and Backus, S. (1998) Prevalence of freshwater flocculation in cold regions: a case study from the Mackenzie River Delta, Northwest Territories, Canada. *Arctic*, **51**, 155–164.

Gatley, S., Somerville, I., Morris, J.H., Sleeman, A.G. and Emo, G. (2005) *Geology of Galway-Offaly. 1:100,000 scale, Bedrock Geology Map Series, Sheet 15.* Geological Survey of Ireland, Dublin.

Hambley, G.W. and Lamoureux, S.F. (2006) Recent summer climate recorded in complex varved sediments, Nicolay Lake, Cornwall Island, Nunavut, Canada. *J. Paleolimnology*, **35**, 629–640.

Knight, J., Coxon, P., McCabe, A.M. and McCarron, S.G. (2004) Pleistocene glaciations in Ireland. In: *Quaternary Glaciations – Extent and Chronology. Part 1: Europe* (Eds J. Ehlers and P.L. Gibbard), pp. 183–191.

Lambert, A. and Hsü, K.J. (1979) Non-annual cycles of varve-like sedimentation in Walensee, Switzerland. *Sedimentology*, **26**, 453–461.

Lamoureux, S.F. and Gilbert, R. (2004) A 750-yr record of autumn snowfall and temperature variability and winter storminess recorded in the varved sediments of Bear Lake, Devon Island, Arctic Canada. *Quatern. Res.*, **61**, 134–137.

Lindeberg, G. and Ringberg, B. (1999) Image analysis of rhythmites in proximal varves in Blekinge, southeastern Sweden. *GFF*, **121**, 182–186.

Long, M.M. and O'Riordan, N.J. (2001) Field behaviour of very soft clays at the Athlone embankments. *Géotechnique*, **51**, 293–309.

Malmaeus, J.M. (2004) Variation in the settling velocity of suspended particulate matter in shallow lakes, with special implications for mass balance modelling. *Int. Rev. Hydrobiol.*, **89**, 426–438.

McCabe, A.M. (1987) Quaternary deposits and glacial stratigraphy in Ireland. *Quatern. Sci. Rev.*, **6**, 259–299.

McCabe, A.M. (1996) Dating and rhythmicity from the last deglacial cycle in the British Isles. *J. Geol. Soc. London*, **153**, 499–502.

McCabe, A.M. and Clark, P.U. (1998) Ice-sheet variability around the North Atlantic Ocean during the last deglaciation. *Nature*, **392**, 373–377.

Meehan, R. (2004) Evidence for several ice marginal positions in east central Ireland, and their relationship to the Drumlin Readvance Theory. In: *Quaternary Glaciations – Extent and Chronology. Part 1: Europe* (Eds J. Ehlers and P.L. Gibbard), pp. 193–194.

Mitchell, F.M. (1986) *The Shell Guide to Reading the Irish Landscape*. Country House, Dublin, 228 pp.

O'Brien, N.R. and Pietraszek-Mattner, S. (1998) Origin of the fabric of laminated fine-grained glaciolacustrine deposits. *J. Sed. Res.*, **68**, 832–840.

Parthenaides, E. (1991) Effect of bed shear stresses on the deposition and strength of deposited cohesive muds. In: *Microstructure of Fine-Grained Sediments* (Eds R.H. Bennett, W.R. Bryant and M.H. Hulbert), pp. 175–183. Springer-Verlag, New York.

Ridge, J.C. (2003) The last glaciation of the northeastern United States: a combined varve, paleomagnetic and calibrated ^{14}C chronology. *New York State Museum Bulletin*, **497**, 15–45.

Ringberg, G. (1984) Cyclic lamination in proximal varves reflecting the length of summers during Late Weichsel in southernmost Sweden. In: *Climatic changes on a yearly to millennial basis* (Eds N.-A. Mörner and W. Kárlen), pp. 57–62, Reidel Publishing Co.

Ringberg, B. and Erlström, M. (1999) Micromorphology and petrography of Late Weichselian glaciolacustrine varves in southeastern Sweden. *Catena*, **35**, 147–177.

Scourse, J.D., Hall, I.R., McCave, I.N., Young, J.R. and Sugdon, C. (2000) The origin of Heinrich layers: evidence from H2 for European precursor events. *Earth Planet. Sci. Lett.*, **182**, 187–195.

Shaw, J. and Archer, J.J. (1978) Winter turbidity current deposits in Late Pleistocene glaciolacustrine varves, Okanagan Valley, British Columbia, Canada. *Boreas*, **7**, 123–130.

Shaw, J., Gilbert, R. and Archer, J.J. (1978) Proglacial lacustrine sedimentation during winter. *Arctic Alpine Res.*, **10**, 689–699.

Smith, N.D. (1978) Sedimentation processes and patterns in a glacier-fed lake with low sediment input. *Can. J. Earth Sci.*, **15**, 741–756.

Smith, N.D. (1981) The effect of changing sediment supply on sedimentation in a glacier-fed lake. *Arctic Alpine Res.*, **13**, 75–82.

Smith, N.D. and Ashley, G.M. (1985) Proglacial lacustrine environments. In: *Glacial Sedimentary Environments* (Eds G.M. Ashley, J. Shaw and N.D. Smith). SEPM Short Course, **16**, 135–215.

Tsai, C.-H., Iacobellis, S. and Lick, W. (1987) Flocculation of fine-grained lake sediments due to a uniform shear stress. *J. Great Lakes Res.*, **13**, 135–146.

Van der Meer, J.J.M. and Warren, W.P. (1997) Sedimentology of late glacial clays in lacustrine basins, Central Ireland. *Quatern. Sci. Rev.*, **16**, 779–791.

Waelbroeck, C., Dupless, J.-C., Michel, E., Labeyrie, L., Paillard, D. and Duprat, J. (2001) The timing of the last deglaciation in North Atlantic climate records. *Nature*, **412**, 724–727.

Warren, W.P. and Ashley, G.M. (1994) Origins of the ice-contact stratified ridges (eskers) of Ireland. *J. Sed. Res.*, **A64**, 433–449.

Woodward, J.C., Porter, P.R., Lowe, A.T., Walling, D.E. and Evans, A.J. (2002) Composite suspended sediment particles and flocculation in glacial meltwaters: preliminary evidence from Alpine and Himalayan basins. *Hydrol. Process.*, **16**, 1735–1744.

Anatomy and facies association of a drumlin in Co. Down, Northern Ireland, from seismic and electrical resistivity surveys

BERND KULESSA*[1], GORDON CLARKE†, DAVID A. B. HUGHES† and S. LEE BARBOUR‡

*School of the Environment and Society, University of Wales Swansea, Singleton Park, Swansea, SA2 8PP, UK
†School of Planning, Architecture, and Civil Engineering, Queen's University Belfast, Belfast, BT9 5AG, UK
(e-mail: gordon.clarke@qub.ac.uk)
‡Department of Civil and Geological Engineering, University of Saskatchewan, Saskatoon, Saskatchewan, SK S7N 5A9, Canada

ABSTRACT

Seismic refraction and electrical resistivity geophysical techniques were used to reconstruct the internal architecture of a drumlin in Co. Down, Northern Ireland. Geophysical results were both validated and complemented by borehole drilling, flow modelling of ground water, and geological mapping. The geophysical anatomy of the drumlin consists of five successive layers with depth including; topsoil, partially saturated and saturated glacial tills, and weathered and more competent greywacke bedrock. There are numerous, often extensive inclusions of clay, sand, gravel, cobbles, and boulders within the topsoil and the till units. Together geophysical and geotechnical findings imply that the drumlin is part of the *subglacial lodgement, melt-out, debris flow, sheet flow* facies described by previous authors, and formed by re-sedimentation and streamlining of pre-existing sediments during deglaciation of the Late Devensian ice sheet. Seismic refraction imaging is particularly well suited to delineating layering within the drumlin, and is able to reconstruct depths to interfaces to within ± 0.5 m accuracy. Refraction imaging ascertained that the weathered bedrock layer is continuous and of substantial thickness, so that it acts as a basal aquifer which underdrains the bulk of the drumlin. Electrical resistivity imaging was found to be capable of delineating relative spatial changes in the moisture content of the till units, as well as mapping sedimentary inclusions within the till. The moisture content appeared to be elevated near the margins of the drumlin, which may infer a weakening of the drumlin slopes. Our findings advocate the use of seismic refraction and electrical resistivity methods in future sedimentological and geotechnical studies of internal drumlin architecture and drumlin formation, owing particularly to the superior, 3-D spatial coverage of these methods.

Keywords Drumlin, stratigraphy, formation, geophysics, seismic refraction, electrical resistivity.

INTRODUCTION

During the Late Devensian glaciation in Northern Ireland, ice flowed offshore through coastal embayments away from the ice divide in the Lough Neagh region (Fig. 1), terminating at marine margins (e.g. McCabe, 1987). Following the Last Glacial Maximum, drumlinisation in Northern Ireland occurred on strong ice-moulded bedrock (e.g. Eyles & McCabe, 1989) during the *Belderg* Stadial (~18–16.6 ^{14}C kyr BP) (e.g. Knight, 2002). Drumlins formed by mass and debris flow together

with lodgement processes in a hydrologically active subglacial environment beneath warm ice, and can be classified into five facies associations which include (e.g. McCabe & Dardis, 1989):

Facies Association 1: 'Forms with a core of older drift' (predating 30 ka BP); occur near major divides of ice dispersion in the Late Pleistocene;
Facies Association 2: 'Overridden ice-marginal subaqueous facies'; occurs along the central and northern coast of western Ireland and is characterised by interbedded diamictons, muds, sands, and gravels;

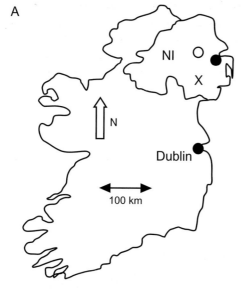

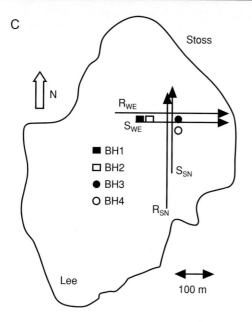

Facies Association 3: 'Subglacial lodgement, melt-out, debris flow, sheet flow facies'; particularly common in Ireland and characterised by drumlins composed of till and commonly inter-till sands beds and/or stratified diamictons;

Facies Association 4: 'Subglacial channel stratified facies'; occur in subglacial tunnel-type valleys and are generally cored by sequences of sands and gravels;

Facies Association 5: 'Lee-side stratified facies'; occurs throughout the Irish drumlin belt and is '. . . characterised by proximal to distal sediment transformations from massive to stratified diamicts which are interbedded with a wide range of sand and gravel lithofacies . . .' (McCabe & Dardis, 1989).

The internal architecture of drumlins, as represented by these facies, is of interest because it is a common record of past glaciological conditions and processes operating at the base of ice masses. From an applied engineering perspective, knowledge of drumlin architecture is pre-requisite to the successful design of geotechnical cuttings, e.g. for road construction. Unfortunately, reconstruction of drumlin stratigraphy often suffers from poor spatial coverage because suitably large exposures are commonly not available (e.g. Menzies, 1987), and borehole logging is limited to one spatial dimension. Geophysical investigations of internal drumlin architecture are very sparsely documented. Nonetheless, existing work is encouraging, suggesting that particularly seismic and electrical resistivity surveys are well suited for identification of tills (e.g. Birch, 1989) and reconstruction of internal layering (e.g. Sutinen, 1985; Sharpe *et al.*, 2004) in drumlins and other glacial deposits.

In 2004–05 a substantial cutting was excavated through a large drumlin in Co. Down, Northern

Fig. 1 (A) Location of field site, marked by 'X', in Northern Ireland (NI). The open and full circles in NI respectively mark the locations of Lough Neagh and Belfast City. (B) Photograph of drumlin, view is to the north. The area of proposed excavation prior to road construction is marked by stripped topsoil, as indicated by lighter grey colours on the eastern side of the photograph than on the western side. (C) Schematic layout of seismic refraction (S_{WE}, S_{SN}) and electrical resistivity (R_{WE}, R_{SN}) profiles, together with borehole (BH) locations. The centre of the drumlin coincides approximately with the location of BH1, and the stoss and lee sides of the drumlin are labelled.

Ireland (Fig. 1). Extensive seismic refraction and electrical resistivity geophysical data were collected prior to excavation in order to reconstruct the internal architecture of the drumlin. These data are reported here, together with direct field observations and geotechnical borehole logs recorded during subsequent drilling. These studies aimed to answer four particular questions:

1 What value can geophysical surveying add to more traditional methods of geological field mapping and geotechnical site investigation?
2 How accurate are geophysical reconstructions of internal drumlin architecture, as compared with borehole logs and direct field observation?
3 How well suited are seismic and electrical resistivity surveys for reconstructing internal layering within the drumlin and the physical properties of the sediments it is composed of?
4 What are the implications of the geophysical and geotechnical findings for existing models of drumlin formation?

It is hoped that the present study will increase interest in the use of geophysical methods in the investigation of glacial sediments.

FIELD SITE AND METHODS

Field site

The field site is a large drumlin located approximately 1 km south of Loughbrickland, Co. Down (Fig. 1A). The drumlin covers an area of approximately 0.75 km^2 and is part of a swarm aligned approximately north-south, consistent with ice flow away from the Late Devensian ice divide at Lough Neagh towards the Carlingford Lough coastal embayment (e.g. McCabe, 1987; Eyles & McCabe, 1989; Knight, 2002). The present study was prompted by plans to construct a dual carriageway along the main road between Belfast and Dublin (A1) in the Loughbrickland area. Prior to borehole drilling and excavation as part of the proposed cutting (Fig. 1B), several seismic refraction and electrical resistivity geophysical profiles were acquired to reconstruct the spatial distribution of vadose zone thickness and depth to bedrock across the drumlin (O'Loughlin, 2003; Sexton, 2003).

The locations of these profiles were chosen so that they coincided with the area of the proposed cutting and could be used to assist with the siting of four boreholes drilled subsequent to the geophysical surveys. The principles of both geophysical methods are well established and field equipment and interpretation software is widely available. Only a brief introduction to either technique is therefore given below, and the interested reader is referred to popular textbooks for further details (e.g. Reynolds, 1997; Sharma, 1997).

Seismic refraction surveys

The seismic waves were generated using a sledgehammer impacting on a metal plate coupled to the ground surface. This is standard practice in near-surface applications and indeed compares favourably with alternative seismic sources (e.g. van der Veen *et al.*, 2000). Seismic waves were expected to travel through the body of the drumlin, refracting at major interfaces, such as the phreatic surface within the drumlin and the till-bedrock interface at its base.

A commercially available 24-channel seismic system was used to record seismic arrivals at the ground surface. The system consisted of 24 vertical 40 Hz geophones connected to a *Geometrics Geode* seismograph, which in turn was linked to a laptop computer. The *OYO SeisImager* software, version 2.20, was used for picking of refracted arrivals in the recorded seismograms, estimation of velocity models using delay times, and for tomographic reconstruction of layer seismic velocities and thicknesses. The *SeisImager* software is widely used in near-surface applications of the seismic refraction method (e.g. Sheehan *et al.*, 2005). The tomographic reconstruction iteratively updates a user-specified initial model, consisting of layer velocities and thicknesses, until a minimum, non-linear least-squares misfit is obtained between observed and calculated seismic velocities.

Several seismic refraction profiles with a geophone spacing of 3 m were acquired using nine forward or reverse shots per profile. Shot points were located at successively greater in-line distances from the geophone array. This yielded total profile lengths of approximately 170 m. Profiles were aligned either south-north parallel to the long axis of the drumlin, or west-east parallel to the short axis of the drumlin. These refraction profiles were acquired for use in reconstructing the internal architecture of the drumlin. Shorter seismic refraction profiles with a geophone spacing of 1 m were also acquired, again using several, laterally offset forward or reverse shots. These shorter profiles (total length: ~85 m) served for accurate estimation of very near-surface seismic velocities, and for improving the user-specified initial model for tomographic inversion of the longer seismic refraction profiles (above). Respectively one long refraction profile in the south-north direction (profile S_{SN}) and in the west-east direction (profile S_{WE}) are reported here (Fig. 1C), together with a short refraction profile centred on the midpoint of the long refraction profile S_{WE}.

Electrical resistivity surveys

Electrical resistivity surveys create an electrical current in the ground using two dedicated electrodes, and measure the resulting electrical potential field using a further two electrodes. By combining a large number of such four-pole measurements at regular intervals along a profile, the bulk electrical resistivity distribution (inverse of bulk electrical conductivity) in the subsurface along the profile can be reconstructed. It was expected that the interfaces between (i) the partially saturated and the saturated tills, and (ii) the overlying till and the underlying greywacke bedrock could be detected based on noticeable contrasts in bulk resistivity between these layers.

A commercially available electrical resistivity imaging system, the *IRIS Syscal R1* with 48 nodes (electrodes), was used in the present study. The stainless steel electrodes were inserted into the ground, and connected to an imaging cable, at regular spacing along a straight line. The cable was connected to the *Syscal R1* resistivity meter, which was programmed by the user to switch between a large number of electrical four-poles. Data acquisition was rapid and automated, with the data being initially stored in the meter and later downloaded to a PC once acquisition completed. The inversion code DCIP2D (e.g. Li & Oldenburg, 2000) was used to tomographically reconstruct the bulk subsurface resistivity distribution. The reconstruction process involves iterative updating of a *current* model, calculated from the field data together with the

measurement errors, until the least-squares misfit between observed and calculated data is minimised (e.g. Li & Oldenburg, 2000). The current model is then assumed to correspond as closely to the actual bulk resistivity distribution in the subsurface as can feasibly be reconstructed from the field data.

Several electrical resistivity profiles with an electrode spacing of 5 m, yielding total profile lengths of 235 m, were acquired in the same area as the seismic refraction surveys. Profiles were aligned either south-north parallel to the long axis of the drumlin, or west-east parallel to the short axis of the drumlin. These four resistivity profiles were acquired for use in reconstructing the internal architecture of the drumlin. Each of the long profiles was repeated with a smaller electrode spacing of 1 m, yielding shorter profile lengths of 47 m. The shorter profiles were used to estimate the very near-surface electrical resistivity distribution, which allowed estimation of the thickness of the partially saturated zone, as well as improvement of the tomographic reconstruction of the long profiles. Respectively one long resistivity profile in the south-north direction (profile R_{SN}) and in the west-east direction (profile R_{SN}) are reported here (Fig. 1C), together with a short resistivity profile centred on the midpoint of the long profile R_{SN}.

Borehole logs and other relevant field observations

During borehole drilling, geological materials encountered with depth were logged by the on-site crew. Upon completion of drilling, falling-head permeability tests were conducted to measure bulk hydraulic conductivity. Vibrating-wire sensors were installed in all four boreholes to continuously monitor the long-term response of pore water pressure to rainfall events and stress relief during excavation. Excavation of a south-north section of the drumlin as part of road construction (Fig. 1B) provided an exposure along which the bedrock elevation could be established in the direction of the roadway.

RESULTS

Inverted, long seismic refraction and electrical resistivity profiles in the south-north and west-east directions are illustrated in Figs 2 and 3. These

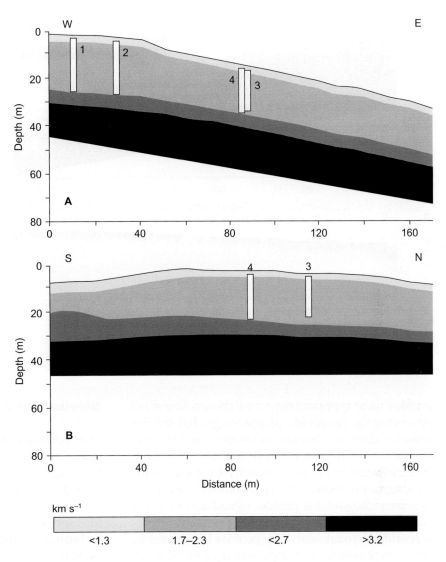

Fig. 2 Representative long seismic refraction profiles. (A) Profile S_{WE}. (B) Profile S_{SN}. Approximate boreholes locations are shown for reference, labelled 1, 2, 3, 4. Note that the seismic velocities of topsoil and partially saturated till combine to produce an averaged value of <1.3 km s^{-1}. See Fig. 1 for a plan view of profile and borehole locations.

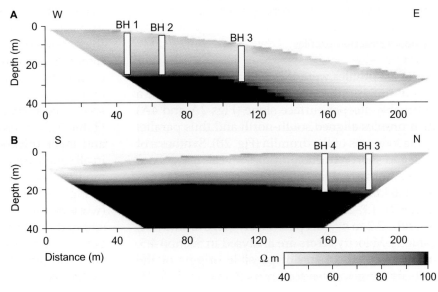

Fig. 3 Representative long electrical resistivity profiles. (A) Profile R_{WE}. (B) Profile R_{SN}. Approximate boreholes (BH) locations are shown for reference. See Fig. 1 for a plan view of profile and borehole locations.

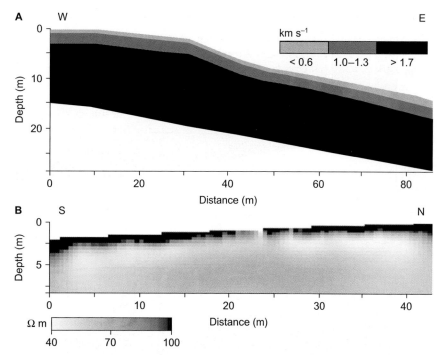

Fig. 4 Representative short seismic refraction and electrical resistivity profiles. (A) Short seismic refraction profile centred the midpoint of the long profile S_{WE} (see Fig. 2A). (B) Short electrical resistivity profile centred the midpoint of the long profile R_{SN} (see Fig. 3B). Note that in (B) the resistivities of topsoil and partially saturated till combine to produce an averaged value of >100 Ω m. See Fig. 1 for a plan view of profile and borehole locations.

profiles were representative and chosen based on three criteria: (a) profile direction, guided by the need to illustrate changes in seismic velocity and electrical resistivity along both the long and short axes of the drumlin; (b) proximity to a maximum number of boreholes; and (c) avoidance of repetition (including further profiles would add no extra value). Similarly, the inverted, short seismic refraction and electrical resistivity profiles illustrated in Fig. 4 were respectively representative of all short seismic and resistivity profiles acquired.

Seismic refraction profiles

West-east profiles were aligned parallel to the short axis of the drumlin, and therefore typically have a noticeably steeper surface slope (Figs 2A and 4A) than profiles aligned south-north and thus parallel to the long axis of the drumlin (Fig. 2B). Synthesis of long and short refraction profiles reveals a sequence of five horizontal layers, having seismic velocities of <0.6 km s^{-1} (shallowest layer), 1.0–1.3 km s^{-1} (layer 2), 1.7–2.3 km s^{-1} (layer 3), 2.7 km s^{-1} (layer 4) and >3.2 m s^{-1} (deepest layer) (Figs 2 and 4A; Table 1). Velocity errors are analysed in Section 4.5, including a discussion of possible origins of the velocity ranges given for layers 2 and 3 (Table 1).

Electrical resistivity profiles

The west-east resistivity profiles are aligned parallel to the short axis of the drumlin, and therefore have a noticeably steeper surface slope (Fig. 3A) than the profiles aligned south-north (Figs 3B and 4B). Synthesis of long and short resistivity profiles reveals the presence of three dominant layers including; a shallow layer of >100 Ω m, an intermediate layer of <50 Ω m, and a deep layer of >100 Ω m (Figs 3 and 4A, Table 1). As discussed in more detail in Section 4.3, comparison of layer thicknesses with those obtained from refraction imaging suggests that the shallow resistivity layer corresponds to a combination of seismic layers 1 and 2, and the intermediate and deep resistivity layers correspond to seismic layers 3 and 5 (Table 1). Uncertainties in estimates of resistivity and layer thickness are analysed below. Notably, in all long west-east profiles resistivities of the intermediate layer were lower downslope (eastern end in Fig. 3A) than nearer the centre of the drumlin (western side in Fig. 3A). Many localised anomalies situated within the shallow and the intermediate layers were also found. These anomalies variably were less or more resistive than the dominant layer resistivity.

Table 1 Geophysical anatomy of drumlin. Seismic velocities are accurate to within ±0.05 km⁻¹, and electrical resistivities to within ±4 Ω m (Section 4). The resistivities of layers 1 and 2 combine to produce an averaged value of >100 Ω m (Fig. 4b). The layer of weathered greywacke could not be detected using the electrical resistivity surveys, and is therefore marked by '?'.

	Description	Velocity (km s^{-1})	Resistivity (Ω m)
Layer 1	Topsoil	<0.6	} >100
Layer 2	Partially saturated till	1.0–1.3	
Layer 3	Saturated till	1.7–2.3	<50
Layer 4	Weathered greywacke	~2.7	?
Layer 5	Competent greywacke	>3.2	>100

Borehole logs and other relevant field observations

The two western (boreholes 1 and 2) and eastern (boreholes 3 and 4) logs were characterised by similar lithologies respectively (Fig. 5). In the western boreholes, topsoil approximately 0.5 m thick is underlain by glacial tills of varying consistency and colour, although differences in these properties are small (Fig. 5A). The total thickness of the till units varied between 23.5 m in borehole 1 and 22.5 m in borehole 2. In both boreholes a thin (~0.5 m) band of sandy gravel is sandwiched between the till and the underlying greywacke bedrock (Fig. 5A). Since the boreholes only marginally penetrated bedrock, neither thickness nor spatial extent of the weathered greywacke can be ascertained from the borehole logs. Boreholes 3 and 4 were located east of boreholes 1 and 2 (Fig. 1C), approximately 50–80 m further away from the centroid of the drumlin. The thickness of topsoil (~0.5 m) and sequence of glacial till units in the eastern boreholes (Fig. 5B) were similar to those in the western boreholes (Fig. 5A). Till thickness totalled 18.5 m and 20 m in boreholes 3 and 4 respectively, and was therefore smaller than in the western boreholes, albeit not substantially. By comparison with the western boreholes, no layer of sandy gravel was found in the eastern boreholes, but the latter consistently terminated in weathered greywacke bedrock (Fig. 5).

Further evidence of internal drumlin architecture comes from field observation conducted once excavation had begun (Fig. 1B), exposing topsoil, till profiles, and in some areas bedrock. Next to the

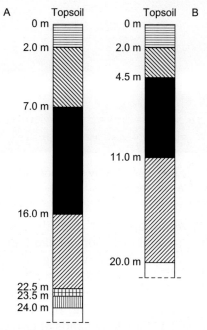

Stiff grey brown gravelly mottled silt with some cobbles

Very stiff grey sandy gravelly silt with some cobbles

Very stiff dark grey sandy gravelly silt with some cobbles and boulders

Very stiff grey sandy gravelly silt with some cobbles and boulders

Band of dense grey and fine medium and coarse sandy gravel with some cobbles

Completely to highly weathered grey shale rock

Highly to moderately weathered grey shale rock

Fig. 5 Representative boreholes logs as recorded during drilling by the on-site crew. (A) Borehole 2. (B) Borehole 4. See Fig. 1 for a plan view of borehole locations.

Fig. 6 Example of a massive clay inclusion, showing one of authors (~2 m tall) for scale.

dominant layering apparent from the four borehole profiles, several inclusions of sand, gravel or clay, in some cases mixed to varying proportions, were observed to be embedded within the tills. Such inclusions were often several metres in vertical and horizontal extent (Fig. 6). Several field drains, exposed as thin, long bands of gravel, were also observed close to the surface of the drumlin. Detailed results from falling-head permeability tests, pore water pressure monitoring, and other geotechnical tests will be reported elsewhere. Insitu permeability tests revealed that the hydraulic conductivity of the till units is very small (~10^{-9}–10^{-8} m s^{-1}) compared to that of the underlying weathered bedrock (hydraulic conductivity was too large to be measured reliably). Monitored water levels allowed the seasonal fluctuations of the thickness of the partially saturated zone to be tracked within the upper till units. The 2-D, south-north profile of bedrock elevation, reconstructed post-excavation, revealed that total thickness of topsoil and till units was several metres less

near the southern margin of the survey area than at the northern margin. Groundwater discharge was observed to be common near the toe of the drumlin.

INTERPRETATION

Calibration of geophysical data

Geophysically-derived layering is conveniently calibrated using the borehole logs (Fig. 5) together with other field observations, as summarised in Table 1. Layers 1, 2, and 3 in the seismic refraction and electrical resistivity data correspond to topsoil and partially saturated and saturated glacial tills. Layer 4 corresponds to weathered greywacke bedrock, and in contrast to the seismic refraction data cannot be distinguished reliably in the electrical resistivity data (Table 1). In the absence of any other plausible explanation, it is inferred that that layer 5 most likely corresponds to relatively competent bedrock. The interface between weathered bedrock (lower seismic velocity of approximately 2.7 km s^{-1}) and more competent bedrock (higher seismic velocity of >3.2 km s^{-1}) thus gives rise to a strong seismic refraction.

Neither partially saturated (1.0–1.3 km s^{-1}) nor saturated (1.7–2.3 km s^{-1}) tills could be assigned a typical seismic velocity (Table 2). Instead, velocity was found to increase progressively with depth in these layers, which is interpreted to be due to increasing material bulk density with depth. A similar increase in seismic velocity in a high-resolution, cross-borehole seismic profile of a section of glacial till was observed at a local test site (Mulholland, 2004).

Topsoil and glacial till units

Following calibration, the seismic refraction data in particular allow multi-dimensional extrapolation of lithology across the drumlin, thus complementing results from more traditional geotechnical site investigation well. Topsoil thickness is largely constant at approximately 0.5 m within the survey area (Fig. 2), and the overall thickness of the partially saturated zone (including topsoil) is typically 2–2.5 m (Fig. 4A). Noticeable spatial changes in the thickness of topsoil and the partially saturated zone appear to occur only near the southern (Fig. 2B) and eastern (Fig. 4A) margins of the survey area.

Table 2 Comparison of depths to bedrock (from ground surface using a topsoil thickness of 0.5 m), as taken from borehole (BH) logs and seismic refraction (S_{WE}, S_{SN}) and electrical resistivity profiles (R_{WE}, R_{SN}) (Sections 4.3 and 4.5).

	Log (m)	S_{WE} (m)	S_{SN} (m)	R_{WE} (m)	R_{SN} (m)
BH 1	24.5	24 ± 0.5	–	23 ± 1.5	–
BH 2	23.5	23.5 ± 0.5	–	23 ± 1.5	–
BH 3	19	19.5 ± 0.5	20.5 ± 0.5	–	20.5 ± 1.5
BH 4	20.5	19.5 ± 0.5	21 ± 0.5	–	21.5 ± 1.5

These apparent changes must be considered with great caution since seismic data coverage near the margins of the survey area is poor compared to more central portions. The changes may thus well be artefacts produced by mathematical extrapolation during data inversion. Any minor changes in thickness of topsoil and partially saturated zone (Figs 2 and 4A) must similarly be considered with caution since they typically lie within the range of uncertainty of depth estimates.

Seismic refraction (Fig. 2) and electrical resistivity (Fig. 3) data consistently suggest that the thickness of the saturated zone is greatest near the centre of the drumlin, thinning towards the southern and eastern margins of the survey area. These changes in thickness are generally consistent with the borehole logs (Fig. 5) and the 2-D south-north profile of bedrock elevation reconstructed post excavation. The direction of thinning of the saturated tills is, thus, aligned in the direction of former ice flow.

Importantly from a geotechnical perspective, electrical resistivities of saturated tills were found to decrease away from the centre of the drumlin (Fig. 3A). As discussed in popular geophysical textbooks (e.g. Reynolds, 1997; Sharma, 1997), the most prominent factors that could generate a decrease in resistivity within a given material are increases in moisture content, porosity, cementation of grains, conductivity of the waters filling the pores, or sediment clay content. Consistent downslope variations in till composition were not observed, and in particular neither porosity, cementation of grains, water conductivity, or clay content changed consistently in the study area. The most likely explanation for the observed, downslope decrease in resistivity is therefore a gradual increase in till moisture content.

The marked difference in hydraulic conductivity between the till units and the weathered bedrock was modelled to produce nearly vertical groundwater flow down through the drumlin to the underlying aquifer. Groundwater flow direction was modelled to be reversed near the toe of the drumlin, where strong upward gradients were predicted to cause groundwater discharge at the ground surface. Field observations confirm that such discharge indeed occurs at the toe of the drumlin. The upward hydraulic gradients are interpreted to decrease the effective stress near the toe of the drumlin, resulting in the inferred increase in moisture content and decrease in resistivity in this area. This is important since softening of tills due to a decrease in effective stress may decrease the stability of the drumlin slopes.

Depth to bedrock

Seismic and electrical resistivity estimates of depth to bedrock are summarised in Table 2 together with actual depths taken from the borehole logs. For convenient visual comparison, boreholes are superimposed on seismic and resistivity images in Figs 2 and 3.

For seismic profiles S_{WE} and S_{SN} the sudden jump in velocity from a maximum of approximately 2.3 km s^{-1} in layer 3 to approximately 2.7 km s^{-1} in layer 4 (Table 1, Fig. 2) is inferred to indicate the transition from dense, fully saturated tills to weathered bedrock. Within the given ranges of uncertainty, depths to bedrock estimated for both seismic profiles (S_{WE} and S_{SN}) match actual depths in all boreholes closely (Table 2). Notably, the thickness of weathered bedrock (2.7 km s^{-1}) appears to suddenly increase near the southern margin of survey area (Fig. 2B). Since seismic data coverage near the margins of the survey area is poor compared to more central portions), this observed increase is probably an artefact produced by mathematical extrapolation during data inversion.

For resistivity profiles R_{WE} and R_{SN} depths were determined by calculating vertical resistivity gradients at the locations closest to the boreholes (Figs 1C and 3), and taking the depth corresponding to the largest increase in resistivity as being equal to depth to bedrock. Within the given ranges of uncertainty, depths to bedrock estimated for both resistivity profiles match actual borehole depths closely (Table 2).

Drumlin facies and formation

In elucidating past glaciological conditions and processes operating at the base of the Late Devensian ice sheet in Northern Ireland, it is interesting to classify the drumlin surveyed here within the framework of the five distinct facies associations identified by McCabe & Dardis (1989). The collective evidence from borehole logs (Fig. 5), seismic (Figs 2 and 4A) and electrical resistivity (Figs 3 and 4B) surveys, and associated field observations (Fig. 6) is entirely consistent with the criteria defining Facies Association 3, but violates at least one criterion defining each of the other four facies associations so that the drumlin cannot be part of:

• Facies Association 1 because it is not located near a major ice divide and apparently does not have a core of older drift;
• Facies Association 2 because it is located in the east rather than the west of Ireland and apparently lacks evidence of interbedded diamictons, muds, sands, or gravels;
• Facies Association 4 because it is not located in a subglacial tunnel-type valley and sequences of sands and gravels are not apparent;
• Facies Association 5 because there is no apparent evidence for interbedded sequences of massive or stratified diamicts and sand or gravel lithofacies.

We therefore infer that the drumlin is part of the *subglacial lodgement, melt-out, debris flow, sheet flow* facies (Facies Association 3, McCabe & Dardis, 1989). Synthesis of the findings from the present study with the detailed facies model of McCabe & Dardis (1989) reveals that:

• Several stratified units of poorly sorted till/ diamicton are present, thinning and generally dipping towards the south/south-east, and thus in the general direction of former ice flow in this area. These units are believed to have formed by re-sedimentation of material originally deposited subglacially by debris flow.
• Near the centre of the drumlin, a horizontally extensive (>20 m in diameter), thin (~0.5 m) inclusion of sand and gravel is sandwiched between till and weathered bedrock. Such basal deposits are typical of sheet flow events during transient, enhanced flow of subglacial melt water.
• Numerous, often extensive sediment inclusions of varying geometries are present within the till units, consisting of various mixtures of clay, sand, gravel, cobbles, and boulders. These inclusions probably originated at different times by transient subglacial debris and sheet flow events.
• Weathered, hydraulically highly permeable bedrock is present beneath the saturated till units, forming a continuous layer that is typically >5 m thick. Allowing for temporal changes in the thermal regime at the base of the former ice sheet, weathering processes likely included bed fracture and formation of Kamb-type cavities and/or Nye-type meltwater channels at different times (e.g. Knight, 2002).

The main phase of original sedimentation is believed to have occurred prior to drumlinisation in this area. Slow or negligible ice sheet motion is pre-requisite to sedimentation of several tens of metres of subglacial debris (e.g. McCabe & Dardis, 1989), which likely occurred particularly during the *Glenavy* Stadial (~25–18 ^{14}C kyr BP; e.g. Knight, 2002). It has been speculated that drumlins were later streamlined by fast ice flow during the *Belderg* Stadial (~18–16.6 ^{14}C kyr BP; e.g. Knight, 2002), and further that fast ice flow eventually lead to the rapid disintegration of the Late Devensian ice sheet (e.g. Eyles & McCabe, 1989).

Uncertainty in geophysical estimates

Two main lines of evidence can be used to assess the accuracy of geophysical estimates of subsurface layer geometries and physical properties (i.e. seismic velocity and electrical resistivity). First, an initial assessment of uncertainty can be obtained by comparing geophysically derived layer thicknesses or depths to bedrock with actual thicknesses and depths taken from the borehole logs, as implied e.g. by Table 2. However, this shows only an approximate indication of geophysical accuracy since boreholes were often located several metres to >10 m away from the geophysical profiles (Fig. 1C). Second, estimates of uncertainty can be obtained by systematically comparing geophysically derived layer geometries and physical properties at the cross-over points between seismic profiles or between electrical resistivity profiles (Fig. 1C). This has the advantage that geophysical estimates should match between any two profiles at their cross-over point so that quantitative estimation of relative errors is possible. However, it has the disadvantage that absolute errors cannot be estimated since a ground-truthing data set (as derived e.g. from borehole logs) is not considered.

Application of standard statistical techniques of error estimation to all available data for cross-over points (including all additional geophysical data not presented here) reveals that seismic velocities are accurate to within ±0.05 km s^{-1}, and depths to interfaces derived from seismic refraction data are accurate to within ±0.5 m. Estimates of electrical resistivity were found to be accurate to within ±4 Ω m, and depths to interfaces derived from electrical resistivity data are accurate to within ±1.5 m. Within the given ranges of uncertainty, comparison of geophysically-derived depths to bedrock with actual depths taken from boreholes logs (Table 2) was found to be satisfactory in all cases.

SYNTHESIS AND CONCLUSIONS

Synthesis of results

The geophysical anatomy of the drumlin consists of five distinct layers successively including; topsoil (~0.5 m thick), partially saturated glacial tills (~1.5–2 m thick), fully saturated glacial tills (up to >20 m thick), weathered greywacke bedrock (>5 m thick), and more competent greywacke bedrock. This succession of layers is confirmed by borehole logs and direct field observations. There are numerous, often extensive, inclusions of clay, sand, gravel, cobbles, and boulders included within the topsoil and the till units.

Novelty value of the present study

It is now possible to answer the three key questions that motivated the present study:

1 The key additional value of geophysical surveying is its superior spatial coverage, often supporting reconstruction of subsurface properties in three spatial dimensions at metre-scale spatial resolution. This is particularly desirable in sedimentological reconstructions of drumlins since suitably large outcrops for field mapping are commonly rare, constituting the *drumlin problem* (e.g. Menzies, 1987; McCabe & Dardis, 1989). The present study presents a particularly desirable scenario: initially geophysical data are calibrated at point locations in terms of drumlin stratigraphy using the borehole logs and other field observations, and subsequently these data are used to extrapolate changes in stratigraphic layering across the drumlin. Here, additional value is also added to the information from borehole logs since:

(a) Drilling through bedrock is logistically and financially particularly expensive, such that boreholes were terminated when weathered bedrock was reached. In contrast, relatively cheaply acquired seismic refraction data could map both thickness and spatial extent of the weathered layer, which is geotechnically particularly relevant.

(b) Low electrical resistivities are diagnostic of increased till moisture content near the toe of the drumlin, indicating strong upward gradients in water flow which potentially increase the risk of landslides. Detecting and delineating such zones of enhanced moisture content using traditional, invasive techniques of geotechnical site investigation is logistically and financially more expensive.

2 Geophysical estimates of depths to interfaces were found to be accurate to within ±0.5 m and ±1.5 m for seismic refraction and electrical resistivity data, respectively. Within these ranges of uncertainty, geophysically derived depths to bedrock matched those taken from borehole logs very well (Table 2). Further case studies must confirm whether such encouraging results are a fortunate exception in the present case, or indeed typically the norm in seismic refraction and electrical resistivity reconstructions of drumlin anatomy.

3 Since the seismic refraction method is subject to considerably less depth uncertainty than the electrical resistivity method (±0.5 m vs. ±1.5 m), it is relatively well suited for mapping of layering within the drumlin. This inference is confirmed by the fact that electrical resistivity data could neither detect nor delineate the layer of weathered bedrock, and could not distinguish between topsoil and partially saturated tills. In contrast, electrical resistivity is much better suited to delineating lateral changes in till moisture content and mapping inclusions within topsoil and till than seismic refraction imaging.

4 Together geophysical and geotechnical findings imply that the drumlin is part of the *subglacial lodgement, melt-out, debris flow, sheet flow* facies as described by McCabe & Dardis (1989). The drumlin is inferred to have formed by re-sedimentation and streamlining of pre-existing sediments, controlled primarily by debris and sheet water flows at the base of fast-flowing ice during deglaciation of the Late Devensian ice sheet (e.g. McCabe & Dardis, 1989; Knight, 2002).

Future work

The findings presented here generally advocate the use of seismic refraction and electrical resistivity methods in reconstructing the internal architecture of glacial sediments. This has triggered profound interest in conducting further extensive studies at a number of other sites in Northern Ireland. Motivated particularly by the success of the present study, it is planned to use seismic and electrical geophysical methods as a standard tool of site investigation in the future.

ACKNOWLEDGEMENTS

Gordon Clarke acknowledges a DEL Ph.D. scholarship, and Lee Barbour an EPSRC Visiting Fellowship. The authors would particularly like to thank Eamon O'Loughlin and Michael Sexton who spent many hours in the field in summer 2003 collecting the geophysical data. The Geophysical Inversion Facility of the University of British Columbia provided a free academic license for the code DCIP2D, which is gratefully acknowledged. The Construction and Road Services of Northern Ireland provided support in many different ways, which made this study possible. We thank Neil Glasser and John Woodward for their detailed comments on the original version of this manuscript.

REFERENCES

Birch, F.S. (1989) A geophysical study of Quaternary sediments near the Late Pleistocene marine limit in Epping, New Hampshire. *Northeast Geol.*, **11**, 124–132.

Eyles, N. and McCabe, A.M. (1989) The Late Devensian (<22,000 BP) Irish Sea Basin: The sedimentary record of a collapsed ice sheet margin. *Quaternary Sci. Rev.*, **8**, 307–351.

Knight, J. (2002) Bedform patterns, subglacial meltwater events, and Late Devensian ice sheet dynamics in north-central Ireland. *Global Planet. Change*, **35**, 237–253.

Li, Y. and Oldenburg, D.W. (2000) 3-D inversion of induced polarisation data. *Geophysics*, **65**, 1931–1945.

McCabe, A.M. (1987) Quaternary deposits and glacial stratigraphy in Ireland. *Quaternary Sci. Rev.*, **6**, 259–299.

McCabe, A.M. and Dardis, G.F. (1989) A geological view of drumlins in Ireland. *Quaternary Sci. Rev.*, **8**, 169–177.

Menzies, J. (1987) Towards a general hypothesis on the formation of drumlins. In: *Drumlin Symposium* (Eds J. Menzies and J. Rose), pp. 9–24. Balkema, Rotterdam.

Mulholland, P. (2004) *Cross-borehole seismic transmission surveys at the EERC test site, Queen's University Belfast.* Unpublished B.S. thesis, Queen's University Belfast, Belfast, Ireland.

O'Loughlin, E. (2003) *The use of seismic refraction for bedrock detection at an engineering site, Loughbrickland, County Down.* Unpublished M.S. thesis, Queen's University Belfast, Belfast, Ireland.

Reynolds, J.M. (1997) *An introduction to applied and environmental geophysics.* John Wiley & Sons.

Sexton, M. (2003) *The use of electrical resistivity surveying as a complementary technique for traditional geotechnical investigations.* Unpublished M.S. thesis, Queen's University Belfast, Belfast, Ireland.

Sharma, P.V. (1997) *Environmental and Engineering Geophysics.* Cambridge University Press.

Sharpe, D., Pugin, A., Pullan, S. and Shaw, J. (2004) Regional unconformities and the sedimentary architecture of the Oak Ridges Moraine area, southern Ontario. *Can. J. Earth Sci.*, **41**, 183–198.

Sheehan, J.R., Doll, W.E. and Mandell, W.A. (2005) An evaluation of methods and available software for seismic refraction tomography analysis. *J. Environ. Eng. Geoph.*, **10**, 21–34.

Sutinen, J. (1985) Application of radar, electrical resistivity, and seismic soundings in the study of morainic landforms in northern Finland. In: *INQUA Till Symposium*, Geological Survey of Finland, Special Paper, **3**, 65–75.

Van der Veen, M., Bueker, F., Green, A.G. and Buness, H.A. (2000) Field comparison of high-frequency seismic sources for imaging shallow (10–250 m) structures. *J. Environ. Eng. Geoph.*, **5**, 39–56.

The Newbigging esker system, Lanarkshire, Southern Scotland: a model for composite tunnel, subaqueous fan and supraglacial esker sedimentation

MATTHEW R. BENNETT*, DAVID HUDDART† and GEOFFREY S.P. THOMAS‡

*School of Conservation Science, University of Bournemouth, Talbot Campus, Fern Barrow, Poole, BH12 5BB, UK
(e-mail: mbennett@Bournemouth.ac.uk)
†School of Education, Community and Social Science, Liverpool John Moores University, I.M. Marsh Campus, Barkhill Road, Liverpool L17 6BD, UK
‡Department of Geography, University of Liverpool, Liverpool L69 3BX, UK

ABSTRACT

The Newbigging esker system shows an ordered variation in stratigraphy and sedimentology within three morphological and spatially distinct landform assemblages: a single linear ridge, interrupted and terminated by shallow fans along its length; a series of multiple sub-parallel ridges and shallow fans, and a complex multi-ridge structure. The sedimentology of the system indicates a composite origin and the sedimentary architecture is consistent with the progressive infilling of a large lake basin by successive shifts in the sediment input as the ice margin retreated during the Late Devensian glacial stage. Quarry A consists of stacked overlapping, sand-dominated, wedges formed during a single episode of fan sedimentation as a result of flow expansion at the exit of a tunnel into the lake during a short ice-front still-stand. Quarry B consists of a boulder-rich tunnel-fill deposited in a single episode at the cessation of a high-magnitude discharge event. Quarry C, in contrast, was deposited in multiple episodes and consists of an upwards depositional transition from coarse tunnel deposition, into a subaqueous fan fronting the tunnel portal, and then into a supraglacial outwash sandur deposited in an ice-walled trough created by tunnel unroofing. This case study adds to the depositional models available with which to interpret complex esker and kame assemblages, both in the Scottish Highlands and elsewhere in the world. It also illustrates the need for caution when interpreting such assemblages due to their potential for polygenesis.

Keywords Eskers, Scotland, subaqueous fans, supraglacial sedimentation.

INTRODUCTION

In recent decades the morphology and sedimentology of Pleistocene eskers have received much attention in Britain (Thomas, 1984, Terwindt & Augustinus, 1985; Gray, 1991; Auton, 1992; Huddart & Bennett, 1997; Owen, 1997; Thomas & Montague, 1997), Ireland (Warren & Ashley, 1994; Delaney, 2002), North America (Cheel, 1982; Shreve, 1985; Henderson, 1987; Shulmeister, 1989; Gorrell & Shaw, 1991; Spooner & Dalrymple, 1993; Brennand, 1994; Brennand & Shaw, 1996) and Scandinavia (Hedbrand & Amark, 1989; Mäkinen, 2003). At the same time there have been only a few investigations into eskers forming at modern glacier margins (Price, 1966, 1969; Howarth, 1971; Gustavson & Boothroyd, 1987; Boulton & Van der Meer, 1989; Huddart et al., 1998). Consequently, the origin of eskers remains poorly understood and frequently controversial, especially in terms of their palaeohydrology, depositional processes and environments of deposition. This is particularly true of complex esker systems that consist of multiple, anastomosing or braided, ridges that have previously been interpreted by some authors as indicative of subglacial water flow while others have favoured their formation with supraglacial systems (Warren & Ashley, 1994). The aim of this paper is to help resolve this debate by contributing a detailed case study of one complex esker system in Southern Scotland.

It is worth first clarifying the characteristics of the three main types of esker. Most common are narrow single ridges displaying a winding planform and periodically spaced topographic highs and lows. They occur in lengths from a few hundred metres to hundreds of kilometres, in widths from a few tens of metres to a few hundred and are mostly orientated parallel to regional ice flow directions. Internally they are composed of subhorizontally bedded, often very coarse gravel, frequently deformed at the margins, and are interpreted as backfilling in single subglacial tunnels in thin ice towards the outer margin of the ice-sheet (Warren & Ashley, 1994).

A second type of esker is beaded, with ridges consisting of long, narrow segments punctuated by wide mounds, often in repeated down-ice sequence (Warren & Ashley, 1994). The mounds, or beads, are commonly composed of finer grained gravel, sand and silt deposited in a fan radiating outwards from the terminus of the up-ice ridge segment. They have been interpreted as subaqueous fans, alluvial fans or deltas. In each case they mark deposition at ice-sheet margins where sediment discharges from sub-glacial tunnel exits into the immediate proglacial environment (Banerjee & McDonald, 1975; Saunderson, 1977; Cheel, 1982; Thomas, 1984; Henderson, 1987; Diemer, 1988; Hedbrand & Amark, 1989; Mäkinen, 2003).

A third type of esker consists of multiple, anastamosing or braided, ridges. They are commonly shorter, usually less than a few kilometres, and much wider, up to a kilometre, than single ridge or beaded eskers. The best examples in Britain occur at Carstairs in central Scotland (Huddart & Bennett, 1997; Thomas & Montague, 1997) and at Flemington in north-east Scotland (Auton, 1992). This type of esker is less well documented and their origin is controversial. One hypothesis is that they form subglacially as a response to catastrophic flooding in rare, high-magnitude events that either destroy or significantly modify previous subglacial drainage systems (Gorrel & Shaw, 1991; Brennand, 1994; Fard, 2003). An alternative explanation involves deposition in large supraglacial stream channels (Bennett and Glasser, 1996). These are open channel systems, flowing on the surface of

the ice and laterally constrained by marginal ice-walls. They have been observed on the margin of Holmstrombreen in Svalbard (Boulton and Van der Meer, 1989), at the margins of the Woodworth Glacier (Flint, 1970) and the Casement Glacier, Alaska (Price, 1966), and form when flood events cause abandonment of braided stream channels. The sediment-fill in abandoned channels inhibits ablation of underlying ice, but debris-free ice on either side of the channel will melt more rapidly. The result is a ridge of ice-cored channel-fill sediment, standing on the surface of the glacier. Successive shifts in the course of supraglacial streams may generate a series of such ridges and as the ice margin ablates they become superimposed upon underlying substrate as a series of braided ridges. A modern analogue of this model has been provided by Huddart *et al.* (1998) from Svalbard.

Two independent investigations (Huddart & Bennett, 1997; Thomas & Montague, 1997), have invoked a supraglacial model to explain the complex, anastamosing, sub-parallel ridges at Carstairs, Southern Scotland. In the absence of sections through the ridges, however, both papers concentrated on the morphology of the ridges and the sedimentology of peripheral features, which consists mainly of sandur, subaqueous fan and delta sediments. Although differences in emphasis occur, especially the role of large, pro-glacial lake basins in triggering esker sedimentation (Thomas & Montague, 1997), both contributions argue that the ridges were formed as a response to supraglacial sedimentation.

The Carstairs ridges form part of a belt of glacial depositional landforms and sediments that extends from Douglas in the west for almost 100 km north-eastwards towards Edinburgh (Fig. 1). Thomas and Montague (1997) argued that the deposits of this belt formed in an interlobate sediment sink that collected drainage from the uncoupling margins of Scottish Highland and Southern Uplands ice-lobes during Late Devensian deglaciation. Similar interlobate zones, each associated with extensive ice-marginal sedimentation, have been identified during retreat of the Irish Ice Sheet (Warren & Ashley, 1994), the Finnish portion of the Scandinavian ice-sheet (Punkari, 1995), the Laurentide ice-sheet (Veillette, 1986; Gustavson & Boothroyd, 1987) and a modern analogue from Vegbreen, Svalbard (Huddart *et al.*, 1998). The southern Scottish interlobe includes the classic Carstairs esker system (Merritt *et al.*, 1983;

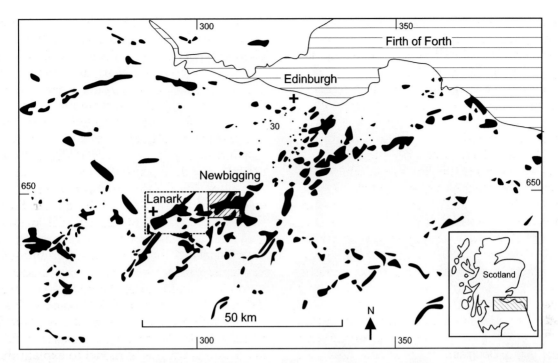

Fig. 1 Distribution of sand and gravel deposits between Lanark and Edinburgh, southern Scotland (after Sutherland, 1991). Cross hatching shows area examined. Pecked box shows adjacent area examined by Thomas and Montague (1997) and Huddart and Bennett (1997). Inset shows location.

Gordon, 1993; Huddart & Bennett, 1997; Thomas & Montague, 1997) and a variety of other esker types in the Douglas valley, to the south-west of Lanark, to the north-east and south-east of Newbigging, between Dolphinton and West Linton and at Romannobridge, together with extensive basins of glaciolacustrine sediment and areas of ice-marginal kame-moraines (Fig. 2).

The ridge system at Newbigging is similar to that at Carstairs, although on a smaller scale, and displays multiple parallel ridges. Whereas the internal architecture of the Carstairs esker ridges is now unexposed, the Newbigging ridge is quarried and allows the structure and internal composition to be examined. The aim of this paper is therefore is to use these exposures to document the morphology and sedimentology of the Newbigging ridge system providing data with which to constrain the formation of similar, multiple ridge esker systems.

GEOMORPHOLOGY AND LANDFORM ASSEMBLAGES

The Newbigging ridge system occupies part of the floor of the South Medwin valley, which drains south-westwards for some 13 km from the flanks of the Pentland Hills towards its junction with the River Clyde (Fig. 2). The valley is two or three kilometres wide but narrows at its western end to a kilometre, where the ridge system occurs. The floor is at an elevation of some 215 m O.D. with steep valley side slopes rising to over 500 m. At the eastern end of the valley, around Garvald, a prominent dry col, the Dolphinton Gap, runs south to take drainage into the River Tweed. A geomorphological map of the area is shown in Figure 3A and is derived from ground mapping at a scale of 1:10,000, supplemented by air photographs. The central portion of the ridge has been almost entirely removed by quarry operations and was mapped from aerial photographs, older 1:10,000 maps and from detail in Jenkins (1991). The area comprises three spatially distinctive, morphological assemblages (Fig. 3B).

Assemblage A: kame topography. This assemblage comprises a morphologically complex set of irregular, low amplitude (<10–15 m) almost randomly distributed ridges, mounds, basins and shallow channels. The assemblage occurs in two separate areas. The first, to the west (A1, Fig. 3B) covers the low, open country south of Newbigging. The

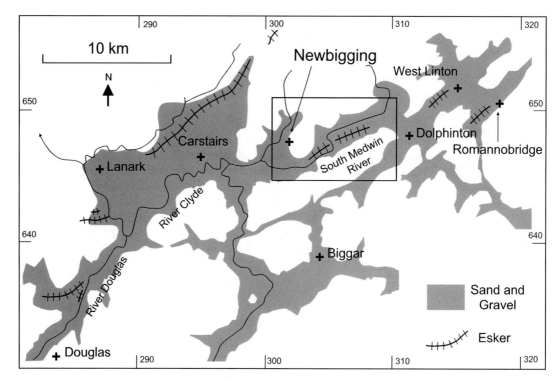

Fig. 2 Distribution of sand and gravel in the area between West Linton and Douglas, showing location of esker systems mentioned in the text and location of the Newbigging area.

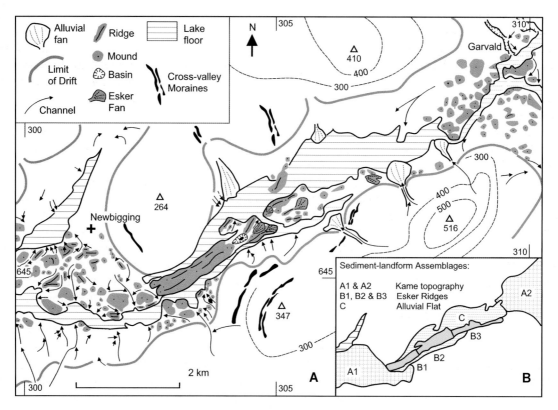

Fig. 3 (A) Geomorphological map of the Newbigging area. (B) Distribution of landform assemblage zones. A: kame topography; B: esker ridges; C: alluvial flat.

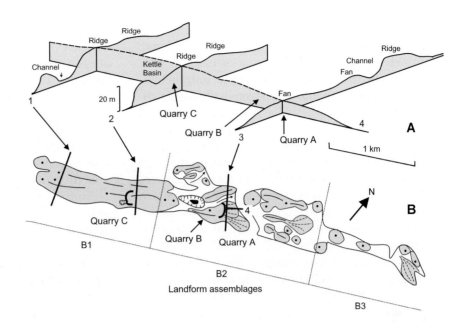

Fig. 4 (A) Cross-profiles through the Newbigging ridge system. (B) Location of cross-profiles and relationship with landform assemblages.

South Medwin River cuts sharply through these ridges and mounds and is flanked by extensive Holocene river alluvium. The second area occupies the eastern portion of the South Medwin valley around Garvald (A2, Fig. 3B). This area is cut through by the South Medwin river as a narrow flat-floored channel that divides west of Garvald. The northern arm carries the present South Medwin river westwards whilst the eastern arm turns southeast, via the deeply entrenched and flat-floored Garvald Burn into the Dolphinton Gap.

Assemblage B: esker ridges. This forms a complex ridge system, some five kilometres long, trending southwest to northeast and running slightly oblique to the centre line of the South Medwin valley (Fig. 3B). At the western end the ridge starts abruptly from within the ridges and mounds of landform assemblage A, south of Newbigging, and occupies the centre of the valley. Its direction progressively shifts eastwards, however, to cross the valley floor and impinge upon its southern flank. The length of the ridge can be divided into three separate sub-assemblages (B1–B3), running west to east.

Sub-assemblage B1. In the west the main ridge complex consists of a very prominent, multiple, linear ridge structure, some two kilometres in length, up to 30 m in height and 500 m wide. Two surveyed cross-profiles (Fig. 4, Profiles 1 and 2) reveal two major ridges separated by either a deep channel

(Profile 1) or kettle basin (Profile 2). The ridge to the north is broad and composed of two subsidiary ridges, whilst that to the south is narrow and sharp-crested.

Sub-assemblage B2. Eastwards the main ridge complex extends a further two kilometres but is lower in height and consists of two or three discontinuous and slightly bifurcating linear ridges, up to 15 m high, separated by, or passing eastwards into, shallow east-sloping fans. Each fan has a steep slope to the west and a gentle slope to the east. Two surveyed profiles are shown in Figure 4 (Profiles 3 and 4). Profile 3 runs normal to the ridge trend and shows two, shallow, intersecting fans separated from a sharp-crested ridge to the north by a channel. Profile 4 runs down the front face of the southern-most fan.

Sub-assemblage B3 Further to the east, the main ridge system breaks up into a single line of low amplitude, often well-separated, irregular ridges and mounds, over a kilometre in length and terminated by a large, radially sloping fan that grades out into alluvial sediment on the floor of the South Medwin valley.

Assemblage C: alluvial flat. This assemblage flanks both sides of the main esker ridge system, but particularly to the north, and consists of extensive flat, ill-drained terrain at an approximate general elevation of 215 m OD (Fig. 3B), underlain by Holocene peat and alluvium.

In addition to these three major assemblages there are a number of other significant landforms. Prominent amongst them are a series of five alluvial fans that drain from deeply entrenched bedrock gullies cut into the steep valley side of the South Medwin valley (Fig. 3A). In all cases the toes of the fans merge out into the alluvial flat valley floor. On both valley side slopes are a series of low-amplitude moraine ridges that run downslope, oblique to the contours, in an arcuate pattern (Fig. 3A). Rarely more than a few metres in height, they are composed of bouldery diamicton on bedrock and occur either singly or as double, sub-parallel ridges. They die out downslope at the junction with kame topography on the valley floor. At a number of locations fragmentary valley-side terraces occur at elevations of approximately 235 m OD, some 20 m above the alluvial floor. Many of the numerous rock or drift cut drainage channels (Fig. 3A) feeding into the valley floor terminate down-slope at approximately this elevation (McMillan et al., 1981).

STRATIGRAPHY

McMillan et al., (1981) listed the details of sixteen boreholes in the Newbigging area and their locations are shown in Figure 5, together with vertical log profiles. Most boreholes reached at least 15 m in depth and many penetrated bedrock. The relatively close spacing of the boreholes allows reconstruction of the bedrock surface, the extent and thickness of sedimentary units, the relationship between them, and the association between sediments and overlying landforms.

Figure 6 is a fence diagram of the distribution of the major sediment types through the Newbigging area, derived from the borehole records. The axis of the diagram is parallel to the centre line of the South Medwin valley with offsets across the wider portions of the valley floor. The pattern shows that the central area of the valley, underlying the main esker ridge and the areas of alluvial sediment peripheral to it, forms a shallow, overdeepened rock basin. This basin is bounded up- and down-valley by rock bars underlying the two areas of kame topography (landform assemblages A1 & A2). A down-valley rock bar to the west is particularly prominent with a bedrock high of 218 m OD near

the centre of the valley (Borehole 2, Fig. 6). To the east of the main esker ridge system the basin deepens and bedrock lies below 185 m OD in boreholes 10 and 14 (Fig. 6) in the centre of the basin. The bedrock surface then rises rapidly to the east to reach 215 m OD in borehole 16 (Fig. 6), located in the centre of the kame topography of landform assemblage A2 around Garvald. The maximum proven amount of overdeepening in the basin is thus of the order of 30–40 m.

From the borehole records, four major glacigenic sediment types can be distinguished. Diamicton occurs across much of the floor, especially its eastern half, and across the foot of valley side-slopes, as a mantle lying directly above bedrock (Fig. 6). Deeply buried by younger sediment, the diamicton averages 3–4 m in thickness, but up to a maximum of 6 m occurs in the centre of the basin (Borehole 7, Fig. 6). It is not exposed at the surface but drilling records describe it as a stiff, reddish brown to brown clay, with ill-sorted angular to well-rounded clasts up to cobble size (McMillan et al., 1981). The diamicton is interpreted as a subglacial till. Gravel, ranging from granule to cobble and occasionally boulder size, underlies most of the main esker ridge system and the narrow western portion of the basin. The eastern portion, in contrast, is relatively gravel free except for tongues of gravel associated with alluvial fans that impinge into the basin on its southern side (Borehole 12, Fig. 6). Further gravel underlies the kame topography at the eastern end of the basin. Coarse and medium *sand* occurs extensively in the central and eastern portions of the basin in sequences up to 15 m thick (Borehole 10, Fig. 6), but is rare in the western portion beneath the main esker ridges. The gravels and coarse and medium sands are together interpreted as glacio-fluvial deposits. *Fine sand, laminated silt and clay*, sometimes containing isolated 'oversize' clasts, occur in sequences up to 10 m thick throughout the basin, but particularly in the central and eastern portions (Borehole 13 and 14, Fig. 6). They are interpreted as glaciolacustrine sediments (McMillan et al., 1981).

The overall stratigraphic succession in the basin shows a simple upward-fining passage from diamicton, through gravel and sand into silts and clays. Superimposed upon this is a down-basin transition from gravel-dominant in the west, beneath

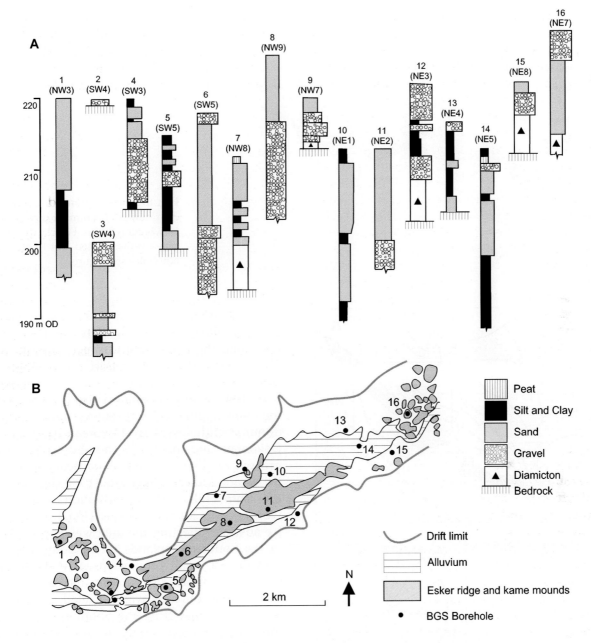

Fig. 5 Boreholes in the Newbigging area. (A) Borehole logs. Borehole records from McMillan *et al.* (1981). Numbers in brackets are British Geological Survey borehole reference numbers for Sheet NT04. (B) Borehole locations showing relationship to main ridge and alluvial sediment.

the main esker-ridge, to sand- and silt-dominant in the east, beneath the flat floor of the valley. This is consistent with the progressive filling of a shallow lake basin by sediment input from the west, which delivered coarser sediment to the more proximal locations and finer sediment to the more distal.

PROVENANCE AND ICE DIRECTION

The solid geology of the district around Newbigging is shown in Figure 7A. The area to the south-east of the Southern Upland Fault is dominated by Lower Palaeozoic greywackes with subordinate shale, chert and conglomerate; that to the north-west

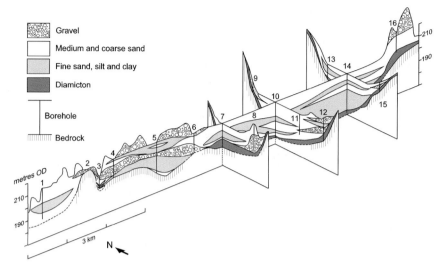

Fig. 6 Fence diagram showing distribution of sediment through the Newbigging basin. For clarity some minor units omitted. For location of boreholes see Figure 5.

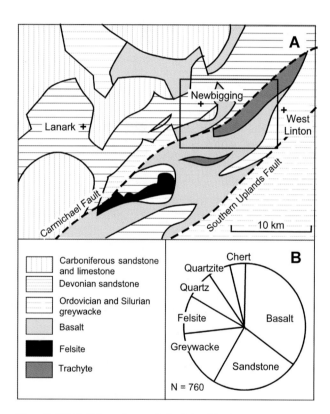

Fig. 7 (A) Solid geology of the Newbigging area. (B) Provenance distribution of gravel clasts from the Newbigging ridge. Combined data from authors and McMillan *et al.* (1981).

by Devonian conglomerates and sandstones, Carboniferous sandstones, shales and limestones and extrusions of andesite, trachyte, rhyolite and basalt. A large intrusion of Devonian felsite outcrops in the Tinto Hills, ten kilometres south-west

of Newbigging (McMillan *et al.*, 1981; Shaw & Merritt, 1981). The South Medwin valley itself lies along the line of the Carmichael Fault, with Carboniferous basalts and trachytes to the south-east and Devonian sandstones to the north-west.

A clast lithological analysis from three sites in the Newbigging esker system shows that the dominant lithologies are Devonian basalts (35%), Devonian sandstones (23%) and Lower Palaeozoic greywacke (15%) (Fig. 7B). A particularly distinctive erratic indicator is felsite (10%) derived from the Tinto Hills. These distributions confirm the views of McCall and Goodlet (1952) that the last ice advance affecting the area was from a source in the Southern Uplands to the southwest. The presence of small proportions of meta-quartzite (6%) in all samples also supports the views of these authors that the area was glaciated by Highland ice prior to that from the Southern Uplands as this rock type occurs only north of the Highland Boundary Fault to the north-west. Clast lithology, together with the orientation of streamlined drift landforms, the direction of striae (McMillan *et al.*, 1981), the flow direction of meltwater channel systems (Sissons, 1961), the arcuate pattern of moraines and the north-east fining of sediment, all confirm that the most recent ice moved from the south-west.

SEDIMENTOLOGY

Extensive quarrying through the Newbigging ridge system permits the sedimentology and structure of the deposits to be investigated in some detail. A recently abandoned quarry (Quarry A) occurs in the central portion of the ridge and a major working quarry (Quarry C) lies some 800 m to the west (Fig. 4). The area between has largely been removed but detail of former exposure (Quarry B) is given in Jenkins (1991). In each quarry continuous serial sections were surveyed and a series of vertical log profiles recorded (Fig. 8). The lithofacies classification used was modified from Eyles *et al.* (1983) and Eyles and Miall (1984) and the facies codes, descriptions and proportion of total bed thickness logged in each exposed quarry are given in Table 1. Type facies are illustrated in Figures 9 and 10.

Quarry A

Quarry A (NT045460) occurs within esker landform sub-assemblage B2 and the surrounding morphology consists of two discontinuous, slightly bifurcating linear ridges, 15 m high, separated by, or passing eastwards into, a series of shallow, wedge-shaped fans, sloping east. The quarry is cut at right angles to the axis of the southern most fan. Figure 8 shows a serial section of the face and three logs measured through it.

The sequence is dominated by sand and includes trough cross-stratified sand (St), horizontally stratified sand (Sh), parallel-laminated sand (Sl) and rippled sand (Sr) facies that, together, make up

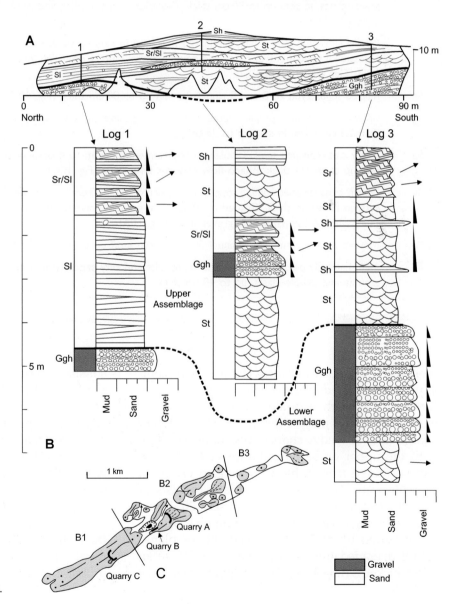

Fig. 8 Quarry A, Newbigging. (A) Serial section. (B) Vertical log profiles. (C) Location of quarry and relationship to landform assemblages.

Table 1 Lithofacies types, descriptions and percentage thickness of sediment types derived from vertical profile logs in quarries A and C. Facies codes after Eyles *et al.* (1983)

Code	Description	% Quarry A	% Quarry C
Gm	*Massive Gravel.* Boulder, cobble and pebble gravel in laterally persistent, stacked sets up to 4.5 m thick. Erosional bases (Fig. 9A).	1.7	21.7
Ggc	*Graded gravel (Channelled).* Pebble gravel grading upwards into coarse and medium sand, in sets up to 1.2 m thick, with erosionally channelled bases (Fig. 9D).	–	10.9
Ggh	*Graded gravel (horizontal).* Pebble gravel grading upwards into granule gravel in sets up to 3 m thick. Each set composed of numerous graded sub-sets. Horizontally stratified. Conformable bases. (Fig. 9A).	10.0	22.8
Sh	*Horizontally stratified sand.* Coarse to medium sand, with isolated pebble gravel clasts, in sets up to 80 cm thick. Occasional thin grading and rippled units (Fig. 9C).	3.3	6.3
Sl	*Parallel laminated sand.* Thinly laminated fine sand in sets up to 90 cm thick. Occasional pebble gravel clasts and granule gravel lines. Commonly caps facies Ggc (Fig. 10 B & C).	19.2	10.2
Sr	*Rippled sand.* Type A, B and S ripples in sets up to 80 cm thick (Fig. 10 B).	38.8	6.2
Sg	*Graded sand.* Thin, multiple coarse to fine sand graded co-sets, usually less than 2.5 cm thick, in sets up to 75 cm.	3.1	5.9
Sp	*Planar cross-stratified sand.* Tabular cross-stratified, medium to coarse sands in sets up to 60 cm thick. Palaeo-flow towards 040–060° (Fig. 10C).	–	2.1
St	*Trough cross-stratified sand.* Troughs up to 60 cm wide, 20 cm deep in multiple sets up to 80 cm thick. Graded upwards from granule gravel to medium sand, with occasional pebble gravel clasts (Fig. 10A).	22.7	1.9
Sm	*Massive sand.* Asymmetric channel fills in coarse sand up to 40 cm thick.	–	0.6
F	*Mud.* Laterally persistent, laminated sandy silts and silty sands with occasional granule and pebble clasts, up to 40 cm thick.	1.2	4.0

more than 80% of log thickness (Table 1). Horizontally stratified graded gravel (Ggh) contributes a further 10%. Architecturally, two separate sediment assemblages may be defined. Each consists of a large lens, or wedge shaped, packet of sediment with the upper on-lapping the lower to the east and separated from it by a low-angled erosional disconformity. The lower assemblage consists of trough cross-stratified sand (St), in sets, 20–50 cm thick, showing current directions running east. This is succeeded by multiple, stacked sets of planar stratified, graded gravels (Ggh), averaging 30–40 cm in set thickness (Log 3, Fig. 8). The upper assemblage

consists predominantly of trough cross-stratified sand (St), rippled sand (Sr), parallel laminated fine sand and silt (Sl) and laminated muds (F) in a generally fining upwards succession (Logs 1 & 2, Fig. 8). The muds contain sporadic gravel clasts and clusters (Fig. 10C) showing characteristic dropstone structures (Thomas & Connell, 1985). Palaeocurrent indicators, derived from ripples, show palaeo-transport towards the east or east-north-east.

The morphology into which this quarry is cut, the architecture of the sediment packages, the sediment types and structures, and the occurrence of dropstones, all suggests deposition in a subaqeous

Fig. 9 Sedimentary facies types: 1. (A) Massive gravel (Gm) below trowel and graded gravel (Ggh) above (Quarry C). (B) Graded gravel (Ggh) interbedded with finer units (Quarry C). (C) Horizontally stratified medium/coarse sand (Sh) overlying planar cross-stratified coarse sand (Sp) (Quarry A). (D) Graded gravel (Ggc) with channelled base, grading up to parallel laminated sand (Sl) (Quarry C).

Fig. 10 Sedimentary facies types: 2. (A) Trough cross-stratified coarse sand and granule gravel (St) (Quarry A). (B) Parallel-laminated sand (Sl) and rippled sand (Sr) on fan distal slope (Quarry A). (C) Parallel-laminated fine sand (Sl) and silt with dropstones in distal fan (Quarry A). (D) Small-scale faulting indicating meltout of underlying ice blocks (Quarry C). (E) Part of the core of boulder gravels (Quarry B). (F) Dropstones and dropstone pod in silty fine sands, silts and parallel-laminated and rippled fine sands (Quarry B).

fan. In this environment sedimentation occurs when a subglacial tunnel abuts directly into standing water. Rapid reduction in flow velocity causes immediate deposition of the coarser-grained sediment and a fan is built outward from the tunnel exit (Rust & Romanelli, 1975; Banerjee & MacDonald, 1975; Thomas, 1984; Cheel, 1982; Henderson, 1987; Diemer, 1988; Warren & Ashley, 1994; Paterson & Cheel, 1997; Russell & Arnott, 2003). Such fans are dominated by mass-flow, underflow and suspension sedimentation, and are characterised by normally-graded gravels in proximal environments and trough cross-stratified sands, parallel-laminated sands and rippled sands in distal environments. They commonly show rapid lateral and down-current facies transitions due to dispersion of flow entering the lake and to periodic variation in sediment input arising from seasonal fluctuations in meltwater flow. In this case the development of two laterally off-lapping packages, separated by a disconformity, are probably indicative of subaqueous fan-switching in response to a high-magnitude discharge event.

The trough cross-stratified sands (St) at the base of the lower package represent the migration of sinuous and straight-crested dune bedforms on the floor of distributary channels or troughs (Shaw, 1975). Their upward passage into graded gravel sets (Ggh) indicates an increase in flow velocity, a change to hyperconcentrated debris flow and a more proximal location. Gravel deposition occurred as cohesionless clast dispersion supported by fluid turbulence at the base of individual turbidite flows (Postma & Roep, 1985; Postma, 1986). The absence of reversed grading at the base of the flow deposits suggests a low slope (<10°) to the fan (Walker, 1975).

The overall fining-upwards succession in the upper package, from trough cross-stratified sands (St) at the base, through thinly parallel-laminated sand (Sl) into rippled fine sand (Sr) indicates declining underflow and an increase in deposition from suspension, caused by increased distality from the retreating tunnel mouth. The alternation between cross-stratified sand (St) and horizontally stratified sand (Sh) and between rippled sand (Sr) and parallel-laminated sand (Sl) demonstrates that flow was unsteady, with periods of high flow velocity alternating with phases of decreased flow velocity when dunes could form and migrate.

Quarry B

Along the axis of the main ridge formerly connecting Quarry A to Quarry C, but now mostly removed, Jenkins (1991) described a core of 'Boulder Drift'. This was 10–20 m wide, at least 5 m thick and extended a few hundred metres east towards the apex of the fan exposed in Quarry A. The 'Boulder Drift' consisted of very poorly sorted, large cobble and boulder gravel, up to a maximum clast size of 1.5 m, set in a matrix of granule gravel, fine sand and silt (Fig. 10E & F). The boulders were described as erosively scoured into underlying sands, with lateral contacts often steep or overhanging, and overlain by laminated silts and fine sands containing outsize clasts. Although largely structureless, some sections displayed a crudely defined, low-angled cross-bedding, on a set scale of 2 m, running parallel to the ridge and indicating flow to the north-east.

Similar boulder gravels have been observed in other ice-marginal settings. They have been interpreted as a response to dumping of supraglacial debris into proximal outwash sediment (Rust & Romanelli, 1975); to the slumping of supraglacial debris into an ice-contact subaqeous environment (Eyles *et al.*, 1987); to rapid reduction in flow competence at a tunnel exit (McCabe & O'Cofaigh, 1994); to the reworking of basally-sheared debris by englacial streams (Aitken, 1995); and to tunnel deposition (Warren & Ashley, 1994). Similar gravels have locally been described by Goodlet (1964) and Laxton (1980) from a steep-sided, double ridge in the Carstairs esker system. The occurrence of these boulder gravels at a location immediately up-ice of the apex of the subaqueous fan of quarry A, and along the line of a former single esker ridge connecting it, suggests that they were deposited either within a feeder tunnel, or at its immediate exit, by loss of flow competence on entering standing water. The occurrence above them of sands and silts with outsize clasts is indicative of the rapid onlap of more distal suspension sediment on the fan as the tunnel portal retreated.

Quarry C

Quarry C (NT033452) is located towards the western end of the ridge system, within landform assemblage B1, a little east of cross-profile 2

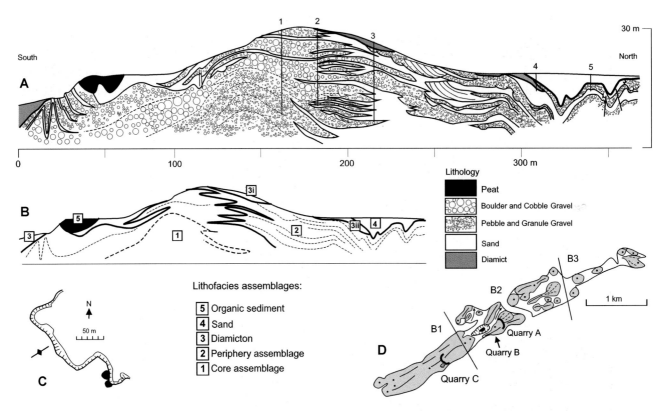

Fig. 11 Quarry C, Newbigging. (A) Serial section showing general distribution of sediment types and location of vertical profile logs. Vertical exaggeration × 3. (B) Distribution of principal facies assemblages. For discussion see text. (C) Map of exposed faces. (D) Location of Quarry C.

in Figure 4. At this point the main ridge is 500 m wide and 30 m high, with steeply sloping margins, particularly to the south. The surface is compound and consists of two major parallel ridges, separated by a number of enclosed, peat-filled kettle basins and shallow channels. The quarry cuts at right angles through the southernmost ridge and a serial section, some 350 m long, based on field mapping and a photo-montage is shown in Figure 11. Five vertical log profiles taken through the exposure are shown in Figure 12. Gravels predominate, with massive gravel (Gm) and graded gravel (Ggh & Ggc), together making up more than 55% of logged thickness (Table 1). Because of the generally coarse nature of the sediments, and the dominance of massive or graded gravel units, palaeocurrent indicators are rare and of insufficient number to establish clear flow directions. Trough cross-stratified sand sets, rippled sands and rare planar cross-bedded sands, however, show flow directions in an arc ranging from north-north-east to south-east, with a mean direction running

approximately sub-parallel to the trend of the main ridge, towards 050°.

Sedimentary assemblages

The overall internal architecture comprises an arched core of gravel, some 100 m wide by 30 m high, coarsening upwards to the ridge crest and southern margin and fining laterally to the northern margin. The axial plane of the arch is not coincident with the crest of the main ridge, but is 50 m to the south (Fig. 11A). The trend of the axis, however, is parallel with the ridge direction. Both limbs of the arch are symmetrical and dip outwards between 10 and 15° in the core, but flatten towards the margins. The sequence can be divided into two major and three minor facies assemblages, the relationships between which are shown in Figure 11B (assemblages 1–5).

Assemblage 1: Core assemblage. This assemblage comprises the sediment exposed in the core of the arch and on its southern limb (Unit 1, Fig. 11B). It

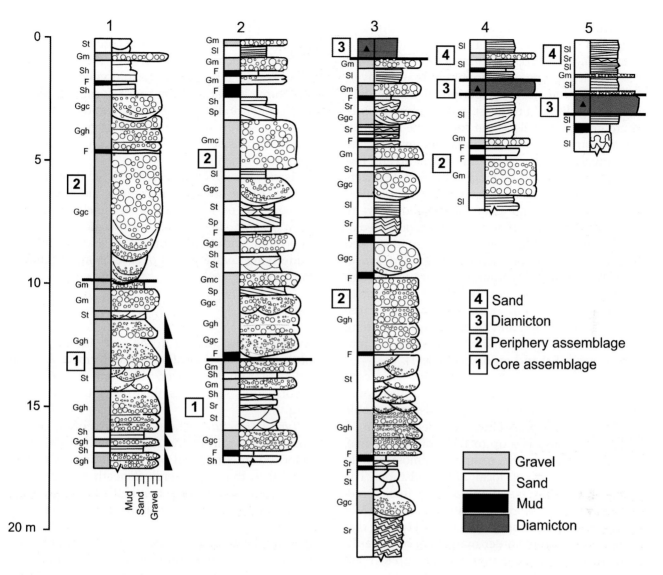

Fig. 12 Vertical log profiles through Quarry C. For location of logs see Figure 12. For description of facies codes see Table 1.

is dominantly composed of coarse, graded gravel, in either channels (Ggc) or horizontal sheets (Ggh), and coarse, massive gravel (Gm). Minor proportions of parallel-laminated sand (Sl), horizontally stratified sand (Sh), rippled sand (Sr) mud (F) and graded sand (Sg) also occur (Table 1). In the core, an overall, upward shift of facies type from horizontally stratified, graded gravel (Ggh), through channelled, graded gravel (Ggc) into massive gravel (Gm) is discernible (Log 3, Fig. 12). The horizontally stratified gravel (Ggh) comprises pebble gravel grading upwards into granule gravel in sets up to 3 m thick. Each set is composed of numer-

ous graded sub-sets. The channelled gravel (Ggc) comprises pebble gravel grading upwards into granule gravel and coarse and medium sand, in sets up to 1.2 m thick. Each set has an erosionally channelled base and some internal trough cross-stratification. Channels are in the order of 5–10 m in width. The massive gravel (Gm) comprises boulder, cobble and pebble gravel in laterally persistent, stacked sets up to 4.5 m in thickness. Each set is separated from the next above either by an erosional disconformity or by thin, laterally persistent clays and parallel-laminated silts (F). The Gm gravels are particularly well developed in the upper part of the southern limb of the arch, where they are commonly composed of boulder gravel.

Assemblage 2: Peripheral assemblage. This comprises the thickening sequence of sands and gravels exposed on the northern periphery of the arch (Unit 2, Fig. 11B). In contrast to the southern limb, where graded gravels in the core of the arch pass upwards and outwards into massive gravels, the core gravels on the northern limb break up and fine laterally over a short distance into discrete sheets of finer gravel separated by gravelly sands, parallel-laminated sands, and large sandy and gravelly troughs. The assemblage is dominantly composed of massive, pebble and cobble gravel (Gm) in stacked sets; large troughs of channelled, graded gravel (Ggc) that fine laterally from pebble gravel through granule gravel into pebbly sand; horizontally stratified sands (Sh); occasional planar cross-stratified sands (Sp); and frequent upward-fining transitions from parallel-laminated sand (Sl) into rippled sand (Sr) and mud (F) (Table 1).

Assemblage 3: Diamicton assemblage. This consists of two types. The first occurs draped over the crest of the main ridge and is disconformable upon the gravels of assemblage 1 below (Unit 3i, Fig. 11B). It comprises up to 1.5 m of very coarse, cobble to boulder gravel, composed almost entirely of angular blocks of local red sandstone. The matrix is variable, but dominantly a fine, sometimes laminated, sandy silt showing evidence of soft-sediment deformation. The second type of diamicton occurs on the outer margins of both flanks of the main ridge and comprises units up to 1.5 m in thickness (Unit 3ii, Fig. 11B). It consists of a poorly sorted sand and gravel set in a matrix of mud and showing poorly defined internal bedding. On the northern slope this diamicton lies conformably over the sands and gravels of assemblage 2 but is progressively overlain by a thick succession of sands (assemblage 4) farther down slope (Logs 4 and 5, Fig. 12). On the southern slope a thick sequence of crudely bedded, pebbly diamicton lies unconformably upon strongly folded boulder gravels of assemblage 1 and thickens rapidly down the slope.

Assemblage 4: Sand assemblage. On the northern margin of the ridge, the sands and gravels of assemblage 2 are deformed into a series of low amplitude folds. The folds are partially truncated by a disconformable sequence of parallel laminated sand, rippled sand and occasional muds that form a series of basin fills within the core of underlying synform structures (Unit 4, Fig. 11B). Upwards, however, the influence of the intervening antiform structures is lost and the sands form a conformable sequence dipping north-east at low angle and thinning south-west against the ridge margin. On the upslope side, these sands locally overlie diamicton of assemblage 3.

Assemblage 5: Organic assemblage. Beneath the surface of the southern margin of the main ridge a sequence of up to 2.2 m of black peats and grey organic silts fill two semi-coalescent basins deposited within the core of a shallow fold in underlying coarse gravel of assemblage 1 (Unit 5, Fig. 11B).

Structure

Except for the arching in the core, the main part of the sequence is relatively undeformed. At both ridge margins, however, there are a variety of deformation structures that fall into two types (Fig. 13)

Type 1 structures. These structures occur on the southern margin of the ridge only and are characterised by structural orientation approximately normal to the ridge trend. Beneath and immediately to the south of the two kettle basins, rapidly alternating sequences of massive pebble to boulder gravel (Gm), and thin parallel-laminated sands (Sl), are thrown into a series of tight, upright folds with a vertical amplitude of some 15 m and with limbs dipping up to 75° (Fig. 13). The axial trend of the folds is approximately normal to the trend of the main ridge (Fig. 13). The upper part of each fold is truncated by the unconformable base of overlying diamicton, dipping south. The geometry of these folds suggests that they were formed by active ice bulldozing marginal sediment as their stress direction, approximately south-south-east to north-north-west, is coincident with ice directions inferred from provenance, ridge-trend and meltwater-channel direction.

Type 2 structures. These structures occur at both margins, are smaller than the type 1 and are characterised by structural orientation parallel, or sub-parallel to the ridge trend. They occur on two scales. On the larger scale they include a series of folds exposed on the northern margin (Fig. 13). At

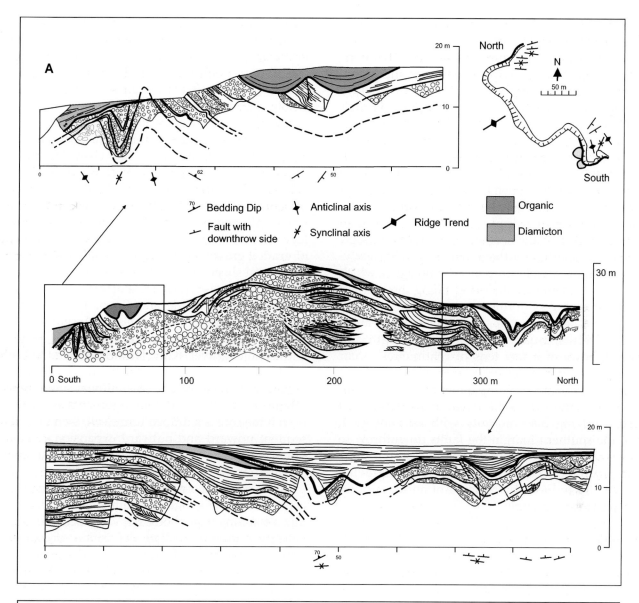

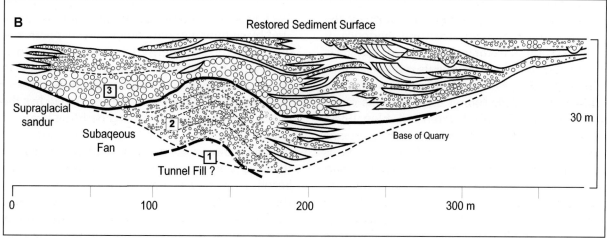

Fig. 13 The structure of the Newbigging ridge in Quarry C. (A) Tectonic structure with enlarged sections at the northern and southern end as shown in the inset. (B) Section (A) restored to original horizontal depositional surface to show distribution of tunnel, subaqueous fan and supraglacial sandur facies. Later organic sediments, diamictons, distal fan sediments and deformation structures removed for clarity. Vertical exaggeration of × 3.

this location assemblage 2 comprises approximately co-equal sets of massive pebble gravel (Gm) and parallel-laminated medium sand (Sh). From the crest of the main ridge these sediments dip at low angle to the north-east but are then tilted sharply downwards to the north at an angle of 40°. Further down-dip they are progressively deformed into a series of open, upright folds of low amplitude, striking approximately east-west, sub-parallel to the trend of the main ridge. This trend, and the low amplitude and open form of the folds, suggest that they were formed as sag basins, generated by melt of underlying dead-ice or the removal of an ice support from the northern margin of the ridge. This is supported by the geometry of the overlying, disconformable units of assemblage 4 which shows parallel-laminated sand filling the cores of the underlying synforms.

On the smaller scale, a number of high-angle normal faults, usually less than 2 m in length, and with throws of a few tens of centimetres, occur on both margins. On the northern margin these faults are orientated sub-parallel to the fold axes, show downthrow to the north and terminate at the overlying disconformity with assemblage 4. On the southern margin the faults downthrow to the south and are orientated parallel to the ridge margin. In both cases these faults suggest gravitational slip induced by removal of an ice support from the ridge margin.

Interpretation of Quarry C

Three models can explain the sedimentary sequences exposed in Quarry C. In a 'tunnel' model, sediment should comprise gravel deposited in a closed conduit. The continuous, steep-sided form of the main ridge, together with structural evidence of laterally confining, ice-marginal support, suggests tunnel deposition. Four arguments militate against this interpretation, however. First is the width of the ridge. Most single tunnel esker systems range fom 10 to 100 m in width (Warren & Ashley, 1994); at more than 500 m the Newbigging ridge seems too wide to have been structurally supported as a single tunnel, especially in thin, decaying, marginal ice. Second, the arched bedding in the core of the ridge, though architecturally characteristic of tunnel-fill that was subsequently warped by removal of marginal ice support, occupies less than

a fifth of the exposed ridge width, and there is no evidence for more than one tunnel. Third, normal faulting, usually indicative of the removal of lateral ice support, does not occur at the margins of the arched core, but at the margins of the ridge as a whole. Removal of marginal support therefore occurred after deposition of the whole complex, not after the deposition of the core assemblage. Fourth, the sediments are not characteristic of tunnel fill, which is normally composed of sub-horizontally bedded, longitudinally stacked, sheets of poorly sorted, clast-supported boulder to cobble gravel (Warren & Ashley, 1994). The core assemblage (Assemblage 1) is predominantly composed of graded gravel (Ggc & Ggh), in either persistent horizontal sheets or broad, cross-cutting and overlapping channels which are characteristic of mass-flow deposits (Postma and Roep, 1985; Postma, 1986). The peripheral assemblage (Assemblage 2) is predominantly composed of massive pebble and cobble gravel (Gm), horizontally stratified sands (Sh), and parallel-laminated sand (Sl) which are more characteristic of glaciofluvial successions (Thomas et al., 1985). All four arguments imply that even if the core is a draped tunnel-fill, then the subsequent upward and outward expansion of peripheral sedimentation has some other depositional explanation.

In a 'sub-aqueous' fan model, sedimentation occurs when a subglacial tunnel abuts directly into a standing water body. Rapid reduction in flow velocity causes immediate deposition of coarser-grained sediment and a fan is built out-ward from the tunnel exit (Rust & Romanelli, 1975; Banerjee & MacDonald, 1975; Thomas, 1984; Cheel, 1982; Henderson, 1987; Diemer, 1988; Paterson & Cheel, 1997; Russell & Arnott, 2003). Such fans are dominated by sediment gravity flows and suspension fall-out and are characterised by massive to graded gravels in proximal environments with a general basinward transition to traction deposits. Although the core assemblage (Assemblage 1) matches these characteristics, and the arched bedding bears similarity to a fan in strike section, other arguments militate against a full acceptance of this model. First, the ridge does not display a characteristic fan form. Second, the wider sedimentary sequences above the core, in the peripheral facies assemblage 2, are not characteristic of subaquous fan sedimentation. Thus, even if the core facies represent an original subaquous fan, issuing from a tunnel a little distance up-current from the present-day quarry face, then the wider spread of subsequent sedimentation requires some other explanation.

In a 'supraglacial' model, sedimentation takes place in a trough formed on the surface of the ice. As water enters the trough from a tunnel, pipe-flow gives way to open-channel flow, and normal glaciofluvial sedimentation results in an ice-walled trough that extends down-ice to the margin. During active sedimentation, the sedimentary body geometry would form a large trough, with the upper, depositional surface horizontal. As the ice melts down, however, and the base of the trough reaches the underlying substrate, this geometry would be inverted to form a broadly draped arch of sediment; thinner, and hence lower, at the margins, and thicker and higher, in the centre. This process would be accompanied by deformation at the margins of the trough as supporting walls wasted away. The overall architecture provides some support for this model, as does the structural evidence. Thus, the series of shallow, sag folds and accompanying normal faults structurally orientated parallel to the ridge indicate adjustment to decay and removal of lateral ice-support.

Sedimentary sequences deposited in large, open, supraglacial troughs are not well documented, although they are likely to be similar to those observed in proximal proglacial environments. These are characterised by deposition of migrating bars comprising predominantly massive gravel bar-core facies; planar cross-bedded sand and gravel, bar-front migration facies; trough cross-stratified sand and gravel channel facies; parallel-laminated and rippled sand channel-floor or bar-top facies; and a low occurrence of rippled sands, and laminated and massive mud-overbank flood facies (Ashley, 1975, Thomas *et al.*, 1985). These bear similarity to the peripheral assemblage of sediments (Assemblage 2), but are dissimilar to the core assemblage of sediments (Assemblage 1). Clearly, this model alone, whilst possibly explaining the upper, peripheral assemblage, does not explain the core assemblage.

A composite model

For the reasons given above, none of the three models provides a satisfactory explanation for

all the sedimentary characteristics seen in Quarry C, although each has components present. The sequence is best explained as a composite, and a model of the likely development is shown in Figure 14. This involves three stages and identifies a transition from esker tunnel-fill into a subaqueous fan fronting the tunnel portal and then into a supraglacial outwash sandur deposited in an ice-walled trough created by tunnel unroofing.

Stage A – Tunnel. At this stage rapid retreat from the previous still-stand position and fan lead to the continued deposition of a single tunnel-bound ridge towards Quarry C (Fig. 14A). No tunnel facies were recorded in Quarry C, although the core facies (Assemblage 1) appears to be draped over a narrow sediment body located below the floor of the quarry. Given that Borehole 6 (Fig. 5), located some 100 m east of the quarry, failed to reach bedrock at 195 m OD and that the base of the quarry is at approximately 205 m OD, there is at least 10 m of unexposed sediment below the quarry floor that could comprise the tunnel facies.

Stage B – Subaqueous fan. Assuming that the ice margin stabilised close to the location of Quarry C, a further subaqeous fan developed at the tunnel portal and buried the underlying tunnel facies (Fig. 14B). The resulting sequence of sediments, dominated by coarse graded gravels, in either channels or sheets, is represented by the core assemblage exposed in Quarry C.

Stage C – Supraglacial sandur and subaqeous fan. During continued still-stand at the position of Quarry C the ice margin thinned and the feeding tunnel became unroofed to form a short ice-walled channel on the surface of the ice (Fig. 14C). Unroofing probably occurred as a response to both normal lowering of the ice surface, and the control placed on the height of the feeding tunnel due to the prominent rock bar beneath the ice to the rear of the margin. The unroofed channel would have abutted directly into the fronting lake margin. Consequently, a broad ice-walled trough containing supraglacial sandur sediment developed above the water level and a large subaqueous fan below it. Thus, the upward passage from the core assemblage to the peripheral assemblage in Quarry C represents the break-point between

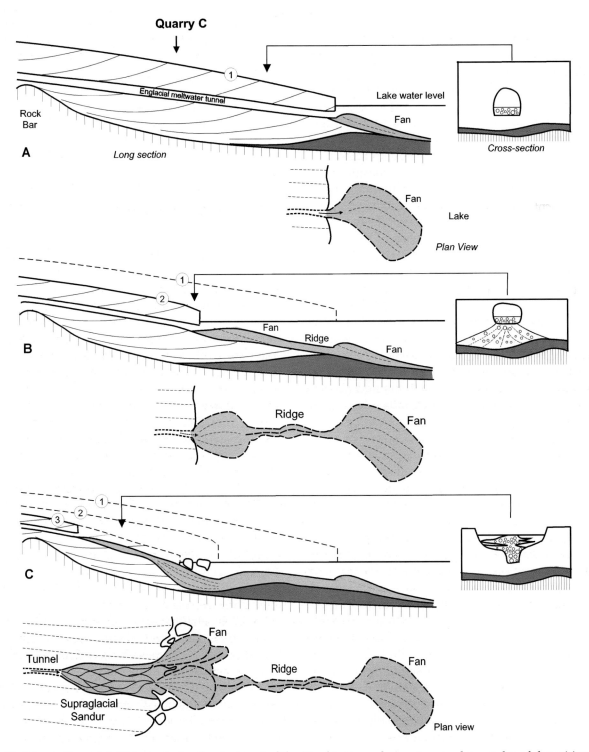

Fig. 14 Schematic model of stages in the development of the Newbigging esker system to show styles of deposition. For explanation see text.

supraglacial, fluviatile sedimentation and sub-aqeous fan sedimentation. Rapid reduction in flow velocity on entering the lake would cause immediate deposition of the coarser-grained sediment in the proximal parts of the fan by mass-flow and underflow, and deposition of the finer-grained sediment in the distal parts of the fan by underflow and suspension. Such sediment commonly shows rapid lateral and down-current facies transition due to dispersion of flow entering the lake and to periodic variations in sediment input arising from seasonal and other episodic fluctuations in flow. This is consistent with the sedimentology of the core facies assemblage (Assemblage 1), which is dominated by graded gravels deposited in shallow troughs and laterally persistent sheets. As the fan prograded, sedimentation would have been gradually replaced upwards by more shallow water facies and ultimately by subaerial outwash sedimentation. This is consistent with the sedimentology of the peripheral facies assemblage (Assemblage 2), which is dominated by gravel channels and bars. Figure 13B shows a restoration of the supraglacial sandur to its original horizontal depositional surface and the upward relationship between the tunnel, fan and supraglacial sandur facies.

As the ice retreated from this prominent still-stand position, removal of lateral support from the trough caused deformation at the margin. In addition small ice-marginal fluctuations appear to have slightly tectonised the southern margin of the sediment body. During and after this process reworking on the subaqueous slopes of the fan resulted in the deposition of diamictons as subaqueous flow tills. As the ice retreated to the west, and the confining ice both surrounding and beneath the supraglacial sandur/subaqeous fan complex melted, sedimentation became more distal. This led to the deposition of discrete lobes of sand across the partially abandoned, but still submerged proximal fan, by underflow and suspension. This is consistent with the sedimentology of the sand facies (Assemblage 4), which is dominated by parallel-laminated sand, rippled sand and occasional mud. These distal fan sediments partially blanket and infill tectonically collapsed lateral margins of the main ridge.

PALAEOGEOGRAPHICAL RECONSTRUCTION

From the morphological, stratigraphic and sedimentological data, and the interpretation of the ridge structure as resulting from an ordered sequence of esker depositional sub-environments, it is possible to devise a palaeogeographic reconstruction of the events that occurred during the development of the Newbigging esker ridge system. A number of these events are illustrated in Figure 15.

Stage A. Retreat to the south west of the Late Devensian Southern Uplands ice-sheet progressively uncovered a bedrock topography comprising northeast- to southwest-orientated rock ridges and intervening overdeepened basins. At one stage in deglaciation, a lobate ice-margin occupied the South Medwin valley and terminated at its eastern end around Garvald (Stage A, Fig. 15). Proglacial meltwater flow ran northeastwards as a braided outwash stream system that turned south through the Dolphinton Gap to drain towards the headwaters of the present River Tweed. This outwash system deposited large thicknesses of sediment across the valley floor and underlying buried ice. As the ice-margin retreated this outwash surface was progressively abandoned, the underlying buried ice melted out and a complex kame topography resulted.

Stage B. On retreat from stage A, an overdeepened rock basin, generated by earlier glacial erosion, was uncovered and a small ice-frontal lake was impounded, trapped against the rock-bar at Garvald (Stage B, Fig. 15). At its maximum the lake filled to a height of at least 235 m O.D. This is confirmed by the occurrence of probable lake-shoreline terraces and by the termination of valley-side drainage channels and marginal moraine systems at this height. The maximum water depth in the lake was of the order of 60 m. On filling, the lake over-flowed eastwards by rapidly cutting through the fronting barrier of ice-marginal landforms and sediment at Garvald. As the ice-margin retreated westwards across the lake basin, subglacial drainage entered the lake via a tunnel with an exit below water-level. At the tunnel exit, water-flow decelerated, resulting in rapid deposition of bed-load and the formation of a fronting subaqeous fan (Fan F1, Stage B, Fig. 15). The size of this fan implies that the ice margin remained

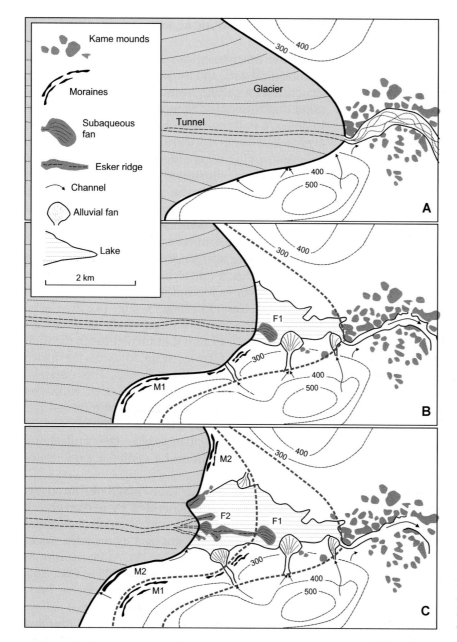

Fig. 15 Palaeogeographic reconstruction of stages A, B, C, D and E in the development of the Newbigging esker system. For explanation see text.

relatively stationary at this position for some time. This is confirmed by the occurrence of a series of small, arcuate moraines running obliquely down the southern valley wall towards the position of the fan, but terminating at the lake shore-line (Moraines M1, Stage B, Fig. 15).

Stage C. Retreat from stage B caused abandonment of Fan F1 and its replacement upcurrent by the rapid uncovering of tunnel-fill sediment, now marked by a discontinuous, single ridge structure. A further period of retreat and subsequent still-

stand is indicated by a second fan developed at the western end of this ridge (Fan F2, Stage C, Fig. 15). At this position there is a significant change in the ridge morphology from a single ridge system with fans down-current, to a double ridge system with intervening fans up-current. Thus, at this stage the development of Fan F2 is accompanied on both its northern and southern margin by adjacent single ridges. This implies multiple stream exits at the margin at this time; probably caused by sub-glacial tunnel branching immediately up-ice of the

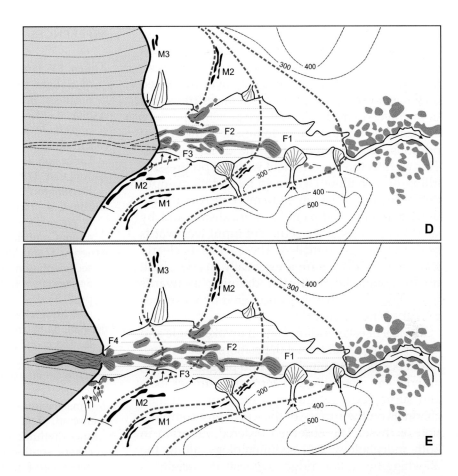

Fig. 15 (*cont'd*)

margin, possibly generated by a high-magnitude flood event. This episode of still-stand is accompanied by further arcuate moraine ridges developed across both the northern and southern valley walls (Moraines M2, Stage C, Fig. 15).

Stage D. Between stages C and D retreat was rapid as, upcurrent, the two ridges and the fan at stage C are replaced by three parallel, single ridges; again suggestive of subglacial tunnel branching. Immediately upcurrent, a further stage of still-stand may be identified by the occurrence of two adjacent fans (Fans F3, Stage D, Fig. 15). These are accompanied, to the north, by a single ridge; suggesting further sub-glacial tunnel branching. Interpretation of this stage is assisted by detail from Quarry A, which shows the development of two packages of off-lapping sediment stacked one above the other and indicative of switching across a subaqueous fan surface directed to the east.

Stage E. On retreat from stage D, Fans F3 were abandoned. The tunnel feeding these fans, exposed

in Quarry B, and comprising coarse boulder gravels, was then exhumed as a single ridge structure. A temporary still-stand in retreat buried the tunnel facies and built a large subaqeous fan in front of the tunnel portal (Fan F4, Stage E, Fig. 15). With thinning of the ice the tunnel became unroofed and a supraglacial outwash sandur was deposited in a widening ice-walled trough that fed the fan. With subsequent retreat a large, multiple ridge system was left behind as a consequence of differential melt beneath braided channel systems.

CONCLUSION

The glaciological context of the Newbigging ridge system shows that it was built during a stage in the retreat of the Late Devensian Southern Uplands ice sheet as directional indicators all confirm ice-flow from south west to north east. The morphological context indicates that the ridge system consists of three morphologically and spatially distinct landform assemblages that merge into one another in the down-ice direction. The stratigraphic context confirms that the ridge system was associated with the filling of an exhumed, glacially over-deepened rock basin through a down-basin facies transition from gravel-dominant beneath the proximal portions of the ridge, through sand-dominant in the central portions, to silt-dominant in the distal portions. This architecture is consistent with the progressive filling of a shallow lake basin by sediment input delivering coarse sediment to more proximal locations and finer sediment to more distal, and by successive shifts in the sediment input westwards as the ice margin retreated. These contexts clearly show that the ridge form is essentially an esker, built up by subglacial meltwater draining eastwards into a lake basin that was uncovered by westward retreat of the ice margin.

Models of esker sedimentation invoke single and multiple subglacial tunnel deposition, subaqueous-fan deposition at tunnel exits into proglacial lakes, alluvial fan deposition at terrestrial tunnel exits and supraglacial sandur deposition in ice-walled channels on the surface of the ice; or any combination of these. The observed ordered variation in both morphology and sedimentology along the Newbigging esker ridge suggest its origin is composite, with the style of deposition changing significantly during retreat. Thus, the sequence exposed in each of the three quarries in the esker are radically different and imply different esker sub-environments. Quarry A provides evidence for distal subaqueous fan sedimentation whilst Quarry B provides evidence for boulder-rich subglacial tunnel fill. Quarry C, in contrast, is composite and identifies a transition during a single stage of retreat still-stand from esker tunnel-fill into a subaqueous fan fronting the tunnel portal into a supraglacial outwash sandur deposited in an ice-walled trough created by the tunnel unroofing.

This case study demonstrates that complex, multiple-ridge eskers may form by a combination of both subglacial and surpraglacial processes, although the complexity of morphology is a function of the supraglacial rather than subglacial sedimentation. The composite nature of the landform assemblage at Newbiggin cautions against making over-simplistic interpretations on the basis of limited sedimentological data when make palaeo-glacialogical inferences.

REFERENCES

Aitken, J.F. (1995) Lithofacies and depositional history of a Late Devensian ice-contact deltaic complex, northeast Scotland. *Sed. Geol.*, **99**, 11–130.

Ashley, G.M. and Warren, W. (1995) Irish eskers – origins of ice-contact stratified deposits. INQUA Commission Ireland 1995 Symposium and Field Excursion, *Formation and Properties of Glacial Deposits*.

Auton, C.A. (1992) Scottish landform examples – 6 the Flemington Eskers. *Scot. Geog. Mag.*, **108**, 190–196.

Auton, C.A., Firth, C.R. and Merritt, J.W. (1990) *Beauly to Nairn Field Guide*, Quaternary Research Association, Cambridge, 149 pp.

Banerjee, I. and McDonald, B.C. (1975) Nature of esker sedimentation. In: *Glaciofluvial and Glaciolacustrine Sedimentation* (Eds A.V. Jopling and B.C. McDonald), pp. 133–154. Society of Economic Paleontologists and Mineralogists Special Publication, **23**.

Bennett, M.R. and Glasser, N.F. (1996) *Glacial Geology: Ice Sheets and Landforms*. Wiley, Chichester, 364 pp.

Boulton, G.S. and Van De Meer, J. (1989) *Preliminary report on an expedition to Spitzbergen in 1984 to study glaciotectonic phenomena* (Glacitecs '84). University of Amsterdam, Amsterdam.

Brennand, T.A. (1994) Macroforms, large bedforms and rhythmic sedimentary sequences in subglacial eskers, south-central Ontario: implications for esker genesis and meltwater regime. *Sed. Geol.*, **91**, 9–55.

Brennand, T. and Shaw, J. (1996) The Harricana glacio-fluvial complex, Abiti region, Quebec: its genesis and implications for meltwater regime and ice-sheet dynamics. *Sed. Geol.*, **102**, 221–262.

Cheel, R.J. (1982) The depositional history of an esker near Ottawa, Canada. *Can. J. Earth Sci.*, **19**, 1417–1427.

Delaney, C. (2002) Sedimentology of a glaciofluvial land-system, Lough Ree area, Central Ireland: implications for ice margin characteristics during Devensian glaciation. *Sed. Geol.*, **149**, 111–126.

Diemer, J.A. (1988) Subaqueous outwash deposits in the Ingraham Ridge, Chazy, New York. *Can. J. Earth Sci.*, **25**, 1384–1396.

Eyles, N., Eyles, C.H. and Miall, A.D. (1983) Lithofacies types and vertical profile models: an alternative approach to the description and environmental interpretation of glacial diamict and diamictite sequences. *Sedimentology*, **30**, 393–410.

Eyles, N., Clark, B.M. and Clague, J.J. (1987) Coarse-grained sediment gravity flow facies in a large supraglacial lake. *Sedimentology*, **34**, 193–216.

Eyles, N. and Miall, A.D. (1984) Glacial facies. In: Facies Models (Ed. R.G. Walker). *Geoscience Canadian Reprint Series*, **1**, 15–38.

Fard, A.M. (2003) Large dead-ice depressions in flat-topped eskers: evidence of a Preboreal Jökulhlaup in the Stockholm area, Sweden. *Global Planet. Change*, **35**, 273–295.

Flint, R.F. (1970) *Glacial and Quaternary Geology*. New York. 892 pp.

Goodlet, G.A. (1964) The kamiform deposits near Carstairs, Lanarkshire. *Bull. Geol. Surv. Great Brit.*, **21**, 175–196.

Gordon, J.E. (1993) Carstairs Kames. In: *Quaternary of Scotland* (Eds J.E. Gordon, J.E. and D.G. Sutherland), pp. 544–549. Geological Conservation Review Series 6, Chapman and Hall, London.

Gorrell, G. and Shaw, J. (1991) Deposition in an esker bead and fan complex, Lanark, Ontario, Canada. *Sed. Geol.*, **72**, 285–314.

Gray, J.M. (1991) Glaciofluvial landforms. In: *Glacial deposits in Great Britain and Ireland* (Eds J. Ehlers, P.L. Gibbard and J. Rose), pp. 443–453. Balkema, Rotterdam.

Gustavson, T.C. and Boothroyd, J.C. (1987) A depositional model for outwash sediment sources, and hydrologic characteristics, Malaspina Glacier, Alaska: A modern analog of the south-eastern margin of the Laurentide Ice Sheet. *Bull. Geol. Soc. Am.*, **99**, 292–302.

Hedbrand, M. and Amark, M. (1989) Esker formation and glacier dynamics in eastern Skane and adjacent areas, southern Sweden. *Boreas*, **18**, 67–81.

Henderson, P.J. (1987) Sedimentation in an esker system influenced by bedrock topography near Kingston, Ontario. *Can. J. Earth Sci.*, **25**, 987–999.

Howarth, P.J. (1971) Investigations of two eskers at eastern Breidermerkurjökull, Iceland. *Arctic Alpine Res.*, **3**, 305–318.

Huddart, D. and Bennett, M.R. (1997) The Carstairs Kames (Lanarkshire, Scotland): morphology, sedimentology and formation. *Journal of Quatern. Sci.*, **12**, 467–484.

Huddart, D., Bennett, M.R. and Glasser, N.F. (1998) Morphology and sedimentology of a High-Arctic esker complex: Vegbreen, Svalbard. *Boreas*, **28**, 253–273.

Jenkins, K.A. (1991) *The origin of eskers and fluvioglacial features: an example from the area south of the Pentland Hills*. Unpublished report, Department of Geology and Geophysics, University of Edinburgh, 36 pp.

Laxton, J.L. (1980) A method for estimating the grading of boulder and cobble grade material. *Report Instit. Geol. Sci.*, **80/1**, 31–35.

Mäkinen, J. (2003) Time-transgressive deposits of repeated depositional sequences within interlobate glaciofluvial (esker) sediments in Köyliö, SW Finland. *Sedimentology*, **50**, 327–360.

Merritt, J.W., Laxton, J.L., Smellie, J.L. and Thomas, C.W. (1983) Summary assessment of the sand and gravel resources of south-east Strathclyde, Scotland. *British Geological Survey Technical Report*, WF/83/6.5.

McCabe, A.M. and O'Cofaigh, C. (1994) Sedimentation in a subglacial lake, Enniskerry, eastern Ireland. *Sed. Geol.*, **91**, 57–95.

McCall J. and Goodlet, G.A. (1952) Indicator stones from the drift of Midlothian and Peebles. *Trans. Edinb. Geol. Soc.*, **14**, 401–409.

McMillan, A.M., Laxton, J.L. and Shaw, A.J. (1981) The sand and gravel resources of the country around Dolphinton, Strathclyde Region, and West Linton, Borders Region. Description of 1:25,000 resource sheets NT04 and 14 and parts of NT05 and 15. *Mineral Assessment Report of the Institute of Geological Sciences*, 62 pp.

Owen, G. (1997) Origin of an esker-like ridge – erosion or channel-fill? Sedimentology of the Monington 'Esker' in southwest Wales. *Quatern. Sci. Rev.*, **16**, 675–684.

Paterson, J.T. and Cheel, R.J. (1997) The depositional History of the Bloomington Complex, An Ice-Contact Deposit in the Oak Ridges Moraine, Southern Ontario, Canada. *Quatern. Sci. Rev.*, **16**, 705–719.

Postma, G. (1986) Classification of sediment gravity-flow deposits based on flow conditions during sedimentation. *Geology*, **14**, 291–294.

Postma, G. and Roep, Th.B. (1985) Resedimented conglomerates in the bottomsets of Gilbert-type gravel deltas. *J. Sed. Pet.*, **55**, 874–885.

Price, R.J. (1966) Eskers near Casement Glacier, Alaska. *Geografiska Annaler*, **48A**, 111–125.

Price, R.J. (1969) Moraines, sandur, kames and eskers near Breidermerkurjökull, Iceland. *Trans. Instit. Brit. Geogr.*, **46**, 17–43.

Punkari, M. (1997) Glacial and fluvioglacial deposits in the interlobe areas of the Scandinavian Ice Sheet. *Quatern. Sci. Rev.*, **16**, 741–754.

Russell, H.A.J. and Arnott, R.W.C. (2003) Hydraulic-jump and hyperconcentrated-flow deposits of a glacigenic subaqueous fan: Oak Ridges Moraines, southern Ontario, Canada. *J. Sed. Res.*, **73**, 887–905.

Rust, B.R. and Romanelli, R. (1975) Late Quaternary subaqueous outwash deposits near Ottowa, Canada. In: *Glaciofluvial and Glaciolacustrine Sedimentation* (Eds A.V. Jopling and B.C. McDonald), pp. 177–192. Society of Economic Paleontologists and Mineralogists Special Publication **23**.

Shaw, A.J. and Merritt, J.W. (1981) The sand and gravel resources of the country around Biggar, Strathclyde region: Description of 1:25,000 sheets NS93 and NT03 and parts of NS92 and NT02. *Mineral Assessment Report of the Institute of Geological Sciences*, 95 pp.

Shaw, J. (1975) Sedimentary successions in Pleistocene ice-marginal lakes. In: *Glaciofluvial and Glaciolacustrine Sedimentation* (Eds A.V. Jopling and B.C. McDonald), pp. 281–303. Society of Economic Paleontologists and Mineralogists Special Publication **23**.

Shreve, R.L. (1985) Esker characteristics in terms of glacier physics, Katahdin esker system, Maine. *Bull. Geol. Soc. Am.*, **96**, 639–646.

Shulmeister, J. (1989) Flood deposits in the Tweed esker (southern Ontario, Canada*). Sed. Geol.*, **65**, 153–163.

Sissons, J.B. (1961) A subglacial drainage system by the Tinto Hills, Lanarkshire. *Trans. Edinb. Geol. Soc.*, **18**, 175–192.

Spooner, I.S. and Dalrymple, R.W. (1993) Sedimentary facies relationships in esker-ridge/esker-fan complexes, south-eastern Ontario: application to the exploitation for asphalt blending sand. *Quatern. Int.*, **20**, 81–92.

Terwindt, J.H.J. and Augustinus, P.G.E.F. (1985) Lateral and longitudinal successions in sedimentary structures in the Middle Mause esker, Scotland. *Sed. Geol.*, **45**, 161–188.

Thomas, G.S.P. (1984) Sedimentation of a sub-aqueous esker-delta at Strabathie, Aberdeenshire. *Scot. J. Geol.*, **20**, 9–20.

Thomas, G.S.P. and Connell, R. (1985) Iceberg drop, dump and grounding structures from the Pleistocene glaciolacustrine sediments, Scotland, *J. Sed. Pet.*, **55**, 243–249.

Thomas, G.S.P., Connaughton, M. and Dackombe, R.V. (1985) Facies variation in a Late Pleistocene supraglacial outwash sandur from the Isle of Man. *Geol. J.*, **20**, 193–213.

Thomas, G.S.P. and Montague, E. (1997) The Morphology, Stratigraphy and Sedimentology of the Carstairs Esker, Scotland, U.K. *Quatern. Sci. Rev.*, **16**, 661–674.

Veillette, J.J. (1986) Former southwesterly ice flows in the Abitibi-Timiskaming region: implications for the configuration of the Late Wisconsinan ice sheet. *Can. J. Earth Sci.*, **23**, 1724–1741.

Walker, R.G. (1975) Generalised facies models for resedimented conglomerates of turbidite association. *Bull. Geol. Soc. Am.*, **86**, 737–748.

Warren, W.P. and Ashley, G.M. (1994) Origins of the ice-contact stratified ridges (eskers) of Ireland. *J. Sed. Res.*, **64(A)**, 433–449.

The age and origin of the Blakeney esker of north Norfolk: implications for the glaciology of the southern North Sea Basin

S.J. GALE *and* P.G. HOARE

School of Geosciences, The University of Sydney, Sydney, New South Wales 2006, Australia (e-mail: sgale@mail.usyd.edu.au)

ABSTRACT

The age and origin of the Blakeney ridge of eastern England have been debated for over a century. The feature is shown to be an esker formed in the late stages locally of a glaciation of Middle Quaternary age. The weight of evidence indicates that the esker was associated with the ice that laid down the Marly Drift of central north Norfolk, although this cannot be demonstrated conclusively. The flows that formed the Blakeney esker were subglacial. They took place from northwest to southeast along the length of the ridge, with at least the final phase occurring under hydrostatic pressure. It is likely that the esker developed during a glacial stillstand and under steep hydraulic gradients close to the ice sheet terminus. Although the esker conduit appears to have functioned largely as a Röthlisberger channel, there is evidence of simultaneous incision of the glacial bed. A valley is cut into bedrock beneath the esker. The esker and the buried valley may form the distal components of one of the broadly north–south aligned subglacial valleys that were trenched across the southern North Sea Basin in Middle Quaternary times. The complex assemblage of proglacial glaciofluvial landforms found to the south of the esker may represent the products of transport along the subglacial valley. The ice lobe that occupied the southern North Sea Basin may have left similar buried valley–proglacial fan complexes elsewhere along the northern fringe of East Anglia.

Keywords East Anglia, North Sea Basin, Blakeney, Quaternary, glaciation, Marly Drift.

INTRODUCTION

The Blakeney esker is part of a complex assemblage of Quaternary landforms and drift deposits found within and around the lower Glaven valley in northern East Anglia, eastern England (Fig. 1). These have been the subject of debate for more than a century (Holmes, 1883, 1884), and both their origin and their exact age are still disputed (Booth in Lawson and Allen, 2000; Hamblin, 2000; Hamblin *et al.*, 2000, 2003).

The landforms and the drift rest at unknown depth on Upper Chalk of Late Cretaceous age (British Geological Survey, 1985). The surface of the Chalk is entrenched by a broadly north–south oriented valley within which are preserved deposits of 'sand, gravel and clay' (Woodland, 1970) (Fig. 1B). Blanketing the area and concealing the palaeovalley and its fill are highly calcareous tills known as Marly Drift or, where they are disturbed, as Contorted Drift (Lyell, 1863; Wood and Harmer, 1868, 1869; Wood, 1880; Reid, 1882; Woodward, 1884, 1885; Boswell, 1914, 1935; Straw, 1965; Banham *et al.*, 1975; Perrin *et al.*, 1979; Hoare and Gale, 1986; Ehlers *et al.*, 1987, 1991; Fish *et al.*, 2000; Fish and Whiteman, 2001; Pawley *et al.*, 2004).

The Marly Drift was assigned to the Anglian (or its equivalent) Stage by Solomon (1932), Baden-Powell and Moir (1942) and Baden-Powell (1948a, 1948b) (see also Harmer, 1904, 1905, 1909). Most recent commentators have implicitly supported this thesis by attributing all pre-Devensian East Anglian tills to this stage (Sparks and West, 1964, 1972; Turner, 1973; Shotton *et al.*, 1977; Perrin *et al.*, 1979; Bowen *et al.*, 1986; Hart and Peglar, 1990; Ehlers *et al.*, 1991; Whiteman, 1995; Bowen, 1999). Assuming this attribution to be correct, the only lithostratigraphical evidence of a post-Hoxnian, but pre-Ipswichian, glaciation of East

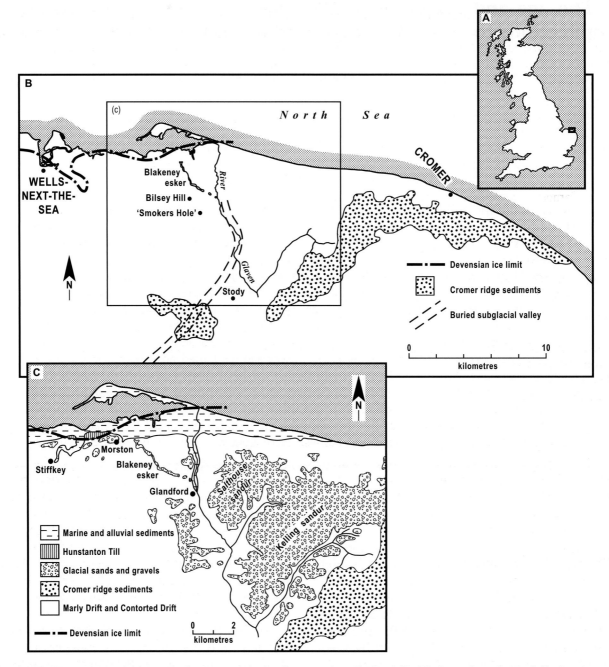

Fig. 1 (A) Great Britain showing the location of the study area in north Norfolk. (B) Central north Norfolk showing the sediments of the Cromer ridge (Straw, 1965), the buried subglacial valley of the Glaven (Woodland, 1970), the Devensian ice limit (Straw, 1960) and the sites referred to in the text. (C) The Quaternary geology of the Glaven valley (Gale *et al.*, 1988).

Anglia comes from Tottenhill in west Norfolk. Here glaciofluvial sands and gravels overlie the marine beds of the Nar Valley Clay Member (Gibbard *et al.*, 1991, 1992; Lewis and Rose, 1991). Uranium-series dating and amino acid analyses indicate that the sands and gravels are younger than Oxygen Isotope Stage 9 (Rowe *et al.*, 1997; Scourse *et al.*, 1999), though other amino acid determinations suggest an age more recent than Oxygen Isotope Stage 11 (Bowen *et al.*, 1989; Ashton *et al.*, 1995).

By contrast, Straw has argued for a Wolstonian (or its equivalent) age for the Marly Drift, proposing that the maximum extent of Wolstonian ice lay across the eastern part of central Norfolk (Straw, 1965, 1967, 1973, 1974, 1979a, 1979b, 1983, 1984; Anon, 1970). Cox and Nickless (1972, 1974), Bristow and Cox (1973a, 1973b) and Cox (1981) went further and assigned a Wolstonian age to all pre-Devensian tills in East Anglia (see also Ranson, 1968). However, neither Laban et al. (1984), British Geological Survey and Rijks Geologische Dienst (1986), Cameron et al. (1987), Long et al. (1988) nor Beets et al. (2005) were able to find evidence of glacial deposition this far south in the North Sea Basin during the Wolstonian Stage.

On the basis of its particle-size distribution, mineralogy, chemistry and colour, the Marly Drift has been divided into four lithological types (Banham et al., 1975; Perrin et al., 1979):

1 Lowestoft-type: Lowestoft Till enriched with Chalk.
2 Cromer-type: Cromer Till enriched with Chalk.
3 Intermediate-type: a mixture of Lowestoft-type and Cromer-type.
4 Reconstituted Chalk.

The ice that laid down the Lowestoft Till is regarded as having entered the area from the north, northwest and west, whilst that which deposited the Cromer Tills is thought to have advanced from the northwest and northeast (Banham, 1975; Perrin et al., 1979; Ehlers et al., 1991; Hart and Boulton, 1991).

Ehlers et al. (1987) identified two superimposed Marly Drift units in central Norfolk. The lower was considered to have been deposited by ice advancing from the southwest, whilst the upper was thought to have been laid down by ice moving from the north. Ehlers et al. (1987, 1991) found no trace of meltwater deposition or ice melt between the two units and concluded that the tills probably represent different advances of the same ice sheet. They further argued that the Marly Drift is not an individual stratigraphic unit, but a Chalk-rich facies of the 'normal' Anglian tills (Ehlers et al., 1987), a thesis earlier advanced by Reid (1882), Woodward (1884), Banham et al. (1975) and Perrin et al. (1979). These conclusions were reiterated by Ehlers et al. (1991), who proposed that the lower unit of the Marly Drift represents a Chalk-rich

equivalent of the Lowestoft Till carried by ice that advanced from the southwest, and that the upper unit was laid down by a separate advance of the same ice sheet from between northwest and north.

Until recently, the Marly Drift deposits were ascribed to a single cold stage and for most workers the only debate turned on whether deposition occurred contemporaneously or whether the Lowestoft Till and the Cromer Tills were laid down at different times within the same stage (see Ehlers et al., 1991; Hart and Boulton, 1991; and references therein). Of late, however, members of the British Geological Survey's Northern East Anglia Project have argued that several of the glaciogenic units generally assigned to the Anglian should be ascribed to other cold stages. Despite being in the enviable position of knowing that '... the glacial stratigraphy has finally been resolved ...' (Hamblin, 2000), the group has rapidly and repeatedly revised its scheme. In its most recent version, the Contorted Drift, Third Cromer Till and the 'Cromer Till' variant of the 'Marly Drift' are placed in a post-Anglian Sheringham Cliffs Formation, although it is conceded that the stratigraphic position of other facies of the Marly Drift in the Glaven valley and west Norfolk is unclear (Lee et al., 2004).

Overlying these highly calcareous tills in the Glaven valley area is a complex series of deposits, composed largely of sand and gravel. These make up the Blakeney esker, as well as features interpreted as kames, kame terraces and sandar by Woodward (1884), Solomon (1932), West (1957), Straw (1960, 1965, 1967, 1973), Sparks and West (1964), Gale and Hoare (1986) and Gray (1988, c. 1994, 1997). Most workers have implicitly assumed that these landforms are of the same age as the deposits of which they are composed. That this may not necessarily be the case should be borne in mind when assessing the chronology of these features.

Solomon (1932) and Baden-Powell and Moir (1942) assigned the sand and gravel of the Blakeney esker to the Wolstonian (or its equivalent) Stage. West (1958), Woodland (1970), Thornton and Cox (in Boulton et al., 1984) and, speculatively, Booth (in Lawson and Allen, 2000) suggested that the landform is of Devensian age, although West (1961) later revised his assessment and proposed that the feature dates from the Wolstonian. A Wolstonian age was also propounded by Sparks and West

(1964, 1972), West (1967), Straw (1965, 1967, 1973), Turner (1973) and, somewhat guardedly, by Shotton *et al.* (1977) and Boulton (in Boulton *et al.*, 1984). Most recently, the British Geological Survey also assigned the esker and the nearby kames to the penultimate cold stage (the Wolstonian of other workers). The deposits of which the esker and kames are composed were grouped with the Briton's Lane Sand and Gravel Member that occupies much of the higher ground on and adjacent to the Cromer ridge, a complex ice-marginal feature that lies southeast of the esker (Fig. 1b) (Hamblin, 2000; Hamblin *et al.*, 2000; Hamblin *et al.*, 2003; Moorlock *et al.*, 2002). Although most workers have maintained that the sand and gravel and the immediately underlying tills were deposited during separate cold stages, Straw (1960, 1965, 1967, 1973, 1979b, 1983) argued that they are penecontemporaneous. Similarly, Slater (1923) believed the cannon-shot gravels (the spherical, flint-rich gravels of, *inter alia*, the Cromer ridge and the Blakeney esker [Solomon, 1932]) and the Marly Drift to be of the same age.

Although the Blakeney ridge was referred to as an esker by early workers (Holmes, 1883, 1884; Woodward, 1884), their use of the term was very different to that of modern researchers. Slater (1923) was the first to interpret the ridge deposits as the product of glaciofluvial deposition in a subglacial conduit. Subsequently, three further hypotheses have been advanced to explain the formation of the ridge. These are that it is morainic in origin (West, 1957); that it has been eroded out of a much larger spread of outwash sands and gravels, and now assumes a linear form by chance (Straw, 1960, 1965, 1973; Hamblin *et al.*, 2000); and that it represents the fill resulting from subaerial meltwater flow to the northwest along intersecting rectilinear fractures in an ice sheet (Sparks and West, 1964; West, 1967). Sparks and West (1964, 1970) also suggested that the ridge may be a composite feature, partly crevasse filling, partly esker, though still the product of northwesterly flows. West (1957), too, hinted at evidence of flows to the north and west at the northwestern end of the ridge.

We report here the results of investigations of the sedimentology and stratigraphy of the Blakeney esker and its underlying Quaternary deposits. The aim of this work has been to understand the origin and age of this enigmatic feature. Further work has been undertaken since our initial studies (Gale and Hoare, 1986; Hoare and Gale, 1986), and site and sample numbers in the preliminary accounts are not necessarily the same as those employed in this final report.

THE GEOMORPHOLOGY OF THE BLAKENEY ESKER

The Blakeney esker (national grid reference: TG 04) forms a distinctive, 2.9 km long, northwest–southeast oriented ridge, standing up to 20 m above the undulating surface of the surrounding countryside and varying in width from 40 to 170 m (Figs 1, 2). The feature has a more or less rectilinear plan form (although ploughing may have straightened the edges of an originally more serpentine shape [Sparks and West, 1964]). Though in places the esker has been entrenched, its crest increases irregularly in altitude in a southeasterly direction (Fig. 3). The feature lies, with apparent disregard for the underlying topography, across a surface underlain by a suite of glaciogenic deposits. Parts of the ridge have been worked for sand and gravel (Ordnance Survey sheets show that excavations began prior to 1886), and there are relatively large pits at TG 025426, between TG 027424 and TG 031422, and near its northwestern end.

METHODS

The ridge was mapped at a scale of 1:10 560 with the aid of vertical and oblique aerial photographs (Fig. 2). Every exposure of Quaternary sediment was investigated and recorded. At the majority of the sites only a single unit is seen. However, sites 1, 2, 8 and 15 each reveals a succession of strata and these were used to establish the stratigraphy of the deposits within and beneath the ridge.

Gravel fabrics were determined by establishing the orientation of all clasts of *a*-axis >20 mm within a limited area of the exposure. Twenty-five gravels were measured at each site. In the case of the largely clast-supported gravels of Unit 5, the inclination and declination of the maximum projection plane of the gravels were noted. The fabrics of the matrix-supported gravels of Units 4 and 6 were determined by recording the inclination and declination of the axis of dip (whether *a* or *b*) of

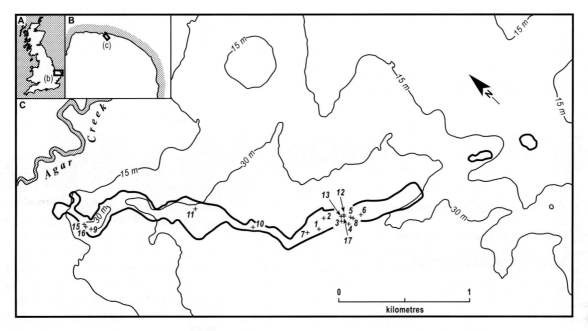

Fig. 2 The Blakeney esker showing the location of the sites investigated. Mapping based on field surveys at 1:10 560 scale and aerial photograph analysis at ~1:10 000 scale.

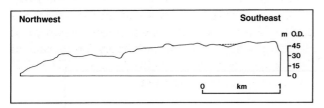

Fig. 3 The long profile of the Blakeney esker (Sparks and West, 1964).

each clast. Sites close to the former ground surface were avoided in order to reduce the chance of measuring fabrics that had been disturbed by weathering.

The orientation of cross-beds within the sands of Unit 5 was determined by recording the angle and direction of dip of all visible bedding surfaces within a single cross-set. Between six and 25 measurements were made at each site.

The location and extent of the gravel fabric and sand cross-bed measurement sites are shown in Figures 2, 4 and 5.

After cleaning the sections, samples of each of the six major units were taken by collecting the entire volume of material within an area not exceeding 0.55 m^2 and no deeper than 0.4 m. The positions and dimensions of the samples taken

from sites 1 and 2 are shown in Figures 4 and 5. Given the coarse nature of many of the sediments, strenuous efforts were made to collect sufficient material to ensure representative sampling of their coarsest fractions (Gale and Hoare, 1992, 1994) (Table 1).

Colour was determined on freshly exposed materials using *Munsell Soil Color Charts*. The colour of samples of Units 1 and 3 after laboratory air-drying was measured in the same way. Unless otherwise stated, all colours reported in the text are those of the material in its field condition.

Particle-size distribution was established using sieves and calibre plates at quarter phi-unit intervals and by the pipette method of sedimentation analysis (Gale and Hoare, 1991). Particle-size descriptions follow the terminology of Wentworth (1922) and textural terms are those proposed by Folk (1954). The lithology of the 4.0–5.7 mm fraction of samples 1:2, 2:3, 2:7 and 9, the 5.7–8.0 mm fraction of samples 1:1, 1:2, 2:3, 2:7, 9, 10 and 11, the 8.0–16.0 mm fraction of samples 2:7, 6D, 7, 9 and 11, and the 16.0–32.0 mm fraction of samples 6A, 9, 10 and 11 was also determined.

The shape of those flint clasts in the 22.5–27.0 mm fraction of sample 9 was measured following the procedures described by Gale and Hoare (1991).

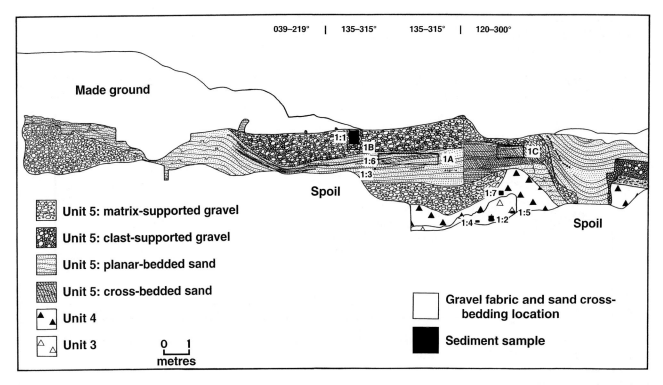

Fig. 4 The stratigraphy of site 1, Blakeney esker.

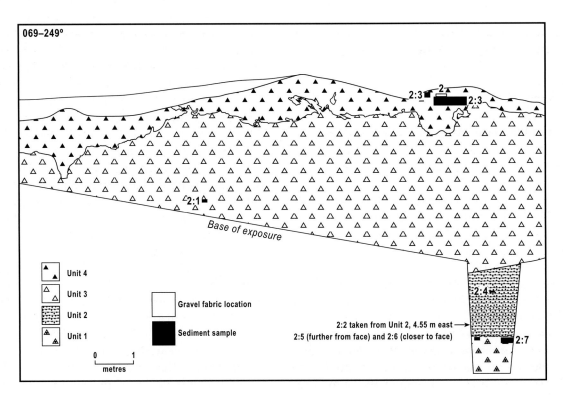

Fig. 5 The stratigraphy of site 2, Blakeney esker.

Table 1 Summary particle-size distribution statistics (calculated following the procedures of Folk and Ward, 1957) and textural classification (following the definitions of Folk, 1954) of samples from Units 1 to 6

Unit	Sample	Sample size (g)	Texture[a]	Graphic mean (Φ-units)	Inclusive graphic standard deviation (Φ-units)	Inclusive graphic skewness (Φ-units)	Graphic kurtosis (Φ-units)
6	6A	45 765	sG	−3.68	2.55	0.49	0.71
6	6D	142 480	msG	−3.63	2.94	0.67	0.94
6	7	51 741	sG	−3.23	2.72	0.45	0.89
6	8	31 761	sG	−2.37	2.47	0.40	0.56
6	9	66 669	G	−4.63	2.84[b]	0.74[b]	1.83[b]
6	10	92 688	sG	−3.25	2.79	0.29	0.69
6	11	111 517	sG	−2.87	3.29	0.50	0.58
5	1:1	49 188	G	−3.18	2.81	0.70	1.13
5	1:3	64.72	S	1.45	0.41	0.04	0.96
5	1:6	154.76	S	1.16	0.38	0.12	0.92
4	2:3	5659	msG	0.67	4.93[b]	0.13[b]	1.78[b]
3	1:2	4553	gM	5.95	4.12[b]	0.04[b]	0.87[b]
2	2:2	not available	S	2.31	0.55	0.05	1.09
2	2:4	194.45	S	2.32	0.54	0.09	1.08
1	2:7	2849	gM	6.64[b]	7.10[b]	0.44[b]	1.19[b]

[a] G: gravel; sG: sandy gravel; msG: muddy sandy gravel; S: sand; gM: gravelly mud.
[b] Calculated using extrapolated percentile(s).

The surface characteristics of flint clasts in the 5.7–8.0 mm fraction of samples 1:1 and 17, and the 13.2–16.0 mm fraction of sample 10 were also recorded.

Sub-samples of the <2.0 mm fraction of material collected from Units 1, 3 and 4 were ground to a powder and their calcium carbonate-equivalent content determined gasometrically using a Bascomb calcimeter (Bascomb, 1961; Gale and Hoare, 1991).

RESULTS AND INTERPRETATION

Unit 1

The lowest bed observed consists of an extremely poorly sorted, yellowish brown to very pale brown, gravelly mud. This is exposed at sites 2, 15 and 16 and has a maximum observed thickness of >0.92 m. The gravel assemblage is dominated by Chalk and flint (Tables 2, 3). The calcium carbonate-equivalent content of the <2.0 mm fraction is around 25% (Table 4).

The extremely poor sorting and wide particle-size range of the deposit (Fig. 6, Table 1) are characteristic of a diamicton, but are not necessarily diagnostic of a glacial origin. However, the interpretation of this material as a till is supported by the presence of striated and facetted gravels, and by the exotic clasts in the fine gravel fractions (Tables 2, 3). Although other mechanisms are possible, the generally massively bedded nature of the deposit points to its likely accumulation by lodgement.

Unit 2

Unconformably overlying Unit 1 at site 2 (Fig. 5) and site 15 is a sequence of planar-bedded, moderately sorted, light yellowish brown to brownish yellow sand, with a maximum observed thickness of 1.87 m. The sand contains occasional lenses of laminated silt. The unit has a sharp and undulating erosional contact with the subjacent bed. The particle-size characteristics of the sand accord with those of many fluvial sediments (Gale

Table 2 The lithology of the 4.0–5.7 mm fraction of samples from Units 1, 3, 4 and 6: percentage frequency by number

Lithology	Unit 1, sample 2:7	Unit 3, sample 1:2	Unit 4, sample 2:3	Unit 6, sample 9
Chalk	78.4	75.8	0.4	
Flint	12.4	18.9	91.2	85.8
Sandstone	3.8	0.4	2.1	11.0
Calcareous till	0.3	1.2	2.1	
Oölitic limestone		0.4		
Quartzite	1.7	0.4		0.4
Igneous			0.3	0.2
Vein quartz	3.4	2.0	3.8	2.6
Carbonate precipitate		0.4		
Shell		0.4		
Unidentifiable			0.1	
n	291	244	1143	508

Table 3 The lithology of the 5.7–8.0 mm fraction of samples from Units 1, 3, 4, 5 and 6: percentage frequency by number

Lithology	Unit 1, sample 2:7	Unit 3, sample 1:2	Unit 4, sample 2:3	Unit 5, sample 1:1	Unit 6, sample 9	Unit 6, sample 10	Unit 6, sample 11
Chalk	72.5	71.7					
Flint	24.2	20.7	94.7	85.5	96.2	92.9	93.3
Sandstone		3.3	3.2	3.1	2.1	3.1	4.8
Calcareous till			0.8				
Oölitic limestone				0.4			
Carboniferous limestone		1.1					
Quartzite	1.3			3.0		0.6	0.6
Igneous		1.1		1.3			
Vein quartz	1.3	2.2	1.4	6.2	1.7	3.1	1.3
Carbonate precipitate	0.7						
Other				0.6		0.4	
n	153	92	506	1347	238	1038	315

and Hoare, 1991) (Fig. 6, Table 1), whilst planar-bedding is not uncommon in fluvial environments (Allen, 1983). Given the nature of the under- and overlying beds, this sediment is almost certainly the product of glaciofluvial deposition.

Unit 3

At sites 2 and 15, the extremely poorly sorted, very pale brown, gravelly mud of Unit 3 displays a sharp, erosional contact with Unit 2 (Fig. 5). Restricted outcrops of a very similar deposit are also found at sites 1, 12, 13 and 17, in locations where, prior to commercial excavation, they must have been overlain by Units 4, 5 and/or 6. The position of each of these exposures close to and at the same height as Unit 3 at sites 2 and 15 strongly suggests that these outcrops also reveal representatives of Unit 3. The unit reaches a maximum thickness of 4.34 m. Although the mud is texturally similar to

Table 4 Calcium carbonate-equivalent content of the <2.0 mm fraction of samples from Units 1, 3 and 4

Unit	Sample	Calcium carbonate-equivalent (% by mass)		Number of subsamples analysed
		Mean	Range	
4	1:7	0.1	0.1–0.2	3
4	2:3	2.6	2.5–2.7	3
3	1:4	52.7	52.4–53.1	3
3	1:5	52.8	52.7–52.9	2
3	2:1	69.2	69.0–69.5	2
3	12	70.2	68.9–71.6	3
3	13:1	67.9	67.0–69.3	3
3	13:2	67.3	66.3–68.1	3
1	2:5	24.5	24.3–24.6	3
1	2:6	24.6	24.4–24.9[a]	4

[a] A fifth subsample yielded an inexplicably low value of 18.4%.

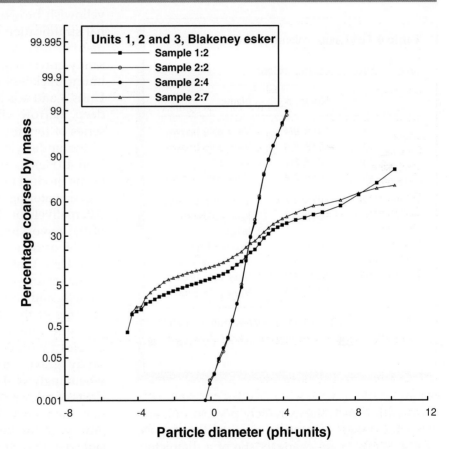

Fig. 6 The particle-size distribution of samples from Units 1 (sample 2:7), 2 (samples 2:2 and 2:4) and 3 (sample 1:2), Blakeney esker.

that of Unit 1 (Fig. 6, Table 1), the two beds may apparently be distinguished by their contrasting matrix calcium carbonate-equivalent content (Table 4) and air-dry colour (Table 5). Yet the field-moist colours of the two units are similar (Table 6).

Pertinently, the occurrence of sharp-edged, boulder-sized inclusions of white reconstituted Chalk within very pale brown deposits at site 13 (Table 6) suggests that colour provides an unreliable guide to stratigraphic position.

Table 5 Air-dry colours of samples from Units 1 and 3

Unit	Sample	Munsell colour	
		Notation	Name
3	1:4	5Y 8/2	White
	1:5	10YR 8/2	White
	2:1	5Y 8/2	White
	12	2.5Y 8/2	White
	13:1	2.5Y 8/2	White
	13:2	2.5Y 8/2	White
1	2:6	10YR 7/3	Very pale brown
	2:7	10YR 5/6–	Yellowish brown–
		10YR 6/6	brownish yellow

Table 6 Field-moist colours of Units 1 and 3

Unit	Site	Munsell colour	
		Notation	Name
3	1	10YR 8/4	Very pale brown
	13	10YR 7/4	Very pale brown
	13[a]	10YR 8/1	White
	15	10YR 7/4	Very pale brown
1	2	10YR 5/6–	Yellowish brown–
		10YR 6/4	light yellowish
			brown
	2:4	10YR 6/6	Brownish yellow
	2:5[b]	10YR 5/6	Yellowish brown
	15	10YR 7/4	Very pale brown
	16	10YR 7/4	Very pale brown

[a] Boulder-sized inclusions of reconstituted Chalk.
[b] Air-dried sample moistened with distilled water.

As with Unit 1, the extremely poor sorting and the wide range of particle sizes of the deposit (Fig. 6, Table 1) are characteristic of a diamicton. The interpretation of this material as a till is supported by the presence of striated and facetted gravels, and of exotic clasts in the fine gravel fractions (Tables 2, 3).

Although generally massively bedded, some parts of Unit 3 at site 2 possess sub-horizontal structures. These may be the product of a range of processes including progressive accretion, shearing, the drawing out of crushed Chalk clasts and post-depositional unloading. In some places, these features are truncated by, and must therefore predate, the entrenchment of the top of the unit. Elsewhere, they are developed in sympathy with the topography of the surface of the bed and may represent unloading or deformational features. The existence of massive bedding and the possible presence of accretion and shearing structures are suggestive of accumulation by lodgement.

Unit 4

Overlying Unit 3 at sites 1 and 2 (Figs 4, 5) is a texturally variable, yellowish brown to dark yellowish brown deposit, possessing some internal stratification and composed of silty sand and matrix-supported, extremely poorly sorted, muddy sandy gravel (Fig. 7, Table 1). Unit 4 reaches a maximum thickness of 1.74 m. The junction between Units 3 and 4 is sharp and complex, consisting of deep, narrow channels cut into Unit 3, and a series of tongues, flames and detached fragments of the underlying till drawn up into the body of Unit 4 (Figs 4, 5). The flames and fragments may be the product of diapirism resulting from the loading, dewatering and thaw consolidation of Unit 3. Alternatively or additionally, they may be the result of the reworking of material from Unit 3 during the deposition of Unit 4, incorporating sediments from Unit 3 into the overlying bed.

The texture of Unit 4 at site 2 varies both horizontally and vertically. Silty sand occurs at the base of some of the channels, whilst elsewhere in the exposure rounded (and, less frequently, angular) cobbles are supported within a silty sand to muddy sandy gravel matrix. The gravels consist overwhelmingly of flint, although there are occasional clasts of calcareous till (Tables 2, 3). These till clasts may have been incorporated into the deposit by erosion of the underlying unit, but it is also possible that they are the product of diapirism. Exotic gravels make up >6% of the 4.0–5.7 mm fraction and >4% of the 5.7–8.0 mm fraction (Tables 2, 3).

The local channelling of the bed of Unit 4 is strongly indicative of fluid flow. Furthermore, the

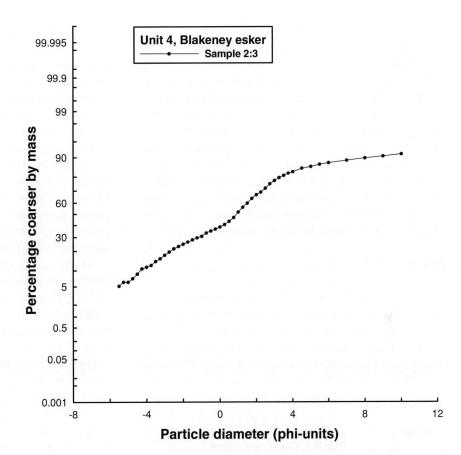

Fig. 7 The particle-size distribution of sample 2:3 from Unit 4, Blakeney esker.

absence of fines from some parts of the unit and the occasional internally stratified nature of the deposit provide convincing evidence that the material has been affected by water action. The fabric of the sediment is dominated by *a*-axis dips, characteristic of gravels transported in suspension in fluids (Rees, 1968; Allen, 1982) (Fig. 8). Nevertheless, the

extremely poorly-sorted nature of much of the sediment indicates that transport has generally resulted in little winnowing and that the material was derived from a similarly poorly-sorted source. Taken together, these features are indicative of fluid flows capable of scouring the substrate and transporting the entire load supplied to them. The small amount of sorting and stratification that can be identified may have occurred on the declining limb of one or more flow pulses.

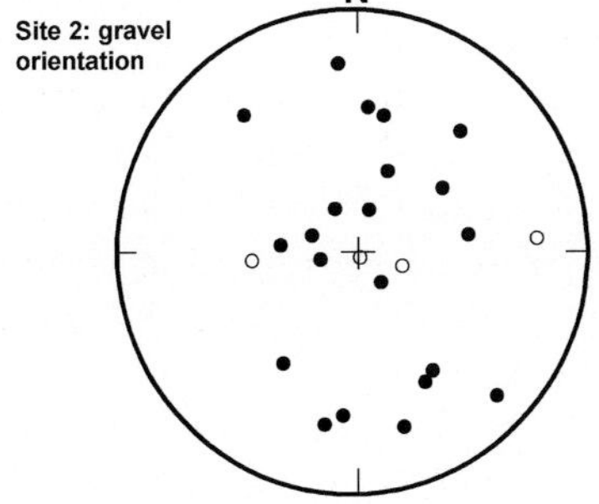

Unit 4 may have formed when the ice sheet was uncoupled locally from its bed at times of active subglacial meltwater flow. Relatively thin, coarse-grained, spatially disjunct patches of sediment may have been released from the basal ice and lowered onto the glacial bed with little modification. Energetic fluid flows might have transported this material to low pressure or topographically low areas; or it may have flowed into such locations beneath the weight of the ice. Saturated sediments are subject to syn- and post-depositional disaggregation and re-sedimentation (Menzies and Shilts, 1996), and these mechanisms

Fig. 8 The orientation of gravels of *a*-axis >20 mm, Unit 4, site 2, Blakeney esker: stereographic equal-angle projection. Closed circles represent *a*-axis dip, open circles represent *b*-axis dip.

may have given rise to the local absence of fines and the occasional internal stratification that may be recognised in Unit 4. However, Gray's (1988) contention that Unit 4 accumulated as a subglacial meltout till *sensu stricto* is difficult to reconcile with the contrasting lithological composition of the highly calcareous subjacent till (presupposing these units to be penecontemporaneous) (Tables 2 to 4). Assuming that the carbonate-rich Unit 3 accumulated by lodgement, the almost complete absence of calcium carbonate from the matrix and gravel fractions of Unit 4 is quite unlike that to be anticipated in a lodgement till–meltout till couplet.

Unit 5

At site 1, Unit 4 is overlain unconformably by yellow, reddish yellow and strong brown sand containing frequent stringers and lenses of very poorly sorted gravel (Fig. 4). The unit has a maximum thickness in excess of 9.65 m. The sand is typically of medium grade, is well-sorted, displays planar-bedding and tabular and trough cross-stratification, and includes occasional small lenses of coarse sand. The sand has been subject to small-scale normal and reverse faulting, and contains many examples of minor loading structures. The gravel largely forms a series of lenses, though lag deposits a single clast thick also occur. Some of the gravel lenses possess a basal layer of coarser clasts, again probably a channel lag. The lenses themselves are composed mainly of clast-supported gravel with a matrix of sandy gravel. Occasionally the gravel exhibits matrix support. The gravel possesses a bimodal particle-size distribution, with peaks in the coarse pebble and medium sand fractions (sample 1:1: Fig. 9, Table 1).

At site 1, Unit 5 fills channels up to several metres deep incised into Unit 4 (Fig. 4). At TG 030422, Gray (1988, *c.* 1994, 1997) recorded Unit 5 infilling similar channels tens of metres wide and several metres deep entrenched into till (probably Unit 3).

Unit 5 can be traced laterally to sites 3, 4 and 5, and on to site 8. In these sections, the unit is made up largely of well sorted, medium sand displaying

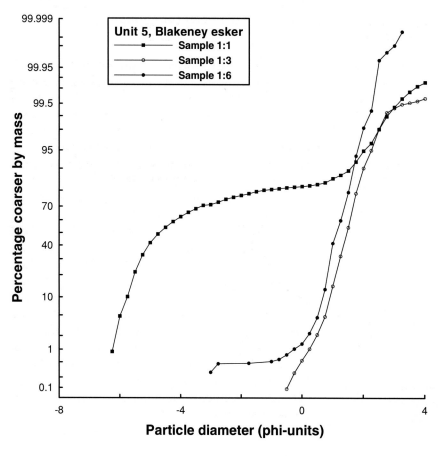

Fig. 9 The particle-size distribution of samples from Unit 5, Blakeney esker.

planar-bedding and tabular and trough cross-stratification. The sand contains only occasional stringers of gravel, and gravel lenses are absent.

The presence of lenses of gravel and coarse sand, tabular and trough cross-bedded sand, lag gravel at the base of channels, and channels scoured into the underlying bed is compelling evidence that Unit 5 was laid down under conditions of fluid flow. The marked imbrication of the maximum projection plane of the gravel (Fig. 10) is indicative of bedload transport of the sediments (Potter and Pettijohn, 1963; Johansson, 1965; Walker, 1975).

This interpretation is supported by the evidence of gravel lag deposits and by the occurrence of clast-supported gravel in the lenses. Similarly, the particle-size distributions of the sand and gravel are compatible with a fluvial origin for the unit (Gale and Hoare, 1991) (Fig. 9, Table 1). The deposits of Unit 5 at site 1 include a vertical sequence of four gravel lenses and a deep, sand-filled channel (Fig. 4). The complex sequence of cut-and-fill episodes that these features represent was probably the result of rapid, lateral shifts of the channels as the sediments accumulated.

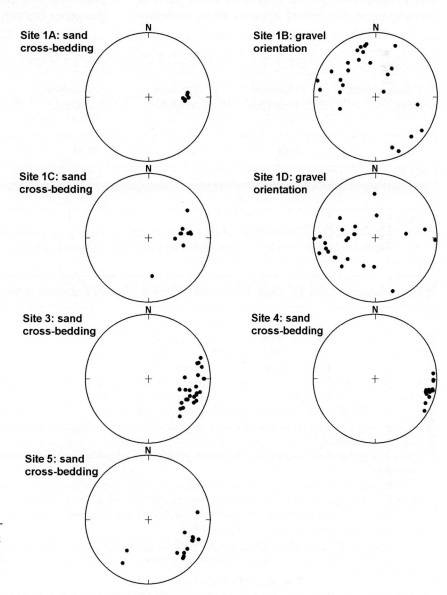

Fig. 10 The orientation of sand cross-bedding and the maximum projection plane of gravels of *a*-axis >20 mm, Unit 5, Blakeney esker: stereographic equal-angle projection.

The imbrication of the gravel from sites 1B and 1D, and the orientation of cross-beds in the sand at sites 1A, 1C, 3, 4 and 5 are shown in Figure 10. These reveal a consistent palaeoflow direction along the line of the ridge from west to east and from northwest to southeast.

The gravel fraction of Unit 5 is dominated by flint, Chalk is absent and exotic clasts make up >14% of the 5.7–8.0 mm fraction (Table 3).

Unit 6

Overlying the sand of Unit 5 at site 8 is a >7.38 m thick sequence of yellowish red to brownish yellow, matrix-supported, sandy gravel. In some parts of this exposure, the gravel appears to be conform-able with the subjacent sand of Unit 5; in other parts, however, the base of Unit 6 cuts across the bedding of the underlying unit. Elsewhere along the ridge, the gravel of Unit 6 infills channels cut into till (Sparks and West, 1964). The gravel extends to the present surface of the ridge at site 8. Unit 6 also forms the uppermost sedimentary unit in all other cases where the materials making up the top of the ridge can be seen.

Unit 6 is generally massively bedded and displays inverse grading at a number of sites (Fig. 11). The gravel clasts are made up largely of rather spherical, angular to sub-rounded flint (Tables 2, 3, 7). Significant proportions of the flints possess a patina on one face and secondary rounding along the other fractured edges (Table 8). The clasts are

Table 7 The mean shape of flint clasts in the 22.5–27.0 mm fraction of sample 9 from Unit 6

Oblate–prolate index	Maximum projection sphericity	Modified Wadell roundness	Krumbein's intercept sphericity	Krumbein's visual roundness	Number of clasts in fraction
−0.14	0.68	0.09	0.71	0.41	78

Table 8 The surface characteristics of flint clasts in the 5.7–8.0 mm fraction of sample 17 from Unit 3, the 5.7–8.0 mm fraction of sample 1:1 from Unit 5 and the 13.2–16.0 mm fraction of sample 10 from Unit 6: percentage frequency of flint clasts by number

Unit	Sample	Class 1	Class 2	Class 3	Class 4	No. of flint clasts in fraction	Total no. of clasts in fraction
6	10	0.0	83.8	14.9	1.4	74	76
5	1:1	0.0	84.8	15.2	0.0	197	215
3	17	26.9	37.2	20.6	15.4	78	not available

Class 1: cortex[a] has survived on part of surface, remaining parts may or may not be patinated[b].
Class 2: no cortex; the surface of the clast possesses an inherited patina, although the clast may have been fractured subsequently and some rounding or other modification of new faces and edges may have occurred.
Class 3: surface completely free of cortex or patina; some rounding or other modification of new faces and edges may have occurred.
Class 4: angular clast, a small part of which possesses a pitted surface, remaining parts may or may not be patinated.
[a] 'Flint nodules extracted directly from chalk bedrock possess an irregular, pitted surface or cortex. This cortex is usually coated by a white layer of either imperfectly formed or partially decomposed flint . . . a few millimetres in thickness . . . This coating is relatively soft, and is therefore rapidly lost once the flint has been removed from the chalk by natural agencies.' (Gibbard, 1986).
[b] Weathering of flint initially produces a bleached, often blotchy surface colouring and a 'polished' or 'varnished' surface. As weathering progresses, however, a porous, often brown-coloured surface rind or 'patina' develops (Gibbard, 1986).

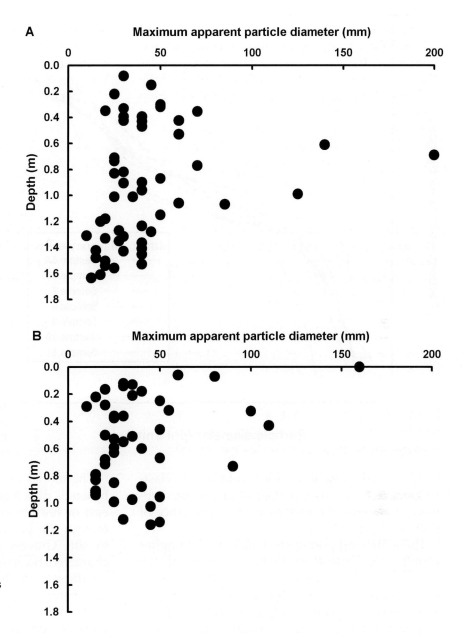

Fig. 11 Maximum apparent particle diameter plotted as a function of depth below the top of Unit 6 at sites (A) 8 and (B) 10. All clasts of diameter >20 mm were measured.

matrix-supported and the sediments typically form sandy gravel (Table 1). They possess markedly bimodal particle-size distributions with peaks in the coarse pebble and medium to fine sand grades; there is a notable absence of granules and coarse sands (Fig. 12). The deposits are very poorly sorted and very positive (fine)-skewed (Table 1). It should be stressed, however, that these characteristics are, at least in part, statistical artefacts resulting from the application of measures devised for use with unimodal data sets to bimodal distributions.

Three samples of Unit 6 (9, 10 and 11) were selected in order to determine the lithology of the

5.7–8.0 mm fraction (Table 3). The three samples are statistically closely comparable, irrespective of whether the entire assemblage or the durable component (flint, quartzite and vein quartz) only is considered (entire: $\chi^2 = 9.03$, $df = 4$, $\alpha = 0.01$; durable: $\chi^2 = 4.25$, $df = 2$, $\alpha = 0.01$). Justification for focusing attention on these most obdurate lithologies is provided by Gale and Hoare (1991). The samples may therefore be regarded as derived from the same background population and may be treated as a single bulk sample of the unit.

It will be clear from the foregoing that Moorlock *et al.*'s (2002) description of this unit as composed

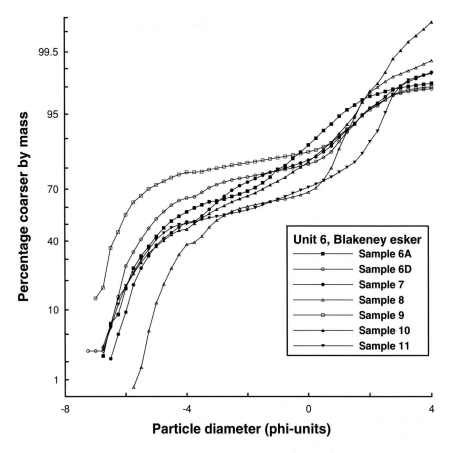

Fig. 12 The particle-size distribution of samples from Unit 6, Blakeney esker.

'. . . predominantly of rounded, cobble-size, clast-supported, flint gravel devoid of any exotic pebbles . . .' is (with the exception of the presence of flint) wrong on all counts.

The matrix support of the Unit 6 gravel implies simultaneous transport and deposition of the coarser and finer components of the material. Transport must therefore have been by some process capable of maintaining the large particles in suspension in fluid flow. Possible mechanisms include conventional fluid turbulence, dispersive pressure (Bagnold, 1954), grain contact (Pierson, 1981), Bernouillian lift (Fisher and Mattinson, 1968; Mattinson and Fisher, 1970), boundary lift (Rubinow and Keller, 1961; Saffman, 1965) and buoyancy enhancement in high-density suspensions (Rodine and Johnson, 1976; Hampton, 1979; Pierson, 1981). The existence of inverse grading suggests the operation of dispersive stresses (Bagnold, 1954), kinematic sieving (Middleton, 1970), the formation of pressure gradients due to streamline deflection around particles (Mattinson and Fisher, 1970) or some similar mechanism (but see Legros, 2002).

The fabrics of the Unit 6 gravel are dominated by *a*-axis dips (Fig. 13). Most possess a clear west–east or northwest–southeast orientation (Fig. 13), though the preferred inclinations vary from site to site. Gravels carried in suspension in fluids characteristically possess *a*-axis fabrics parallel to the direction of flow, usually (but not always) with an upstream inclination (Johansson, 1965; Rees, 1968; Allen, 1982). The flows that laid down Unit 6 therefore appear to have taken place along the line of the ridge. More specifically, we interpret the flow as having occurred from northwest to southeast. Support for this conclusion comes from two sources. First, a flow from northwest to southeast along the line of the ridge is identical to that determined for Unit 5 conformably beneath the gravels. Secondly, Banham (1977) recorded cross-strata dipping to the southeast in Wiveton Downs pit (TG 031422) towards the southeastern end of the esker.

Evidence apparently contradictory to the picture of southeasterly flow comes from West (1957), Straw (1960) and Sparks and West (1964), who noted that the bedding (or perhaps the gravels) of Unit 6

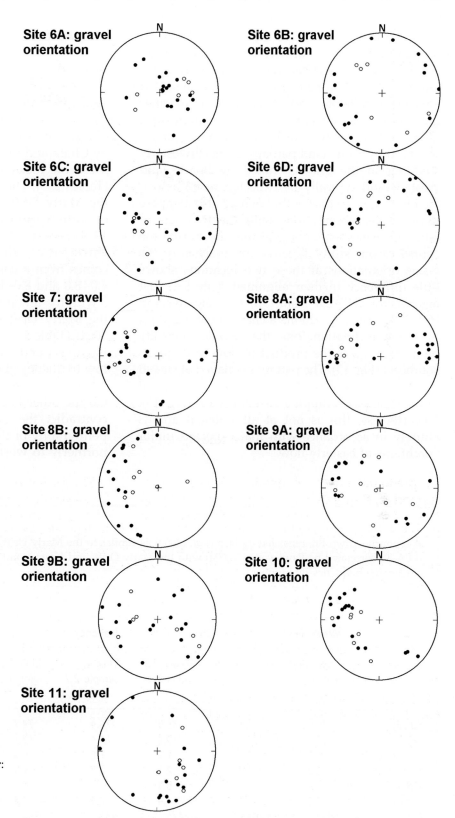

Fig. 13 The orientation of gravels of *a*-axis >20 mm, Unit 6, Blakeney esker: stereographic equal-angle projection. Closed circles represent *a*-axis dip, open circles represent *b*-axis dip.

at the northwestern end of the esker (TG 016436) dipped to the north, northwest and west (though Sparks and West's [1964] report is open to other interpretations). None of these authors presented any measurements and it is unclear whether their observations were of real or apparent dips. More pertinently, the northward-dipping gravels observed by Straw (and perhaps the northwesterly 'bedding in the gravels' noted by Sparks and West) would typically be expected to be associated with flow to the south. We may also compare these earlier observations with the two sets of fabric measurements made at this site during the course of our study. Eigenvector analysis of the *a*-axis orientations at these two locations shows little difference in their alignment (site 9A, first eigenvector: 117–297°; site 9B, first eigenvector: 099–279°). Site 9A exhibits a characteristic pattern of *a*-axis inclination, one that would typically be interpreted as the product of flow to the east-southeast (Fig. 13). The pattern displayed at site 9B is more equivocal, with an approximately equal number of *a*-axes dipping east and west (Fig. 13). Nevertheless, this is not at all unusual and is entirely in accordance with that expected under conditions of easterly flow.

DISCUSSION

Units 1–3

Two calcareous tills and an intervening unit of glaciofluvial sand lie beneath the Blakeney ridge. The gravel assemblage of the tills is dominated by Chalk and flint, with low percentages of other Mesozoic rocks, of quartzite and of vein quartz (Tables 2, 3). The tills are closely comparable with the Marly Drift that blankets much of this part of north Norfolk (Banham *et al.*, 1975; Perrin *et al.*, 1979; Fish *et al.*, 2000; Fish and Whiteman, 2001; Pawley *et al.*, 2004). Support for this assessment comes from a comparison of the lithology of the 4.0–8.0 and 8.0–16.0 mm fractions of Units 1 and 3 with the few published counts on material from the Marly Drift (Fish *et al.*, 2000; Pawley *et al.*, 2004) (Table 9). The source of most of the gravels in Units 1 and 3 is likely to be very local, and the low frequency of exotic clasts does not permit any reliable assessment of the direction of transport. On the other hand, the gravel lithology does not contradict the evidence of other work on the petrography of the Marly Drift, which suggests a northerly to westerly provenance (Banham *et al.*,

Table 9 The lithology of the 4.0–8.0 mm and/or 8.0–16.0 mm fractions of samples 4ac and 6 from units 4 and 6, Weybourne Town Pit, regarded by Fish *et al.* (2000) as similar to the Marly Drift, the Weybourne Town Till (LFA-3), correlated by Pawley *et al.* (2004) with the Marly Drift, and Units 1 and 3 of the Blakeney esker: mean percentage frequency by number

| Lithology | 4.0–8.0 mm fraction | | | | | 8.0–16.0 mm fraction | |
| | Weybourne | | | Blakeney | | | |
	Town Pit 4ac	Town Pit 6	Town Till (LFA-3)	Unit 1, sample 2:7	Unit 3, sample 1:2	Weybourne Town Till (LFA-3)	Blakeney Unit 1, sample 2:7
Chalk	80.0	81.1	75.8	76.4	74.7	76.3	75.8
Flint	15.4	15.3	19.4	16.4	19.3	22.1	19.7
Quartzite	0.5	0.2	1.0	1.6	0.3	0.4	1.5
Vein quartz	2.2	1.4	2.1	2.7	2.1	0.4	3.0
Chert			0.3			0.4	
Other	2.5	2.1	1.4	2.9	3.6	0.4	
n	1053	1284	1775	444	336	262	66

1975; Fish *et al.*, 2000; Fish and Whiteman, 2001; Riding, 2001; Pawley *et al.*, 2004).

Banham *et al.* (1975) noted two sections in which the generally darker, less calcareous, Cromer-type Marly Drift passes up into the generally paler, more calcareous, Lowestoft-type, but none in which the reverse succession could be seen. Although there is some evidence of a similar succession beneath the Blakeney ridge, we hesitate to assign Units 1 and 3 to these easy lithostratigraphic categories given the considerable variation in colour and, to a lesser extent, carbonate content within each of these units. In addition, the gravel lithologies of Units 1 and 3 are statistically closely comparable, irrespective of the size fraction investigated or whether the entire assemblage or the durable component only is considered (4.0–5.7 mm entire: $\chi^2 = 7.27$, $df = 4$, $\alpha = 0.01$; 4.0–5.7 mm durable: $\chi^2 = 4.03$, $df = 1$, $\alpha = 0.01$; 5.7–8.0 mm entire: $\chi^2 = 2.53$, $df = 2$, $\alpha = 0.01$; 5.7–8.0 mm durable: $\chi^2 = 0.18$, $df = 1$, $\alpha = 0.01$). This being the case, the sub-ridge sequence may have been the product of a single ice sheet that deposited the lower till and the overlying glaciofluvial sands and was then reactivated to deposit the upper till. It is also possible that the sands represent deposition during the early part of the phase of reactivation. Alternatively, but perhaps less likely given the similarity of the two tills, the stratigraphy may indicate the retreat and advance of different ice bodies separated by a phase of sub-aerial deposition. Whatever explanation is invoked, the presence of glaciofluvial deposits between the two Marly Drifts contradicts the observations of Ehlers *et al.* (1987, 1991) that there is no evidence of meltwater deposition or ice melt between the older and the younger Marly Drift units.

Since the gravel assemblages of Units 1 and 3 may be regarded as derived from the same background population, the data may be grouped to improve the sample size and treated as a single bulk sample of the till.

Units 4 to 6

Overlying the lower part of the succession along the line of the Blakeney ridge is a series of deposits laid down by glacially related fluid flows. Some evidence of the penecontemporaneity of these units and the underlying till was provided by Straw (1965), who noted the interdigitation of the basal

layers of the Blakeney ridge sands and gravels with Marly Drift towards the northwestern end of the ridge (TG 017438). This site is no longer exposed, unfortunately, so it is not possible to establish whether the interfingering is a primary depositional feature or the product of processes such as ice thrusting or diapirism. It is noteworthy, however, that P.H. Banham (pers. commun., 1993) recorded the diapiric intrusion of Marly Drift into the overlying sands and gravels at this site. Interestingly, Marly Drift is also found interfingered with deposits very similar to those of Units 4–6 within a few kilometres of the Blakeney ridge at Bilsey Hill (TG 024415), 'Smoker's Hole' (TG 032405, though this is not the true Smoker's Hole) and Stody (TG 056346) (Ehlers *et al.*, 1987) (Fig. 1b).

Measurements of palaeoflow directions in Units 5 and 6 show a concordance of alignment (Figs 10, 13), whilst measurements of the channels within which these units were deposited reveal a close correspondence with flow direction in the overlying beds (Figs 10, 14). Since we should expect that flow directions in the channels would be constrained by the channels themselves, the accordance of fabric and channel data in one part of the esker enhances our confidence in the fabric measurements and strengthens our belief that the fabrics provide a reliable measure of flow direction elsewhere. This entire assemblage of palaeoflow data offers strong support for the thesis that

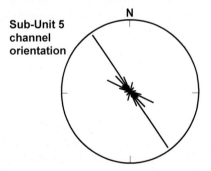

Fig. 14 The orientation of the channels beneath the Blakeney ridge within which Unit 5 was deposited. The data represent the percentage of total channel length within 10° class intervals. The measurements were abstracted from the survey made by Gray (1997) in the area between sites 1 and 8 (Fig. 2). Orientations expressed as a function of the surface area or volume of the channels would yield even more highly clustered distributions. Measured channel length = 723 m.

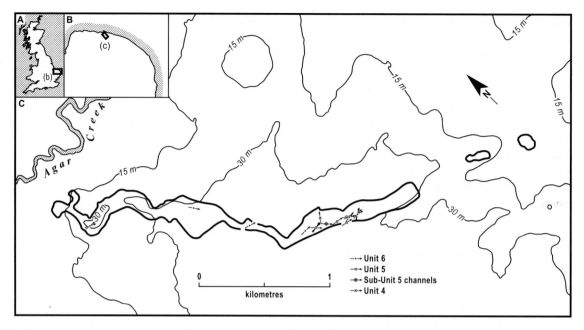

Fig. 15 The Blakeney esker showing palaeoflow directions during Units 4, 5 and 6 times. The resultant flow directions are derived from eigenvector analysis of a range of orientation data and represent the direction of the first eigenvector. The cross-bedding within the sand of Unit 5 dips downstream. The maximum projection planes of the clast-supported gravel of Unit 5 dip upstream. The sub-Unit 5 channels are aligned parallel to the direction of flow. The interpretation of flow direction in these features is based on that determined for Unit 5, which infills the channels. The *a*-axes of the matrix-supported gravel of Units 4 and 6 are also characteristically aligned parallel to the direction of flow. During the period of deposition of these units, flow is interpreted as having been generally to the east. This assessment is based on the flow directions reconstructed for Unit 5, which overlies Unit 4 and conformably underlies Unit 6.

Units 5 and 6, and perhaps Unit 4, were deposited by successive episodes of flow along essentially the same path.

The flows that deposited Unit 4 incised channels into the underlying till, whilst those that laid down Unit 5 entrenched deep channels into both Unit 4 and the till. The flows that deposited Unit 6 cut into both the till and Unit 5 (they are also likely to have entrenched Unit 4, though no direct evidence of this has yet been found). The period represented by Units 4 to 6 must therefore have been characterised by repeated incision and reworking of sediment along the line of the esker.

The origin of the Blakeney ridge

The correspondence between the palaeoflow direction of the sands and gravels that form the Blakeney ridge and the alignment of the ridge itself makes the assertion of Straw (1960, 1965, 1973) and more recently Hamblin *et al.* (2000) that the ridge is erosional in origin extremely difficult to accept

(Fig. 15). It also makes a morainic genesis, as proposed by West (1957), but later abandoned, highly unlikely. Furthermore, the clear evidence of a south-easterly flow of water in an apparently upslope direction (Fig. 3) and diametrically opposed to that suggested by Sparks and West (1964, 1970) undermines their suggestion that the ridge represents crevasse-fill deposits resulting from northwesterly open channel flow along intersecting rectilinear fractures in an ice sheet. Instead, the concordance between the palaeoflow direction in the sands and gravels, the alignment of the sediment body and the orientation of the ridge, as well as the morphology of the ridge and its relationship to the underlying topography strongly suggest that the feature is an esker.

Support for this assessment comes from several quarters. First, although matrix-supported gravel of the sort that makes up Unit 6 may be deposited under a range of conditions, materials of identical character have been widely recognised as common components of eskers (Saunderson, 1977; Ringrose,

1982; Lindström, 1985; Gorrell and Shaw, 1991). Secondly, the low variability of the palaeocurrent indicators with respect to the plan form of the deposit (Fig. 15) is strongly indicative of unidirectional flow laterally constrained by the walls of a conduit. Thirdly, uphill flow can have taken place only under hydrostatic pressure and with full-pipe discharge. Such conditions are unlikely to have existed hereabouts except beneath or within a glacier. Fourthly, flows in a southeasterly direction are in accordance with the glaciohydraulic gradients likely to have existed near the southern edge of an ice mass occupying the North Sea Basin.

The environment of esker formation

The Blakeney esker is part of a complex assemblage of glaciofluvial landforms, including outwash plains and kame-like mounds and ridges, that dominates the landscape of the lower Glaven valley (Fig. 1c). The formation of eskers at the margins of ice sheets (Shreve, 1985; Hambrey, 1994) and the limited potential for the preservation of glaciofluvial landscapes beneath advancing ice strongly suggests that this assemblage is the product of glacial stillstand or stagnation. This may have been either the maximum stage of a readvance, a major halt during retreat or an episode of downwasting *in situ* (Boulton *et al.*, 1984).

Although Hart and Boulton (1991) have argued that the Blakeney esker formed englacially, there is compelling evidence that the flows that formed the esker must have been subglacial. Thus, the base of Unit 4 infills channels scoured into the underlying till and the base of Unit 5 infills channels scoured into the underlying Unit 4. In addition, Sparks and West (1964) observed channels cut into the till and infilled by Unit 6 gravel, whilst channels developed in the till surface and infilled by the sand and gravel of Unit 5 may be traced for up to 375 m parallel to the line of the ridge (Gray, 1988, *c.* 1994, 1997). Furthermore, although Price (1966) has shown that ridge forms can be preserved as ice melts out from beneath englacial esker gravels, it is difficult to conceive of materials being let down an appreciable distance without intense disruption of sedimentary structures. The sediments of Unit 5 display evidence of minor faulting (Fig. 4), but this is explicable in terms of the loss of support caused by the melting of the pipe

walls and is clearly not the product of englacial sediments being released onto the till surface. Indeed, there is no evidence of large-scale collapse in any part of Units 4 to 6.

Gray (*c.* 1994, 1997) and Bennett and Glasser (1997) interpreted the features cut into Unit 3 as Nye channels. Gray (*c.* 1994, 1997) argued that the channels were subsequently over-filled by sands and gravels to produce the Blakeney ridge. He suggested that flows thus changed from erosional to depositional, and that a single major channel of Röthlisberger-type developed. The incised channels are not strictly of Nye-type, however, since these develop only on rigid beds (Nye, 1973; Knight, 1999). Walder and Fowler (1994) and Clark and Walder (1994) considered the situation of water flowing at the interface between the ice and a saturated deformable substrate (till). Where water cannot escape into an underlying aquifer, they proposed that two types of conduit would develop. These cannot exist simultaneously and the type that forms depends on the hydraulic gradient (largely determined by the ice surface slope) and the rheology of the ice and till. Under deformable bed conditions and low hydraulic gradients, the drainage network should consist of 'canals' incised into the subglacial sediment with a more or less flat roof of ice. Flow will be relatively sluggish with velocities of ~0.1–0.3 m s^{-1}. Unless the sediment is extremely cohesive, these channels will be much wider than deep. The drainage system should consist of channels distributed more or less uniformly over the glacial bed, perhaps connected in a braided fashion. By contrast, where the substrate is rigid or where the ice has a deformable bed and hydraulic gradients are steep (~10^{-1}), a classical Röthlisberger channel will form. The drainage system will consist of an arborescent network with relatively few trunk channels along which flows occur at velocities of ~1–3 m s^{-1}.

Using this model as their basis, Clark and Walder (1994) argued that eskers (characterised by high velocity flow along trunk conduits) are the product of sedimentation along Röthlisberger channels and that these are likely to form only where the glacial bed is free of or discontinuously covered by sediment, or where the sediment does not pervasively deform under the stresses applied by the ice. Clearly, these conditions cannot have existed at Blakeney, a point made by Hart (1996) in her criticism of Clark and Walder's (1994) model. On the other hand, it is very likely that the Blakeney esker formed at the margin of the ice sheet, where hydraulic gradients are likely to have been of the order of 10^{-1} (Wright, 1973; Mathews, 1974; Augustinus, 1999; Weidick *et al.*, 2004) and where the model predicts that Röthlisberger channels would develop irrespective of the bed material. The simultaneous entrenchment of the glacial bed under these circumstances is entirely to be expected, given the relatively easily erodible nature of Unit 3, the bed upon which flow would largely have taken place.

The massively bedded, matrix-supported gravel of Unit 6 was transported in suspension under conditions of high boundary-shear stress. A range of transport mechanisms can give rise to deposits of this sort. It is thus not possible to discriminate between open and closed channel sediment transport from the evidence of the sediments alone. However, a clue to the environment of deposition of this gravel comes from the overall increase in height of the unit from northwest to southeast along the line of the ridge; that is, in a downstream direction. Such an increase in altitude could be the result of sediments being let down from within or upon the ice onto a northwest-sloping surface. Nevertheless, as shown above, there is no evidence to support this hypothesis. A down-ice increase in ridge altitude could also occur if the sediments were time transgressive, in other words, if they had been laid down with decreasing sediment yield at different stages as the ice retreated. But there is no indication that Unit 6 (the upper part of the ridge) represents anything but a single-stage feature; it possesses no evidence of beading, for example. Instead, it is highly likely that the southeasterly increase in height of Unit 6 is a primary depositional element. Since the flows that laid down this unit were directed to the southeast, this can only mean that the deposition of Unit 6 occurred under hydrostatic pressure and therefore under pipe-full conditions.

The maximum particle sizes of the sediments that make up Units 5 and 6 are very similar (Figs 9, 12), possibly evidence for source control on the range of particle sizes available for entrainment. Notwithstanding this, such particles were moved as bedload during Unit 5 times and carried in suspension during Unit 6 times. The increase in flow power during Unit 6 times by comparison with Unit 5 may have been caused by an increase in the hydraulic gradient of the glaciohydrological system. Equally, however, it may have been a result of conduit constriction, either by plastic deformation as an adjustment to a reduction in water pressure in the conduit, or by freezing on the conduit walls and roof, or because of deposition on the bed of the conduit.

Most esker paths run approximately parallel to the general direction of former ice movement (Shreve, 1972, 1985), particularly when deposition takes place from full-pipe flows (Hooke, 1984), as appears to have been the case here. The esker is likely to represent the last stages of glaciation in the area (see, for example, Shreve, 1985; Hambrey, 1994) and therefore the latest phases of deposition. Its southeasterly orientation is in broad accord with the petrographic evidence for the direction of movement of the Marly Drift ice. This may provide some support for the thesis that Units 1 to 3 and 4 to 6 date from the same cold stage.

Sediment provenance

The gravel lithologies of Units 1 and 3 are statistically indistinguishable (4.0–5.7 mm entire: $\chi^2 = 7.27$, $df = 4$, $\alpha = 0.01$; 4.0–5.7 mm durable: $\chi^2 = 4.03$, $df = 1$, $\alpha = 0.01$; 5.7–8.0 mm entire: $\chi^2 = 2.53$, $df = 2$, $\alpha = 0.01$; 5.7–8.0 mm durable: $\chi^2 = 0.18$, $df = 1$, $\alpha = 0.01$), as are those of Units 4 and 6 (4.0–5.7 mm durable: $\chi^2 = 0.28$, $df = 1$, $\alpha = 0.01$; 5.7–8.0 mm durable: $\chi^2 = 3.36$, $df = 1$, $\alpha = 0.01$). By contrast, the gravel of Unit 5 bears no comparison with that of either Unit 4 (5.7–8.0 mm entire: $\chi^2 = 38.97$, $df = 2$, $\alpha = 0.01$; 5.7–8.0 mm durable: $\chi^2 = 33.89$, $df = 1$, $\alpha = 0.01$) or Unit 6 (5.7–8.0 mm entire: $\chi^2 = 74.60$, $df = 2$, $\alpha = 0.01$; 5.7–8.0 mm durable: $\chi^2 = 51.63$, $df = 1$, $\alpha = 0.01$), but is closely comparable with that of the underlying tills (Unit 5 vs Unit 1, 5.7–8.0 mm durable: $\chi^2 = 0.07$, $df = 1$, $\alpha = 0.01$; Unit 5 vs Unit 3, 5.7–8.0 mm durable: $\chi^2 = 0.12$, $df = 1$, $\alpha = 0.01$; Unit 5 vs Units 1 and 3, 5.7–8.0 mm durable: $\chi^2 = 0.04$, $df = 1$, $\alpha = 0.01$). This implies that the deposits of Unit 5 were either reworked from the tills or derived from the same source as the tills within the ice. This assessment is supported by the relatively high proportions of igneous and oölitic limestone clasts within Unit 5. Both these lithologies occur in the tills, but are almost entirely absent from Units 4 and 6 (Tables 2, 3). The distinction between the tills and Unit 5, and Units 4 and 6 is supported by the comparison of their durable components (Tables 10, 11) (*contra* Straw, 1965, 1967).

By contrast, the surface characteristics of the flints from Units 5 and 6 are closely comparable, and are significantly different to those of Unit 3 (Table 8). The high frequency of Class 1 flints (those possessing some surviving cortex) in the till of Unit 3 implies that gravels were derived more or less directly from the bands of nodular flint that are abundant throughout the Upper Chalk. By contrast, the absence of Class 1 clasts in Units 5 and 6 strongly suggests that the hydraulically-transported sediments in the esker did not come directly from the erosion of the underlying Chalk. It is possible that any cortex that the flints possessed prior to their entrainment by the meltwater was removed during transport under highly energetic flow conditions. If this had been the case, however, any inherited patination is also likely to have been destroyed. Yet patination remains intact in 16.2% (Unit 6) and 17.8% (Unit 5) of the gravels.

The presence in Units 5 and 6 of patinated clasts possessing freshly-fractured faces, some of which have been rounded or otherwise modified during transport, strongly suggests that the patination is inherited and not the product of post-depositional weathering. The Class 2 gravels are thus likely to have been reworked from pre-existing deposits of weathered flint. These may have been the marine and estuarine sediments that were deposited in the North Sea Basin during the Early Quaternary or the fluvial and coastal gravels deposited in and around the North Sea during times of lowered sea level prior to the earliest glaciation of the region (Hey, 1976; Funnell *et al.*, 1979; Green and McGregor, 1990; Rose *et al.*, 1996, 1999, 2001). This material would also have been available for entrainment by the ice, perhaps explaining its significant presence in Unit 3.

Unit 5 thus appears to be dominated by weathered flint, probably from the North Sea Basin, with accessory exotic gravels either reworked from the tills or derived from the same source as the tills within the ice. Units 4, 5 and 6 are all trenched

Table 10 The lithology of the durable component (flint, quartzite, vein quartz) of the 4.0–5.7 mm fraction of samples from Units 1, 3, 4 and 6: percentage frequency by number

Lithology	Unit 1, sample 2:7	Unit 3, sample 1:2	Unit 4, sample 2:3	Unit 6, sample 9
Flint	70.6	88.5	95.9	96.7
Quartzite	9.8	1.9		0.4
Vein quartz	19.6	9.6	4.1	2.9
n	51	52	1086	451

Table 11 The lithology of the durable component (flint, quartzite, vein quartz) of the 5.7–8.0 mm fraction of samples from Units 1, 3, 4, 5 and 6: percentage frequency by number

Lithology	Unit 1, sample 2:7	Unit 3, sample 1:2	Unit 4, sample 2:3	Unit 5, sample 1:1	Unit 6, sample 9	Unit 6, sample 10	Unit 6, sample 11
Flint	90.2	90.5	98.6	90.3	98.3	96.2	98.0
Quartzite	4.9			3.1		0.6	0.7
Vein quartz	4.9	9.5	1.4	6.6	1.7	3.2	1.3
n	41	21	486	1275	233	1002	300

into the tills at some points along their flow paths beneath the esker. However, this is unlikely to have provided the source of the exotic gravels in Unit 5 (and such gravels are rare in Unit 4 and 6 deposits) since Chalk, which dominates the till assemblages, is effectively absent from Units 4 to 6 (Tables 2, 3).

The provenance of the sediments thus appears to have changed over time, from glacially quarried material during Units 1 and 3 times to deposits of weathered Cenozoic flints during Units 4, 5 and 6 times. On the evidence of their surface texture and ostensibly 'well-rounded' shape, it has long been argued that the Unit 6 gravels have been reworked from beach deposits (Straw 1965, 1967; Ranson cited in Banham 1977; Gibbard, 1986). Attractive though this argument is, the beach gravel thesis has three flaws. First, the percussion marks that are found on some of the Unit 6 gravels and which have been interpreted as the product of beach processes, also appear to develop in other environments (Gale and Hoare, 1991). It is considered that insufficient is known of the genesis of these features for them to be used as reliable environmental indicators. In any case, the incidence of percussion marks is much less than is implied by previous workers. Of the 74 flints examined in the 13.2–16.0 mm fraction of Unit 6 (Table 8), only one shows signs of 'percussion'. Secondly, beach gravels are typically well or very well sorted (Gale and Hoare, 1991). If such gravels had been transported by glacially related fluid flows capable of entraining a relatively wide range of particle sizes, we might have anticipated that the resultant sediments would have largely retained their beach character. Yet the gravels of Unit 6 are very to extremely poorly sorted (Table 1) and are most unlikely to represent reworked beach deposits alone. Thirdly, the mean shape of the esker gravels (Table 7) lies firmly within the fluvial environmental envelope defined by the oblate–prolate index and maximum projection sphericity (Dobkins and Folk, 1970; Gale and Hoare, 1991). Furthermore, the angular to sub-rounded shape of the Unit 6 flints is similar to that of flints from fluvial environments and contrasts with the dominantly rounded shape of flints from beach and shallow marine environments (Bridgland and D'Olier, 1995; Bridgland et al., 1997; Bridgland, 1999, 2000). It would thus appear more likely that the esker gravels were derived from sources of weathered flint other than those possibly represented by beach deposits.

Woodland (1970) has proposed that a major subglacial conduit lay along the line of the modern Glaven valley (Fig. 1b). He speculated that flow along this feature was to the north, but it is impossible to demonstrate this with confidence since the deepest point lies mid-way along the known length of the channel, since the boreholes that provide the evidence for the buried valley do not necessarily intersect the valley axis and since the subglacial stream thought to have formed this feature is likely to have had an up-and-down long profile. More pertinently, we know that similar, broadly north–south aligned subglacial valleys were trenched across the southern North Sea Basin in Anglian times (see, for example, Balson and Cameron, 1985; British Geological Survey and Rijks Geologische Dienst, 1986; Long et al., 1988; Wingfield, 1989, 1990; Balson and Jeffery, 1991; and Cameron et al., 1992) (Fig. 16). We hypothesise that these offshore valleys may have extended onshore, perhaps becoming the tunnel valleys identified by Woodland. The lower parts of the offshore valleys are infilled by poorly bedded, coarse-grained material that may represent glaciofluvial, glaciolacustrine or glaciomarine sediments, till or slump deposits (Balson and Cameron, 1985; British Geological Survey and Rijks Geologische Dienst, 1986; Long et al., 1988; Balson and Jeffery, 1991; Cameron et al., 1992). This may be similar to the material known to infill the onshore palaeovalley.

From the evidence of their size and spacing, it would appear that the subglacial valleys were stable, long-lived features. Indeed, Ehlers and Linke (1989) have argued that, although such valleys may have been formed by catastrophic outbursts of meltwater, they must have been resculptured by glacial ice and enlarged by subsequent meltwater flows. These flows took place along stable paths whose locations were determined by the hydraulic gradient of the ice sheet. In addition, we suggest that the existence of linear zones of deeper ice along the subglacial valleys would have acted to focus englacial and subglacial meltwaters, thereby enhancing and maintaining these flow lines. We speculate that the esker represents the final episode of deposition along the flow path that cut the Glaven palaeovalley (Fig. 1b). At first sight, the discrepancy between the size of

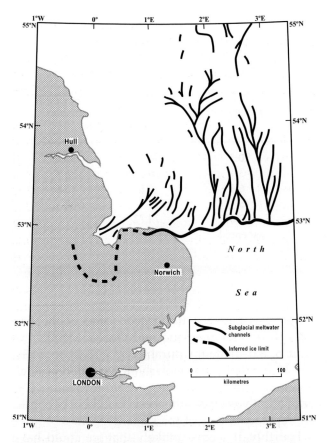

Fig. 16 The palaeogeography of eastern England and the southern North Sea Basin during the maximum stage of a readvance or a major halt during the retreat of the Anglian ice. The pattern of subglacial meltwater channels and the inferred ice limit are based on the distribution of the Swarte Bank Formation (Cameron *et al.*, 1992) and the location of the Cromer ridge (Fig. 1B).

have been capable of transporting weathered flint from the sea floor into north Norfolk. By contrast, much of the till upon which the esker lies appears to have been locally derived.

Such a model may also provide an explanation for the frequent occurrence in the Glaven valley region of those deposits of spherical, patinated flints known as the cannon-shot gravels (Wood and Harmer, 1872; Moir, 1928; Solomon, 1932; Baden-Powell and Moir, 1942; Straw, 1965; Green and McGregor, 1990). These may have been entrained from the floor of the North Sea and carried along the conduit system to the ice front where they were reworked into the glaciofluvial deposits of the lower Glaven valley. The model may also explain the occurrence of those large, fan-like deposits composed in part of cannon-shot gravels known as the Salthouse and Kelling sandar (Fig. 1c). These may represent the outwash fans formed at the mouth of the tunnel valley. Evidence in support of a genetic link between the sandar and the sub-esker tills comes from the presence of large masses of diamicton in the Salthouse and Kelling deposits. Such deposits are likely to have been reworked, at least in part, into the glaciofluvial gravels with which they are interbedded. Comparison of the lithology of the Salthouse sandur gravels with that of the gravels that make up Units 1, 3 and 5 reveals that they are closely similar (Table 12), supporting the suggestion that they are representatives of the same episode of deposition.

The age of the Blakeney esker and its associated sediments

There has been considerable speculation concerning the exact age of the Blakeney esker, the sediments of which it is composed and the tills that underlie it. Most of this is based on lithostratigraphic comparison with other Quaternary units in northern East Anglia. These comparisons cannot therefore be regarded as necessarily of chronostratigraphic significance. Nevertheless, a number of statements can be made regarding the stratigraphic position of the esker deposits. First, if the esker gravels do indeed interfinger with the underlying Marly Drift (Straw, 1965), this may suggest that Unit 6 and either Unit 1 or Unit 3 are penecontemporaneous. Units 4 and 5 (and possibly Unit 2) must therefore date from the same episode.

the esker and that of the underlying valley may appear to indicate that the features are unlikely to be genetically related. However, Shreve (1972) has argued that wide tunnel valleys and small eskers may form with no change in discharge, and Boulton and Hindmarsh (1987) have demonstrated that tunnel valleys may be excavated by relatively small streams. Indeed, there are numerous examples of eskers associated with, at the mouths of and developed along the axes of tunnel valleys and deep troughs cut by subglacial meltwaters (Michalska, 1969; Wright, 1973; Krüger, 1983; Mooers, 1989, 1990). We therefore suggest that the esker may represent the distal end of a glacial conduit system that originated within the North Sea Basin (Fig. 16), and that such a conduit would

Table 12 Statistical comparison of the lithology of the durable component (flint, quartzite, vein quartz) of the 5.7–8.0 mm fraction of the Salthouse sandur with that of Units 1 and 3, and 4, 5 and 6 of the Blakeney esker using the χ^2 test. Degrees of freedom = 1 in all cases. The Salthouse sandur sample comes from TG 054416

Lithology	Units 1 and 3, samples 2:7 and 1:2	Unit 4, sample 2:3	Unit 5, sample 1:1	Unit 6, samples 9, 10 and 11	Salthouse sandur
Flint	90.3	98.6	90.3	96.9	88.0
Quartzite	3.2		3.1	0.5	6.4
Vein quartz	6.5	1.4	6.6	2.6	5.6
n	62	486	1275	1535	987
χ^2	0.11	45.02	2.67	74.72	
H$_0$ at 0.01 significance level	accepted	rejected	accepted	rejected	

This contradicts those workers (including, most recently, Hamblin, 2000; Hamblin *et al.*, 2000, 2003; and Moorlock *et al.*, 2002) who have maintained that Units 4 to 6 represent the deposits of a later cold stage than the underlying tills.

Secondly, although it is not possible to determine in which cold stage these deposits were formed, it is unlikely that they are of Devensian age, as suggested (but rapidly abandoned) by West (1958), and by Woodland (1970) and Booth (in Lawson and Allen, 2000). This is because the maximum extent of Devensian ice in East Anglia lies a kilometre northwest of the northernmost end of the esker (Fig. 1), despite the evidence-free assertions to the contrary of Moorlock *et al.* (2002). Furthermore, the Devensian tills and the Marly Drift differ markedly in lithology (except in rare cases [Gale and Hoare, unpublished data]). More equivocally, till lithologically similar to the Marly Drift, and thus comparable with the Marly Drift facies of Units 1 and 3, is stratigraphically separated from Devensian till by warm-stage beach deposits at Morston (TF 98714406), 3 km to the west of the esker (Gale *et al.*, 1988).

Thirdly, in the most recent version of its scheme, the British Geological Survey has sought to correlate the esker gravels with the Briton's Lane Sand and Gravel Member, claimed to be of penultimate glacial age (Hamblin *et al.*, 2003). In Table 13, the lithologies of the durable components of gravels from Briton's Lane are compared with those of Unit 6 of the Blakeney esker. There are significant differences between the gravel suites at the two sites (Briton's Lane Sand and Gravel Member vs Unit 6, 8.0–16.0 mm durable: $\chi^2 = 12.48$ to 54.50, $df = 1$, $\alpha = 0.01$), particularly the relatively high quartzite and low flint counts at Briton's Lane. This dramatically weakens the thesis that these two units are representatives of the same deposit.

Fourthly, it seems unlikely that ice could have advanced across the Blakeney esker without, at the very least, dramatically modifying its form. The well-preserved character of both its geomorphology and its sediments suggest this cannot have happened. If this is the case, the esker must date from the late stages of the last glaciation locally.

CONCLUSIONS

The Blakeney ridge, a landform whose origin and age have been debated for over a century, is shown here to be an esker formed in the late stages locally of a glaciation of Middle Quaternary age. The weight of evidence indicates that the esker was associated with the ice that laid down the Marly Drift of central north Norfolk, although this cannot be demonstrated conclusively. A sequence of glacially related fluid flow deposits, glaciofluvial sand and gravel, and matrix-supported gravel has been preserved within the esker, although the sedimentary record is almost certainly highly incomplete as a result of major flushing and erosional events in the tunnel.

Table 13 The lithology of the durable component (flint, quartzite, vein quartz and chert) of the Briton's Lane Sand and Gravel Member and Unit 6 of the Blakeney esker: percentage frequency by number. Briton's Lane gravel data from Moorlock *et al.* (2000, 2002)

| Lithology | 8.0–16.0 mm fraction | | | | | 16.0–32.0 mm fraction | | | | |
| | Briton's Ln | Blakeney | | | | Briton's Ln | Blakeney | | | |
	Sand and Gravel Member	Unit 6, sample 6D	Unit 6, sample 7	Unit 6, sample 9	Unit 6, sample 11	Sand and Gravel Member	Unit 6, sample 6A	Unit 6, sample 9	Unit 6, sample 10	Unit 6, sample 11
Flint	86.8	98.3	98.9	99.0	100.0	90.3	100.0	100.0	99.8	99.8
Quartzite	9.3					8.9				
Vein quartz	1.7	1.7	1.1	1.0					0.2	0.2
Chert	2.2					0.9				
n	1651	346	459	105	153	113	582	342	633	636

The flows that formed the esker were subglacial. They took place from northwest to southeast, with at least the final phase occurring under hydrostatic pressure. It is likely that the esker developed during a phase of stillstand and under steep hydraulic gradients close to the ice sheet terminus. Although the conduit may have functioned largely as a Röthlisberger channel, there is evidence of simultaneous entrenchment of the relatively easily erodible glacial bed. The pattern of southeasterly flow along the esker is indicative of the general direction of former ice movement. This is in broad accord with other evidence for the direction of movement of the Marly Drift ice.

The weathered flints that make up the bulk of the esker gravels are likely to have been reworked from weathered Early and Middle Quaternary deposits on the floor and margins of the North Sea Basin. The underlying tills, by contrast, are largely the product of glacial quarrying of the Upper Chalk. It is possible that the esker and the major subglacial conduit that lies along the line of the modern Glaven valley represent the distal component of the broadly north–south oriented subglacial valleys that were trenched across the southern North Sea Basin in Middle Quaternary times. The complex assemblage of proglacial glaciofluvial landforms found to the south of the esker may represent the products of transport along these valleys. The ice lobe that advanced into the southern North Sea Basin may have left similar buried valley–proglacial fan complexes elsewhere along the fringe of northern East Anglia.

ACKNOWLEDGEMENTS

We should like to thank Rodger Connell for innumerable discussions, some of which were related to the geology of the Blakeney esker, Dr Anna Wilson for assistance with fieldwork, Professor Philip Gibbard for thoughtful comments on flint provenance and Dr Jim Riding for providing us with a copy of his unpublished report on till palynology.

REFERENCES

Allen, J.R.L. (1982) *Sedimentary Structures Their Character and Physical Basis.* Elsevier, Amsterdam, 593 + 663 pp.

Allen, J.R.L. (1983) River bedforms: progress and problems. In: *Modern and Ancient Fluvial Systems* (Eds J.D. Collinson and J. Lewin), *Int. Assoc. Sedimentol. Spec. Publ.*, **6**, 19–33.

Anon. (1970) Locality 3 landforms in the Holt–Blakeney area. In: *Norwich Easter–1970* (Ed. G.S. Boulton), pp. 27–31. Quaternary Research Association, Norwich.

Ashton, N.M., Bowen, D.Q. and Lewis, S.G. (1995) Discussion on excavations at the Lower Palaeolithic site at East Farm, Barnham, Suffolk, 1989–1992. *J. Geol. Soc. London*, **152**, 571–574.

Augustinus, P.C. (1999) Reconstruction of the Bulgobac glacial system, Pieman River basin, western Tasmania. *Aust. Geogr. Stud.*, **37**, 24–36.

Baden-Powell, D.F.W. (1948a) Long-distance correlation of boulder clays. *Nature*, **161**, 287–288.

Baden-Powell, D.F.W. (1948b) The Chalky Boulder Clays of Norfolk and Suffolk. *Geol. Mag.*, **85**, 279–296.

Baden-Powell, D.F.W. and Moir, J.R. (1942) On a new Palaeolithic industry from the Norfolk coast. *Geol. Mag.*, **79**, 209–219.

Bagnold, R.A. (1954) Experiments on a gravity-free dispersion of large solid spheres in a Newtonian fluid under shear. *Proc. Roy. Soc. London A*, **225**, 49–63.

Balson, P.S. and Cameron, T.D.J. (1985) Quaternary mapping offshore East Anglia. *Modern Geol.*, **9**, 221–239.

Balson, P.S. and Jeffery, D.H. (1991) The glacial sequence of the southern North Sea. In: *Glacial Deposits in Great Britain and Ireland* (Eds J. Ehlers, P.L. Gibbard and J. Rose), pp. 245–253. Balkema, Rotterdam.

Banham, P.H. (1975) Glacitectonic structures: a general discussion with particular reference to the contorted drift of Norfolk. In: *Ice Ages: Ancient and Modern* (Eds A.E. Wright and F. Moseley), pp. 69–94. Seel House, Liverpool.

Banham, P.H. (1977) Wiveton Downs gravel pits, Blakeney. (TG 031 422). In: *East Anglia International Union for Quaternary Research X Congress 1977 Guidebook for Excursions A1 and C1* (Ed. R.G. West), pp. 31–32. Geo Abstracts, Norwich.

Banham, P.H., Davies, H. and Perrin, R.M.S. (1975) Short field meeting in north Norfolk, 19–21 October 1973. *Proc. Geol. Assoc.*, **86**, 251–258.

Bascomb, C.L. (1961) A calcimeter for routine use on soil samples. *Chemistry and Industry*, **1961**, 1826–1827.

Beets, D.J., Meijer, T., Beets, C.J., Cleveringa, P., Laban, C. and van der Spek, A.J.F. (2005) Evidence for a Middle Pleistocene glaciation of MIS 8 age in the southern North Sea. *Quatern. Int.*, **133–134**, 7–19.

Bennett, M.R. and Glasser, N.F. (1997) *Glacial Geology: Ice Sheets and Landforms.* John Wiley, Chichester, corrected edn, 364 pp.

Boswell, P.G.H. (1914) On the occurrence of the North Sea Drift (Lower Glacial), and certain other brick-earths, in Suffolk. *Proc. Geol. Assoc.*, **25**, 121–153.

Boswell, P.G.H. (1935) Geology of the Norwich district. In: *A Scientific Survey of Norwich and District* (Ed. R.H. Mottram), *Report of the British Association for the Advancement of Science Annual Meeting, 1935 Norwich September 4–11*, Appendix, 49–59.

Boulton, G.S., Cox, F.C., Hart, J.K. and Thornton, M.H. (1984) The glacial geology of Norfolk. *Bull. Geol. Soc. Norfolk*, **34**, 103–122.

Boulton, G.S. and Hindmarsh, R.C.A. (1987) Sediment deformation beneath glaciers: rheology and geological consequences. *J. Geophys. Res.*, **92**, 9059–9082.

Bowen, D.Q. (1999) On the correlation and classification of Quaternary deposits and land–sea correlations. In: *A Revised Correlation of Quaternary Deposits in the British Isles* (Ed. D.Q. Bowen), *Geol. Soc. Spec. Rep.*, **23**, 1–9.

Bowen, D.Q., Hughes, S., Sykes, G.A. and Miller, G.H. (1989) Land–sea correlations in the Pleistocene based on isoleucine epimerization in non-marine molluscs. *Nature*, **340**, 49–51.

Bowen, D.Q., Rose, J., McCabe, A.M. and Sutherland, D.G. (1986) Correlation of Quaternary glaciations in England, Ireland, Scotland and Wales. *Quatern. Sci. Rev.*, **5**, 299–340.

Bridgland, D.R. (1999) Analysis of the raised beach gravel deposits at Boxgrove and related sites. In: *Boxgrove A Middle Pleistocene Hominid Site at Eartham Quarry, Boxgrove, West Sussex* (Eds M.B. Roberts and S.A. Parfitt), *English Heritage Archaeological Report*, **17**, 100–111.

Bridgland, D.R. (2000) Discussion: the characteristics, variation and likely origin of the Buchan Ridge Gravel. In: *The Quaternary of the Banffshire Coast & Buchan: Field Guide* (Eds J.W. Merritt, E.R. Connell and D.R. Bridgland), pp. 139–143. Quaternary Research Association, London.

Bridgland, D.R. and D'Olier, B. (1995) The Pleistocene evolution of the Thames and Rhine drainage systems in the southern North Sea Basin. In: *Island Britain: a Quaternary Perspective* (Ed. R.C. Preece), *Geol. Soc. Spec. Pub.*, **96**, 27–45.

Bridgland, D.R., Saville, A. and Sinclair, J.M. (1997) New evidence for the origin of the Buchan Ridge Gravel, Aberdeenshire. *Scot. J. Geol.*, **33**, 43–50.

Bristow, C.R. and Cox, F.C. (1973a) The Gipping Till: a reappraisal of East Anglian glacial stratigraphy. *J. Geol. Soc. London*, **129**, 1–37.

Bristow, C.R. and Cox, F.C. (1973b) Additional note. In: *A Correlation of Quaternary Deposits in the British Isles* (Eds G.F. Mitchell, L.F. Penny, F.W. Shotton and R.G. West), *Geol. Soc. London Spec. Rep.*, **4**, 9, 12.

British Geological Survey (1985) *East Anglia Sheet 52°N–00°.* 1:250 000 Series. Solid Geology. British Geological Survey, Keyworth.

British Geological Survey and Rijks Geologische Dienst (1986) *Indefatigable Sheet 53°N–02°E.* 1:250 000 Series. Quaternary Geology. British Geological Survey, Keyworth and Rijks Geologische Dienst, Haarlem.

Cameron, T.D.J., Crosby, A., Balson, P.S., Jeffery, D.H., Lott, G.K., Bulat, J. and Harrison, D.J. (1992) *United Kingdom Offshore Regional Report: the Geology of the Southern North Sea.* Her Majesty's Stationery Office, London, 152 pp.

Cameron, T.D.J., Stoker, M.S. and Long, D. (1987) The history of Quaternary sedimentation in the UK sector of the North Sea Basin. *J. Geol. Soc. London*, **144**, 43–58.

Clark, P.U. and Walder, J.S. (1994) Subglacial drainage, eskers, and deforming beds beneath the Laurentide and Eurasian ice sheets. *Geol. Soc. Am. Bull.*, **106**, 304–314.

Cox, F.C. (1981) The 'Gipping Till' revisited. In: *The Quaternary in Britain Essays, Reviews and Original Work on the Quaternary Published in Honour of Lewis Penny on his Retirement* (Eds J. Neale and J.R. Flenley), pp. 32–42. Pergamon, Oxford.

Cox, F.C. and Nickless, E.F.P. (1972) Some aspects of the glacial history of central Norfolk. *Bull. Geol. Surv. Great Brit.*, **42**, 79–98.

Cox, F.C. and Nickless, E.F.P. (1974) The glacial geomorphology of central and north Norfolk: comment. *East Midland Geogr.*, **6**, 92–95.

Dobkins, J.E. and Folk, R.L. (1970) Shape development on Tahiti-Nui. *J. Sed. Petrol.*, **40**, 1167–1203.

Ehlers, J., Gibbard, P.L. and Whiteman, C.A. (1987) Recent investigations of the Marly Drift of northwest Norfolk, England. In: *Tills and Glaciotectonics* (Ed. J.J.M. van der Meer), pp. 39–54. Balkema, Rotterdam.

Ehlers, J., Gibbard, P.L. and Whiteman, C.A. (1991) The glacial deposits of northwestern Norfolk. In: *Glacial Deposits in Great Britain and Ireland* (Eds J. Ehlers, P.L. Gibbard and J. Rose), pp. 223–232. Balkema, Rotterdam.

Ehlers, J. and Linke, G. (1989) The origin of deep buried channels of Elsterian age in Northwest Germany. *J. Quatern. Sci.*, **4**, 255–265.

Fish, P.R. and Whiteman, C.A. (2001) Chalk micropalaeontology and the provenancing of Middle Pleistocene Lowestoft Formation till in eastern England. *Earth Surf. Proc. Land.*, **26**, 953–970.

Fish, P.R., Whiteman, C.A., Moorlock, B.S.P., Hamblin, R.J.O. and Wilkinson, I.P. (2000) The glacial geology of the Weybourne area, north Norfolk: a new approach. *Bull. Geol. Soc. Norfolk*, **50**, 21–45.

Fisher, R.V. and Mattinson, J.M. (1968) Wheeler Gorge tubidite–conglomerate series, California; inverse grading. *J. Sed. Petrol.*, **38**, 1013–1023.

Folk, R.L. (1954) The distinction between grain size and mineral composition in sedimentary-rock nomenclature. *J. Geol.*, **62**, 344–359.

Folk, R.L. and Ward, W.C. (1957) Brazos River bar: a study in the significance of grain size parameters. *J. Sed. Petrol.*, **27**, 3–26.

Funnell, B.M., Norton, P.E.P. and West, R.G. (1979) The Crag at Bramerton, near Norwich, Norfolk. *Phil. Trans. Roy. Soc. London B*, **287**, 489–534.

Gale, S.J. and Hoare, P.G. (1986) Blakeney ridge sands and gravels. In: *The Nar Valley & North Norfolk. Field Guide* (Eds R.G. West and C.A. Whiteman), pp. 91–95. Quaternary Research Association, Coventry.

Gale, S.J. and Hoare, P.G. (1991) *Quaternary Sediments: Petrographic Methods for the Study of Unlithified Rocks.* Belhaven, London, 323 pp.

Gale, S.J. and Hoare, P.G. (1992) Bulk sampling of coarse clastic sediments for particle-size analysis. *Earth Surf. Proc. Land.*, **17**, 729–733.

Gale, S.J. and Hoare, P.G. (1994) Reply: Bulk sampling of coarse clastic sediments for particle-size analysis. *Earth Surf. Proc. Land.*, **19**, 263–268.

Gale, S.J., Hoare, P.G., Hunt, C.O. and Pye, K. (1988) The Middle and Upper Quaternary deposits at Morston, north Norfolk, U.K. *Geol. Mag.*, **125**, 521–533.

Gibbard, P.L. (1986) Flint gravels in the Quaternary of southeast England. In: *The Scientific Study of Flint and Chert: Proceedings of the Fourth International Flint Symposium held at Brighton Polytechnic 10–15 April 1983* (Eds G. de G. Sieveking and M.B. Hart), pp. 141–149. Cambridge University Press, Cambridge.

Gibbard, P.L., West, R.G., Andrew, R. and Pettit, M. (1991) Tottenhill, Norfolk (TF 636115). In: *Central East Anglia & the Fen Basin. Field Guide* (Eds S.G. Lewis, C.A. Whiteman and D.R. Bridgland), pp. 131–143. Quaternary Research Association, London.

Gibbard, P.L., West, R.G., Andrew, R. and Pettit, M. (1992) The margin of a Middle Pleistocene ice advance at Tottenhill, Norfolk, England. *Geol. Mag.*, **129**, 59–76.

Gorrell, G. and Shaw, J. (1991) Deposition in an esker, bead and fan complex, Lanark, Ontario, Canada. *Sed. Geol.*, **72**, 285–314.

Gray, J.M. (1988) Glaciofluvial channels below the Blakeney esker, Norfolk. *Quatern. Newsl.*, **55**, 9–12.

Gray, J.M. (*c.* 1994) The Blakeney Esker, Norfolk: conservation and restoration. In: *Conserving our Landscape* (Eds C. Stevens, J.E. Gordon, C.P. Green and M.G. Macklin), pp. 82–86. Proceedings of the Conference Conserving our Landscape: Evolving Landforms and Ice-age Heritage, May 1992, Crewe.

Gray, J.M. (1997) The origin of the Blakeney Esker, Norfolk. *Proc. Geol. Assoc.*, **108**, 177–182.

Green, C.P. and McGregor, D.F.M. (1990) Pleistocene gravels of the north Norfolk coast. *Proc. Geol. Assoc.*, **101**, 197–202.

Hamblin, R.J.O. (2000) A new glacial stratigraphy for East Anglia. *Mercian Geol.*, **15**, 59–62.

Hamblin, R.J.O., Moorlock, B.S.P. and Rose, J. (2000) A new glacial stratigraphy for eastern England. *Quatern. Newsl.*, **92**, 35–43.

Hamblin, R.J.O., Moorlock, B.S.P., Rose, J. and Lee, J.R. (2003) New developments on the Quaternary of Norfolk. *Mercian Geol.*, **15**, 232–233.

Hambrey, M.J. (1994) *Glacial Environments*. UCL Press, London, 296 pp.

Hampton, M.A. (1979) Buoyancy in debris flows. *J. Sed. Petrol.*, **49**, 753–758.

Harmer, F.W. (1904) The Great Eastern Glacier. *Geol. Mag.*, **5(1)**, 509–510.

Harmer, F.W. (1905) The Great Eastern Glacier. *Report of the Seventy-Fourth Meeting of the British Association for the Advancement of Science held at Cambridge in August 1904*, 542–543.

Harmer, F.W. (1909) The Pleistocene Period in the eastern counties of England. *Geology in the Field. The Jubilee Volume of the Geologists' Association (1858–1908)*, 103–123.

Hart, J.K. (1996) Subglacial deformation associated with a rigid bed environment, Aberdaron, North Wales. *Glacial Geology and Geomorphology*, **1996**, rp01.

Hart, J.K. and Boulton, G.S. (1991) The glacial drifts of northeastern Norfolk. In: *Glacial Deposits in Great Britain and Ireland* (Eds J. Ehlers, P.L. Gibbard and J. Rose), pp. 233–243. Balkema, Rotterdam.

Hart, J.K. and Peglar, S.M. (1990) Further evidence for the timing of the Middle Pleistocene Glaciation in Britain. *Proc. Geol. Assoc.*, **101**, 187–196.

Hey, R.W. (1976) Provenance of far-travelled pebbles in the pre-Anglian Pleistocene of East Anglia. *Proc. Geol. Assoc.*, **87**, 69–81.

Hoare, P.G. and Gale, S.J. (1986) Blakeney and Salthouse (TG 04). In: *The Nar Valley & North Norfolk. Field Guide* (Eds R.G. West and C.A. Whiteman), pp. 74–94. Quaternary Research Association, Coventry.

Holmes, T.V. (1883) On eskers or kames. *Geol. Mag.*, **2(10)**, 438–445.

Holmes, T.V. (1884) On eskers or kames. *Proc. Norwich Geol. Soc.*, **1**, 263–272.

Hooke, R. LeB. (1984) On the role of mechanical energy in maintaining subglacial water conduits at atmospheric pressure. *J. Glaciol.*, **30**, 180–187.

Johansson, C.E. (1965) Structural studies of sedimentary deposits orientation analyses, literature digest, and field investigations. *Geol. Fören. Stockh. Förh.*, **87**, 3–61.

Knight, P.G. (1999) *Glaciers*. Stanley Thornes, Cheltenham, 261 pp.

Krüger, J. (1983) Glacial morphology and deposits in Denmark. In: *Glacial Deposits in North-West Europe* (Ed. J. Ehlers), pp. 181–191. Balkema, Rotterdam.

Laban, C., Cameron, T.D.J. and Schüttenhelm, R.T.E. (1984) Geologie van het Kwartair in de zuidelijke bocht van de Noordzee. *Meded. Werkgr. Tert. Kwart. Geol.*, **21**, 139–153.

Lawson, T.J. and Allen, P. (2000) QRA Annual Field Meeting – Norfolk and Suffolk University of East Anglia, 10th–14th April, 2000. *Quatern. Newsl.*, **91**, 15–22.

Lee, J.R., Booth, S.J., Hamblin, R.J.O., Jarrow, A.M., Kessler, H., Moorlock, B.S.P., Morigi, A.N., Palmer, A., Pawley, S.M., Riding, J.B. and Rose, J. (2004) A new stratigraphy for the glacial deposits around Lowestoft, Great Yarmouth, North Walsham and Cromer, East Anglia, UK. *Bull. Geol. Soc. Norfolk*, **53**, 3–60.

Legros, F. (2002) Can dispersive pressure cause inverse grading in grain flows? *J. Sed. Res.*, **72**, 166–170.

Lewis, S.G. and Rose, J. (1991) Tottenhill, Norfolk (TF 639120). In: *Central East Anglia & the Fen Basin. Field Guide* (Eds S.G. Lewis, C.A. Whiteman and D.R. Bridgland), pp. 145–148. Quaternary Research Association, London.

Lindström, E. (1985) The Uppsala esker: the Åsby-Drälinge exposures. *Striae*, **22**, 27–32.

Long, D., Laban, C., Streif, H., Cameron, T.D.J. and Schüttenhelm, R.T.E. (1988) The sedimentary record of climatic variation in the southern North Sea. *Phil. Trans. Roy. Soc. London B*, **318**, 523–537.

Lyell, C. (1863) *The Geological Evidences of the Antiquity of Man with Remarks on Theories of the Origin of Species by Variation*. John Murray, London, 520 pp.

Mathews, W.H. (1974) Surface profiles of the Laurentide ice sheet in its marginal areas. *J. Glaciol.*, **13**, 37–43.

Mattinson, J.M. and Fisher, R.V. (1970) Impossibility of Bernoulli pressure forces on particles suspended in boundary layers: reply. *J. Sed. Petrol.*, **40**, 520–521.

Menzies, J. and Shilts, W.W. (1996) Subglacial environments. In: *Past Glacial Environments: Sediments, Forms and Techniques* (Ed. J. Menzies), pp. 15–136. Butterworth–Heinemann, Oxford.

Michalska, Z. (1969) Problems of the origin of eskers based on the examples from central Poland. *Geographia Polonica*, **16**, 105–119.

Middleton, G.V. (1970) Experimental studies related to problems of flysch sedimentation. In: *Flysch Sedimentology in North America* (Ed. J. Lajoie), *Geol. Assoc. Can. Spec. Pap.*, **7**, 253–272.

Moir, J.R. (1928) Palaeolithic implements from the Cannon-shot gravel of Norfolk. *Proc. Prehist. Soc. East Anglia*, **6**, 1–11.

Mooers, H.D. (1989) On the formation of the tunnel valleys of the Superior lobe, central Minnesota. *Quatern. Res.*, **32**, 24–35.

Mooers, H.D. (1990) A glacial-process model: the role of spatial and temporal variations in glacier thermal regime. *Geol. Soc. Am. Bull.*, **102**, 243–251.

Moorlock, B.S.P., Booth, S.J., Fish, P.R., Hamblin, R.J.O., Kessler, H., Riding, J.B., Rose, J. and Whiteman, C.A. (2000) Briton's Lane Gravel Pit, Beeston Regis (TG 168 415). In: *The Quaternary of Norfolk and Suffolk: Field Guide* (Eds S.G. Lewis, C.A. Whiteman and R.C. Preece), pp. 115–117. Quaternary Research Association, London.

Moorlock, B.S.P., Hamblin, R.J.O., Booth, S.J., Kessler, H., Woods, M.A. and Hobbs, P.R.N. (2002) *Geology of the Cromer District – a Brief Explanation of the Geological Map. Sheet Explanation of the British Geological Survey. 1:50 000 Sheet 131 Cromer (England and Wales).* British Geological Survey, Keyworth, 34 pp.

Nye, J.F. (1973) Water at the bed of a glacier. In: *Symposium on the Hydrology of Glaciers. Int. Assoc. Sci. Hydrol. Publ.,* **95**, 189–194.

Pawley, S.M., Rose, J., Lee, J.R., Moorlock, B.S.P. and Hamblin, R.J.O. (2004) Middle Pleistocene sedimentology and lithostratigraphy of Weybourne, northeast Norfolk, England. *Proc. Geol. Assoc.,* **115**, 25–42.

Perrin, R.M.S., Rose, J. and Davies, H. (1979) The distribution, variation and origins of pre-Devensian tills in eastern England. *Phil. Trans. Roy. Soc. London B,* **287**, 535–570.

Pierson, T.C. (1981) Dominant particle support mechanisms in debris flows at Mt Thomas, New Zealand, and implications for flow mobility. *Sedimentology,* **28**, 49–60.

Potter, P.E. and Pettijohn, F.J. (1963) *Paleocurrents and Basin Analysis.* Springer, Berlin, 296 pp.

Price, R.J. (1966) Eskers near the Casement Glacier, Alaska. *Geogr. Ann.,* **48A**, 111–125.

Ranson, C.E. 1968. An assessment of the glacial deposits of north-east Norfolk. *Bull. Geol. Soc. Norfolk,* **16**, 1–16.

Rees, A.I. (1968) The production of preferred orientation in a concentrated dispersion of elongated and flattened grains. *J. Geol.,* **76**, 457–465.

Reid, C. 1882. The geology of the country around Cromer. *Memoirs of the Geological Survey. England and Wales.* Her Majesty's Stationery Office, London, 143 pp.

Riding, J.B. (2001) A palynological investigation of till samples from the Sheringham area, north Norfolk. *Brit. Geol. Surv. Internal Rep.,* IR/01/155, 11 pp.

Ringrose, S.M. (1982) Depositional processes in the development of eskers in Manitoba. In: *Research in Glacial, Glacio-Fluvial, and Glacio-Lacustrine Systems* (Eds R. Davidson-Arnott, W. Nickling and B.D. Fahey), pp. 117–137. Geo Books, Norwich.

Rodine, J.D. and Johnson, A.M. (1976) The ability of debris, heavily freighted with coarse clastic materials, to flow on gentle slopes. *Sedimentology,* **23**, 213–234.

Rose, J., Gulamali, N., Moorlock, B.S.P., Hamblin, R.J.O., Jeffery, D.H., Anderson, E., Lee, J.A. and Riding, J.B. (1996) Pre-glacial and glacial Quaternary sediments, How Hill near Ludham, Norfolk, England. *Bull. Geol. Soc. Norfolk,* **45**, 3–28.

Rose, J., Lee, J.A., Candy, I. and Lewis, S.G. (1999) Early and Middle Pleistocene river systems in eastern England: evidence from Leet Hill, southern Norfolk, England. *J. Quatern. Sci.,* **14**, 347–360.

Rose, J., Moorlock, B.S.P. and Hamblin, R.J.O. (2001) Pre-Anglian fluvial and coastal deposits in Eastern England: lithostratigraphy and palaeoenvironments. *Quatern. Int.,* **79**, 5–22.

Rowe, P.J., Richards, D.A., Atkinson, T.C., Bottrell, S.H. and Cliff, R.A. (1997) Geochemistry and radiometric dating of a Middle Pleistocene peat. *Geochim. Cosmochim. Acta,* **61**, 4201–4211.

Rubinow, S.I. and Keller, J.B. (1961) The transverse force on a spinning sphere moving in a viscous fluid. *J. Fluid Mech.,* **11**, 447–459.

Saffman, P.G. (1965) The lift on a small sphere in a slow shear flow. *J. Fluid Mech.,* **22**, 385–400.

Saunderson, H.C. (1977) The sliding bed facies in esker sands and gravels: a criterion for full-pipe (tunnel) flow? *Sedimentology,* **24**, 623–638.

Scourse, J.D., Austin, W.E.N., Sejrup, H.P. and Ansari, M.H. (1999) Foraminiferal isoleucine epimerization determinations from the Nar Valley Clay, Norfolk, UK: implications for Quaternary correlations in the southern North Sea basin. *Geol. Mag.,* **136**, 543–560.

Shotton, F.W., Banham, P.H. and Bishop, W.W. (1977) Glacial–interglacial stratigraphy of the Quaternary in Midland and eastern England. In: *British Quaternary Studies Recent Advances* (Ed. F.W. Shotton), pp. 267–282. Clarendon, Oxford.

Shreve, R.L. (1972) Movement of water in glaciers. *J. Glaciol.,* **11**, 205–214.

Shreve, R.L. (1985) Esker characteristics in terms of glacier physics, Katahdin esker system, Maine. *Geol. Soc. Am. Bull.,* **96**, 639–646.

Slater, G. (1923) May 21st. Report by G. Slater. In: Whitsuntide excursion to the Cromer and Norwich districts. May 18th–23rd, 1923 (P.G.H. Boswell and G. Slater). *Proc. Geol. Assoc.,* **34**, 227–230.

Solomon, J.D. (1932) The glacial succession on the north Norfolk coast. *Proc. Geol. Assoc.,* **43**, 241–271.

Sparks, B.W. and West, R.G. (1964) The drift landforms around Holt, Norfolk. *Inst. Brit. Geogr. Trans. Pap.,* **35**, 27–35.

Sparks, B.W. and West, R.G. (1970) Landforms in the Holt–Blakeney area. In: *Norwich Easter–1970* (Ed. G.S. Boulton), p. 27. Quaternary Research Association, Norwich.

Sparks, B.W. and West, R.G. (1972) *The Ice Age in Britain.* Methuen, London, 302 pp.

Straw, A. (1960) The limit of the 'Last' Glaciation in North Norfolk. *Proc. Geol. Assoc.,* **71**, 379–390.

Straw, A. (1965) A reassessment of the Chalky Boulder Clay or Marly Drift of north Norfolk. *Z. Geomorphol.,* N.F., **9**, 209–221.

Straw, A. (1967) The Penultimate or Gipping Glaciation in north Norfolk. *Trans. Norfolk Norwich Nat. Soc.,* **21**, 21–24.

Straw, A. (1973) The glacial geomorphology of central and north Norfolk. *East Midland Geogr.*, **5**, 333–354.

Straw, A. (1974) The glacial geomorphology of central and north Norfolk: reply. *East Midland Geogr.*, **6**, 95–98.

Straw, A. (1979a) Eastern England. In: *Eastern and Central England* (A. Straw and K.M. Clayton), pp. 1–139. Methuen, London.

Straw, A. (1979b) The geomorphological significance of the Wolstonian glaciation of eastern England. *Trans. Inst. Brit. Geogr.*, N.S. **4**, 540–549.

Straw, A. (1983) Pre-Devensian glaciation of Lincolnshire (eastern England) and adjacent areas. *Quatern. Sci. Rev.*, **2**, 239–260.

Straw, A. (1984) Reply to comments. *Quatern. Sci. Rev.*, **3**, ix–x.

Turner, C. (1973) Eastern England. In: *A Correlation of Quaternary Deposits in the British Isles* (Eds G.F. Mitchell, L.F. Penny, F.W. Shotton and R.G. West), *Geol. Soc. London Spec. Rep.*, **4**, 8–18.

Walder, J.S. and Fowler, A. (1994) Channelized subglacial drainage over a deformable bed. *J. Glaciol.*, **40**, 3–15.

Walker, R.G. (1975) Conglomerate: sedimentary structures and facies models. In: *Depositional Environments as Interpreted from Primary Sedimentary Structures and Stratification Sequences. SEPM Short Course*, **2**, 133–161.

Weidick, A., Kelly, M. and Bennike, O. (2004) Late Quaternary development of the southern sector of the Greenland Ice Sheet, with particular reference to the Qassimiut lobe. *Boreas*, **33**, 284–299.

Wentworth, C.K. (1922) A scale of grade and class terms for clastic sediments. *J. Geol.*, **30**, 377–392.

West, R.G. (1957) Notes on a preliminary map of some features of the drift topography around Holt and Cromer, Norfolk. *Trans. Norfolk Norwich Nat. Soc.*, **18(5)**, 24–29.

West, R.G. (1958) The Pleistocene epoch in East Anglia. *J. Glaciol.*, **3**, 211–216.

West, R.G. (1961) The glacial and interglacial deposits of Norfolk. In: *The Geology of Norfolk. Trans. Norfolk Norwich Nat. Soc.*, **19**, 365–375.

West, R.G. (1967) The Quaternary of the British Isles. In: *The Quaternary, Volume 2* (Ed. K. Rankama.), pp. 1–87. Interscience, New York.

Whiteman, C.A. (1995) Processes of terrestrial deposition. In: *Modern Glacial Environments: Processes, Dynamics and Sediments* (Ed. J. Menzies), pp. 293–308. Butterworth–Heinemann, Oxford.

Wingfield, R.T.R. (1989) Glacial incisions indicating Middle and Upper Pleistocene ice limits off Britain. *Terra Nova*, **1**, 538–548.

Wingfield, R.T.R. (1990) The origin of major incisions within the Pleistocene deposits of the North Sea. *Mar. Geol.*, **91**, 31–52.

Wood, S.V. (1880) The Newer Pliocene period in England. *Q. J. Geol. Soc. London*, **36**, 457–528.

Wood, S.V. and Harmer, F.W. (1868) Abstract of a paper on 'The glacial and post-glacial structure of Norfolk and Suffolk.' *Geol. Mag.*, **5**, 452–456.

Wood, S.V. and Harmer, F.W. (1869) On the glacial and postglacial structure of Norfolk and Suffolk. *Report of the Thirty-Eighth Meeting of the British Association for the Advancement of Science; held at Norwich in August 1868*. Notices and Abstracts of Miscellaneous Communications to the Sections, 80–83.

Wood, S.V. and Harmer, F.W. (1872) An outline of the geology of the Upper Tertiaries of East Anglia. In: *Supplement to the Monograph of the Crag Mollusca, with Descriptions of Shells from the Upper Tertiaries of the East of England. Vol. III. Univalves and Bivalves. With an Introductory Outline of the Geology of the Same District, and Map*, ii–xxxi. The Palæontographical Society, London.

Woodland, A.W. (1970) The buried tunnel-valleys of East Anglia. *Proc. Yorks. Geol. Soc.*, **37**, 521–578.

Woodward, H.B. (1884) The geology of the country around Fakenham, Wells, and Holt. *Memoirs of the Geological Survey. England and Wales*. Her Majesty's Stationery Office, London, 57 pp.

Woodward, H.B. (1885) The Glacial Drifts of Norfolk. *Proc. Geol. Assoc.*, **9**, 111–129.

Wright, H.E. (1973) Tunnel valleys, glacial surges, and subglacial hydrology of the Superior Lobe, Minnesota. In: *The Wisconsinan Stage* (Eds R.F. Black, R.P. Goldthwait and H.B. Willman), *Geol. Soc. Am. Mem.*, **136**, 251–276.

Sediments and landforms in an upland glaciated-valley landsystem: upper Ennerdale, English Lake District

DAVID J. GRAHAM† and MICHAEL J. HAMBREY*

*Centre for Glaciology, Institute of Geography and Earth Sciences, University of Wales, Aberystwyth, SY23 3DB, UK
(e-mail: D.J.Graham@lboro.ac.uk)
†Department of Geography, Loughborough University, Loughborough, Leicestershire, LE11 3TU, UK

ABSTRACT

The genesis of moraines associated with British glaciers of Younger Dryas age has proved controversial in recent years. A number of alternative hypotheses exist and, whilst it is generally accepted that such features are polygenetic in origin, some workers have argued that not all of the proposed mechanisms are valid. This paper seeks to explore these issues, using a case study from the English Lake District. A landsystems approach is adopted, integrating information at a variety of spatial scales to explain the development of the sediment-landform associations in upper Ennerdale. The evidence suggests that landform development resulted from a combination of ice-marginal deposition and englacial thrusting. It is probable that thrusting resulted from flow compression against a reverse bedrock slope, combined with the confluence of ice from two separate source areas. It is argued that, whilst englacial thrust moraines may not be commonly associated with British Younger Dryas glaciers, under certain conditions englacial thrusting is an important process in landform development.

Keywords Glacial geomorphology, glacial landsystems, moraine, Younger Dryas, Loch Lomond Stadial, English Lake District, glacial processes.

INTRODUCTION

The suite of landforms and sediments found at glacier margins varies according to thermal regime, relief and tectonic setting. Comparison between modern environments and palaeoenvironments is essential if reliable reconstructions of past glacier extent and conditions are to be derived. By comparing the sediments and landform morphology with modern glaciers, this paper aims to reconstruct the Younger Dryas (Loch Lomond stade) glacial conditions in part of the Lake District, a classic area for the study of upland glaciation in the UK. This area, which is of moderate relief (700–900 m), is typical of many tectonically stable mountain belts, and lacks the precipitous slopes that are common in tectonically active regions.

Apparently chaotic associations of moraine mounds and ridges are characteristic of many former Younger Dryas glaciers in the UK. They are commonly known as 'hummocky moraine' (Sissons, 1979), but the term 'moraine-mound complex' is preferred because of the genetic connotations associated with the former (following Bennett et al., 1996b). The significance of such moraine-mound complexes has been controversial (Bennett, 1994). Early workers applied Alpine analogues and interpreted all moraines as indicative of ice-marginal positions (e.g. Charlesworth, 1956). In the 1960s and 1970s, the Sissons school interpreted moraine-mound complexes as products of large-scale areal stagnation, similar to that described from Scandinavia and North America (Sissons, 1980; Gray, 1982). Later, Eyles (1983) applied an Icelandic analogue and proposed that the features result from incremental stagnation of a glacier margin undergoing active recession. In the 1990s, work by Benn (1990; 1992) and

Bennett (1991; Bennett and Boulton, 1993a) led to the conclusion that moraine-mound complexes are polygenetic in origin, representing ice-marginal landforms associated with active glacier recession and oriented perpendicular to ice flow, subglacial bedforms oriented parallel to ice flow, and small areas of stagnation topography with no discernable organisation. The ice-marginal landforms were used by Bennett (1991) to reconstruct the pattern of deglaciation in the northwest Highlands of Scotland. Subsequently, by analogy with landforms developed at polythermal glaciers in Svalbard, it has been proposed that some moraine-mound complexes in Britain may be englacial thrust moraines (Hambrey et al., 1997). Such features are oriented approximately perpendicular to glacier flow, but are associated with the internal structure of the glacier and do not represent ice-marginal positions. However, the inference of thrusting has proved controversial. Firstly, the interpretation by Bennett et al. (1998) of a classic moraine-mound complex in Glen Torridon, NW Scotland as a thrust-moraine complex, was refuted by Wilson & Evans (2000). Then, Lukas (2005) stated that the thrust hypothesis for British Younger Dryas moraines was universally incorrect, although his study was based on a quite different set of morainic landforms in Sutherland, NW Scotland (Graham et al., 2007). These differences of opinion highlight the need to investigate Younger Dryas moraines on a case-by-case basis, combining detailed analysis of sedimentary facies and landform morphology.

The changing interpretations of British moraine-mound complexes illustrate how the sediment-landform associations that record the presence and nature of glaciers in the past can be interpreted in different ways. Even at contemporary glaciers, where landform-generating processes may be observed and the sedimentological and morphological products of these processes are well preserved and exposed, different researchers often cannot agree. This is to be expected, given the extremely complex interaction of glacial, fluvial, gravity-driven and, commonly, aeolian and lacustrine or marine processes operating. In the palaeolandform record, where the morphology may have been subject to thousands of years of postglacial modification and the sedimentary exposure is commonly poor, the challenges are even greater.

Partially as a response to this complexity, much recent glacial sedimentological and geomorphological work has adopted a landsystems approach (Evans, 2003a). This approach recognises that the landscape is composed of a range of related elements at a variety of spatial scales, and that understanding the significance of individual elements requires an appreciation of their environmental context. This framework enables the significance of individual landscape elements to be understood in the context of the wider landscape associated with a particular glacier system. Further, the approach facilitates the creation of a range of process-form models relating particular styles of glaciation and their characteristic products at a range of spatial scales (Benn and Evans, 1998). Such models may be applied to assist researchers in interpreting landforms of uncertain origin in both the contemporary and ancient landform record.

A particular benefit of the landsystems approach is that the application of a particular model at a field site does not require full information about the nature of the sediments and landforms present. Provided that the information available is sufficient to allow confident application of an existing landsystem model, inferences may be made about the nature of any missing information. So for example, in the absence of extensive sedimentary exposure within a landform, inferences may be made about the likely nature of the constituent facies and their architecture based on the application of a model that fits the available evidence. Conversely, a danger of the landsystems approach is that it is possible to over interpret a landscape, applying a particular model where the evidence is insufficient to do so and leading to inappropriate conclusions being drawn (Benn and Evans, 2004).

This paper adopts a hierarchical approach to the interpretation of the glacial history of a single drainage basin. At the smallest scale, the individual sedimentary facies identified in the valley are described and interpreted. At a larger scale, the individual sediment-landform associations are described and alternative hypotheses for their formation evaluated in the light of the available evidence. Finally, the evidence is combined to develop a process-form model that represents the most likely history of glaciation in the valley and permits an assessment to be made of the wider significance of the study.

ENNERDALE

Ennerdale is a classic over-deepened glacial trough located in the western part of the English Lake District (Fig. 1), carved out of Ordovician volcanic rocks (the Borrowdale Volcanic Group) and late Caledonian granitic rocks (the Ennerdale Granophyre) in the upper part of the valley, and mudstone and sandstone of the Ordovician Skiddaw Group flanking the lake of Ennerdale Water. The central part of the Lake District is characterised by a generally radial pattern of glacial troughs, to which Ennerdale conforms, running approximately from the south-east to the north-west in its upper parts. The valley begins abruptly in the cliffs of Great Gable (899 m), which dominate its head (Fig. 2). The upper valley is encircled by the peaks of Pillar (892 m) and Kirk Fell (802 m) to the southwest and Brandreth (715 m), Haystacks (582 m) and High Crag (704 m) to the northeast. The lower part of the valley is occupied by the glacially carved rock basin containing 4-km long Ennerdale Water, and shows a well-developed suite of landforms associated with the recession of the last (Devensian) British Ice Sheet to occupy the region (Huddart, 1967).

In common with many of the Lake District valleys, there is evidence that a valley glacier sub-

sequently occupied the upper part of Ennerdale. Although there is no direct dating control, this glacier is assumed to be of Younger Dryas age by analogy with other valleys containing similar evidence, and for which dating control is available (Pennington, 1978). The Ennerdale moraines, which extend for about 3 km from the head of the valley, have attracted interest since the late 19th century, when Ward (1873) described them as 'perhaps the finest examples of [a] large series of moraines' in the Lake District (p. 427). Manley (1959) mapped a glacier in the upper part of Ennerdale and attributed it to the Younger Dryas, although the map was at a small scale and no geomorphological evidence was presented. The most detailed examination of the valley landforms was presented by Sissons (1980), who mapped a glacier on the basis of 'remarkable hummocky moraine' (p. 23). Sissons identified two suites of mounds on the northern valley side. Those lower down the valley (described here as the valley-side subdued ridges, A_3) were described as 'smooth mounds of drift' (p. 19), and were inferred to be of pre-Younger Dryas age. Those higher up the valley (described here as the valley-side moraine-mound complex, A_2) are smaller with steeper faces, and were inferred to be of Younger Dryas age, although there is no direct dating control. The boundary

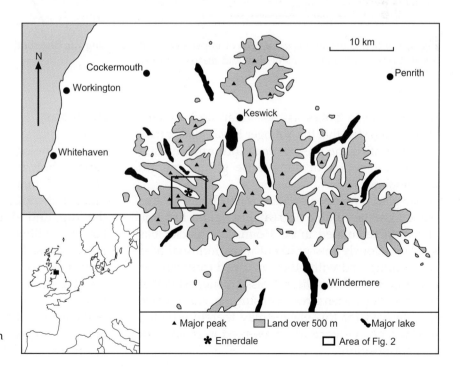

Fig. 1 The location of the field site in the English Lake District.

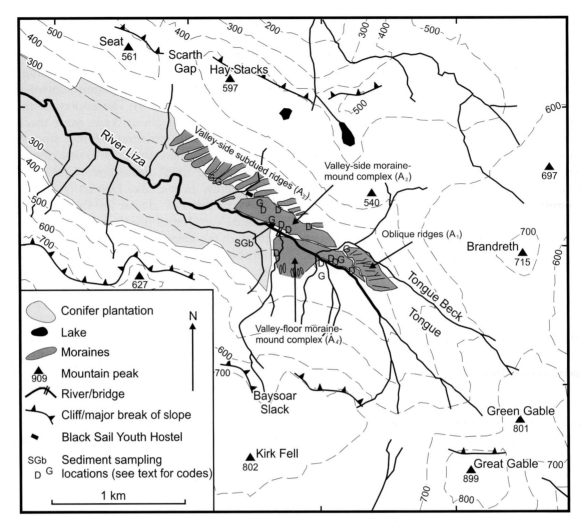

Fig. 2 Outline geomorphological map of upper Ennerdale showing the distribution of the key sediment-landform associations and the locations of clast and matrix samples.

between these zones is not clear and Sissons (1980) suggested that the Younger Dryas glacier may have modified the pre-existing landforms making the determination of the precise glacier margin problematic.

The glacier mapped by Sissons (1980) had an area of 1.28 km², a volume of 0.049 km³ and a firn line elevation calculated to be 465 m. The glacier has an anomalously low gradient in its terminal area when compared with others mapped from similar evidence in the Lake District. Sissons suggested that this may have been a result of surging, and the firn line derived for this glacier may therefore be too low, but no evidence was presented in support of this supposition.

METHODS

Geomorphological mapping was undertaken from aerial photographs and validated in the field. Structural mapping of the valley-side moraine-mound complex was undertaken by measuring the dip and direction of dip of the up-valley facing rectilinear faces and the orientation of the ridge crests. There are few extensive sedimentary exposures in the Ennerdale moraines, so it was not possible to observe the facies architecture of the landform-sediment associations. Observations were made, and clast and matrix samples taken, at all natural exposures encountered. The grain-size distribution of the matrix samples was determined by sieve and SediGraph 5100 analyses in the laboratory. Clast-roundness measurements were made using the Powers scale and plotted as histograms. Measurements of clast *a*-, *b*- and *c*-axes were made with a ruler and clast shape data plotted on Sneed and Folk ternary diagrams (Sneed and Folk, 1958; Benn and Ballantyne, 1993), using the software of Graham and Midgley (2000a). Covariant plots of the C_{40} index (proportion of clasts with a c:a axial ratio of less than or equal to 0.4) and RA index (proportion of angular and very angular clasts) were used to aid discrimination of debris transport pathways (Benn and Ballantyne, 1994). The proportions of striated and faceted clasts were also noted. A digital elevation model was generated using total-station derived data.

SEDIMENTARY FACIES

Lithofacies descriptions

The physical properties of the lithofacies observed in upper Ennerdale are described below and summarised in Table 1 and Figure 3. The sampling locations are indicated on Figure 2.

Gravel (G)

This facies occurs in exposures throughout the study area. The clast and matrix characteristics of this facies form a continuum with the diamicton facies (D) described below and they are differentiated on the basis of texture.

Diamicton (D)

This facies is similar in many respects to the gravel (G) facies and is found throughout the study area. It appears massive, although exposures rarely extend for more than a few metres.

Bedded sand and gravel (SGb)

This facies occurs in a single small exposure in the valley-side moraine-mound complex. It occurs in association with the diamicton (D) facies, but the exposure is too limited to establish the relationship between the facies. The facies consists of predominantly coarse and very coarse sand and fine pebbles, with centimetre-scale bedding. Clasts are predominantly rounded and subrounded and have no facets or striations.

Control sample: River channel (CS1)

Examples of this facies were obtained from the River Liza by the footbridge and from Tongue Beck.

Control sample: Scree (CS2)

This facies was collected from scree on the northern valley side, above the valley-side moraine-mound complex (A_2).

Lithofacies interpretations

Gravel (G) and diamicton (D)

These facies are both interpreted predominantly as basally transported glacial sediment, the principal

Table 1 Summary of the physical properties of the facies observed in Ennerdale. Textural descriptions based on the classification of Moncrieff (1989) for poorly sorted sediments.

Facies	n (clast samples)	Textural description	C_{40} index	RA index	Faceted clasts	Striated clasts
Gravel (G)	7	Muddy sandy-gravel	8–21 (mean = 14.5)	4–18 (mean = 10)	0–8%	4–24%
Diamicton (D)	10	Clast-rich sandy diamicton	8–35 (mean = 19)	0–36 (mean = 9)	0–8%	0–22%
River channel (CS1)	2	–	35, 40	0	0	0
Scree slope (CS2)	2	–	80, 86	66, 70	0	0

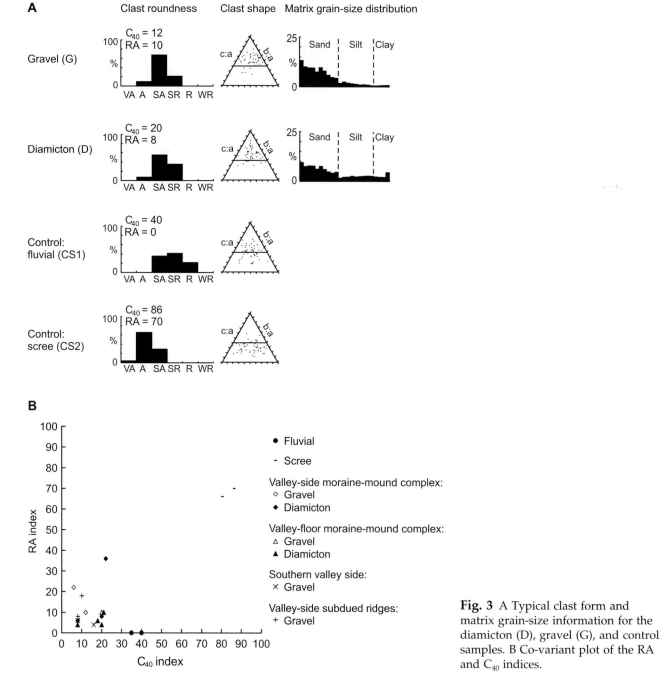

Fig. 3 A Typical clast form and matrix grain-size information for the diamicton (D), gravel (G), and control samples. B Co-variant plot of the RA and C_{40} indices.

difference between them being the proportion of gravel. Unambiguous evidence that these facies have been subjected to traction-zone transport at the glacier bed is the presence of faceted and striated clasts (Boulton, 1978). The poorly sorted sediment, with significant proportions of fine particles in the matrix grain-size distribution, supports a subglacial origin and reflect crushing and abrasion at the glacier bed (Boulton, 1978; Kirkbride, 1995). All

of the samples show peaks in clast roundness in the subangular class and intermediate C_{40} indices, commonly reported for subglacially transported sediment (Bennett *et al.*, 1997). A co-variant plot of the RA and C_{40} indices, on which it is impossible to separate the facies, provides support for a common interpretation for the gravel and diamicton facies (Fig. 3B). Both facies may be clearly differentiated from the fluvial and scree control samples. It is not possible to determine whether the sediments derived these characteristics from transport by a Younger Dryas glacier, or whether they represent reworked material from earlier glaciations.

Bedded sand and gravel (SGb)

The bedded nature of the sediment, absence of fines and degree of sorting suggests a fluvial origin for this facies (Maizels, 1995). The small size of the exposure precluded the possibility of determining the nature of larger-scale structures within the deposit and it is not possible to determine a more specific origin.

River channel samples (CS1)

The fluvial samples are typical of mountain stream sediments. Most of the fine-grained material has been removed, and the clasts are clearly distinguishable from the other facies by clast shape and roundness. The most likely source of the sediment is the moraines through which the streams flow, which are composed of gravel (G) and diamicton (D). The significant proportion of subangular clasts reflects this origin and suggests the sediment has been subjected to limited fluvial transport distances.

Scree samples (CS2)

These samples are typical of the Ennerdale scree deposits, two-thirds of the clasts being angular or very angular. However, the proportion of angular clasts is low compared with many active scree deposits. This probably represents edge-rounding during a prolonged period of weathering processes. The facies is clearly distinguishable from all of the other facies identified in Ennerdale on the basis of clast roundness and shape.

Synthesis

The sediments within the upper Ennerdale moraines appear to have been subject to active basal glacial transport, with some fluvial modification. There is little evidence of a significant passively transported component. These observations are consistent with the general topographic setting of upper Ennerdale, with few steep slopes capable of supplying significant volumes of supraglacial debris. The sedimentary evidence tends to support the hypothesis that the last glacier in upper Ennerdale was characterised by a warm-based thermal regime in which meltwater facilitated sliding over, and deformation of, the bed. However, the possibility that the sediments gained their physical properties during the previous glacial episode and were subjected to minimal modification during the last glacial period cannot be ruled out. Whichever is the case, the last glacier in upper Ennerdale apparently did not carry a significant supraglacial debris load.

SEDIMENT-LANDFORM ASSOCIATIONS

Four sediment-landform associations record the presence of a glacier in the upper part of Ennerdale.

Sediment-landform association descriptions

Oblique ridges on the Tongue (A₁)

A series of subparallel, generally subdued, ridges run obliquely down the south-west facing slope of the Tongue (Figs. 2 and 4A). Nine clear, fairly continuous, ridges can be identified on aerial photographs and from a distant vantage point (Fig. 4), but identification of individual ridges at close quarters in the field is difficult because the slope on which they are located is steep. The ridges appear to bifurcate in places and sometimes die out. Their height was estimated to be between 2 m and 4 m. The ridges have a slightly asymmetric cross-profile, with a steeper down-valley face. Measurements from aerial photographs indicated that the ridges are generally oriented between 265° and 085°. There are no sedimentary exposures in the ridges so determination of their constituent facies was not possible. Occasional, predominantly angular, blocks lie on the surface of the ridges.

Valley-side moraine-mound complex (A₂)

An extensive moraine-mound complex dominates the northern side of upper Ennerdale (Figs. 2 and 4B). Down valley, and higher up on the valley side, the complex has a diffuse boundary with the valley-side subdued ridges (A₃). The moraine-mound complex is separated from the valley-floor moraines (A₄) by the River Liza and Tongue Beck, although the forms immediately adjacent to the river are morphologically similar to the valley-floor moraines.

The moraine-mound complex consists of mounds and ridges rising up to 17 m from the surrounding topography (Fig. 5). The overall impression is of a complex morphology, but close examination reveals clear structure within the landform-sediment association. The organisation is strongest immediately up-valley of the Youth Hostel, where chains of mounds form ridges up to 250 m long. The overall appearance of these ridges is akin to a string of beads, with the slight depressions in the long profile between each mound being well elevated above the surrounding topography (Fig. 5). Although the ridges formed by the chains of mounds are commonly continuous for considerable distances, it is also common for individual mounds to be offset from the trend of the ridge crest, and the ridges do not form an anastomosing network. The ridges are orientated sub-parallel to the long-axis of the valley, trending slightly towards its centre down-valley. With distance up-valley the organisation becomes increasingly chaotic, until distinct ridges can no longer be traced. Where the mounds and ridges are widely spaced, flat boggy ground lies between them. Bedrock is exposed between the mounds in several places, most extensively directly north of Tongue Beck where the mounds appear to lie directly on bedrock. Elsewhere, the proximity of bedrock to the surface could not be directly determined, but exposures between the mounds and in the bed of the River Liza indicate that bedrock is rarely at a depth of more than a few metres between the mounds.

Many of the mounds have rectilinear faces dipping predominantly towards the south (oblique to the valley side which dips to the southwest in this area). Of 74 mounds examined, 54 (73%) had rectilinear faces. Slope dips ranged from 20° to 38° with a mean of 28° (Fig. 6). The direction of dip of the rectilinear faces varies systematically through the moraine-mound complex. Whilst there is a considerable scatter, there is a general trend for the dips to be generally to the south in the down-valley part, turning towards the southwest in the up-valley part of the complex (the trend of the

Fig. 4 (A) The oblique ridges on the Tongue (A₁) viewed from the southern side of the valley. Mounds of the valley-floor moraine mound complex (A₄) are visible in the centre-left of the photograph. (B) The valley-side moraine-mound complex (A₂) viewed from the head of Ennerdale. The structure of the complex becomes increasingly ordered down-valley (further away from the photographer). (C) View from within the valley-side moraine-mound complex (A₂) looking up-valley along the rectilinear faces of several mounds. The imbricate relationships between individual mounds are clear.

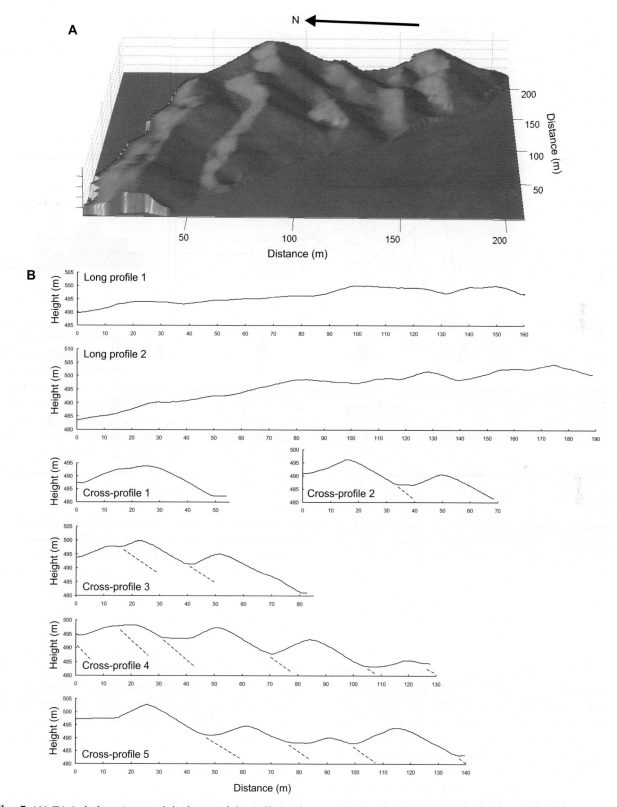

Fig. 5 (A) Digital elevation model of part of the valley-side moraine-mound complex (A$_2$). Faces to the right are commonly rectilinear, whilst those to the left are less regular. (B) Profiles along and across the ridge crests in the valley-side moraine-mound complex (A$_2$). Heights and distances in metres relative to an arbitrary datum. Dashed lines indicated inferred tectonic contacts.

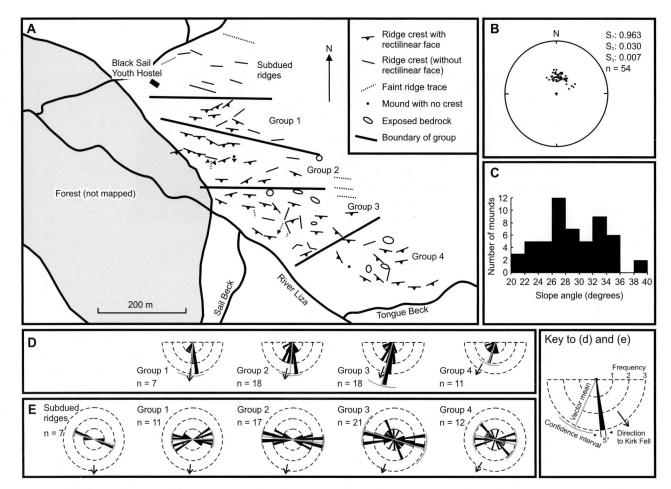

Fig. 6 (A) Structural map of the valley-side moraine-mound complex (A_2) and valley-side subdued ridges (A_3). Ridge-crest orientations and directions of dip of rectilinear faces are marked. (B) Schmidt lower-hemisphere equal-area stereographic projection of the dip and direction of dip of rectilinear faces in the valley-side moraine-mound complex (A_2). (C) Frequency of dips of rectilinear faces in the valley-side moraine-mound complex (A_2). (D) Rose diagrams illustrating the direction of dip of rectilinear faces in the valley-side moraine-mound complex (A_2) at 5° intervals. The vector mean, its confidence interval and the direction to Kirk Fell are marked. (E) Rose diagrams illustrating the orientations of ridge crests in the valley-side moraine-mound complex (A_2) and valley-side subdued ridges (A_3) at 5° intervals. The vector mean, its confidence interval and the direction to Kirk Fell are marked. The number of mounds in each diagram varies between D and E because not all mounds have distinct ridge crests or rectilinear faces.

valley axis is southeast-northwest). This general trend is reflected in the strikes of the ridge crests (although the orientation of individual mounds is not always perpendicular to the dip of the recti-linear slopes), with the additional notable factor being an increase in variability up-valley. The north-facing sides of the mounds are generally irregular and rectilinear slopes are absent.

Exposures in the moraine mounds principally reveal the diamicton (D) facies, although the gravel (G) facies is also represented. The bedded sand and gravel (SGb) facies occurs in a single exposure

and is associated with the diamicton facies, but their geometrical relationship could not be established. No systematic spatial variations in the charac-teristics of the sediments could be determined. Angular boulders are abundant on the surface of the mounds.

Valley-side subdued ridges (A_3)

A series of subdued ridges are present on the valley side, down-valley of the valley-side moraine-mound complex. These ridges are clearly visible on aerial photographs and from a distance in the field, but are less easily identified close up. The landform association is most clear on the valley side beneath Scarth Gap (Fig. 2), but can also be identified in places on the valley side above the moraine-mound complex. As noted above, there is no clear boundary between the valley-side moraine-mound complex and the valley-side subdued ridges. However, they can be differentiated on the basis of: (i) their more subdued, whaleback, morphology; (ii) continuous ridge crests (breached occasionally by streams); (iii) absence of rectilinear faces; and (iv) consistently different ridge-crest orientations (Fig. 6).

Seven ridges, up to 200 m in length, are identifiable on the ground. The amplitude of the ridges is variable, but in the region of 50 m, and they rise up to 6 m above the surrounding topography. Two small exposures reveal examples of the gravel (G) and diamicton (D) facies.

Valley-floor moraine-mound complex (A₄)

A moraine-mound complex is present on the floor of upper Ennerdale from below the Tongue to a little way into the conifer plantation (Figs 2 and 4A). The moraine-mound complex consists of numerous mounds and short-crested ridges rising up to 4 m from the surrounding topography. The mounds commonly rise sharply from flat boggy ground and their true relief may be significantly larger than the apparent relief. The boundary between the oblique ridges on the Tongue (A₁) and the valley-floor moraine-mound complex is sharp (Fig. 4A), but it is unclear whether this is a result of a real difference in landform type, or infilling by sediment. In contrast, the boundary between the valley-floor (A₄) and valley-side (A₂) moraine-mound complexes is diffuse.

The morphology of the moraine-mound complex appears to be chaotic, the orientations and heights of individual mounds being inconsistent and their distribution irregular. However, on the southern side of the valley the complex appears to merge into a series of six small ridges (Fig. 2).

Exposures in the mounds themselves are limited to small fluvial and footpath scrapings. The sediments are predominantly diamictons (facies D), but the gravel (G) facies are also represented. Dispersed, predominantly angular, blocks lie on the surface of the mounds.

Possible origins of the sediment-landform associations

Many models have been proposed to explain the formation of British Younger Dryas moraines, and it is widely accepted that they are polygenetic in origin (Benn, 1996). These models fall into four classes:

1 deformation of sediment at the glacier bed (e.g. Benn, 1992; Bennett, 1995);
2 deposition at the glacier margin during active ice-margin recession (e.g. Benn, 1992; Bennett and Boulton, 1993b; Lukas, 2005);
3 deposition of englacial and supraglacial debris during ice stagnation (e.g. Eyles, 1983; Benn, 1992);
4 deposition of debris entrained along englacial thrusts (Hambrey *et al.*, 1997; Bennett *et al.*, 1998; Graham and Midgley, 2000b).

The key properties of these models, and their ability to explain the four glacial sediment-landform associations in Ennerdale, are assessed in the following sections.

Subglacial deformation

Bedforms may be formed beneath warm-based glaciers at a variety of spatial scales and are commonly associated with Scottish Younger Dryas glaciers (Benn, 1992; Bennett, 1995). They are commonly classified on the basis of their length and elongation ratio as flutes, megaflutes and drumlins (Rose, 1987). Constituent sediments are variable, reflecting the particular material over which the glacier flowed, but commonly consist of sheared basal till (Benn, 1992). Their morphology is highly varied and may reflect more than one glacial episode. They may be older landforms reworked by subglacial processes during subsequent glacier advances, or subglacial landforms submerged by supraglacial sediment during deglaciation (Bennett, 1995). However, typical bedforms have a low stream-lined morphology oriented parallel to glacier flow.

In upper Ennerdale, three of the sediment-landform associations are possible candidates for a subglacial origin. The valley-side subdued ridges (A₃) do not have the classic streamlined form associated with drumlins and flutes, but their whaleback morphology may result from the subglacial remoulding of earlier landforms. The sedimentology and gross morphology of the valley-side moraine-mound complex (A_2) is consistent with a subglacial origin. Bedforms of this scale occur at contemporary glacier margins, but are generally smoothly streamlined and often taper down-glacier (Krüger and Thomsen, 1984). Relict subglacial features in Scotland are generally low and broad (Bennett, 1995). In Ennerdale, the sharp and undulating ridge crests, with offset peaks and common rectilinear faces, are difficult to explain using a subglacial deformation model.

The valley-floor moraine-mound complex (A_4) also lacks the classic streamlined morphology commonly associated with subglacial landforms. The inconsistent orientation of the ridge crests militates against a subglacial origin, although the constituent sediments are compatible with it.

Active ice-margin recession

Moraines form at active glacier margins by a variety of processes, the most common being dumping and washing of sediment from the glacier surface to form dump moraines and ice-contact fans, and the small-scale pushing and squeezing of sediment by the ice (Benn and Evans, 1998). The composition of ice-marginal moraines is highly variable, reflecting the transport history of the sediment and the method by which it was deposited. However, a number of common elements characterise the morphology of each of the ice-marginal moraine types. The plan-form of the moraines reflects that of the ice margin at which they formed and the spatial distribution of debris within the glacier (Bennett and Boulton, 1993a). The volume of sediment within the moraines reflects the balance between the sediment flux and the stability of the location of the ice margin (Bennett and Glasser, 1996). Chains of mounds are common along the frontal margins where the debris flux varies across the glacier margin. The relative stability of the lateral margins of glaciers is conducive to the formation of large dump moraines (Eyles, 1979; Small, 1983). Because they form over long periods, the morphology of lateral moraines is controlled by the location and gradient of the ice margin, and is insensitive to short-term variations in the spatial distribution of debris within the glacier.

As ice-marginal moraines tend to be reworked by glacier advances, and have little topographic

expression when formed during continuous glacier recession, they are characteristic of punctuated ice-margin recession, with each moraine ridge representing the location of a still-stand or minor readvance (Evans, 2003b). Bifurcating ridge crests are common where the rate of ice-margin recession is variable along the ice front. The cross-profile of ice-marginal moraines is strongly controlled by the depositional process. Where the moraines are ice-supported during deposition, ice-contact faces tend to rest at the angle of repose; however, they may be subject to considerable paraglacial reworking. Where ice-marginal pushing and squeezing is important, the ice-contact face is commonly gently dipping (Price, 1970). The distal faces of push and squeeze moraines tend to be steep, whilst ice-contact fans have comparatively gentle distal slopes (Lukas, 2005).

In upper Ennerdale, all of the sediment-landform associations are potential candidates for an ice-marginal origin. The oblique ridges on the Tongue (A_1) are sub-parallel and have anastomosing ridge crests, which are key diagnostic indicators of formation at an ice margin. Their valley-side location and orientation oblique to the axis of the valley also support an ice-marginal origin. The absence of sedimentary exposures in this sediment-landform association makes the determination of a more specific origin difficult.

The valley-side location and planform morphology of the valley-side subdued ridges (A_3) is also consistent with an ice-marginal origin, but their precise origin is difficult to determine in the absence of good sedimentary exposure. Their comparatively subdued morphology (relative to the other sediment-landform associations in Ennerdale) suggests they may be older than the other landforms in the valley, or that they may have been over-ridden by a later glacier advance. Sissons (1980) argued that both were the case and that the diffuse boundary between the valley-side moraine-mound complex (A_2) and the subdued ridges may have resulted from modification of the latter by a Younger Dryas glacier. Without direct dating control, it is likely that the relative ages of these landforms will remain uncertain.

The valley-side moraine-mound complex bears some resemblance to the lateral dump moraines commonly associated with relatively high-relief mountain environments (Benn et al., 2003). Nested sets of moraines may record successive ice-margin positions (Boulton and Eyles, 1979). Models of dump moraine formation in lateral positions stress the importance of the delivery of passively transported sediment from the glacier surface (Boulton and Eyles, 1979; Small, 1983). Whilst such moraines may contain significant proportions of passively transported sediment, particularly in more frontal positions (Matthews and Petch, 1982; Benn and Ballantyne, 1994), the proportion of passively transported sediment observed in the Ennerdale moraine-mound complex (as indicated by the proportion of angular and very angular clasts) are much lower than the proportions observed in Scottish Younger Dryas lateral moraines (Benn, 1989), and towards the lower end of the range observed at contemporary Norwegian glaciers (Matthews and Petch, 1982; Benn and Ballantyne, 1994). Futhermore, lateral moraines tend to reflect the planform morphology of the ice-margin and generally decrease in elevation in a down-glacier direction (Bennett and Glasser, 1996). These characteristics contrast with observations in Ennerdale, where the crest-lines of the moraines are undulating and individual mounds are commonly offset from the general trend-line of the ridge. Finally, although lateral moraines are commonly of considerably different sizes on opposite sides of a valley (e.g. Benn, 1989), they are generally expected to occur in pairs, one on each side of the valley. Although precise matching of the number of moraines on each side of a valley is often not possible, in upper Ennerdale moraines appear to be entirely absent on the southern side of the valley. Overall, these characteristics strongly suggest that the valley-side moraine-mound complex did not form as nested lateral dump moraines.

An alternative hypothesis is that the valley-side moraine-mound complex resulted from ice-marginal pushing and squeezing. This seems unlikely because such landforms commonly have shallow ice-proximal slopes and it is difficult to explain why the moraines should be so well developed on the northern side of the valley and absent elsewhere. It has recently been argued that the majority of Scottish moraine-mound complexes represent suites of nested ice-contact fans (Lukas, 2005). However, the shallow distal slopes associated with such features are in marked contrast to the steep distal slopes in the valley-side moraine-mound complex (A_2).

Groups of moraine mounds that have no clear organisation when viewed from the ground have been used by various workers to delimit former ice extent when their distribution is mapped from aerial photographs (e.g. Benn, 1992; Bennett, 1994; McDougall, 2001). In the case of the valley-floor moraine-mound complex (A_4), the distribution of mounds does not suggest an ice-marginal origin, except on the lower slopes of Kirk Fell where six sub-parallel ridges are present. The relation between these ridges, whose sub-parallel nature does indeed suggest an ice-marginal origin, and the remainder of the moraine-mound complex is unclear. It is possible that they actually represent the result of two distinct sets of processes.

Ice stagnation

Ice-stagnation describes the situation where ice no longer slides and deforms internally and may occur either when heavily debris-covered ice becomes dynamically separated from the main glacier body, or as a result of a decrease in surface gradient (e.g. following a surge or a period of negative mass balance). When stagnation occurs, any debris within the glacier cannot be delivered to the glacier margin and so is deposited *in situ* as the ice melts. This process may occur at a variety of scales, from complete glacier tongues, through narrow debris-covered bands close to the glacier margin (Eyles, 1983), to the melting of small cores of ice trapped beneath sediment (Benn, 1992). During deposition, the sediment is commonly subject to repeated remobilisation as the surface topography varies as a result of differential ablation and may result in topographic inversion (Eyles, 1979). The resulting sediment-landform association (ablation moraine) is commonly characterised by a chaotic morphology with little or no discernable structure, although marked spatial variations in debris concentration may be weakly preserved in the resulting moraine morphology (Boulton, 1972; Benn and Evans, 1998). The nature of the constituent sediments is extremely diverse and depends on both their transport history and the process by which they were deposited. The structures within the sediments commonly reflect the importance of water in re-distributing the sediment. Typical sediments include those associated with debris flows, meltwater streams and resedimentation into small water bodies, and there may be evidence of deformation structures (Boulton, 1972; Eyles, 1979).

In upper Ennerdale, both the valley-side (A_2) and valley-floor (A_4) moraine-mound complexes are possible candidates for an ice-stagnation origin. The clear morphological structure within the other sediment-landform associations militates against a stagnation origin. The apparently chaotic morphology of the valley-floor moraine-mound complex (A_4) is entirely consistent with a stagnation origin, although the close association with the six sub-parallel ridges on the slopes below Kirk Fell suggests that there may be some morphological organisation within the complex that is concealed beneath the boggy sediments between them. The constituent sediments are poorly exposed so the internal structure of individual mounds cannot be determined. Where they are exposed, the constituent sediments are interpreted to have been subject to subglacial transport. This does not preclude an ice-stagnation origin for the mounds, but requires that the sediment was elevated into the body of the glacier prior to deposition.

The valley-side moraine-mound complex (A_2) has a much stronger spatial organisation than that on the valley floor. This is not inconsistent with an ice-stagnation origin and may be explained by the distribution of debris septa within the glacier. However, such structure is generally weakly preserved after deglaciation (Benn and Evans, 1998), and the overall topography would be exceptional for a stagnation origin. In particular, it is difficult to explain the presence of systematically orientated rectilinear faces on many of the mounds. Nevertheless, the presence of bedded sand and gravel within one exposure is entirely consistent with the remobilisation processes that are commonly associated with the formation of ablation moraine. As for the valley-floor moraines, if formation of these moraines was associated with ice-stagnation processes, the basally transported sediment that appears to be their most common constituent would have required elevation into the body of the glacier so that ice could be trapped beneath it.

Englacial thrusting

Thrusting is a common process in the lower reaches of glaciers where they are subjected to longitudinal flow compression, and is particularly widespread in glaciers of polythermal character. In Svalbard polythermal glaciers, thrusting takes place at the ice margin, englacially and even proglacially, especially when the glacier is dynamic and advancing. Entrainment of large volumes of subglacial or basal glacial debris accompanies this process (Hambrey et al., 1997; Hambrey et al., 1999; Bennett, 2001; Glasser and Hambrey, 2001). Ice-marginal thrusting involving substantial volumes of debris has also been documented in temperate Icelandic glaciers, and locally in alpine temperate glaciers (Goodsell et al., 2005). Release of the debris produces moraine-mound complexes that maintain their integrity in zones of permafrost, albeit modified by slope processes (Hambrey and Huddart, 1995; Huddart and Hambrey, 1996; Bennett et al., 1999).

The resulting moraine complexes consist of consistently aligned, imbricate, mounds and ridges. The form of individual moraines is variable, ranging from ridges up to hundreds of metres long to almost conical mounds; most are a few metres in height (Bennett et al., 1998). Rectilinear faces, typically dipping at 30°, are characteristic of the ice-proximal faces and represent the geometry of the original thrust plane. Distal faces are often steeper and irregular, resting at the angle of repose. Thrust-moraine complexes may contain a diverse range of sediment types, reflecting the material over which the glacier flowed. Individual ridges are commonly composed of a single facies or facies association, but may be stacked on another ridge with a quite different character (Glasser and Hambrey, 2003). The ultimate morphology of the moraines following climatic amelioration is strongly controlled by their constituent sediments. Mud-rich moraines tend to be remobilised as gravity-driven flows, whilst thick wedges of well-drained sediments tend to preserve the thrust geometry (Graham, 2002). Bennett et al. (1996b) argued that the final morphology is also dependent on the spacing of thrusts. Where thrusts are closely spaced, there is little disruption to the sediment slabs during deglaciation; where thrusts are widely spaced, slumping and settlement is likely as the ice between them melts. Thus englacial thrust-moraines can be seen as one end of a continuum between thrust-controlled and ablation moraines.

In upper Ennerdale, the valley-floor (A_4) and valley-side (A_2) moraine-mound complexes are potential candidates for a thrust origin, but not necessarily of the Svalbard polythermal type. The broad belt of moraine mounds of variable size, combined with the presence of rectilinear faces with a consistent dip and orientation on the southern side of many of the mounds, is very similar to the morphology of contemporary thrust-moraine complexes. Furthermore, the overall appearance of the moraine-mound complex is suggestive of slabs of sediment that are imbricately stacked upon one another, a relation that is particularly clearly shown in Fig. 4c. Although the evidence of the constituent facies and internal geometry of the moraines in Ennerdale is limited, the presence of sediment inferred to be of basal origin and bedded sands and gravels is consistent with formation by englacial thrusting.

The valley-floor moraine-mound complex (A₄) lacks most of the indicators of a thrust origin (such as rectilinear faces and imbricate structure). However, it is still possible that the moraines have a thrust origin, but that the thrusts were widely spaced and the structure has been destroyed by resedimentation during deglaciation. This supposition is supported by the relatively sparse distribution of the mounds, as well as the transitional nature of the boundary with the valley-side moraine-mound complex with its more typical thrust-moraine morphology.

Synthesis

Based on an evaluation of the strength of the evidence for each of the different models for moraine formation, Table 2 synthesises the likely origins of each sediment-landform association treated in isolation.

It is fairly clear that the oblique ridges (A₁) and valley-side subdued ridges (A₃) represent ice-marginal positions, although the lack of detailed sedimentological evidence has precluded the determination of a more precise origin. The subdued ridges may have obtained their subdued morphology as a result of overriding by a subsequent glacier advance. The origin of the other landforms is more equivocal. Whilst both ice-stagnation and subglacial deformation are possible origins for the valley-side moraine-mound complex (A₂), the evidence is weak. An origin as lateral dump moraines is also possible, but it is difficult to explain the undulating nature of the ridge crests and the total absence of equivalent features on the opposite side of the valley. The support for an englacial thrust origin is not overwhelming, but this hypothesis best fits the available evidence.

Several hypotheses are able to explain the character of the valley-floor moraine-mound complex (A₄). A subglacial origin is possible, but considered unlikely. The available evidence is unable to distinguish between the ice-stagnation and englacial thrusting origin. Given the other evidence (such as ice-marginal moraines) for active deglaciation in upper Ennerdale, it seems unlikely that the valley-floor moraines were associated with large-scale ice-stagnation processes, but localised stagnation cannot be ruled out. The close association and diffuse boundary between the valley-floor and valley-side moraine-mound complexes suggests

Table 2 Summary of the probable origin of each of the sediment-landform associations. Indication of probability ranges from no ticks (very unlikely) to three ticks (most likely)

Sediment-landform association	Possible origin			
	Subglacial deformation	Active ice-margin recession	Ice stagnation	Englacial thrusting
Oblique ridges (A₁)		√√√		
Valley-side moraine-mound complex (A₂)	√	√√	√	√√√
Valley-side subdued ridges (A₃)	√	√√√		
Valley-floor moraine-mound complex (A₄)	√		√√	√√

that they may be genetically related, and lends support to a thrust origin for the valley-floor deposits.

DISCUSSION

Englacial thrusting: a valid mechanism of landform genesis?

Landforms interpreted as relict englacial thrust moraines have now been described in Glen Torridon, Scottish Highlands (Bennett *et al.*, 1998), Cwm Idwal, north Wales (Graham and Midgley, 2000b), and Ennerdale. Although an englacial thrusting origin appears to provide the best fit to the available evidence at each of these sites, a number of workers have disputed the validity of this mechanism in the context of the British Younger Dryas landform record. The origin of the Glen Torridon landforms was disputed by Wilson and Evans (2000), who argued that the landforms represent subglacial bedforms that have been partially reworked from former ice-marginal landforms. However, none of the detailed evidence presented by Bennett *et al.* (1998) in support of an englacial thrust origin was addressed. Indeed, the majority of the evidence that Wilson and Evans (2000) presented is entirely consistent with a thrust origin for the landforms. The only exception is a map of the supposed bed forms, apparently based on aerial photograph interpretation, but there seems little evidence to support a subglacial origin in the images.

Most recently, Lukas (2005) argued, on the basis of detailed sedimentological investigations in the far northwest of Scotland, that Younger Dryas moraine-mound complexes were predominantly formed as a series of nested ice-contact fans. This interpretation is not disputed for these sites. However, the absence of evidence for thrusting at these sites does not mean that thrusting is not a valid mechanism for moraine-mound formation elsewhere (Graham *et al.*, 2007). It has never been proposed that a thrusting origin for moraine-mound complexes is universal, or even particularly common. Indeed, as outlined in the following section, it seems likely that its applicability is restricted to a small number of sites in Britain where very specific conditions prevailed. It is clear that the thrusting model does not fit either the morpho-

logy or sedimentology described by Lukas (2005). Most notably, his description of low-angle ice-distal faces is in marked contrast to the steep distal faces, at the angle of repose, seen at all of the sites where thrusting has been proposed.

The significance of thrust-moraine formation in upper Ennerdale

If the inference that englacial thrusting was responsible for generating some of the sediment-landform associations in upper Ennerdale is correct, the formation and long-term preservation of these landforms has a number of significant implications.

Mapping of thrust-traces on the surfaces of Svalbard glaciers reveals that most thrusts form approximately parallel to the ice-margin, that is perpendicular to the direction of maximum compressive stress (e.g. Hambrey *et al.*, 1999). In Svalbard thrust-moraines, the rectilinear face of each mound appears to record the geometry of the thrust from which the constituent sediment was deposited, and plotting the direction in which these faces dip provides evidence of the direction of maximum local compressive stress in the glacier. A structural map of the rectilinear faces in the Ennerdale valley-side moraine-mound complex suggests a variation in maximum compressive stress from the south in the down-valley part of the moraine-mound complex to the south-west further up-valley (Fig. 6). This pattern of stress is strongly suggestive of ice flow into and across Ennerdale from Kirk Fell. McDougall (2001) mapped a Younger Dryas plateau icefield on the summit of Kirk Fell with an outflow glacier entering Ennerdale *via* Baysoar Slack cirque. Alternatively, a cirque glacier may have formed in Baysoar Slack and been nourished by snow blown from the plateau surface to the south. Additional lines of evidence supporting ice-flow from Kirk Fell are the subdued ridges running up the southern valley-side, which are likely to represent ice-marginal positions. The supposition that a glacier flowed from Kirk Fell also helps to explain the apparently anomalous low gradient of the terminus of the Ennerdale glacier noted by Sissons (1980).

The initiation of thrusting in glacier ice results from longitudinal flow compression that cannot be absorbed by ductile deformation (Bennett *et al.*, 1996b; 1999). In Svalbard polythermal glaciers, this compression results from the transition between warm- and cold-based ice, but a number of other factors may also result in compression. All of the British Younger Dryas thrust moraines so far inferred are stacked against the valley side, and it seems likely that their formation was strongly influenced by compression against a reverse bedrock slope. In Ennerdale, the oblique ridges on the Tongue provide evidence that ice was present in the upper part of Ennerdale, immediately below the slopes of Great Gable, and it seems reasonable to infer that the confluence of ice from Kirk Fell and Great Gable may also have been a contributory factor in generating compressive stresses great enough to initiate thrusting.

It is worthwhile also considering the circumstances in Ennerdale for the thrust moraines to be preserved in the landform record. The preservation of a distinctive thrust-moraine morphology in the landform record requires limited syn- and post-depositional mass movement. Where significant volumes of buried ice are incorporated into the moraines during deposition, the ultimate morphology may be expected to reflect the processes of resedimentation, so the glaciotectonic structure is most likely to be preserved where the volume of sediment in thrusts is large compared with their spacing (Bennett *et al.*, 1996b). If the valley-floor moraine-mound complex was formed from thrust sediment, the lower compressive stresses further away from the valley side may have produced fewer, and more widely spaced, thrusts and explain the poorly preserved tectonic structure. Steeply dipping thrusts also favour preservation as it is the base of the thrust which forms the mound and in steep thrusts more material is left in the base of the mound and less ice is draped with sediment, promoting resedimentation (Bennett *et al.*, 1999). The location of the valley-side moraine-mound complex on well-drained ground away from the valley floor may also have aided its preservation.

Thrusting in Svalbard glaciers is commonly associated with polythermal ice. There is no direct evidence for this in upper Ennerdale, but the formation of thrust-moraines requires that significant quantities of sediment be entrained along englacial thrusts, and a polythermal basal regime may help to explain how this occurred. Whilst warm- and cold-based glaciers tend to have low basal-debris loads (e.g. Chinn and Dillon, 1987;

Kirkbride, 1995), polythermal glaciers may develop very thick, debris-rich, basal ice-layers (in excess of 15 m, e.g. Lawson, 1979). However, such thick layers are associated with glaciers several orders of magnitude larger than that in upper Ennerdale (Kirkbride, 1995). Given the limited distance available for a debris-rich basal ice-layer to form, it is unlikely that sufficient debris could have been entrained in the basal ice-layer to explain the volume of sediment observed in each mound in Ennerdale. Subglacial sediment must, therefore, have been the source for the moraines. The presence of bedrock exposures between the moraine mounds suggests that the sediment–bedrock interface may have acted as a décollement surface, enabling rafts of subglacial debris to be elevated into the glacier. The debris loads of thrusts in warm-based glaciers tend to be much lower than in polythermal glaciers, although there have been few structural studies of non-surge-type warm-based glaciers. It has been suggested that this difference reflects a combination of the freezing-on of rafts of sediment at the thermal boundary and a corresponding increase in longitudinal compression associated with a change from sliding to non-sliding conditions (Hambrey and Huddart, 1995; Bennett, 2001). Therefore, there is some *prima facie* evidence that the entrainment of large volumes of sediment in upper Ennerdale may have been associated with polythermal ice.

A process-form model for upper Ennerdale

In the light of the preceding discussion, and the documentation of the morphology and sedimentology of the landform-sediment associations in Ennerdale, it is possible to propose a process-form model for the Younger Dryas glaciation in the upper valley.

A few thousand years after the complete withdrawal of ice following the Last Glacial Maximum, glaciers became re-established on the Central Fells of the Lake District, including the crags and cirques of Great Gable and Kirk Fell. Ice flowed into and along Ennerdale from these source areas. The maximum extent of the Ennerdale glacier is uncertain, but its limit is likely to be marked by the furthest of the subdued ridges on the northern valley-side. The limit may be obscured within what is currently forested land.

The flow of ice from Kirk Fell across Ennerdale, and its confluence with the main Ennerdale glacier, resulted in longitudinal compression against the northern valley side and the initiation of thrusting within the glacier ice. Debris was entrained along these thrusts and elevated into the body of the glacier. During the subsequent retreat of the glacier, the sediment in these thrusts was lowered onto the valley side, the geometry of the thrust plane being reflected by the dip of the rectilinear slopes of the resulting sediment mounds. The moraine-mound complex formed by this process is likely to reflect thrusting that occurred towards the end of the Younger Dryas glaciation as the influence of the main Ennerdale glacier declined, otherwise the structure would have been modified or destroyed. This also explains why ice-marginal moraines do not overprint or disrupt the pattern, and provides evidence that the recession of the glacier was dynamic during this phase.

The subsequent retreat history of the Ennerdale and Kirk Fell glaciers is uncertain because of the equivocal origin of the valley-floor moraine-mound complex (A_4). In addition, the small, probably ice-marginal, ridges associated with the valley-floor moraine-mound complex (A_4) suggest dynamic recession, but their location and orientation makes it unclear whether they are related to the main Ennerdale or Kirk Fell glacier. If the valley-floor moraine-mound complex represents a stagnation deposit, this may have resulted from the recession of the Kirk Fell glacier. The very low gradient of the lower part of the main Ennerdale glacier was supported by the ice supplied by the Kirk Fell glacier. If this ice supply ceased, the low gradient ice on the floor of the valley may have become dynamically separated from the rest of the main Ennerdale glacier and stagnated *in situ*. If this is the case, no climatic significance can be attached to the stagnation deposit. Alternatively, the complex may provide no evidence about the nature of recession and may be formed of englacially thrust material, the original tectonic structure of which has been disrupted by subsequent glacier flow or reworking during deposition. Whatever the nature of this stage of the recession, the oblique ridges on the Tongue (A_1) appear to record a series of ice-marginal positions of the main Ennerdale glacier, suggesting that the final recession of this glacier was dynamic.

Comparison with Cwm Idwal, North Wales

There is a striking similarity between the topographic setting and sediment-landform associations developed in Ennerdale and those described by Graham & Midgley (2000b) in Cwm Idwal, north Wales. Both sites have: (i) a moraine-mound complex with clear organisation stacked on the valley side; (ii) a more chaotically organised moraine-mound complex on the valley floor; (iii) no clear downvalley ice limits; (iv) a potential glacier source area on the opposite side of the valley to the moraine-mound complexes; and (v) small subparallel ridges beneath the potential source area. This remarkable similarity suggests that the process-form model developed for upper Ennerdale may be more generally applicable at sites where a perhaps extremely dynamic glacier, with its source high on a valley side, is confluent with a glacier at a relatively low altitude in the valley bottom.

Wider significance

The pattern of landform-sediment associations in upper Ennerdale contrasts sharply with the 'glaciated-valley landsystem' model first proposed by Boulton and Eyles (1979). This model is characterised by large volumes of passively transported sediment and the dominant sediment-landform association is large latero-frontal dump moraines. Benn and Evans (1998) distinguished between high- and low-relief glaciated-valley land-systems, with the principal difference being the magnitude of the debris supply from the valley sides. The model of Boulton and Eyles (1979) is appropriate for some tectonically active, high-relief, environments with large fluxes of passively transported sediment, whereas the model developed for upper Ennerdale (and also for Cwm Idwal) may provide an alternative glaciated-valley landsystem for tectonically stable, low-relief, mountain environments in which active transport dominates. The particular suite of sediment-landform associations developed at these sites may be expected where there is a strong cross-valley compressive stress regime.

The recognition of thrust-moraine complexes in the palaeo-landform record is of importance as it has potential implications for the reconstruction of glacier dynamics and palaeoclimate (Hambrey et al., 1997; Bennett et al., 1998; Hambrey and Glasser, 2002). In the case of Ennerdale it appears that the palaeoclimatic implications are limited, as the presence of a reverse bedrock slope and glacier confluence are likely to have been the dominant control on thrusting. The preservation of a glacio-tectonic structure does, however, demonstrate that glacier retreat was dynamic and not dominated by *in situ* stagnation and associated resedimentation. Because they form englacially, thrust-moraine complexes do not provide direct evidence of the position of the ice-margin. Where an ice-marginal position has been inferred from the distribution of moraine-mound complexes, as has often been the case in the reconstruction of British Younger Dryas glaciers (e.g. Sissons, 1980; Gray, 1982), the maximum size of the glacier may be underestimated.

Although the deformation structures present in glacier ice are relatively well understood, the rôle of glacier ice-deformation in landform development has, until recently, been little studied. Recent work on contemporary glaciers in Svalbard (Bennett and Glasser, 1996; Bennett et al., 1996a; 1999; Hambrey et al., 1997; 1999) has begun to redress this, but work on evidence of ice-deformation associated with valley glaciers in the palaeo-landform record has hardly begun. In the British Younger Dryas land-form record, evidence of ice-deformation is limited to the moraine-mound complex interpreted as thrust-moraines in Torridon, Scotland (Bennett et al., 1998; Wilson and Evans, 2000), Ennerdale and Cwm Idwal (Graham and Midgley, 2000b), and some sedimentological evidence of debris eleva-tion in the morainic sediments on Skye, Scotland (Benn, 1990). Although such landforms are appar-ently not abundant, it seems likely that if three englacial thrust-moraine complexes are present in the British Younger Dryas landform record, others remain to be discovered. Similar structures have recently been identified in Pleistocene glacigenic sediments in Poland (Ruszczynska-Szenajch, 2001). With the exception of this Polish work, englacial thrust-moraine complexes in the palaeo-landform record have thus far only been recognised in asso-ciation with cirque and valley glaciers, but there is no reason to believe they may not have formed elsewhere close to the margins of the Pleistocene ice-sheets where hummocky topography is common (e.g. Sollid and Sørbel, 1988; Johnson et al., 1995; Andersson, 1998; Eyles et al., 1999). Where they

do exist, however, their recognition may be prob-lematic because of degradation of the glaciotectonic structure. It is also likely that landforms result-ing from other forms of ice-deformation, such as foliation-parallel ridges and supraglacial debris stripes, may await discovery in the palaeo-landform record (Hambrey and Glasser, 2002).

CONCLUSIONS

The typical Lake District glacial trough of Ennerdale contains one of the finest moraine-mound ('hum-mocky moraine') complexes in England. Inferred to date from the Younger Dryas or Loch Lomond Stade, the complex provides evidence of a multi-plicity of events and a range of glaciotectonic and sedimentary processes. These processes, whilst not necessarily common in Britain, indicate that for certain combinations of glacier dynamics, thermal regime and topography, englacial thrusting is an important process in landform development. More specifically, it is concluded that:

1 Four distinct landform-sediment associations record the presence of a glacier in the upper part of Ennerdale during the Younger Dryas. Ridges run obliquely down an elevated area (the Tongue) in the upper part of the valley. Moraine-mound complexes are developed on the northern valley-side and the floor of the valley. Subdued ridges run down the northern valley-side beyond the moraine-mound complex.
2 The principal sedimentary facies that make up the landforms are gravel and diamicton, interpreted as basal till, and bedded sand and gravel, inferred to be the product of fluvial reworking.
3 Ice-marginal moraines indicate that the recession of the glacier was predominantly active, although the presence of a chaotic moraine-mound complex on the valley floor suggests that ice may have stagnated locally because of dynamic separation of a subsidi-ary glacier from the tongue of the main Ennerdale glacier.
4 There is good morphological and structural evid-ence that debris was entrained along thrusts within the body of a glacier that crossed the valley at right angles to its main axis, and resulted in the deposi-tion of a distinctive sediment-landform association. Although polythermal ice is not ruled out, the prin-cipal contributing factors to thrusting are a glacier confluence and a reverse bedrock slope, which led to

strong longitudinal compression in the ice. Although the inferred subsidiary glacier, originating on Kirk Fell, had previously been placed on a map, this work provides the first published evidence supporting the existence of such a glacier.

5 The thrust-related sediment-landform association in Ennerdale is mirrored by that in Cwm Idwal, Wales. In both cases, a major source area for ice is located along the flanks of the valley. Similar topographic situations occur elsewhere in the British uplands, so it is to be expected that there is scope for discovering other examples.

6 This investigation highlights the importance of sediment, morphology and structure of moraines in establishing the style of glaciation in upland areas, and illustrates the care needed in inferring ice dynamics, glacier thermal regime, and ice limits.

ACKNOWLEDGEMENTS

This work was undertaken whilst D.G. was in receipt of a University of Wales, Aberystwyth postgraduate studentship. Permission to work at the site was granted by the Forestry Commission and the National Trust. Nicholas Midgley and Phillippa Noble provided field assistance. The authors thank David Huddart and an anonymous referee for their helpful and constructive comments.

REFERENCES

Andersson, G. (1998) Genesis of hummocky moraine in the Bolen area, southwestern Sweden. *Boreas*, **27**, 55–67.

Benn, D.I. (1989) Debris transport by Loch Lomond Readvance glaciers in Northern Scotland: basin form and the within-valley asymmetry of lateral moraines. *J. Quatern. Sci.*, **4**, 243–254.

Benn, D.I. (1990) *Scottish Lateglacial Moraines: Debris Supply, Genesis and Significance*. Unpublished PhD thesis, University of St. Andrews.

Benn, D.I. (1992) The genesis and significance of 'hummocky moraine': evidence from the Isle of Skye, Scotland. *Quatern. Sci. Rev.*, **11**, 781–799.

Benn, D.I. (1996) Glacial sedimentological research in Scotland. *Scot. Geogr. Mag.*, **112**, 57–62.

Benn, D.I. and Ballantyne, C.K. (1993) The description and representation of particle shape. *Earth Surf. Proc. Land.*, **18**, 665–672.

Benn, D.I. and Ballantyne, C.K. (1994) Reconstructing the transport history of glaciogenic sediments – a new approach based on the co-variance of clast form indices. *Sed. Geol.*, **91**, 215–227.

Benn, D.I. and Evans, D.J.A. (1998) *Glaciers and Glaciation*. Arnold, London, 734 pp.

Benn, D.I. and Evans, D.J.A. (2004) Introduction and rationale. In: *A Practical Guide to the Study of Glacial Sediments* (Eds D.J.A. Evans and D.I. Benn), pp. 1–10. Arnold, London.

Benn, D.I., Kirkbride, M.P., Owen, L.A. and Brazier, V. (2003) Glaciated valley landsystems. In: *Glacial Landsystems* (Ed. D.J.A. Evans), pp. 372–406. Arnold, London.

Bennett, M.R. (1991) *Scottish 'Hummocky Moraine': its Implications for the Deglaciation of the North West Highlands During the Younger Dryas or Loch Lomond Stadial*. Unpublished PhD thesis, University of Edinburgh.

Bennett, M.R. (1994) Morphological evidence as a guide to deglaciation following the Loch Lomond Readvance: a review of research approaches and models. *Scot. Geogr. Mag.*, **110**, 24–32.

Bennett, M.R. (1995) The morphology of glacially fluted terrain: examples from the Northwest Highlands of Scotland. *Proc. Geol. Assoc.*, **106**, 27–38.

Bennett, M.R. (2001) The morphology, structural evolution and significance of push moraines. *Earth-Sci. Rev.*, **53**, 197–236.

Bennett, M.R. and Boulton, G.S. (1993a) Deglaciation of the Younger Dryas or Loch Lomond Stadial ice-field in the Northern Highlands, Scotland. *J. Quatern. Sci.*, **8**, 133–145.

Bennett, M.R. and Boulton, G.S. (1993b) A reinterpretation of Scottish 'hummocky moraine' and its significance for the deglaciation of the Scottish Highlands during the Younger Dryas or Loch Lomond Stadial. *Geol. Mag.*, **130**, 301–318.

Bennett, M.R. and Glasser, N.F. (1996) *Glacial Geology: Ice Sheets and Landforms*. Wiley, Chichester, 364 pp.

Bennett, M.R., Hambrey, M.J. and Huddart, D. (1997) Modification of clast shape in High-Arctic environments. *J. Sed. Res.*, **67**, 550–559.

Bennett, M.R., Hambrey, M.J., Huddart, D. and Ghienne, J.F. (1996a) The formation of a geometrical ridge network by the surge-type glacier Kongsvegen, Svalbard. *J. Quatern. Sci.*, **11**, 437–449.

Bennett, M.R., Hambrey, M.J., Huddart, D. and Glasser, N.F. (1998) Glacial thrusting and moraine-mound formation in Svalbard and Britain: the example of Coire a'Cheund-chnoic (Valley of a Hundred Hills), Torridon, Scotland. In: *Mountain Glaciation* (Ed. L.A. Owen), pp. 17–34. Wiley & Sons, Chichester.

Bennett, M.R., Hambrey, M.J., Huddart, D., Glasser, N.F. and Crawford, K. (1999) The landform and sediment assemblage produced by a tidewater glacier surge in Kongsfjorden, Svalbard. *Quatern. Sci. Rev.*, **18**, 1213–1246.

Bennett, M.R., Huddart, D., Hambrey, M.J. and Ghienne, J.F. (1996b) Moraine development at the High-Arctic glacier Pedersenbreen, Svalbard. *Geogr. Ann.*, **78A**, 209–222.

Boulton, G.S. (1972) Modern Arctic glaciers as depositional models for former ice sheets. *J. Geol. Soc. London*, **128**, 361–393.

Boulton, G.S. (1978) Boulder shapes and grain-size distributions of debris as indicators of transport paths through a glacier and till genesis. *Sedimentology*, **25**, 773–799.

Boulton, G.S. and Eyles, N. (1979) Sedimentation by valley glaciers: a model and genetic classification. In: *Moraines and Varves* (Ed. C. Schlüchter), pp. 11–23. Balkema, Rotterdam.

Charlesworth, J.K. (1956) The late-glacial history of the Highlands and Islands of Scotland. *Trans. Roy. Soc. Edinb.*, **62**, 103–929.

Chinn, T.J.H. and Dillon, A. (1987) Observations on a debris-covered polar glacier 'Whiskey Glacier', James Ross Island, Antarctic Peninsula, Antarctica. *J. Glaciol.*, **33**, 300–310.

Evans, D.J.A. (Ed.) (2003a) *Glacial Landsystems*. Hodder Arnold, London, 532 pp.

Evans, D.J.A. (2003b) Ice-marginal terrestrial landsystems: active temperate glacier margins. In: *Glacial Landsystems* (Ed. D.J.A. Evans), pp. 12–43. Hodder Arnold, London.

Eyles, N. (1979) Facies of supraglacial sedimentation on Icelandic and Alpine temperate glaciers. *Can. J. Earth Sci.*, **16**, 1341–1361.

Eyles, N. (1983) Modern Icelandic glaciers as depositional models for 'hummocky moraine' in the Scottish Highlands. In: *Tills and Related Deposits* (Eds E.B. Eversen, C. Schlüchter and J. Rabassa), pp. 47–60. Balkema, Rotterdam.

Eyles, N., Boyce, J.I. and Barendregt, R.W. (1999) Hummocky moraine: sedimentary record of stagnant Laurentide Ice Sheet lobes resting on soft beds. *Sed. Geol.*, **123**, 163–174.

Glasser, N.F. and Hambrey, M.J. (2001) Styles of sedimentation beneath Svalbard valley glaciers under changing dynamic and thermal regimes. *J. Geol. Soc. London*, **158**, 697–707.

Glasser, N.F. and Hambrey, M.J. (2003) Ice-marginal terrestrial landsystems: Svalbard polythermal glaciers. In: *Glacial Landsystems* (Ed. D.J.A. Evans), pp. 65–88. Hodder Arnold, London.

Goodsell, B., Hambrey, M.J. and Glasser, N.F. (2005) Debris transport in a temperate valley glacier: Haut Glacier d'Arolla, Valais, Switzerland. *J. Glaciol.*, **51**, 139–146.

Graham, D.J. (2002) *Moraine-mound formation during the Younger Dryas in Britain and the Neoglacial in Svalbard*. Unpublished PhD thesis, University of Wales, Aberystwyth.

Graham, D.J., Bennett, M.R., Glasser, N.F., Hambrey, M.J., Huddart, D. and Midgley, N.G. (2007) 'A test of the englacial thrusting hypothesis of "hummocky" moraine formation: case studies from the northwest Highlands, Scotland': Comments. *Boreas*, **36**, 103–107.

Graham, D.J. and Midgley, N.G. (2000a) Graphical representation of particle shape using triangular diagrams: an Excel spreadsheet method. *Earth Surf. Proc. Land.*, **25**, 1473–1477.

Graham, D.J. and Midgley, N.G. (2000b) Moraine-mound formation by englacial thrusting: the Younger Dryas moraine of Cwm Idwal, North Wales. In: *Deformation of Glacial Materials* (Eds A.J. Maltman, M.J. Hambrey and B. Hubbard), pp. 321–336. Geological Society, London.

Gray, J.M. (1982) The last glaciers (Loch Lomond Advance) in Snowdonia, North Wales. *Geol. J.*, **17**, 111–133.

Hambrey, M.J., Bennett, M.R., Dowdeswell, J.A., Glasser, N.F. and Huddart, D. (1999) Debris entrainment and transfer in polythermal valley glaciers. *J. Glaciol.*, **45**, 69–86.

Hambrey, M.J. and Glasser, N.F. (2002) Development of landform and sediment assemblages at maritime High-Arctic glaciers. In: *Landscapes of Transition: Landform Assemblages and Transformations in Cold Regions* (Eds K. Hewitt, M. Byrne, M. English and G. Young), pp. 11–42. Kluwer, Dordrecht.

Hambrey, M.J. and Huddart, D. (1995) Englacial and proglacial glaciotectonic processes at the snout of a thermally complex glacier in Svalbard. *J. Quatern. Sci.*, **10**, 313–326.

Hambrey, M.J., Huddart, D., Bennett, M.R. and Glasser, N.F. (1997) Genesis of 'hummocky moraines' by thrusting in glacier ice: evidence from Svalbard and Britain. *J. Geol. Soc. London*, **154**, 623–632.

Huddart, D. (1967) The deglaciation of the Ennerdale area: a re-interpretation. *Proc. Cumberland Geol. Soc.*, **2**, 63–75.

Huddart, D. and Hambrey, M.J. (1996) Sedimentary and tectonic development of a high-arctic thrust-moraine complex: Comfortlessbreen, Svalbard. *Boreas*, **25**, 227–243.

Johnson, M.D., Mickelsen, D.M., Clayton, L. and Attig, J.W. (1995) Composition and genesis of glacial hummocks, western Wisconsin, USA. *Boreas*, **24**, 97–116.

Kirkbride, M.P. (1995) Processes of transportation. In: *Modern Glacial Environments: Processes, Dynamics and Sediments* (Ed. J. Menzies), pp. 261–308. Butterworth-Heinemann, Oxford.

Krüger, J. and Thomsen, H.H. (1984) Morphology, stratigraphy, and genesis of small drumlins in front of the glacier Mýdalsjökull, south Iceland. *J. Glaciol.*, **30**, 94–105.

Lawson, D.E. (1979) *Sedimentological analysis of the western margin of Matanuska Glacier, Alaska*. Cold Regions Research and Engineering Laboratory, Hanover, New Hampshire, 122 pp.

Lukas, S. (2005) A test of the englacial thrusting hypothesis of 'hummocky' moraine formation: case studies from the northwest Highlands, Scotland. *Boreas*, **34**, 287–307.

Maizels, J. (1995) Sediments and landforms of modern proglacial terrestrial environments. In: *Modern Glacial Environments: Processes, Dynamics and Sediments* (Ed. J. Menzies), pp. 365–415. Butterworth-Heinmann, Oxford.

Manley, G. (1959) The Late-glacial climate of north-west England. *Liverpool Manchester Geol. J.*, **2**, 188–215.

Matthews, J.A. and Petch, J.R. (1982) Within-valley asymmetry and related problems of Neoglacial lateral moraine development at certain Jotunheim glaciers, southern Norway. *Boreas*, **11**, 225–247.

McDougall, D.A. (2001) The geomorphological impact of Loch Lomond (Younger Dryas) Stadial plateau icefields in the central Lake District, northwest England. *J. Quatern. Sci.*, **16**, 531–543.

Moncrieff, A.C.M. (1989) Classification of poorly sorted sedimentary rocks. *Sed. Geol.*, **65**, 191–194.

Pennington, W. (1978) Quaternary Geology. In: *The Geology of the Lake District* (Ed. F. Moseley), pp. 207–225. Yorkshire Geological Society, Yorkshire.

Price, R.J. (1970) Moraines at Fjallsjökull, Iceland. *Arctic Alpine Res.*, **2**, 27–42.

Rose, J. (1987) Drumlins as part of a glacier bedform continuum. In: *Drumlin Symposium* (Eds J. Menzies and J. Rose), pp. 103–116. Balkema, Rotterdam.

Ruszczynska-Szenajch, H. (2001) Glaciodynamically upthrusted bands of englacially transported debris in the Pleistocene of central Poland. *Sedimentology*, **48**, 585–597.

Sissons, J.B. (1979) The Loch Lomond Stadial in the British Isles. *Nature*, **280**, 199–203.

Sissons, J.B. (1980) The Loch Lomond Advance in the Lake District, northern England. *Trans. Roy. Soc. Edinb. Earth Sci.*, **71**, 12–27.

Small, R.J. (1983) Lateral moraine of Glacier de Tsidjiore Nouve: form, development, and implications. *J. Glaciol.*, **29**, 250–259.

Sneed, E.D. and Folk, R.L. (1958) Pebbles in the lower Colorado River, Texas, a study of particle morphogenesis. *J. Geol.*, **66**, 114–150.

Sollid, J.L. and Sørbel, L. (1988) Influence of temperature conditions in formation of end moraines in Fennoscandia and Svalbard. *Boreas*, **17**, 553–558.

Ward, C. (1873) The glaciation of the northern part of the Lake-District. *Q. J. Geol. Soc. London*, **29**, 422–441.

Wilson, S.B. and Evans, D.J.A. (2000) Coire a'Cheundchnoic, the 'hummocky moraine' of Glen Torridon. *Scot. Geogr. J.*, **116**, 149–158.

Part 4

Pre-Quaternary glacial systems

Grooved and striated pavement of Palaeoproterozoic dolomite overlain by Permo-Carboniferous Karoo tillite, Douglas, Karoo Basin, South Africa. (Photograph by M.J. Hambrey)

Cenozoic climate and sea level history from glacimarine strata off the Victoria Land coast, Cape Roberts Project, Antarctica

P.J. BARRETT

Antarctic Research Centre, Victoria University of Wellington, PO Box 600, Wellington, New Zealand (e-mail: peter.barrett@vuw.ac.nz)

ABSTRACT

This paper reviews the record of past climate and sea level from 34 to 17 Ma provided by continuous core through 1500 m of shallow marine strata off the Victoria Land coast of Antarctica. The site was selected because it is close to the edge of the East Antarctic Ice Sheet. Previous drilling and seismic surveys had suggested a thick seaward-dipping sequence of Oligocene and older age close to the coast. However, the floor of the basin, lower Devonian sandstone, was encountered beneath uppermost Eocene conglomerate. The strata were deposited in a rift basin just seaward of the Transantarctic Mountains c. 20 m.y. after uplift began. Sediment was delivered by rivers and glaciers both from and behind the mountains from the East Antarctic interior. Sediment accumulation was rapid and the record more complete for the period 34 to 31 m.y., but then slowed as basin subsidence declined, leaving major time gaps, but still a representative record of the entire time span. Basin filling kept pace with subsidence. The sedimentary facies – conglomerate, sandstone and mudstone with marine fossils throughout – are typical of the coastal margin of a subsiding sedimentary basin, with the addition of diamictite beds in the upper 900 m recording marine-terminating glaciers that extended periodically beyond the coast. Deposition was characterised by repetitive vertical facies successions of conglomerate and fine sandstone in the lower part of the section and by cyclic facies successions of diamictite, sandstone and mudstone from a few to over 60 m thick in the upper part. These are thought to reflect glacio-eustatic changes in sea level on a wave-dominated coast in concert with advance and retreat of piedmont ice onto the continental shelf, with diamicton and sand (nearshore) grading upwards to mud (shelf) and then to sand (inner shelf to shoreline). Tephra dating of two cycles at 23.98 and 24.22 Ma allows their correlation with 40,000 years cycles in the deep-sea isotope record, and ascribed to eustatic sea-level changes of 30–60 m. This suggests that the cyclicity of the Cape Roberts section reflects the influence of the earth's varying orbital parameters on climate and sea level, with over 50 sedimentary cycles preserved out of the possible 200–400 cycles during this time period. The Cape Roberts section also records a dramatic increase in glacial influence at around 33 Ma, with the proportion of glacial facies ranging from 10 to 30% of the sedimentary section through to 17 Ma. The section also records a progressive shift from chemical to physical weathering (decline in CIA index and increase in % illite-chlorite) from 33 to 25 Ma, and a decline in marine palynomorphs characteristic of fresh melt water mingling over the same interval. The terrestrial pollen record records interglacial climate and indicates this to be cool temperate for the entire period from 34 to 17 Ma though slightly cooler from c. 25 Ma.

Keywords Cape Roberts Project, Cenozoic, palaeoclimate, sequence stratigraphy, glaciomarine facies, Antarctic margin.

INTRODUCTION

Until the early 1970s it was believed that the ice ages spanned just the last 2 million years of earth history – the Quaternary Period (Flint, 1971). That view changed with the drilling of the Ross Sea and Southern Ocean in 1973 by the *Glomar Challenger*, revealing both a physical record of glaciation that began at least 25 million years ago (Hayes and Frakes, 1975) and a proxy record from deep-sea sediments indicating a fall in temperature (and maybe increase in ice volume) both at the Eocene-Oligocene boundary and in the middle Miocene (Shackleton & Kennett, 1975). The Quaternary ice ages had become Cenozoic in their span – at least in the Southern Hemisphere.

In the decade that followed, analysis of seismic sequences from the world's continental margins showed repeated patterns of coastal advance and retreat on time scales of 10^5 to 10^6 years. These changes could be traced back at least to Cretaceous times, with a significant sea-level fall in mid-Oligocene (Vail *et al.*, 1977); more substance was added in a review a decade later by Haq *et al.* (1987). This new view of the way in which strata are deposited and preserved in basins on continental margins, termed sequence stratigraphy (Posamentier *et al.*, 1988), presumed that cyclic fluctuations of tens to hundreds of metres in sea level on time scales ranging from hundreds of thousands to millions of years. Some found flaws in the simple conceptual model from which the early interpretations were made (e.g. Miall, 1991). However, others have shown how the concept can provide a sound basis for interpreting sea-level changes by testing them on well-dated Quaternary strata, representing a period for which the history of eustatic sea level change is reasonably well established (e.g. Carter *et al.*, 1991). In the meantime improvements in chronology of deep-sea cores through both biostratigraphy and magnetostratigraphy have allowed the assembly of a deep-sea oxygen isotope record from many different sites that showed trends in ocean temperature and ice volume (Miller *et al.*, 1987), with an emerging view that the first continent-wide ice sheet formed on Antarctica in earliest Oligocene times (Wise *et al.*, 1991).

The deep-sea isotope community and sequence stratigraphers have both continued to gather their 'far-field' proxy data as a basis for interpreting past climate and sea-level history through Cenozoic (and older) times (e.g. Zachos *et al.*, 2001; Billups & Schrag, 2003; Miller *et al.*, 2005). At the same time a smaller community has been investigating the Antarctic margin itself for records of past climate, some through seismic surveys and offshore drilling under the auspices of the ANTOSTRAT project (Cooper *et al.*, 1991, 1995) and others through investigations on land (reviewed in Barrett, 1996), though land-based investigations have been frustrated by the poorly fossiliferous and fragmentary record of Cenozoic geological history on the Antarctic continent itself.

One location on the Antarctic margin that has proved especially instructive for Cenozoic climate history has been in the southwest corner of the Ross Sea (Fig. 1). Here a series of drilling projects between 1973 and 1999 (reviewed in Hambrey *et al.*, 2002) cored the western margin of the Victoria Land Basin, one of a north-sound-trending troughs that form part of the West Antarctic Rift System (Behrendt, 1999). The most recent of the series, the Cape Roberts Project, resulted in continuous core (95% recovery) that spans the time period from 34 to 17 Ma, and contains a fragmentary, but nevertheless valuable, coastal record of varying climate, ice conditions and sea level representing most of that period (Cape Roberts Science Team, 1998, 1999, 2000). The purpose of this paper is to summarise the results of this work.

The drilling off Cape Roberts took place from a sea-ice platform in three successive field seasons (from late 1997 to late 1999), coring a more-or-less continuous stratigraphic section of 1500 m of uppermost Eocene to lower Miocene strata (34 to 17 Ma). The initial results and scientific reports have been published in a series of special issues of Terra Antartica (http://www.mna.it/english/Publications/TAP/terranta.html#Special). The drilling itself was a significant technical achievement and is documented in Cowie (2002).

This review first outlines the regional setting in which the Cenozoic strata off Cape Roberts were deposited, and the general character and chronology of the sequence. The strata are then described in terms of a small number of characteristic facies, largely organised as facies successions and interpreted in terms of a glacial sequence stratigraphic model. A case is made for the Oligocene-lower

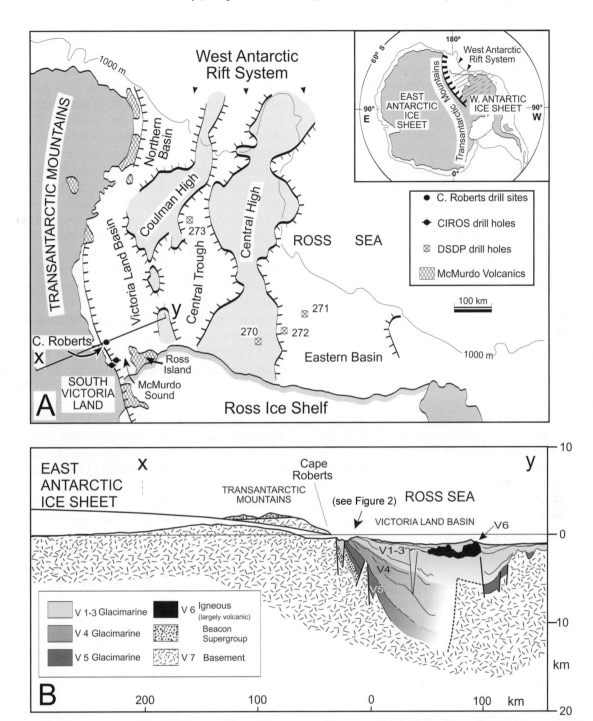

Fig. 1 Setting for Cape Roberts Project drilling on the western margin of West Antarctic Rift System (from Cape Roberts Science Team, 1998). (A) The Ross Sea region, showing the location of the Victoria Land Basin adjacent to the Transantarctic Mountains, and the location of drill sites in the area; 'x–y' shows the section line for B. (B) Cross-section from the East Antarctic interior across the Transantarctic Mountains to the Ross Sea, showing the proximity of the East Antarctic Ice Sheet to the Victoria Land Basin, and hence the potential for strata filling the basin to record climatic and tectonic events from this part of the Antarctic margin. The cross-section has been modified to reflect new knowledge from the Cape Roberts Project, notably the Beacon sandstone flooring the basin, and a new interpretation of the basin fill from Fielding *et al.* (2006) and Wilson *et al.* (unpub. data).

Miocene sequence being deposited in a nearshore marine environment on an open coast with sedimentation strongly influenced by both waves and glaciers discharging into the sea. The cycles record both glacial maxima when ice flowed from the inland ice sheet that lies to the west through the mountains to the coast, at times grounding in the shallow water in the vicinity of the drill site, and periods of glacial retreat, and higher sea level when rivers carried sediment to the coast to be distributed alongshore by waves and currents. Changes in the proportion of glacial facies and in indicators of physical/chemical weathering and temperature, all of which imply a cooling trend, are shown in terms of discrete time periods through the Cape Roberts section, and the implications discussed.

REGIONAL SETTING

Cape Roberts lies on the Transantarctic Mountain Front, a 30-km-wide zone between the rising Transantarctic Mountains to the west and the Victoria Land Basin to the east, and representing the western margin of the West Antarctic Rift System (Fig. 1). The Transantarctic Mountain Front extends for around 1000 km to Cape Adare in the north and over 3000 km to the south, the topographic relief across it typically being around 4000 m. Today the mountains form a significant barrier to the flow of ice through outlet glaciers to the Ross Ice Shelf and Ross Sea.

The firmest indication of the initial growth of the Transantarctic Mountains comes from fission-track data that point to the first significant denudation of the McMurdo sector of the Transantarctic Mountains around 55 Ma, though other sectors of the Transantarctic Mountains record denudation events in the late Cretaceous Period also (Fitzgerald, 1992). The first direct physical evidence of the Transantarctic Mountains as a significant feature comes from the oldest strata cored in the CIROS-1 drill-hole, drilled in 1986 70 km south of Cape Roberts. These include granitic clasts eroded from exposed basement to the west, implying that the Transantarctic Mountains were at least half of their present height, for erosion had even then cut through the more than 2000 m of Devonian-Jurassic Gondwana cover beds to basement (Barrett, 1989).

The Transantarctic Mountains are thought to have risen highest in late Cenozoic time because of the extreme relief they now show between summit and valley floor levels, around 50% more than mountains in temperate regions (Stern *et al.*, 2005). They attributed this increase in relief to middle Miocene cooling that froze mountain tops while outlet glaciers continued to excavate. This same continent-wide cooling would have also reduced sediment supply to the Antarctic continental shelf. Expansions of the Antarctic Ice Sheet have eroded to the shelf edge since its inception in the earliest Oligocene (Hambrey *et al.*, 1992, Anderson, 1999), but the present deep shelf (on average around 500 m) may have resulted from the reduced sediment supply from a largely frozen continent since middle Miocene times.

Today's extreme topography of the Transantarctic Mountains provides a striking contrast to the subdued submarine relief across the Ross Sea, as shown in the recent bathymetric compilation by Davey (2004). Provenance studies from cores taken from the north-south-trending Victoria Land Basin show the basin-fill to have been derived from the adjacent mountains themselves (George, 1989; Sandroni & Talarico, 2001). Eastward-dipping basin geometry and late Cenozoic erosion along the Transantarctic Mountain Front have exposed the oldest strata in the basin in several seismic sections perpendicular to the Transantarctic Mountain Front (Cooper *et al.*, 1995), most notably off Granite Harbour (Barrett *et al.*, 1995; Bartek *et al.*, 1996), the site of the Cape Roberts Project drill sites. Environmental and technical constraints required the selection of three sites in order to core the entire sequence (Figs. 2, 3).

Although the Transantarctic Mountains have been a persistent feature of the region since early Cenozoic time, coastal bathymetry today, with variations in nearshore water-depth from less than 100 m to more than 1000 m, is far more extreme than offshore bathymetric variations during the deposition of the Oligocene–Lower Miocene strata cored off Cape Roberts. The sea-floor relief through that period can be gauged from coast-parallel seismic lines (e.g. NBP9601–93), which passed close to CRP-2/2A (Fig. 4). Resolution is low (~20 m), but the records show persistent stratification parallel to the coast over distances of many kilometres. This contrasts with the channelling many tens of

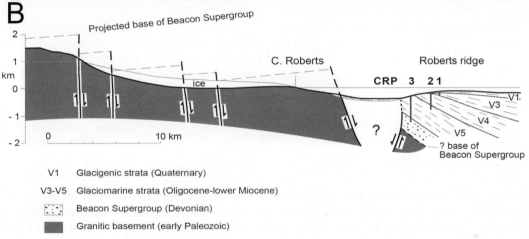

Fig. 2 View north along the Transantarctic Mountains and geological cross-section. (A) Areal photograph (from US Navy photo TMA 2202 F33 141) showing the well-defined topographic boundary between the Transantarctic Mountains and the Ross Sea, as well as the location of the Cape Roberts drill holes and the cross-section below. (B) Geological cross-section (from Cape Roberts Science Team, 2000), showing the geological structure across the Transantarctic Mountain Front (basement faults after Fitzgerald, 1992), and the location and context of the Cape Roberts drill holes near the edge of the Victoria Land Basin.

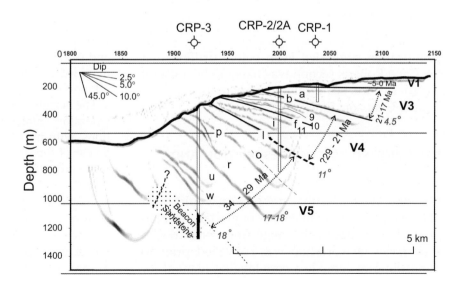

Fig. 3 Seismic stratigraphy of the sequence cored off Cape Roberts (from Henrys *et al.*, 2001), showing drill site locations and chronology.

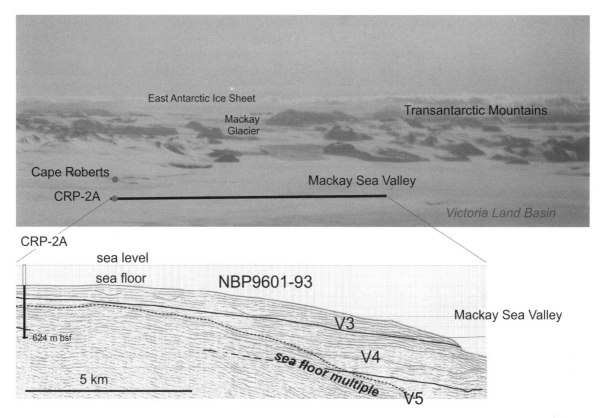

Fig. 4 View of the Transantarctic Mountains from the Ross Sea (US Navy Photo TMA 1558 F33 69). The seismic section beneath (Henrys *et al.*, 2000), which is roughly perpendicular to the regional dip of the strata, shows the parallel stratification and lateral continuity of Oligocene – lower Miocene section cored by CRP2A. This contrasts with the broad channelling many tens of metres deep and hundreds of m across in younger (?late Miocene or Pliocene) strata, and the present day Mackay Sea Valley which is around 800 m deep and 10 km across.

metres deep and hundreds of m across in younger (?late Miocene or Pliocene) strata, and the extreme Quaternary channelling of the Mackay Glacier to form the Mackay Sea Valley.

In summary, the simple seaward-dipping geometry and the coast-parallel persistence of stratification suggested even before drilling that the Cape Roberts section was likely to be a useful recorder of ice, sea level and climate for much of middle Cenozoic time.

BACKGROUND TO CORE DESCRIPTION AND ANALYSIS

With the awareness that the project would most likely be coring a complex mix of glacial and non-glacial strata that might range from terrestrial to shelf facies, a workshop was convened prior to the drilling to compile a core-logging manual for

consistent visual core description (Hambrey *et al.*, 1997), and agreement on the tasks for core-log presentation and analysis. The sedimentological results of the project appear in three main forms

1 The Initial Reports, with a project authorship and completed immediately after the drilling. These included detailed (4 m to a page) core logs, and were intended to be primarily descriptive.
2 The Scientific Reports, with summary descriptions, laboratory analyses and more considered interpretations, with individual authorship, completed within a year of the drilling.
3 Subsequent papers in the open literature.

Glacimarine sediments are inherently varied in texture, composition and sedimentary structures, and while a continuous drill core confers significant advantage in providing a continuous stratigraphic record, it gives no indication of the lateral significance of particular lithologies. Seismic records provide some help in a gross way but only on a scale of tens of metres, being limited by their resolution. As a consequence there were vigorous on-site discussions on criteria for the various facies, and the meaning and significance of many features seen in the core, different people having seen the same feature in different geological contexts. These issues were resolved mostly through discussion, and an agreement to acknowledge differences in interpretation. This took time and a consistent and broadly accepted facies scheme was not established until the Initial Report on CRP-2/2A was produced (Cape Roberts Science Team, 2000). This scheme was tested, and accepted with minor modification for CRP-3 (Cape Roberts Science Team, 2001). It was subsequently used in several papers in the Scientific Reports for CRP-2/2A and CRP-3, as well as papers in the open literature (Powell & Cooper, 2002; Hambrey *et al.*, 2002).

A feature of the project has been the development of a sequence stratigraphic model for explaining the cycles or facies successions that were evident through most of the cored section. Sequence stratigraphy is fundamentally based on changing relative sea level, and requires a degree of wave energy to erode, remove and deposit sand nearshore and mud offshore as rising and falling sea level drives marine transgressions and regressions. Those leading the facies analysis tended to emphasise the glacial aspects of the sediments, whereas those leading the sequence stratigraphy emphasised the role of sea level. These differences are evident in the early reports, but have now been resolved (this paper; Dunbar *et al.*, in press).

The basis of all analyses of the Cape Roberts cores has been the visual core descriptions, compiled at the Cape Roberts camp from the split core face by a team of four sedimentologists led by Ken Woolfe, who also drafted the 4 m-page logs for core from all three sites. These logs were checked, corrected and studied further by a separate team of about six sedimentologists at Crary Lab, McMurdo Station, led for CRP-1 by Mike Hambrey (Cape Roberts Science Team, 1998), CRP-2/2A by Chris Fielding (Cape Roberts Science Team, 1999) and CRP-3 by Malcolm Laird (Cape Roberts Science Team, 2000).

The Initial Report for each site included a summary description of the core in terms of litho-stratigraphical units (LSUs), subunits and facies within subunits. These were determined normally within 48 hours of the core arriving at the camp. The LSUs were a necessary convenience for timely core description, and were useful for summary descriptions for the Initial Reports. However, as drilling proceeded, an appreciation of the whole core grew, a separate sequence stratigraphic terminology was developed that recognised the cyclic nature of the section, with sequence boundaries that did not necessarily correspond with lithostratigraphical units. For those working on cyclostratigraphical aspects of the core, sequence boundaries and the sequence numbering scheme of Fielding *et al.* (2001) became a more convenient reference frame.

The 4 m-per-page logs provide an excellent record, along with the core images available on CD, of the gross lithology, of sedimentary structures and of clasts that feature prominently throughout the sequence. The estimates of mud content, however, can be significantly in error in sediments with a coarse 'tail' – 12 out of 118 samples from CRP-2/2A were described as sandstone but found to have sand contents ranging from only 11 to 39% (Barrett & Anderson, 2000). Where trends in mud content are important for environmental interpretation, then visual core descriptions need to be checked against analytical results.

THE CAPE ROBERTS SECTION

Cape Roberts section is described in terms of 5 sedimentary units, from oldest to youngest:

1 quartz sandstone, 116 m thick, and bearing a very strong similarity to Devonian sandstone cropping out in the adjacent Transantarctic Mountains;
2 conglomerate (uppermost Eocene), 33 m thick;
3 sandstone and minor conglomerate (uppermost Eocene-lower Oligocene), 460 m thick;
4 diamictite, sandstone, mudstone cycles (lower Oligocene-lower Miocene), 1010 m thick;
5 diamicton and minor muddy sand (Pliocene, Quaternary), up to 44 m thick.

A lithologic log is shown as Figure 5, and the facies for each unit or facies association are reviewed below, but first the chronology of the Cenozoic section is reviewed.

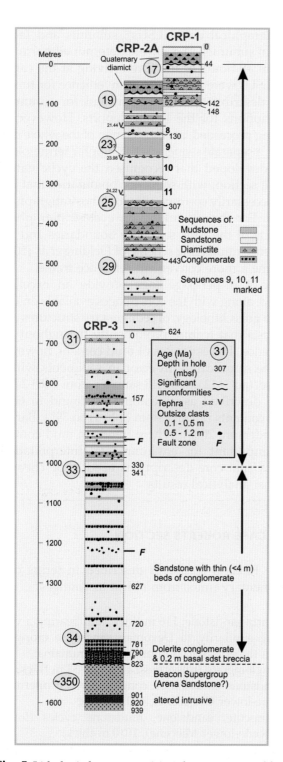

Fig. 5 Lithologic log summarising the section cored by the Cape Roberts Project (adapted from Barrett, 2001). The chronology has been adjusted to conform to the Gradstein *et al.* (2004) timescale (see Naish *et al.*, in press).

Chronology of the CRP Cenozoic section

The Cenozoic section has been dated from bio-stratigraphical and chronostratigraphical datums in conjunction with a high resolution magneto-stratigraphy. The age models developed during and after drilling (Cape Roberts Science Team, 1998, 1999, 2000; Roberts *et al.*, 1998; Lavelle, 1998; Wilson *et al.*, 2000; Hannah *et al.*, 2001) have now been refined, and summarised by Florindo *et al.* (2005). A revised chronology, outlined below, has provided the basis for Table 1 and the ages of ten major lithological divisions (including the basal conglomerate) from 34 to 17 Ma in the Cenozoic pre-Pliocene section.

In particular, an adjustment has been made to the chronology of CRP sequences across the Oligocene–Miocene boundary to conform with recent astrochronological recalibration of this interval (summarised in Gradstein, 2004). The new 'floating' astronomical calibration relies on a statistical match (coherency) between the climate proxy records from ODP Site 1090 and the orbital target curve with a time-step of 2.4 Ma. This precludes an age for the Oligocene–Miocene Boundary of 24.0 Ma as proposed by Wilson *et al.* (2002), and is also inconsistent with the geomagnetic polarity time-scale (GPTS) of Cande & Kent (1995) and Berggren *et al.* (1995).

In the new astrochronology the Ar-Ar tephra ages for normal-polarity sequences 10 and 11 (23.98 and 24.22 Ma) place them in short normal polarity chrons (C7n1n and C7n2n respectively). Consequently, these sequences can be uniquely matched to individual 40-kyr cycles on the composite oxygen isotope stratigraphy of ODP sites 929/1090 (Naish *et al.*, in press) – c. 600-kyrs earlier than the 40-kyr cycle correlation presented by Naish *et al.* (2001a). Sequence 9, whose age is constrained only by the 21.44 Ma ash in Sequence 8 above, could like in either C6n3n or C6n2n. Thus the Oligocene-Miocene boundary at 23.02 Ma could correspond to the uncon-formities below, or above Sequence 9. Here, it is placed at the base of Sequence 8, represent-ing a period of extensive erosion in the western Ross Sea associated with ice sheet expansion equivalent to global sea-level lowering of ~50 m during the earliest Miocene Mi-1 event (Pekar & DeConto, 2006).

Table 1 Main chronological subdivisions and thicknesses of the strata cored in the CRP drillholes (modified from Florindo *et al.*, 2005, following Gradstein *et al.*, 2004). Depths to lithological divisions expressed as mbsf (metres below sea floor) and cmd (cumulative metres drilled through Cenozoic strata below the Miocene-Quaternary boundary in CRP-1 and CRP2/2A). Ages are approximate. See text for explanation

AGE		CRP-1		CRP-2/2A		CRP-3		Lithological subdivisions	
		m bsf	*cmd*	m bsf	*cmd*	m bsf	*cmd*	*Thickness*	*Lithologies*
C	Quaternary 0–2 Ma	0.00 −43.55		5.54 −21.16				44 m	Diamicton Muddy sand/ Sandy mud
	Pliocene ~2–3 Ma			21.16 −26.79					
E									
N	Early Mio 17–19 Ma	43.55 −141.60	0.00 −98.05	26.79 −52.63				98 m	
O	Early Mio 19–23 Ma	141.60 −147.69*	98.05 −104.14*	52.63 −130.27	98.05 −175.69			78 m	
Z	Late Olig 23–25 Ma			130.27 −306.65	175.69 −352.07			176 m	*Cycles of Diamictite, Sandstone and Mudstone*
O	25–29 Ma			306.65 −442.99	352.07 478.41			135 m	
I	29–31 Ma			442.99 −624.15*	478.41 −669.57*			181 m	
C	Early Olig 31–33 Ma					2.80 −329.96	682.37+ −1009.53	340 m+	
	Early Olig 33–34 Ma					329.96 −789.77	1009.53 −1469.34	460 m	Sandstone, minor congl
	Latest Eocene 34 Ma					789.77 −823.10	1469.34 1502.67	33 m	Conglomerate
DEVONIAN ~350 Ma						823.10 −939.42*			Qtz sandstone
to convert m bsf to cmd		For CRP-1 subtract 43.55 m *bottom of hole		For CRP-2A add 45.42 m *bottom of hole		For CRP-3 add 679.57 m *bottom of hole		+ includes 10 m interval between base of CRP-2A and top of CRP-3	

The Eocene–Oligocene boundary is of particular interest because it was about that time that the deep-sea isotope shift marks the development of the first large Antarctic ice sheet (Zachos *et al.*, 1992). The boundary is most likely to lie in or just above a 33-m-thick conglomerate at the base of the Cape Roberts Cenozoic section. The conglomerate is unfossiliferous and unsuitable for magnetostrati-graphy, but the overlying 400 m of sandstone and minor conglomerate include fine beds that show a well-defined polarity zonation (Florindo *et al.*, 2001). These show a largely reversed interval in the upper part that can be identified as C12r on the GPTS (now 31.1 to 33.3 Ma in Gradstein *et al.*, 2004) with the aid of several biostratigraphic datums and Sr isotope ages around 31 Ma in the upper

200 m of CRP-3 (Hannah *et al.*, 2001). Below this lies a normal interval from 340.8 to 627.3 mbsf identified as C13n (now 33.27 to 33.74 Ma in Gradstein *et al.*, 2004), and a further reversed interval below with two thin intervals of normal polarity, regarded as possible cryptochrons (Florindo *et al.*, 2001). The sedimentation rate for that 0.47 Ma interval is 606 m/m.y., which projects the Eocene–Oligocene boundary (33.9 Ma) to be just under 100 m deeper ~725 mbsf. The lower part of the section is significantly more gravelly, implying a higher sedimentation rate, and the boundary may well be deeper still. However for this review it is placed at the base of the sandstone interval in Table 1 and the age rounded to 34 Ma. It is worth noting that although a number of marine palynomorphs were found in samples from the lowest part of this interval (781 and 789 mbsf) they do not contain any elements of the warm late Eocene Transantarctic Flora (Hannah *et al.*, 2001), indicating that the latest Eocene cooling took place before sedimentation began in the Cape Roberts section.

The first well-dated interval above the CRP-3 core begins at an unconformity in CRP-2A at 306.27 mbsf with the appearance of volcanic debris from the McMurdo Volcanic Group. Sanidine was dated from a clast 13 m above the unconformity at 24.98 Ma, and further clasts, as well as tephras (Fig. 5), have been dated from higher in the core (McIntosh, 2000). As noted above, tephras within cycles 11, 10 and 8 have provided chronological pinning points for the magnetostratigraphy in recognizing Milankovitch cyclicity in this interval of CRP-2A (Naish *et al.*, 2001a; Naish *et al.*, in press).

The CRP-2A section below 306.27 mbsf is magnetostratigraphically complex and has no reliable biostratigraphic datums. It represents a long time span (~6 m.y.), but has a significant lithological break at 442.99 mbsf, separating a finer diamictite-poor interval from a diamictite-rich interval above. This boundary also corresponds with seismic reflector 'l' of Henrys *et al.* (2001), shown in Figure 3. The upper part of the lower interval includes a number of miliolid shell fragments in place that have yielded Sr ages in the range 29 to 31 Ma. The interval has therefore been split into a lower 31–29 Ma unit and an upper 29–25 Ma unit, though it is likely that there

is significant time missing at the unconformities at 442.99 and 306.27 mbsf.

The youngest well-dated tephra in the section, the interval from 109 to 114 mbsf in CRP-2A, provides an age of 21.44 Ma. For the ~160 m of lower Miocene strata above this the chronology depends on diatom datums, Sr isotope ages and volcanic clast ages with lower precision. However, this is still sufficient to establish an age of around 17 Ma for the youngest few tens of metres of the Cape Roberts section.

The Oligocene–lower Miocene strata are overlain by a thin cover of soft largely Quaternary diamicton and muddy sand that is 44 m thick in CRP-1 and 27 m thick in CRP-2/2A. The CRP-1 sediments were considered entirely late Quaternary on microfossil evidence (Cape Roberts Science Team, 1998). Subsequently a shell bed from 32 to 34 mbsf has been the subject of detailed study, and is well-dated from magnetostratigraphy, diatom datums and Sr isotopic ratios at 1.1 Ma (Scherer *et al.*, 2006). The 21 m of Late Quaternary diamicton in CRP-2/2A is underlain by 6 m of a sandier diamicton with Pliocene foraminifera that were considered contemporaneous (Webb & Strong, 2000). The post-Oligocene cover at CRP-3 is less than 3 m thick, and none was recovered.

Quartz sandstone (Devonian)

The oldest Cape Roberts core is over 100 m of light reddish brown quartz-cemented quartz sandstone cored form beneath the Cenozoic section in CRP-3. The strata just below and above the contact are shown in Figure 6. The lithology below the contact closely resembles Devonian Taylor Group sandstone 50 km inland and around 3000 m higher in the adjacent Transantarctic Mountains (Turnbull *et al.*, 1994). The sandstone is well sorted, lacking both pebbles and mud, and with grain size varying from medium to very coarse. The bedding in some places is massive and in others well stratified, with small- and medium-scale cross-lamination superimposed on the regional 15° dip. The laminae are typically defined by a single grain-size and in places form low angle ripple structures. Some intervals show colour mottling that might have resulted from soft-sediment deformation or bioturbation. This assemblage of features has been interpreted to

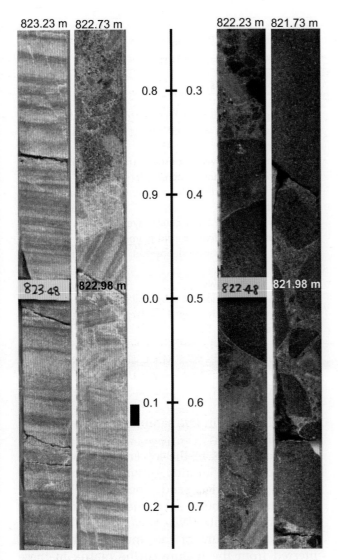

823.23 m   822.73 m   822.23 m   821.73 m

0.8 — 0.3

0.9 — 0.4

823.48   822.98 m   822.48   821.98 m

0.0 — 0.5

0.1 — 0.6

0.2 — 0.7

Fig. 6 The floor of the Victoria Land Basin cored in CRP-3, showing the finely laminated reddish brown quartz sandstone of the lower Taylor Group (Beacon Supergroup) overlain at 823.10 mbsf by angular sandstone talus for around 20 cm and then pebble to boulder conglomerate of Ferrar Dolerite. (Modified from Cape Roberts Science Team, 2000.)

represent a sub-humid to semi-arid continental setting with both fluvial and aeolian activity (Cape Roberts Science Team, 2000, p. 73). The interval also includes a 20-m-thick highly altered basic igneous intrusion. This, along with the quartzose nature of the sandstone, led to the correlation of these beds with the lower part of the Beacon Supergroup (Taylor Group) of early Devonian age.

Note on Cenozoic facies treatment

The characteristics of each facies in the Cape Roberts section are outlined in Appendix 1 (page 286) and summarised in Table 2, with the proportions of each facies from each time period shown in Table 3. The summary is based largely on Cape Roberts Science Team (1998, 1999, 2000), Powell *et al.* (2000, Table 1), and Fielding *et al.* (2000), with textural data from De Santis & Barrett (1998), Barrett & Anderson (2000, Table 2) and Barrett (2001, Table 2). The scheme is based on that of Powell *et al.* (2000) and Fielding *et al.* (2000) for describing and interpreting the CRP-2/2A core, and adopted from a simplified form used by Hambrey *et al.* (2002) for reviewing the late Oligocene and early Miocene glacimarine sedimentation from Ross Sea cores.

Conglomerate (uppermost Eocene)

CRP-3 823–790 mbsf (summarised from Cape Roberts Science Team, 2000)

The oldest bed in the Cenozoic sequence is a 17-cm-thick clast-supported breccia of quartz sandstone like the bedrock beneath (Fig. 6), overlain by 6 cm of matrix-supported conglomerate of the same quartz sandstone, in turn overlain by around 12 m of coarse conglomerate with cobbles and boulders up to 1.9 m long. The clasts are mostly dark Ferrar dolerite, a resistant lithology derived from sills within the Beacon Supergroup. The remainder of the unit comprises pebbly sandstone beds and coarse sedimentary breccia of dolerite clasts, which is also part of a 10-m-thick shear zone that runs through the upper part of this interval.

The basal sandstone breccia is interpreted as talus on sloping Taylor Group sandstone, with the overlying dolerite conglomerate brought in by high gradient streams. The overlying pebbly sandstone and breccia is seen as a mix of fluvial and talus debris. A granitoid clast 8 m above the base indicates that denudation of the Transantarctic Mountains had reached basement by the time this part of the Victoria Land Basin had began to subside. This interval included a mix of clast-supported and matrix-supported conglomerate, but no evidence of glacial influence in the form of facets or striae was found.

Table 2 A summary of the main CRP lithofacies, based on Powell *et al.* (2000, 2001), with a brief interpretation. Appendix 1 provides more detail

Lithofacies	Depositional process
1. Mudstone; minor lonestones	Hemipelagic suspension settling, with iceberg rafting
2. Interstratified sandstone and mudstone	Hemipelagic settling with sediment gravity-flows or wave/current action
3. Poorly sorted (muddy) very fine to coarse sandstone	Sediment gravity-flows or offshore-transition mixing
4/5. Moderately to well sorted, fine to coarse sandstone	Nearshore wave-grading and/or current sorting
6. Stratified diamictite	Subglacial or debris-flow deposition, or heavy rain-out from floating ice
7. Massive diamictite	Subglacial melting of basal debris layer
8. Sandstone-siltstone rhythmite	Suspension settling from turbid plumes
9. Clast-supported conglomerate	Fluvial or shallow marine deposition; discharge from subglacial streams?
10. Matrix-supported conglomerate	Sediment gravity-flows
11. Mudstone breccia	Mass-flow redeposition or subglacial shearing
12. Non-welded lapillistone	Airfall of volcanic ash through water; reworking by currents and gravity flows

Sandstone and subordinate conglomerate (uppermost Eocene-lower Oligocene)

CRP-3 790–330 mbsf (summarised from Cape Roberts Science Team, 2000)

This interval comprises around 460 m of fine- to medium-grain sandstone, with subordinate thin conglomerate beds, some clast-supported and others matrix-supported, and mostly in the lower 200 m. The sandstones range from vaguely stratified to well stratified, mostly parallel-, but with some small and medium scale cross-stratification, and some suggestion of hummocky cross-stratification with rare thin intervals of soft-sediment deformation. No biogenic structures were noted. Units are typically thick-bedded, and commonly show a fining upward trend from a thin conglomerate at the base. This is more evident above 480 mbsf, where Fielding *et al.* (2001) suggested that the beds show a simple cyclicity that might lend itself to sequence stratigraphic analysis.

Biogenic indicators are extremely rare, but clearly indicate a marine depositional environment, with a modiolid mussel at 781 mbsf and a gastropod mould at 359 mbsf, and marine paly-nomorphs occur in five samples between 789 and 440 mbsf.

The sandstones in the lower 200 m were initially described as 'muddy', giving way up section to clean well-sorted sandstones above 580 mbsf. The 'mud' has been subsequently identified as diagenetic smectite (Wise *et al.*, 2001). The post-depositional nature of the 'mud' has also been inferred from the texture of the matrix in these samples, which have virtually no silt but 'clay' forming between 10 and 21% of the sample (Barrett, 2001). These sediments were originally designated as facies 3, but as the 'mud' is post-depositional then they should be included in facies 5, as has been done in Table 2.

The conglomerate clasts are pebble to cobble-sized, and mostly a mix of pre-Devonian granitoids and Jurassic Ferrar Dolerite, sills of which intrude the Beacon Supergroup. In addition to the conglomerate beds the interval also includes a number of isolated out-sized clasts (more than 0.1 m long and more than 100 times larger than the enclosing grain size). Seven of these were striated, the lowest only a few metres above the basal conglomerate. Striated clasts were also found in the conglomerate beds (Atkins, 2001).

This interval was interpreted by most as representing inner shelf sedimentation above fair-weather wave-base with ice berg influence but no direct evidence of an ice margin close to the site (Cape Roberts Science Team, 2000). A minority view considered the interval to have been dominated by deposition from high-density sediment gravity flows in a deep water base-of-slope setting. However, in view of the stratal geometry and the shallow water character of the overlying strata this seems now most unlikely.

Diamictite, sandstone, mudstone (Oligocene-lower Miocene)

CRP-3 above 330 mbsf, CRP-2A up to 27 mbsf, CRP-1 up to 44 mbsf

This interval includes a range of shallow marine facies with a strong glacial influence and exhibiting a well-developed cyclicity. The evidence for the glacial influence comes from several indicators, including the diamictic texture of a significant proportion of this interval, based on both visual core description and textural analysis, common out-sized clasts, and the faceting (between 10 and 50%) and striae (around 5%) on clasts in all facies (Atkins, 2001). No trend through time was seen in clast surface features. Atkins (2001) noted that clasts averaged subrounded for the older CRP-3 samples, and subrounded to subangular for the younger CRP-2/2A samples, but this simply reflects most older samples coming from conglomerates and younger samples from diamictites. Micromorphological studies of 26 thin sections from diamictites in the upper part of CRP-2A revealed that three showed clear evidence of grounding, with about half the remainder providing some indication of subglacial shearing (van der Meer, 2000).

The interval is just over 1000 m thick, cored with 98% recovery, and has been the object of both detailed and comprehensive sedimentological description and interpretation both in the Initial Reports and subsequent papers by Fielding *et al.* (1998, 2000, 2001), Naish *et al.* (2001a, b) and Powell *et al.* (1998, 2000, 2001). Most of the focus in these papers has been on the cyclic character of this interval, described briefly below, but considered in a following section on sequence stratigraphy. Facies descriptions and proportions may be found in Appendix 1 and Tables 2 and 3. The comment that follows is a summary and interpretive overview.

The lowest part of this interval begins with the first record of grounded ice off shore, massive diamictite with striated clasts at 330 mbsf in CRP-3. Around 10% of the whole interval is diamictite and a similar proportion of the same interval comprises conglomerate transported offshore most likely by a mix of high fluvial discharge and redeposition by waves or sediment gravity-flows beyond. However the most common facies is mudstone (37%). When combined with facies 2, which includes hemipelagic mud as background sedimentation, they record almost half of the section being deposited close to or below wave base. However, a full 20% of the section is facies 5, moderately- to well-sorted sand, very likely wave- or current-sorted in this coastal setting, suggesting that the drill sites site lay at or close to sea level periodically throughout this interval. Facies 1–5 all include lonestones, many faceted and a few striated, that record floating ice during interglacial periods that might have been icebergs from tidewater glaciers, sea-ice or river-ice. The overlying strata all have a higher proportion of glacial facies, and less conglomerate, which virtually disappears above 307 mbsf in CRP-2A, perhaps reflecting less influence from flood discharge. No particular trend is evident in the sandstone and mudstone facies.

The massive diamictites of facies 7 are interpreted as basal glacial debris deposited beneath wet-based glacier ice, implying that the ice front of the time extending beyond the drill site. The stratified diamictites of facies 6, which are commonly interbedded with moderately- to well-sorted sandstone (facies 5) and clast-supported conglomerate (facies 9) may represent deposition in subglacial channels near the ice margin or in grounding line fans beyond. These facies are typically overlain by moderately- to well-sorted sandstones (facies 4/5), becoming poorly sorted (facies 3) before passing into alternating sandstone and mudstone (facies 2) or mudstone (facies 1). Mudstone of facies 1 commonly grades up into sandstones of facies 3 to 5, inferred to represent falling sea level.

This pattern is most obvious in the half dozen cycles that are more than 20 m thick, and for which a simple facies model is shown in Figure 7 for both

Table 3 Percentages of each facies are given for each major lithological division in the section. The section begins with a basal conglomerate and sandstone (facies 4/5 and 9/10) of latest Eocene and earliest Oligocene age in CRP-3, passing up through the cycles of diamictite, sandstone and mudstone (largely facies 1–7) of Oligocene and early Miocene times in CRP-3 and CRP-2A, and is capped by a thin diamictite-dominated Quaternary section in CRP-1. The Quaternary (and Pliocene) strata of CRP-2/2A are similar in facies to the Quaternary strata of CRP-1

Lithofacies	Age in Ma	CRP-3			CRP-2/2A				CRP-1	
		~34	34–33	33–31	31–29	29–25	25–23	23–19	19–17	0–2
1.	Mudstone		2	37	18	17	29	21	26	
2.	Interstrat. sandst/mudstone		2	10	13	16	4	1	19	
3.	Poorly sorted sandstone		[a]	11	17	9	21	12	15	19
4/5.	Mod to well sorted sandst		83	20	37	14	33	37	10	
6.	Stratified diamictite			1	1	16	2	2	3	73
7.	Massive diamictite			9	9	17	10	23	27	
8.	Rhythmite			2		1	1			
9.	Clast-supported congl	41	6	4	2	4		Tr		
10.	Matrix-supported congl	59	7	6	3	2		Tr		
11.	Mudstone breccia						Tr	Tr	1	
12.	Non-welded lapillistone						Tr	3	Tr	
		100	100	100	100		100	100	100	92[b]
		33 m	460 m	327 m	181 m	129 m	176 m	78 m	95 m	19 m

[a] 29% originally assigned to facies 3, but mud diagenetic, so added to facies 5.
[b] 8% lime packstone.

glacial and interglacial times. A perspective view of the Victoria Land coast for both glacial and interglacial times is shown as Figure 8. Detailed studies of the core indicate far more complexity than can be depicted here, particularly with regard to decimetre-scale soft sediment deformation, and in a few places metre–scale brecciated intervals, which have been variously interpreted as a result of slope instability, mass movement and subglacial shearing. Redeposition by sediment gravity flows may well have also been active from time to time. Nevertheless the context of an open coast outlined earlier, and the nature of the diatom flora suggest a relatively nearshore marine environment, with coastal neritic diatoms dominating over open ocean types (Scherer et al., 2000). Indeed in early Miocene times waters were occasionally sufficiently shallow for a significant proportion of benthic diatoms to be deposited (Harwood et al., 1998).

Diamicton, muddy sandstone (Pliocene, Quaternary)

CRP-2/2A above 26.79 mbsf, CRP-1 above 43.55 mbsf

The strata above the lower Miocene section are entirely Quaternary in age at CRP-1. Little was recovered above 19 mbsf, but below this it comprised largely unlithified diamicton with clasts of dolerite and granite up to boulder size, and lesser muddy sand beds with scattered pebbles. The diamictons are presumed to have formed closing to the grounding line of an ice shelf or glacier tongue when ice that was covering the nearby foothills was more extensive than that of today. Powell et al. (1998) provide a useful perspective view of the Victoria Land coast at this time.

A shelly interval between 33.75 and 31.90 mbsf in CRP-1 has yielded a macrofossil fauna of more than 60 species (Taviani et al., 1998, Taviani &

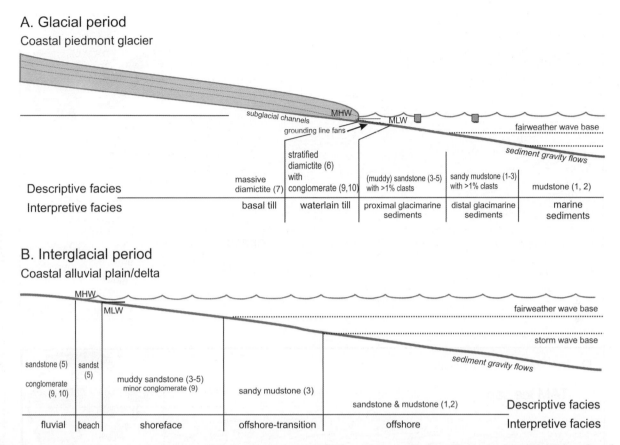

Fig. 7 Facies model for the deposition of sedimentary strata off Cape Roberts during Oligocene (post 33 Ma) and early Miocene times (developed from Hambrey *et al.*, 1989) for glacial (A) and interglacial (B) periods. In both cases wave-and/or current-sorted sand near the shore grade into mud below wave base offshore. During glacial periods glaciers deposited sediment directly over the drillsites as the ice margin advanced beyond them. During interglacial periods sediment was delivered to the coast largely by rivers but striated lonestones suggest some glacial ice still reached the coast in a few places.

Beu, 2003). They are largely molluscs (>40 species) followed by bryozoa (>14 species) and poly-chaetes (3 species). Diatoms are largely species of *Fragiliopsis* and *Thalassiosira*, with few sea ice-related diatoms, indicating ice-free conditions for much of the year (Bohaty *et al.*, 1998). Recent work has shown the age of this interval to correspond to Marine Isotope Stage 31 at 1.07 Ma (Scherer *et al.*, 2006).

Quaternary diamicton with minor sand was encountered in CRP-2, the first hole drilled in 1999, although little was recovered of the top 5 m. The diamicton contains abundant marine benthic Quaternary foraminifera. Below 21 mbsf, how-ever, a 6-m-thick interval of muddy sand includes a 2.4-m-thick bed of diamicton with an *in situ* foraminiferal fauna of 21 marine benthic species

that indicate a Pliocene age (Webb & Strong, 1998). These authors noted deposits of similar age above sea level in lower Taylor Valley and Wright in the Dry Valleys region of adjacent Victoria Land, which implies ~200 m of post-Pliocene uplift across the Transantarctic Mountain Front.

SEQUENCE STRATIGRAPHIC ANALYSIS

Cenozoic glacial and sea-level cycles in the Victoria Land Basin were first recognised in late Oligocene glacigenic strata cored by MSSTS-1, a drill-hole off New Harbour 70 km south of Cape Roberts in 1979 (Barrett, 1986; Barrett *et al.*, 1987). Sub-sequently Hambrey *et al.* (1989) strengthened the concept using a simple facies approach to interpret

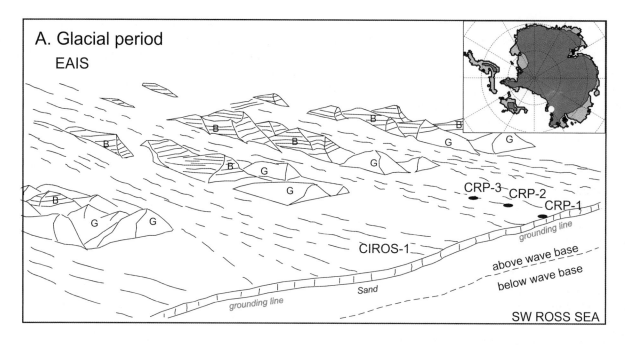

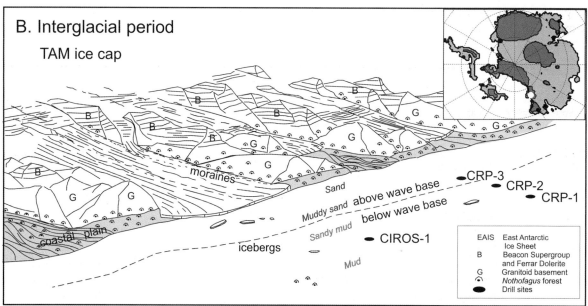

Fig. 8 View of the Victoria Land coast off Cape Roberts during Oligocene (post 33 Ma) and early Miocene times (developed from Hambrey *et al.*, 2002). (A) Glacial period, with an expanded inland ice sheet feeding thin temperate piedmont glaciers depositing sediment to a shallow shelf to be reworked by waves and currents nearshore with mud settling out offshore. (B) Interglacial period, with higher sea level and a much reduced ice sheet. Sediment carried to the coast largely by proglacial rivers. Low woodland beech forest at lower elevations. Insets for A and B show examples of the modelled extent of an ice sheet that might have existed during glacial and interglacial periods in early Oligocene times, representing 21×10^6 cubic km of ice (50 m of sea level equivalent) in A, and 10×10^6 cubic km of ice (24 m of sea level equivalent) in B (DeConto & Pollard, in press).

~300 m of upper Oligocene-lower Miocene strata from the nearby CIROS-1 drill hole in terms of glacial advance and retreat coinciding with sea-level fall and rise. However, the chronology of these strata was not sufficiently well established for correlating the cycles with the deep-sea oxygen isotope curve of Miller *et al.* (1987) and the Haq *et al.* (1987) onlap-offlap curve.

One of the goals of the Cape Roberts Project was to obtain a better-dated record of this time period, and at the same time apply a sequence stratigraphic approach to the Cape Roberts section. This was first achieved by Fielding *et al.* (1998) in the eight lower Miocene cycles from CRP-1, recognising a pattern of unconformity-based diamictite followed by sand and then mudstone as representing a retreat of the ice front over the drill site and a rise in sea level. However they noted that 'it is not generally possible to differentiate between true eustatic signals and local glacial advance/retreat cycles.' Although the chronology of CRP-1 was an improvement over CIROS-1, uncertainties of ~1 m.y. pre-

vented meaningful correlation with the deep-sea oxygen isotope curve then being developed for the Pliocene and potentially older strata with ~0.01 m.y. resolution (Shackleton *et al.*, 1995).

The drilling of CRP-2/2A yielded a section that provided both well-developed cycles and the basis for a chronology that indicated they resulted from orbital forcing. Studies of this core provided more material for analysis, and resulted in an improved sequence stratigraphic model (Fig. 9) containing a *glacial surface of erosion* as an important element (Fielding *et al.*, 2000; Naish *et al.*, 2001a). Three cycles in particular, cycles 9, 10 and 11 in the scheme of Fielding *et al.* (2000), proved to be important because of their thickness and completeness, and also because of a similarity in scale and facies pattern to Plio-Pleistocene cycles of the Wanganui Basin, New Zealand (Fig. 10, Naish & Kamp, 1997), which had already been matched with the Milankovitch cycles of Quaternary deep-sea isotope curve. With the tephra-supported biomagnetostratigraphy, Naish *et al.* (2001a) were

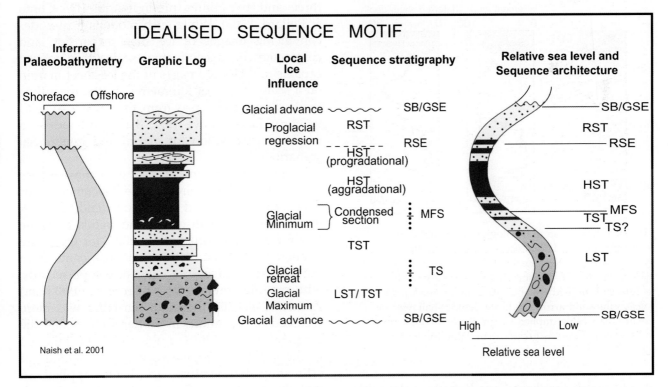

Fig. 9 Sequence stratigraphic model developed for strata cored by CRP-2A from Fielding *et al.*, 2000, and Naish *et al.*, 2001a). SB/GSE – Sequence Boundary-Glacial Surface of Erosion. TS – Transgressive Surface. MFS – Maximum Flooding Surface. RSE – Regressive Surface of Erosion. LST – Lowstand Systems Tract. TST – Transgressive Systems Tract. HST – Highstand Systems Tract. RST – Regressive Systems Tract.

Rangatikei Cycle 6, Wanganui Basin

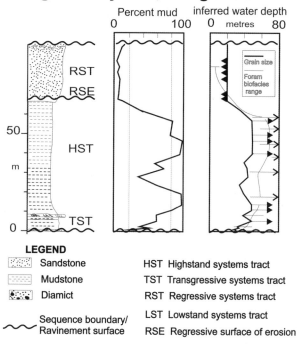

LEGEND

Sandstone	HST	Highstand systems tract
Mudstone	TST	Transgressive systems tract
Diamict	RST	Regressive systems tract
	LST	Lowstand systems tract
Sequence boundary/ Ravinement surface	RSE	Regressive surface of erosion

CRP-2A Cycle 11

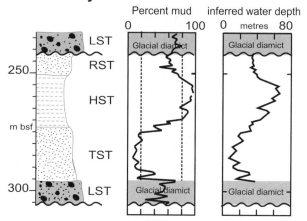

Fig. 10 Comparison of lithologies and water depth interpretations for late Pliocene cycle 6 from the Wanganui Basin, New Zealand, with a latest Oligocene cycle 11 from CRP-2A, showing similar patterns in texture and implied water depth change. For Rangatikei Cycle 6 the solid arrows indicate depths shallower or greater than value implied by grain size. Open arrows indicate water depths greater than value implied by forams.

able to match the three Cape Roberts cycles with cycles in a high resolution deep-sea isotope record from the South Atlantic Ocean (Zachos *et al.*, 1997; Paul *et al.*, 2000). Naish *et al.* (2001b) then used

this indication of orbital forcing as the basis for frequency analysis of the rest of the cycles in the Cape Roberts sequence. The recent revision of the Cenozoic time scale, recalibrating the Oligocene-Miocene boundary, has resulted in a new and improved correlation with particular 40,000 year cycles that correspond to the Ar-Ar ages for each cycle at 23.98 and 24.22 Ma (Naish *et al.*, in press), as noted earlier.

The establishment of the sequence stratigraphic model for the mid-Cenozoic Cape Roberts cycles, and the correlation of particular sedimentary cycles on the Antarctic margin with oxygen-isotope cycles in the deep-sea record, was important because it showed that the Antarctic Ice Sheet in Oligocene and early Miocene times pulsated at Milankovitch frequencies (40,000/100,000 years). This view has been supported by experiments with coupled atmosphere–ice sheet models, which show the Antarctic Ice Sheet varying in volume from 10 to 50 m of sea level equivalent on the same frequencies, although with the greatest sensitivity for modelled atmospheric CO_2 levels between three and two times pre-industrial (De Conto & Pollard, 2003). While such modelling provides convincing images of ice sheet expansion and contraction on a continental scale, it is not yet capable of yielding images of the ice sheet margin on a scale of tens of kilometres. In the meantime the cartoons developed from the review of sedimentary facies above (Fig. 8) are offered as a realistic visualisation for glacial and interglacial scenarios.

Cyclic variations in sea level as a consequence of ice-volume change are now well documented for Quaternary times, and efforts to extend this back in time to track both cycles and trends continue (see for example Miller *et al.*, 2005, for the entire Phanerozoic). Most of this work has come from the analysis of deep-sea oxygen isotope records, with previous reviews by Miller *et al.* (1987) and Zachos *et al.* (2001), though the latter was more focussed on variations in orbital forcing than the temperature-ice volume ambiguity that is inherent in the oxygen isotope record. New methods of estimating temperature are being applied so that the sea level signal can be extracted, with Billups & Schrag (2003) and Lear *et al.* (2004) using the Mg/Ca ratio in carbonate shells as a palaeothermometer. Pekar *et al.* (2002) provided glacio-eustatic estimates by calibrating detrended apparent sea-level amplitudes to $\delta^{18}O$ amplitudes for Oi-events in deep-sea records (Pekar *et al.*, 2002), with Pekar and DeConto (2006) arguing for sea level (and ice-volume) fluctuations of between 30 and 60 m in early Miocene times.

The sequence stratigraphic model for the 54 cycles deposited off Cape Roberts from 31 to 17 million years ago suggests an independent approach to estimating past sea-level change based on the physical characteristics of nearshore sediments. Dunbar & Barrett (2005) studied patterns of sedimentation for coastal nearshore sediments, and reaffirmed the long-standing awareness that where the coast is aggrading, the sea floor sediment texture changes progressively from beach sand to offshore mud (Johnson, 1919; Swift, 1971). More importantly they showed a consistent relationship between mud-percent and water depth, on both seasonal and decadal time scales, and a further relationship between mud percent at a specified depth and wave energy. They concluded that if the wave climate were persistent for nearshore sediments on an open coast then variations in mud content could be used as a proxy for water depth, and if the strength of the wave climate could be estimated then values for water depth could also be determined.

This approach was tested with a late Pleistocene and a late Pliocene cycle from the Wanganui Basin, New Zealand, assuming with strong evidence that wave climate was similar to that of today, and where independent water depths could be estimated from foraminifera. There was close agreement. The textural approach has an advantage in higher resolution for depth trends, though there is lower confidence in depth values on account of the assumption regarding wave climate. Also it cannot register changes in water depths below wave base, where sediment is invariably mud, unless the coast is influenced by strong geostrophic or tidal currents or sediment gravity-flows, which have recognisable characteristics. Preliminary tests of this approach for the Cape Roberts cycles assuming a moderate wave climate indicate water-depth variations of several tens of metres (Fig. 10, Dunbar *et al.*, 2003; see also Dunbar *et al.*, 2007).

The interpretation of the Oligocene-lower Miocene Cape Roberts cycles offered here is based on two key points, the first inferred and the second observed:

1 Sediment was supplied by glaciers and rivers to a subsiding basin from the west across a long, straight open coast, and
2 The geometry of the strata parallel to the coast is persistent and planar.

The sequence stratigraphic interpretation, supported by textural patterns of modern coastal sediments, requires cyclic changes in sea level as the primary control on sedimentary facies architecture. However, without doubt the glacial influence, indicated by the diamictites, sediment gravity flows, sediment deformation and lonestones, is clear and extensive. This interpretation leads to two significant conclusions, one for tectonic history and the other for the value of the Cape Roberts section as a recorder of climate on the Antarctic margin.

1 The well-sorted sandstone facies closely associated with diamictite at many levels through the section indicates a regular return to sea level throughout the 17 m.y. history of the Cape Roberts section. Hence, the initially rapid and then declining net sedimentation-rate throughout the section records the subsidence history of the basin, and indeed this observation has been used in a new view of the tectonic history of the Victoria Land Basin (Fielding *et al.*, 2006; Wilson *et al.*, unpublished data).
2 A total of 54 cycles have been identified through the Cape Roberts section covering the period from 33 to 17 Ma (Fielding *et al.*, 2001; Naish *et al.*, 2001b). However there were most likely between 160 and 400 orbitally forced cycles in that time period, depending on the relative influence of eccentricity (100,000 years) and obliquity (40,000 years). Hence only 15 to 35% of these cycles have been preserved. However, these still have value for the glacial lowstands and interglacial highstands they record at many points through this 16 million year time-span.

CLIMATE TRENDS FROM THE CAPE ROBERTS SECTION

The Cape Roberts section records many cyclic variations in ice extent and sea level, as argued above. Here we review long-term trends in some simple climate proxies for the period represented by Cape Roberts (34 to 17 Ma). The proportion of glacial facies in the section are shown, but the more significant climate proxies are considered

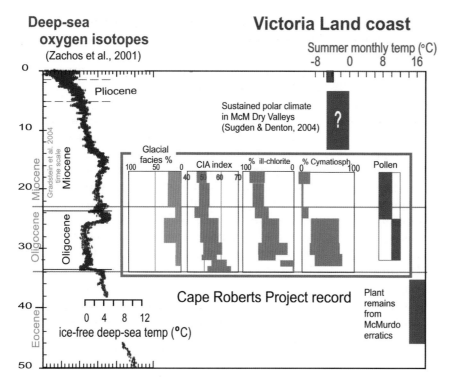

Deep-sea oxygen isotopes (Zachos et al., 2001)

Victoria Land coast

Fig. 11 Trends in climate proxies from the Cape Roberts section for the period from 34 to 17 Ma, compared with the composite deep-sea oxygen isotope curve of Zachos *et al.* (2001). Data were averaged for the time periods set out in Table 1, and are from the following sources: **1** % glacial facies (diamictites, facies 6 and 7) from Table 2; **2** CIA index from Passchier & Krissek (unpublished data); **3** % illite and chlorite in the clay fraction from Ehrmann *et al.* (2005); **4** Fresh water algae (*Cymatosphaera*) from M.J. Hannah (pers. comm., 2006). For sets 2–4 the bar is centred on the mean value, with the width representing ±1 standard deviation. Temperature estimates on the right are for interglacial periods and for mean summer monthly (December-January-February) temperature. For the last two million years it is based on the temperature records from Scott Base, Ross Island, since 1957 (–5°C).

to be the Chemical Index of Alteration (CIA) of Nesbitt & Young (1982), the percentage of illite and chlorite in the clay fraction and the percentage of cymatiosphaerids, marine algae that favour reduced salinities. In each case the data have been averaged for time periods of between one and four million years, and are shown in Figure 11. The climate proxies are compared with estimates of summer monthly temperature from terrestrial palynomorphs (and present-day measurements), and also a widely quoted global climate record for this period, the composite deep-sea oxygen isotope record of Zachos *et al.* (2001).

The oxygen isotope record is influenced largely by deep-sea temperature and the volume of ice on land, but recently Mg/Ca ratios on benthic foraminifera have been found to provide an independent estimate of temperature (Lear *et al.*, 2000, 2004). It is worth noting that both the composite oxygen isotope record and Mg/Ca ratios in the Eocene suggest that temperatures had declined sufficiently by 40 Ma to form small ice sheets on Antarctica (Billups & Schrag, 2003, Fig. 2), and that the dramatic 1 per mil shift in $\delta^{18}O$ at the Eocene-Oligocene boundary most likely resulted from an

increase in ice volume (Lear *et al.*, 2004). Diamictites with exotic clasts at two locations on the Antarctic margin, Prydz Bay at 68°S and 70°E (Hambrey *et al.*, 1991) and Seymour Island at 65°S and 60°W (Ivany *et al.*, 2006) support this view in showing that continental ice reached beyond the present ice limits by earliest Oligocene times, with Prydz Bay diamictites possibly being a little older.

The Cape Roberts section does not provide a comparable record of the onset of Antarctic glaciation, most likely because of the barrier presented by the Transantarctic Mountains, which began rising 20 m.y. earlier (Barrett, 1999). However, the section does record out-sized striated clasts in sand just above the basal conglomerate at ~34 Ma, indicating calving of glacier ice at sea level. In this shallow marine coastal sedimentary section the appearance of diamictites suggests that grounded ice began to extend periodically beyond the Victoria Land coast at ~33 Ma. This continued for the rest of the Oligocene and early Miocene time, with glacial facies comprising 10 to 33% of the total thickness of the section.

The proportion of glacial facies in the section is not easy to interpret and may have as much or more to do with subsidence rates than the duration of ice covering the Cape Roberts sites. However the glacial facies do show that on at least 50 occasions, and most likely on many more, in the period from 33 to 17 Ma grounded ice extended on a broad front into the Ross Sea. We can also speculate that significant unconformities, at ~29, 25 and 23 Ma (442.99, 306.65 and 130.27 mbsf in CRP-2A), for example, represent substantial time loss as a consequence of glacial erosion through more extensive or persistent ice.

The balance between physical and chemical weathering as an indicator of climate through the Cape Roberts section is shown in two ways – through the CIA index and through clay mineralogy. The CIA is calculated from the relative abundances of Al, K, Ca, and Na oxides, and its magnitude increases as the extent of chemical weathering increases. Samples for this purpose were taken from diamictite and mudstone. Values range from ~50 for 'unweathered' feldspar-rich rocks and 70–75 for the 'average shale' to values near 100 for highly weathered sediments. Data were obtained through XRF analysis by Krissek & Kyle (2000, 2001) with further analysis by Passchier (2006), who has also corrected the analyses for biogenic and detrital carbonate.

The oldest strata, from 34 to 33 Ma, were almost entirely sandstone, and lacking mudstone or diamictite for analysis. The overlying finer-grained sediments from 33 to 32 Ma, however, gave CIA values varying considerably in the range between 55 and 65, indicating moderate chemical weathering, moving to between 50 and 60 from 32 to 25 Ma, suggesting a reduction in chemical weathering. Above this from 25 to 17 Ma CIA values showed less variation and centred on a value of around 50. Whether the initial decrease reflects cooling of coastal climate or a rapid but declining sediment accumulation rate during the initial phase of subsidence (Fielding *et al.*, 2006), it is plain that the trend is a declining one and the index itself is low and consistent with a cool to cold climate throughout the time represented by the Cape Roberts section. The range of values obtained from the Cape Roberts section is similar to those reported by Passchier (2004). Their analysis of 32 samples from ancient wet-based glacial deposits found between 1000 and 2500 m asl in the Transantarctic Mountains from 75 to 86°S; they yield a similar range (40–70)

and average (55 ± 7) to the Cape Roberts section. Passchier (2006) has offered a more detailed analysis of these data.

The detrital clay mineral record is a long established climate proxy for tracking long term climate change, and especially in the Antarctic region (Ehrmann *et al.*, 1992). Illite and chlorite are known to result from physical weathering under a cool dry climate, and kaolinite and smectite are normally being derived from chemical weathering under a warm humid climate. Care is required if authigenic clays are present or the strata include easily weathered volcanic detritus. The CRP clay record shows here is summarised from Ehrmann *et al.* (2005).

The basal Oligocene sandstones of the Cape Roberts section have virtually no detrital clay, but as noted earlier they have significant amounts of authigenic smectite, which is of no climatic significance. The interval from 33 to 32 Ma, however, includes fine-grained sediments with much poorly crystallised smectite but still little illite or chlorite (average ~10%). The proportions change dramatically in the following period, from 34 to 33 Ma, to >40% from 32 to 25 Ma, and rise further ~70% from 23 to 17 Ma. These data suggest initially a cool but mild climate, becoming consistently cold from 32 to 25 Ma, perhaps with milder episodes 31 to 29 Ma, and then becoming more frigid from 25 to 17 Ma.

The marine palynomorphs in the Cape Roberts section are common (Hannah *et al.*, 2000, 2001) and include prasinophyte algae, largely the genus *Cymatiosphaera*, well known from the modern Arctic Ocean. There they prefer areas of high freshwater inflow near the mouths of estuaries and fiords (Mudie, 1992). In the open coastal setting off Cape Roberts their abundance is taken to indicate periods of glacial melting that generate extensive freshwater discharges reaching several km offshore. The five palynomorph-bearing samples from 34 to 33 Ma included only one that was productive, possibly because this part of the section is almost entirely sandstone. Above this, however, in the interval representing the period from 33 to 23 Ma, they represent between 30 and 45% of the marine palynoflora, though individual samples vary from a few to almost 90%, suggesting periods of extensive meltwater extending offshore. In the upper part of the section marine palynomorph abundance is high but the proportion of *Cymatiosphaera* is low, and mostly less than 10%, implying reduced melt-water flows and a cooler climate.

The terrestrial palynomorph record is sparse but significant because the assemblages comprise largely modern forms of vegetation. That said, the temperature estimates, although carefully considered are quite speculative, with the detailed reasoning to be found in papers by Raine (1998), Askin & Raine (2000) & Raine & Askin (2001). Further work by Prebble *et al.* (2006) on the detail of two cycles at 31 and 24 Ma confirmed the cooling trend found in the initial work but also found that *Nothofagus* was more persistent in the late Oligocene than previously thought. A goal of this work was to seek changes in the glacial-interglacial pollen assemblage, but these could not be resolved because of a combination of low numbers from the extractions and probable pene-contemporaneous reworking. This suggests that the published assessments that follow are likely to represent an interglacial rather than a glacial climate. This flora consisted of a 'a low-diversity woody vegetation with several species of *Nothofagus* and podocarpaceous conifers' (Raine & Askin, 2001) for the period from 33 to 25 Ma, and from about 25 Ma of a scrub-tundra mosaic with similar *Nothofagus* and podocarpaceous conifers, but an increasing tundra-derived bryophytic [moss] and distinct angiosperm component (Prebble *et al.*, 2006; Raine *et al.*, 2006). The scrub and woodland are compared with the Pacific coast of southern South America, where precipitation is high and mean summer monthly temperatures range from 8 to 12°C. The tundra vegetation represents a colder, periglacial climate, perhaps at higher elevation or in more exposed sites.

To provide context for the temperatures estimated for the Cape Roberts palynomorph record, the present-day mean summer temperature is shown, along with a speculative assessment that the regional climate has been polar for the last 14 Ma based on the persistence of dated geomorphic features in the McMurdo Dry Valleys (Marchant *et al.*, 1996; Sugden & Denton, 2004). Temperatures are also estimated for the period immediately preceding the earliest Cape Roberts record from a more diverse terrestrial palynomorph assemblage that was extracted from middle to late Eocene erratics from southern McMurdo Sound (Askin, 2000).

In summary, the indicators of ice and climate in the Cape Roberts section indicate initially a cold climate for the period from 34 to 33 Ma, with limited glacial influence most likely on account of the barrier presented by the adjacent Transantarctic Mountains. From 33 Ma onwards, however, the cycles of diamictite, sand and mud, left a record of glacial advance and retreat that was synchronised with sea-level fall and rise that is considered a response to huge fluctuations of a dynamic Antarctic ice sheet. The record shows no significant indication of the loss of ice or a clear and persistent warming of the Antarctic region to be expected from the 1 per mil fall at around 25 Ma in the composite deep-sea isotope curve.

Re-analysis of high-latitude ODP sites by Pekar *et al.* (2006) indicates that around this time (24.4 and 23.0 Ma) the Antarctic Ice Sheet was as large as or larger than that of today. They suggested that the isotopic shift is better explained by a strengthening of warmer deep waters originating from the North Atlantic coupled with a reduction in Antarctic deep-water, resulting in a warming of bottom water temperatures in many of the ocean basins, rather than a change in ice volume or surface temperature. They also concluded that ice volume ranged from 50% to 125% of the present-day ice sheet in early Miocene times by calibrating high-resolution records from ODP Sites 1090 and 1218 (Pekar & DeConto, 2006). The Cape Roberts record is consistent with that conclusion.

CONCLUDING REMARKS

The Cape Roberts section represents a collection of fragments of history of the Antarctic margin from the time the ice sheet had just formed until shortly before it became persistent around 14 Ma ago. The sand and minor conglomerate of the first million years is likely to be primarily of regional tectonic interest, with the subsequent period from 33 to 17 Ma of wider climatic interest for its record of the cycles of the continental ice sheet. Although only 20 to 30% of the cycles experienced by the Antarctic margin have been preserved at all, a number have survived with most or all elements intact (e.g. cycle 4 in CRP-1, Cycles 8–11 and 19 in CRP-2A, Cycles 2 and 3 in CRP-3). With their thickness ranging from 26 to 80 m, and considering the time they represent – 40,000 or 100,000 years – they offer obvious potential for high-resolution palaeoclimate studies of the Antarctic margin when the earth was shifting from a high to a low CO_2 world (Pagani *et al.*, 2005).

ACKNOWLEDGEMENTS

I thank the whole Cape Roberts team, scientists, drillers and support staff, for their commitment and effort in extracting, describing and interpreting the cores on which this review is based. In particular, I want to acknowledge the roles of Peter Webb as Crary Laboratory Science Leader, Alex Pyne as Drillsite Science Coordinator, Pat Cooper as Drilling Manager and Jim Cowie as Project Manager. I also thank the science editors, contributors and editorial staff of *Terra Antartica* for the 10 issues that have preserved the primary data from both core-based and geophysical measurements and the sometimes disparate views expressed in trying to reconcile the evidence. The project was also crucially dependent on funding and logistic support from the national programmes of Italy, New Zealand, the United States of America, Germany, Australia, the United Kingdom and the Netherlands.

I am particularly grateful for the advice and critical comments of Rosie Askin, Alan Cooper, Warren Dickinson, Gavin Dunbar, Chris Fielding, Fabio Florindo, Mike Hambrey, Mike Hannah, Dave Harwood, Tim Naish, Steve Pekar, Ian Raine and John Smellie. The preparation of this review was funded by NZ Foundation for Research, Science & Technology Grant COX0410 (Antarctic drilling (ANDRILL) programme).

REFERENCES

Anderson, J.B. (1999) Antarctic Marine Geology. Cambridge University Press, Cambridge. 289 pp.

Armienti, P., Tamponi, M. and Pompilio, M. (2001) Sand provenance from major and trace element analyses of bulk rock and sand grains from CRP-2/2A, Victoria Land Basin, Antarctica. *Terra Antartica*, **8**, 569–582.

Askin, R.A. (2000) Spores and pollen from the McMurdo Sound erratics, Antarctica. In: *Paleobiology and Paleo-environments of Eocene Rocks, McMurdo Sound, East Antarctica* (Eds J.D. Stillwell, and R.M. Feldmann), *AGU Antarctic Research Series*, **76**, 161–181.

Askin, R.A. and Raine, J.I. (2000) Oligocene and Early Miocene terrestrial palynology of the Cape Roberts Drillhole CRP-2/2A, Victoria Land Basin, Antarctica. *Terra Antartica*, **7**, 493–501.

Atkins, C.B. (2001) Glacial influence from clast features in Oligocene and Miocene strata cored in CRP-2/2A and CRP-3, Victoria Land Basin, Antarctica. *Terra Antartica*, **8**, 263–274.

Barrett, P.J. (Ed.) (1986) Antarctic Cenozoic history from the MSSTS-1 drillhole, McMurdo Sound, Antarctica. *NZ DSIR Bulletin*, **237**, 174 p.

Barrett, P.J. (Ed.) (1989) Antarctic Cenozoic history from the CIROS-1 drillhole, McMurdo Sound, Antarctica. *NZ DSIR Bulletin*, **245**, 254 p.

Barrett, P.J. (1996) Antarctic palaeoenvironment through Cenozoic times – a review. *Terra Antartica*, **3**, 103–119.

Barrett, P.J. (1999) Antarctic glacial history over the last 100 million years. *Terra Antartica Reports*, **3**, 53–72.

Barrett, P.J. (2001) Grain-size analysis of samples from Cape Roberts Core CRP-3, Victoria Land Basin, Antarctica, with inferences about depositional setting and environment. *Terra Antartica*, **8**, 245–254.

Barrett, P.J. and Anderson, J. (2000) Grain size analysis of samples from CRP-2A, Victoria Land Basin, Antarctica. *Terra Antartica*, **7**, 245–254.

Barrett, P.J., Elston, D.P., Harwood, D.M., McKelvey, B.C. and Webb, P.-N. (1987) Mid-Cenozoic record of glaciation and sea-level change on the margin of the Victoria Land basin, Antarctica. *Geology*, **15**, 634–637.

Barrett, P.J., Henrys, S.A., Bartek, L.R., Brancolini, G., Busetti, M., Davey, F.J., Hannah, M.J. and Pyne, A.R. (1995) Geology of the margin of the Victoria Land Basin off Cape Roberts, Southwest Ross Sea. In: *Geology and Seismic Stratigraphy of the Antarctic Margin* (Eds A.K. Cooper, P.F. Barker and G. Brancolini), AGU Antarctic Research Series, **68**, 183–207.

Bartek, L.R., Henrys, S.A., Anderson, J.B. and Barrett, P.J. (1996) Seismic stratigraphy of McMurdo Sound, Antarctica: implications for glacially influenced early Cenozoic eustatic change? *Mar. Geol.*, **130**, 79–98.

Behrendt, J.C. (1999) Crustal and lithospheric structure of the West Antarctic Rift System from geophysical investigations – a review. *Global and Planetary Change*, **23**, 25–44.

Berggren, W.A., Kent, D.V., Swisher, C.C., III and Aubry, M.-P. (1995) A revised Cenozoic geochronology and chronostratigraphy. In *Geochronology, Time Scales and Global Stratigraphic Correlation* (Eds W.A. Berggren, D.V. Kent, M.-P. Aubrey and J. Hardenbol), *SEPM (Society for Sedimentary Geology) Special Publication*, **54**, 129–212.

Billups, K. and Schrag, D.P. (2003) Application of benthic Mg/Ca ratios to questions of Cenozoic climate change. *Earth & Planetary Science Letters*, **209**, 181–195.

Bohaty, S.M., Scherer, R.P. and Harwood, D.M. (1998) Quaternary diatom biostratigraphy and palaeo-environments of the CRP-1 drillcore, Ross Sea, Antarctica. *Terra Antartica*, **5**, 431–453.

Cande, S.V. and Kent, D.V. (1995) Revised calibration of the geomagnetic polarity timescale for the late Cretaceous and Cenozoic. *J. Geophys. Res.*, **100**, 6093–6095.

Cape Roberts Science Team (1998) Initial Report on CRP-3. *Terra Antartica*, **5**, 1–187.

Cape Roberts Science Team (1999) Studies from the Cape Roberts Project, Ross, Sea, Antarctica. Initial Report on CRP-3. *Terra Antartica*, **6**, 1–173 with supplement.

Cape Roberts Science Team (2000) Studies from the Cape Roberts Project, Ross, Sea, Antarctica. Initial Report on CRP-3. *Terra Antartica*, **7**, 1–209 with supplement.

Carter, R.M., Abbott, S.T., Fulthorpe, C.S., Haywick, D.W. and Henderson, R.A. (1991) Application of global sea-level and sequence stratigraphic models in southern hemisphere Neogene strata from New Zealand. In: *Sea-level and active plate margins* (Ed. D. MacDonald), *International Association of Sedimentologists Special Publication 12*, pp. 41–65.

Cooper, A.K., Barker, P.F. and Brancolini, G. (1995) *Geology and Seismic Stratigraphy of the Antarctic Margin*, AGU Antarctic Research Series, **68**, 301 pp, atlas, CD-ROMs.

Cooper, A.K., Barrett, P.J., Hinz, K., Traube, V., Leitchenkov, G., Stagg, H.M.J., Meyer, A.W., Davies, T.A. and Wise, S.W. (1991) Cenozoic prograding sequences of the Antarctic continental margin: a record of glacio-eustatic and tectonic events. *Mar. Geol.*, **102**, 175–213.

Cowie, J.C. (2002) Cape Roberts Project Final Report 1995–2001. Antarctica NZ, Christchurch. 134 pp.

Davey, F.J. (2004) Ross Sea Bathymetry scale 1:2000,000. Institute of Geological & Nuclear Sciences, Lower Hutt. Geophysical map.

DeConto, R.M. and Pollard, D. (2003) Rapid Cenozoic glaciation of Antarctica induced by declining atmospheric CO_2. *Nature*, **421**, 245–249.

DeConto, R.M. and Pollard, D. (in press) The influence of sea ice in computer models of the Antarctic Ice Sheet for Oligocene-Miocene times. *Palaeogeography, Palaeoclimatology, Palaeoecology*.

DeSantis, L. and Barrett, P.J. (1998) Grain Size Analysis of Samples from CRP-1. *Terra Antartica*, **5**, 375–382.

Dunbar, G.B., Barrett, P.J. and DeConto, R.M. (2003) Inferring sea level change from sediment texture: application to Oligocene strata off the Victoria Land coast, Antarctica. *Geophysical Research Abstracts*, Vol. 5, 08994.

Dunbar, G.B. and Barrett, P.J. (2005) Estimating palaeobathymetry of wave-graded continental shelves from sediment texture. *Sedimentology*, **52**, 253–269.

Dunbar, G.B., Naish, T., Barrett, P.J., Fielding, C. and Powell, R.D. (in press) Constraining the amplitude of Late Oligocene bathymetric changes in Western Ross Sea during orbitally-induced oscillations in the East Antarctic Ice Sheet: (1) Implications for glacimarine sequence stratigraphic models. *Palaeogeography, Palaeoclimatology, Palaeoecology*.

Ehrmann, W.U., Melles, M., Kuhn, G. and Grobe, H. (1992) Significance of clay mineral assemblages in the Antarctic Ocean. *Mar. Geol.*, **107**, 249–273.

Ehrmann, W.U., Setti, M. and Marinoni, L. (2005) Clay minerals in Cenozoic sediments off Cape Roberts (McMurdo Sound, Antarctica) reveal the palaeoclimatic history. *Palaeogeography, Palaeoclimatology, Palaeoecology*, **229**, 187–211.

Elliott, T. (1986) Siliciclastic shorelines. In: *Sedimentary environments and facies* (Ed. H. Reading), Blackwells, Oxford, 155–188.

Fielding, C.R., Woolfe, K.J., Howe, J.A. and Lavelle, M. (1998) Sequence stratigraphic analysis of CRP-1, Cape Roberts Project, McMurdo Sound, Antarctica. *Terra Antartica*, **5**, 353–361.

Fielding, C.R., Naish, T.R. and Woolfe, K.J. (2000) Facies analysis and sequence stratigraphy of CRP-2/2A, Victoria Land Basin, Antarctica. *Terra Antartica*, **7**, 323–338.

Fielding, C.R., Naish, T.R. and Woolfe, K.J. (2001) Facies architecture of the CRP-3 drillhole, Victoria Land Basin, Antarctica. *Terra Antartica*, **8**, 217–224.

Fielding, C.R., Henrys, S.A. and Wilson, T.J. (2006) Rift history of the western Victoria Land Basin; a new perspective based on integration of cores with seismic reflection data. In: *Antarctica: Contributions to global earth sciences* (Eds D.K. Fütterer, D. Damaske, G. Kleinschmidt, H. Miller and F. Tessensohn), Springer-Verlag, Berlin, 309–318.

Fitzgerald, P.G. (1992) The Transantarctic Mountains of Southern Victoria Land: The application of fission track analysis to a rift shoulder uplift. *Tectonics*, **11**, 634–662.

Flint, R.F. (1971) *Quaternary and Glacial Geology*. Wiley, New York. 892 pp.

Florindo, F., Wilson, G.S., Roberts, A.P., Sagnotti, L. and Verosub, K.L. (2001) Magnetostratigraphy of late Eocene – early Oligocene strata from the CRP-3 core, Victoria Land Basin, Antarctica. *Terra Antartica*, **8**, 599–613.

Florindo, F., Wilson, G., Roberts, A., Sagnotti, L. and Verosub, K. (2005) Magnetostratigraphic chronology of a late Eocene to early Miocene succession from the Victoria Land Basin, Ross Sea, Antarctica. *Global and Planetary Change*, **45**, 207–236.

George, A. (1989) Sedimentary petrology. In: *Antarctic Cenozoic history from the CIROS-1 drillhole, McMurdo Sound, Antarctica* (Ed. P.J. Barrett), DSIR, Wellington, New Zealand. Bull., **245**, 159–168.

Gradstein, F.M., Ogg, J.G. and Smith, A.G. (Eds) (2004) A Geologic Time Scale. Cambridge University Press, Cambridge. 589 pp.

Hambrey, M.J., Barrett, P.J., Hall, K.J. and Robinson, P.H. (1989) Stratigraphy. In: *Antarctic Cenozoic history from the CIROS-1 drillhole, McMurdo Sound, Antarctica* (Ed. P.J. Barrett) DSIR Wellington, New Zealand. Bull. **245**, 23–48.

Hambrey, M.J., Ehrmann, W.U. and Larsen, B. (1991) The Cenozoic glacial record from the Prydz Bay continental shelf, East Antarctica. In: *Scientific results of the Ocean Drilling Program* (Eds J. Barron and B. Larsen), **119**, 77–132.

Hambrey, M.J., Barrett, P.J., Ehrmann, W.U. and Larsen, B. (1992) Cenozoic sedimentary processes on the Antarctic continental shelf: the record from deep drilling. *Zeitschrift für Geomorphologie*, Suppl. Vol. 86, 73–99.

Hambrey, M.J., Krissek, L.A., Powell, R.D. and 7 others (1997) Cape Roberts Project Core Logging Manual. *VUW Antarctic Data Series*, **21**, 89 pp.

Hambrey, M.J., Barrett, P.J. and Powell, R.D. (2002) Late Oligocene and early Miocene glacimarine sedimentation in the SW Ross Sea, Antarctica: the record from offshore drilling. In: *Glacier-Influenced Sedimentation on High-Latitude Continental Margins* (Eds J.A. Dowdeswell and C.O. Cofaigh), *Geol. Soc. London Spec. Publ.*, **203**, 105–128.

Hannah, M.J., Wilson, G.J. and Wrenn, J.H. (2000) Oligocene and Miocene marine palynomorphs from CRP-2/2A, Victoria Land Basin, Antarctica. *Terra Antartica*, **7**, 503–512.

Hannah, M.J., Wrenn, J.H. and Wilson, G.J. (2001) Preliminary report on Early Oligocene and ?Latest Eocene marine palynomorphs from CRP-2/2A, Victoria Land Basin, Antarctica. *Terra Antartica*, **8**, 383–388.

Hannah, M.J., Florindo, F., Harwood, D.M., Fielding, C.R. and CRP Science Team (2001) Chronostratigraphy of the CRP-3 drillhole, Victoria Land Basin, Antarctica. *Terra Antartica*, **8**, 615–620.

Harwood, D.M., Bohaty, S.M. and Scherer, R.P. (1998) Lower Miocene diatom biostratigraphy of the CRP-1 drillcore, McMurdo Sound, Antarctica. *Terra Antartica*, **5**, 499–514.

Haq, B.U., Hardenbol, J. and Vail, P.R. (1987) Chronology of fluctuating sea levels since the Triassic. *Science*, **235**, 1156–1167.

Hayes, D.E. and Frakes, L.A. (Eds) (1975) *Initial Reports of the Deep Sea Drilling Project*, Vol. **28**, U.S. Government Printing Office, 1012 pp.

Henrys, S.A., Bücker, C.J., Bartek, L.R., Bannister, S.A., Neissen, F. and Wonik, T. (2000) Correlation of seismic reflectors with CRP-2/2A, Victoria Land Basin, Antarctica. *Terra Antartica*, **8**, 221–230.

Henrys, S.A., Bücker, C.J., Niessen, F. and Bartek, L.R. (2001) Correlation of seismic reflectors with the CRP-3 drillhole, Victoria Land Basin, Antarctica. *Terra Antartica*, **8**, 127–136.

Ivany, L.C., van Simaeys, S., Domack, E.W. and Samson, S.D. (2006) Evidence for an earliest Oligocene ice sheet on the Antarctic Peninsula. *Geology*, **34**, 377–380.

Johnson, D.W. (1919) *Shore Processes and Shoreline Development*. Wiley, New York, NY, pp. 405–493.

Krissek, L.A. and Kyle, P.R. (2000) Geochemical indicators of weathering, Cenozoic palaeoclimates, and provenance from fine-grained sediments in CRP-2/2A, Victoria Land Basin, Antarctica. *Terra Antartica*, **7**, 281–298.

Krissek, L.A. and Kyle, P.R. (2001) Geochemical indicators of weathering, Cenozoic palaeoclimates, and provenance from fine-grained sediments in CRP-3, Victoria Land Basin, Antarctica. *Terra Antartica*, **8**, 561–568.

Lavelle, M. (1998) Strontium-Isotope stratigraphy of the CRP-1 drillhole, Ross Sea, Antarctica. *Terra Antartica*, **5**, 691–696.

Lear, C.H., Elderfield, H. and Wilson, P.A. (2000) Cenozoic deep-sea temperatures and global ice volumes from Mg/Ca in benthic foraminiferal calcite. *Science*, **287**, 269–272.

Lear, C.H.Y., Rosenthal, Y., Coxall, H.A. and Wilson, P.A. (2004) Late Eocene to early Miocene ice sheet dynamics and the global carbon cycle. *Paleoceanography*, **19**, PA4015, doi: 10.1029/2004PA001039.

Lourens, L., Hilgen, F., Shackleton, N.J., Laskar, J. and Wilson, D. (2004) The Neogene Period. In: *The Geological Timescale* (Eds F.M. Gradstein, J.G. Ogg and A.G. Smith). Cambridge University Press, Cambridge, 409–440.

McIntosh, W.C. (2000) $^{40}Ar/^{39}Ar$ Geochronology of tephra and volcanic clasts in CRP-2A, Victoria Land Basin, Antarctica. *Terra Antartica*, **7**, 621–630.

Marchant, D.R., Denton, G.H., Swisher III, C.C. and Potter Jr, N. (1996) Late Cenozoic Antarctic paleoclimate reconstructed from volcanic ashes in the Dry Valleys region of southern Victoria Land. *Geol. Soc. Amer. Bull.*, **108**, 181–194.

Miller, K.G., Fairbanks, R.G. and Mountain, G.S. (1987) Tertiary oxygen isotope synthesis, sea level history, and continental margin erosion. *Paleoceanography*, **2**, 1–19.

Miller, K.G., Kominz, M.A., Browning, J.V., Wright, J.D., Mountain, G.S., Katz, M.E., Sugarman, P.J., Cramer, B.S., Christie-Blick, N. and Pekar, S.F. (2005) The Phanerozoic record of global sea-level change. *Science*, **310**, 1293–8.

Miall, A.D. (1991) Stratigraphic sequences and their chronostratigraphic correlation. *J. Sed. Petrol.*, **61**, 497–505.

Mudie, P.J. (1992) Circum-Arctic Neogene and Quaternary marine palynofloras: paleoecology and statistical analysis. In: *Neogene and Quaternary dinoflagellate cysts and acritarchs* (Eds M.J. Head and J.H. Wrenn). American Association of Stratigraphic Palynologists Foundation, Dallas, 347–390.

Naish, T.R. and Kamp, P.J.J. (1997) Sequence stratigraphy of sixth order (41 k.y.) Plio-Pleistocene cyclothems, Wanganui basin, New Zealand: A case for the regressive systems tract. *Geol. Soc. Amer. Bull.*, **109**, 978–999.

Naish, T.R., Woolfe, K.J., Barrett, P.J., Wilson, G.S. and 29 others (2001a) Orbitally induced oscillations in the East Antarctic ice sheet at the Oligocene-Miocene boundary. *Nature*, **413**, 719–723.

Naish, T.R., Barrett, P.J., Dunbar, G.B., Woolfe, K.J., Dunn, A.G., Henrys, S.A., Claps, M., Powell, R.D. and Fielding, C.R. (2001b) Sedimentary cyclicity in CRP drillcore, Victoria Land Basin, Antarctica. *Terra Antartica*, **8**, 225–244.

Naish, T.R., Wilson, G.S., Dunbar, G.B. and Barrett, P.J. (in press) Constraining the amplitude of Late Oligocene bathymetric changes in Western Ross Sea during orbitally-influenced oscillations in the East Antarctic Ice Sheet: (2) Implications for global glacio-eustasy. *Palaeogeography, Palaeoclimatology, Palaeoecology.*

Nesbitt, H.W. and Young, G.M. (1982) Early Proterozoic climates and plate motions inferred from major element chemistry of lutites. *Nature*, **299**, 715–717.

Pagani, M., Zachos, J.C., Freeman, K.H., Tipple, B. and Bohaty, S. (2005) Marked decline in atmospheric carbon dioxide concentrations during the Paleogene. *Science*, **309**, 600–603.

Passchier, S. (2004) Variability in geochemical provenance and weathering history of Sirius Group strata, Transantarctic Mountains: implications for Antarctic glacial history. *J. Sed. Res.*, **74**, 607–619.

Passchier, S. (2006) Oligocene-Miocene Antarctic continental weathering trend and ice volume variations from sediment records in the Ross Sea area. *SCAR Open Science Conference, Hobart, Abstracts*, 735.

Paul, H.A., Zachos, J.C., Flower, B.P. and Triparti, A. (2000) Orbitally induced climate and geochemical variability across the Oligocene/Miocene boundary. *Paleoceanography*, **15**, 471–485.

Pekar, S.F. and DeConto, R.M. (2006) High-resolution ice-volume estimates for the early Miocene: evidence for a dynamic ice sheet in Antarctica. *Palaeogeography, Palaeoclimatology, Palaeoecology*, **231**, 101–109.

Pekar, S.F., Christie-Blick, N., Kominz, M. and Miller, K.G. (2002) Calibration between eustatic estimates from backstripping and oxygen isotopic records for the Oligocene. *Geology*, **30**, 903–906.

Pekar, S.F., DeConto, R.M. and Harwood, D.M. (2006) Resolving a late Oligocene conundrum: Deep-sea warming and Antarctic glaciation. *Palaeogeography, Palaeoclimatology, Palaeoecology*, **231**, 29–40.

Posamentier, H.W., Jervey, M.T. and Vail, P.R. (1988) Eustatic controls on clastic deposition I – conceptual framework. In *Sea-level Research: An Integrated Approach* (Eds C.K. Wilgus, B.S. Hastings, C.G. St. C. Kendall, H.W. Posamentier, C.A. Ross and J.C. Van Wagoner). SEPM Spec. Publ., **42**, 109–124.

Powell, R.D. and Cooper, J.M. (2002) A glacial sequence stratigraphic model for temperate glaciated continental shelves. *Geol. Soc. London Special Publication*, **203**, 215–244.

Powell, R.D., Hambrey, M.J. and Krissek, L.A. (1998) Quaternary and Miocene glacial and climatic history of the Cape Roberts drillsite region, Antarctica. *Terra Antartica*, **5**, 341–351.

Powell, R.D., Krissek, L.A. and van der Meer, J.J.M. (2000) Preliminary depositional environmental analysis of CRP-2/2A, Victoria Land Basin, Antarctica: palaeoglaciological and palaeoclimatic inferences. *Terra Antartica*, **7**, 313–322.

Powell, R.D., Laird, M.G., Naish, T.R., Fielding, C.R., Krissek, L.A. and van der Meer, J.J.M. (2001) Depositional environments for strata cored in CRP-3 (Cape Roberts Project), Victoria Land Basin, Antarctica: palaeoglaciological and palaeoclimatic inferences. *Terra Antartica*, **8**, 207–216.

Prebble, J.G., Raine, J.I., Barrett, P.J. and Hannah, M.J. (2006) Vegetation and climate from two Oligocene glacioeustatic sedimentary cycles (31 and 24 Ma) cored by the Cape Roberts Project, Victoria Land Basin, Antarctica. *Palaeogeography, Palaeoclimatology, Palaeoecology*, **231**, 101–109.

Raine, J.I. (1998) Terrestrial palynomorphs from Cape Roberts Project drillhole CRP-1, Ross Sea, Antarctica. *Terra Antartica*, **5**, 539–548.

Raine, J.I. and Askin, R.A. (2001) Terrestrial palynology of Cape Roberts drillhole CRP-3, Victoria Land Basin, Antarctica. *Terra Antartica*, **8**, 389–400.

Raine, J.I., Askin, R.A., Mildenhall, D.C. and Prebble, J.G. (2006) Eocene to Miocene vegetation history and climate, Ross Sea region, Antarctica. *SCAR Open Science Conference, Hobart, Abstracts*, 396.

Roberts, A.P., Wilson, G.S., Florindo, F., Sagnotti, L., Verosub, K. and Harwood, D.M. (1998) Magnetotratigraphy of Lower Miocene strata from the CRP-1 Core, McMurdo Sound, Ross Sea, Antarctica. *Terra Antartica*, **5**, 703–713.

Sandroni, S. and Talarico, F. (2001) Petrography and provenance of basement clasts and clast variability in CRP-3 drillcore (Victoria Land Basin, Ross Sea, Antarctica). *Terra Antartica*, **8**, 449–468.

Scherer, R.P., Bohaty, S.M. and Harwood, D.M. (2000) Oligocene and Lower Miocene Siliceous Microfossil Biostratigraphy of Cape Roberts Project Core CRP-2/2A, Victoria Land Basin, Antarctica. *Terra Antartica*, **7**, 417–442.

Scherer, R.P., Dunbar, R.B., Esper, O., Flores, J.-A. and Gersonde, R. (2006) Southern Ocean and Antarctic nearshore record of an Early Pleistocene Warm Interglacial, MIS-31. SCAR Open Science Conference, Hobart, July, 2006 (abs.).

Shackleton, N.J. and Kennett, J.P. (1975) Paleotemperature history of the Cenozoic and the initiation of Antarctic glaciation: oxygen and carbon isotope analyses in DSDP Sites 277, 279, and 281. In *Initial Reports of the Deep Sea Drilling Project* (Eds J.P. Kennett, and R. Houtz), **29**, 743–755.

Shackleton, N.J., Hall, M.A. and Pate, D. (1995) Pliocene stable isotope stratigraphy of Site 846. In: *Scientific Results of the Ocean Drilling Program* (Eds N.G. Pisias, L.A. Mayer, T.R. Janacek, A. Palmer-Julson and T.H. van Andel), **138**, 337–355.

Smellie, J. (2000) Erosional history of the Transantarctic Mountains deduced from sandstone detrital modes in CRP-2/2A, Ross Sea, Antarctica. *Terra Antartica*, **7**, 545–552.

Smellie, J. (2001) History of Oligocene erosion, uplift and unroofing of the Transantarctic Mountains deduced from sandstone detrital modes in CRP-3 drillcore, Victoria Land Basin, Antarctica. *Terra Antartica*, **8**, 481–490.

Stern, T.A., Baxter, A.K. and Barrett, P.J. (2005) Isostatic rebound due to glacial erosion within the Transantarctic Mountains. *Geology*, **33**, 221–224.

Sugden, D.E. and Denton, G.H. (2004) Cenozoic landscape evolution of the Convoy Range to Mackay Glacier area, Transantarctic Mountains: Onshore to offshore synthesis. *Geol. Soc. Amer. Bull.*, **116**, 840–857.

Swift, D.J.P. (1971) Quaternary shelves and the return to grade. *Mar. Geol.*, **8**, 5–30.

Talarico, F., Sandroni, S., Fielding, C.R. and Atkins, C. (2000) Variability, petrography and provenance of basement clasts from CRP-2/2A drillcore (Victoria Land Basin, Ross Sea, Antarctica). *Terra Antartica*, **7**, 529–544.

Taviani, M. and Beu, A. (2003) The palaeoclimatic significance of Cenozoic marine macrofossil assemblages from Cape Roberts Project drillholes, McMurdo Sound, Victoria Land Basin, East Antarctica. *Palaeogeography, Palaeoclimatology, Palaeoecology*, **198**, 131–143.

Taviani, M., Beu, A. and Lombardo, C. (1998) Pleistocene macrofossils from CRP-1 drillhole, Victoria Land Basin, Antarctica. *Terra Antartica*, **5**, 485–491.

Turnbull, I.M., Allibone, A.H., Forsyth, P.J. and Heron, D.W. (1994) Geology of the Bull Pass-St Johns Range area, Southern Victoria Land, Antarctica, scale 1:50 000. Institute of Geological & Nuclear Sciences, Lower Hutt. Geological map and booklet (52 p.)

Vail, P.R., Mitchum, R.M., Todd, R.G., Widmier, J.M., Thompson III, S., Sangree, J.B., Bubb, J.N. and Hatlelid, W.G. (1977) Seismic stratigraphy and global changes of sea level. *Mem. Amer. Assoc. Petrol. Geol.*, **26**, 49–205.

van der Meer, J.J.M. (2000) Microscopic observations on the upper 300 m pf CRP-2/2A, Victoria Land Basin, Antarctica. *Terra Antartica*, **7**, 339–348.

Webb, P.-N. and Strong, C.P. (1998) Occurrence, stratigraphic distribution and paleoecology of Quaternary foraminifera from CRP-1. *Terra Antartica*, **5**, 455–472.

Webb, P.-N. and Strong, C.P. (2000) Pliocene benthic foraminifera from CRP-2 (Lithostratigraphic Unit 2.2), Victoria Land Basin, Antarctica. *Terra Antartica*, **7**, 453–460.

Wilson, G.S., Bohaty, S.M., Fielding, C.R., Florindo, F., Hannah, M.J., Harwood, D.M., Mcintosh, W.C., Naish, T.R., Roberts, A.P., Sagnotti, L., Scherer, R.P., Strong, C.P., Verosub, K.L., Villa, G., Watkins, D.K., Webb, P.-N. and Woolfe, K.J. (2000) Chronostratigraphy of CRP-2/2A, Victoria Land Basin, Antarctica. *Terra Antartica*, **7**, 647–654.

Wilson, G.S., Lavelle, M., McIntosh, W.C., Roberts, A.P. and 10 others (2002) Integrated chronostratigraphic calibration of the Oligocene-Miocene boundary at 24.0 ± 0.1 Ma from the CRP-2A drill core, Ross Sea, Antarctica. *Geology*, **30**, 1043–46.

Wise, S.W., Jr., Breza, J.R., Harwood, D.M., Wei, W. and Zachos, J. (1991) Paleogene glacial history of Antarctica. In: *Controversies in Modern Geology; Evolution of Geological Theories in Sedimentology, Earth History and Tectonics* (Eds J.A. McKenzie and H. Weissert). Academic Press, London, 133–171.

Wise, S.W., Jr., Smellie, J.L., Aghib, F.S., Jarrard, R.D. and Krissek, L.A. (2001) Authigenic smectite clay coats in CRP-3 drillcore, Victoria Land Basin, Antarctica, as a possible indicator of fluid flow: a progress report. *Terra Antartica*, **8**, 281–298.

Young, G.M. and Nesbitt, H.W. (1998) Processes controlling the distribution of Ti and Al in weathering profiles, siliciclastic sediments and sedimentary rocks. *J. Sed. Research*, **68**, 448–455.

Zachos, J.C., Breza, J. and Wise, S.W. (1992) Early Oligocene ice sheet expansion on Antarctica, sedimentological and isotopic evidence from the Kergulen Plateau. *Geology*, **20**, 569–573.

Zachos, J.C., Flower, B.P. and Paul, H. (1997) Orbitally paced climate oscillations across the Oligocene/Miocene boundary. *Nature*, **388**, 567–570.

Zachos, J., Pagani, M., Sloan, L. and Thomas, E. (2001) Trends, rhythms and aberrations in global climate 65 million years to present. *Science*, **292**, 686–693.

APPENDIX I

SUMMARY DESCRIPTION OF FACIES FROM CENOZOIC STRATA DESCRIBED FROM CRP DRILL CORES

Note: Dispersed clasts or lonestones are a prominent feature throughout Facies 1–5 (mudstone through to sandstone) and facies 8 (rhythmite). They range in size from small pebble to cobble-size, and from angular to rounded. A few are faceted and striated (Atkins, 2001). Lithologies are largely granitoid and dolerite, with lesser amounts of metamorphic and sedimentary types, all sourced from the adjacent Transantarctic Mountains (Talarico et al., 2000, Sandroni & Talarico, 2001).

Facies 1, mudstone with lonestones, is typically a massive slightly sandy clayey silt, although 20% sand is not uncommon. Gravel rarely accounts for more than 0.2%. Fine sand and silt laminae are common and there is little bioturbation. Microfossils include diatoms (common), benthic foraminifera, and both marine and terrestrial palynomorphs, the latter presumably from vegetated ice-free parts of the adjacent coast. The facies also includes a few molluscs.

This facies is interpreted as fine-grained sediment discharged from rivers or glaciers on land and carried offshore to settle out below wave-base along with coarser sediment from floating ice.

Facies 2, interstratified sandstone and mudstone, is characterised by beds of both fine sandstone and mudstone tens of centimentres thick, with sands commonly parallel- and ripple-laminated, and fining upward to mudstones. Some thin intervals show intense soft-sediment deformation.

This facies is interpreted as a mix of hemipelagic sedimentation below wave base, interrupted periodically by occasional low density sediment gravity flows, perhaps triggered by floods or storm events.

Facies 3, poorly sorted muddy very fine to coarse-grained sandstone, varies considerably in texture, although typically it is massive or weakly stratified, with normal grading. Finer beds are ripple-laminated. Thin conglomerates occur at the base of coarse-grained beds. The very poor sorting of this facies has led to several samples visually described as sandstone, analysing as sandy mudstone with 30% fine/very fine sand, 50% silt and 20% clay.

This facies has been interpreted to form from high density sediment gravity flows or by turbid plumes originating from heavy fluvial discharge. Alternatively some intervals could represent mixing between shoreface sand and offshore mud during periods of changing sea level.

Facies 4, a moderately to well-sorted fine sandstone, exhibits planar and low-angle cross-lamination, and in places is interbedded with very fine, medium and coarse sandstones. Recent reviews (e.g. Hambrey *et al.*, 2002) have included it with Facies 5.

Facies 5, a moderately to well sorted stratified or massive, fine to medium-grained sandstone, is typically massive and planar to cross-stratified. Textural analysis shows this facies to be mostly moderately well sorted, but also to have matrix that is far more clay-rich than Facies 1, suggesting that much of the clay migh be of diagenetic origin, and that the facies was better sorted when deposited.

This facies is interpreted as a shoreface deposit influenced primarily by waves and currents.

Facies 6, a stratified diamictite, is a mixture of gravel (>1%), sand and mud, with stratification evident from variations in colour, texture or clast density. It commonly has gravel, sand or silt interbeds, and locally shows strong soft-sediment deformation.

This facies can be interpreted to result from subglacial melt-out through the water column from a floating ice margin. Alternatively, it may have been the product of a debris-flow. This might itself have been fed by subglacial meltwater discharge.

Facies 7, a massive diamictite, is an unstratified mixture of gravel, sand and mud in beds from decimetres to several metres in thickness. Upper and lower contacts are well-defined, though they may be graded. Lower contacts can also be sharp and with indications of loading. Some though not all clast fabrics show a preferred orientation (Atkins, 2001). Samples taken for microtextural analysis by thin section also show rotational structures and other features that indicate that they experienced shearing, possibly by grounded ice (Van der Meer, 2000).

This facies is most likely to have formed near the margin of a temperate glacier from melting of the basal debris layer.

Facies 8, rhythmically interstratified sandstone and siltstone, is characterised by thin silt laminae, thin turbiditic sands and single-grain-thick sand laminae, and is quite unusual, forming <1% of the section.

This facies has been seen to form through suspension settling from turbid plumes accompanied by occasional low density turbidity currents.

Facies 9, a clast-supported conglomerate, is massive, poorly sorted and commonly grades from cobble to small pebble grade, with a matrix of fine- to coarse-grained sand. Clasts are typically moderately to well rounded, but almost invariably include angular clasts too.

The facies is submarine from its close association with other marine facies, but may have formed close to points of fluvial or subglacial meltwater discharge close to or beyond the coast. It could also represent redeposition of such gravels further offshore by sediment gravity flow.

Facies 10, a poorly sorted matrix-supported conglomerate, has the same features as Facies 9, but the clasts are 'floating' in the sandy matrix and it typically has a higher proportion of angular clasts.

It too is interpreted as the product of fluvial or subglacial meltwater discharge, possibly redeposited by sediment gravity flow.

Facies 11, a mudstone breccia, comprises angular fragments of mudstone (with some very fine sandstone) several centimetres across in a matrix of muddy fine sandstone. It was found in just a few places (<0.5% of the section), the most striking being a 4-m-thick bed with clasts up to 75 mm long and intensively brecciated at the base from 311 to 315.5 mbsf in CRP-2A (Cape Roberts Science Team, 2000, Fig. 3.2 k).

These can be interpreted as mass-movement events, induced either by earthquakes or shearing from grounded ice.

Facies 12, comprises phonolitic tephras erupted from ~24 Ma and above ~300 mbsf in CRP-2A (Armienti *et al.*, 2001). Volcanic glass, crystals and clasts can be seen in the core face both dispersed and concentrated as laminae. Several discreet tephra layers were encountered, the most spectacular being a series between 109 and 114 mbsf in CRP-2A, and including a reverse- to normal-graded 1.2-m-thick pumiceous lapillistone. It does not include volcanogenic sandstones with as much as 30% glass and lithic fragments (Smellie, 2000).

This facies is interpreted as volcanic airfall debris deposited through water, in most cases reworked by currents and in some redeposited by sediment gravity-flows. Ar-Ar dating of feldspars and other volcanic material from these deposits, along with some volcanic clasts, has been crucial for the chronology of the period from 25 to 17 Ma (Cape Roberts Science Team, 1998, 1999; Florindo *et al.*, 2005).

Glacial stress field orientation reconstructed through micromorphology and μX-ray computed tomography of till

F. FASANO*, C. BARONI*†, F.M. TALARICO‡, M. BETTUZZI§ *and* A. PASINI§

*Università degli Studi di Pisa, Dip.Scienze della Terra, via S.Maria, 53–56126–Pisa, Italy (e-mail: fasano@unisi.it)
†CNR, Istituto di Geoscienze e Georisorse, Pisa, Italy
‡Università degli Studi di Siena, Dip.Scienze della Terra, via Laterina 8 – 53100 – Siena, Italy
§Università di Bologna, Dipartimento di Fisica, viale C.B. Pichat, 6/2 – 40127 – Bologna, Italy

ABSTRACT

Detailed laboratory analyses largely support field studies on the petrology of glacial materials. In particular, detailed structural analysis of glacial sediments helps reconstruct the presence, orientation and intensity of glacial stress fields. This pilot study compares thin-section micromorphology (two-dimensional observation under a microscope) and μX-ray computed tomography (μXR-CT, three-dimensional observation involving data collection and subsequent processing) to evaluate the potential of the latter technique to yield reliable data. The location and stratigraphy of sampling sites are reported for a complete presentation. Results are encouraging: compared with thin-section micromorphology, μX-ray computed tomography yielded results that are more detailed, reproducible and unaffected by cutting orientation. Moreover, μXR-CT is easily applied to incoherent materials, brittle samples and unique specimens.

Keywords till, XR-CT, micromorphology, glaciotectonic.

INTRODUCTION

This study aims to develop μXR-CT applications in the microstructural investigation of continental glacigenic sediments. The correct genetic interpretation of glacigenic sediments is indispensable for their use as palaeoclimatic indicators and depends on the thorough and accurate collection of textural, microstructural and micromorphological data. These characteristics, which vary considerably in glacial deposits and in different stratigraphic levels within the same unit (van der Meer, 1996), reflect the complex interplay among local ice dynamics, depositional mechanisms and associated post-depositional processes.

For this pilot study samples were selected from the Ricker Hills nunatak, along the inland margin of the Transantarctic Mountains (Victoria Land, Antarctica). It is located between 75°45′ and 75°35′ S and 158°50′ and 159°20′ E. We analyzed two samples collected in the middle of the nunatak at Lake Depression (Fig. 1).

The Ricker Hills geomorphology is complex and records several phases of the glacial history of the East Antarctic Ice Sheet. In particular, palaeo-iceflow directions cannot be easily reconstructed on the basis of field data alone. The Cenozoic 'Ricker Hills Tillite' is exposed in several outcrops and has been mapped by Capponi *et al.* (1999).

To increase our understanding of tillite micro-textures and their use as palaeoenviromental indicators in the Ricker Hills area, we carried out μX-ray tomography analysis of two selected samples. The results of these investigations are here briefly summarised to highlight the potential of this technique for the detailed textural characterisation of tillites. μX-ray tomography is an analytical method that has only recently become available to the geosciences (Ketcham and Carlson, 2001; van Geet *et al.*, 2001), and its application to the

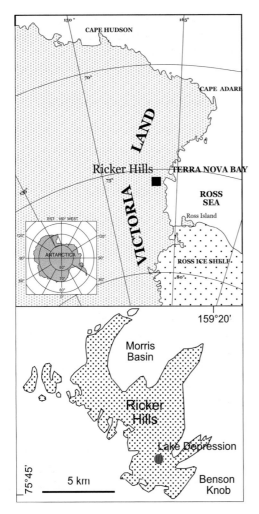

Fig. 1 Location map of the Ricker Hills. The Lake Depression outcrop is indicated.

Samples were impregnated under vacuum with a thin bi-component epoxy resin. Thin sections (9×6 cm) were cut following the methods outlined by Murphy (1986) and van der Meer (1997, 1993). Thin sections were studied using a petrographic microscope with a magnification range of 2× to 40×.

The instrument for μXR-CT analysis is composed of an X-ray source with a microfocus beam (spot size of less than 10 μm), a mechanical system for moving the sample and an XCCD camera as detector. The current is generally set to less than 300 μA, with a voltage of 140 kV. Samples are cores of about 2 cm in diameter and 1.5 cm long. The sample is rotated 360°, and one or more images are collected for every 1° step. All the data are processed through specific software to obtain 1) a 3D dataset, 2) a 3D rendering, and 3) image analysis. Rendering and processing were completed using free software such as MRicro and AMIDE. The image usually obtained is linked to Beer's law:

$$I = I_0\, e^{-\mu\chi}$$

where I_0 is the integral current of incident X-rays, I is the integral current transmitted by the sample, μ is the linear attenuation coefficient of the material and χ is the sample width. There is a direct link between the transmitted current revealed by the detector and the linear attenuation. In positive images, this means that brightness and attenuation are directly linked. In other words, the lighter coloured areas indicate greater attenuation (lower attenuation in negative images).

study of glacigenic sediments has only just begun (Ketcham, 2005).

MATERIALS AND METHODS

The Ricker Hills Tillite is composed of a massive matrix-supported diamictite with a porphyric coarse/fine (bimodal) related distribution, low sorting, low rounding and medium to high angularity. The Tillite is interpreted as a tectomict *sensu* van der Meer *et al.* (2003). In this tectofacies, deformation structures are very common and their orientation is clearly related to stress-field orientation.

SEDIMENTOLOGY

In this study we focus on the 'lake depression' outcrop, since it is a key position in the middle of the Ricker Hills (Fig. 1). Disrupted sandstones of the Beacon Supergroup lie at the base of the profile (Fig. 2). Several high-angle, westward-dipping shear planes were identified; these are generally related to rotational structures marked by a sense of shear to the west. Several injection veins up to some tens of centimetres long and 1–2 cm thick were observed. A massive yellow to greyish-yellow, lithified diamictite overlies the fractured bedrock. It consists of mud-supported clasts (mainly dolerites

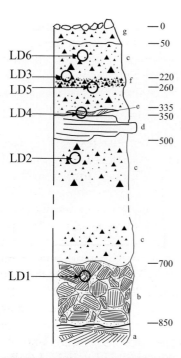

Fig. 2 Lake depression lithological log profile with sampling points. Lithologies: (a) sandstone; (b) disrupted sandstone; (c) diamictite; (d) erratic; (e) laminated lens; (f) clast-supported gravel; (g) uncemented diamictite.

of the Ferrar Supergroup, with some sandstones). The overall texture ranges from sandy to cobbly. The sand fraction is mainly derived from weathering of the Takrouna Sandstone (Beacon Supergroup). In the section described here, this layer is 2 m thick, but it is estimated to reach 8–10 m in the nearby outcrops. A large, erratic dolerite boulder lies within the diamictite. A lens with a laminated structure is present above the boulder. Above the boulder the massive diamictite is interrupted by a 30 cm-thick layer of sorted, roughly stratified gravel. This layer is clast-supported, and the silty matrix is sparse. A yellow, uncemented diamicton containing striated pebbles belongs to a Late Pleistocene glacial drift and forms the upper portion of the outcrop.

In thin section, the diamictite is composed primarily of grains 200–300 µm in size. Quartz is the dominant mineral, with minor lithic fragments of dolerite, plagioclase and K-feldspar. Lithic sandstone fragments are generally scarce and the grain distribution is typically homogeneous. The matrix is mainly composed of quartz fragments and phyllosilicates, and its grain size ranges from silt

to clay. The pale yellow to brown colour indicates oxidation. The coarse fraction is equal to or greater than the fine fraction, and the estimated size limit is 50 µm. Deformational structures such as rotational structures, pressure shadows and grain fracturing are widespread. In particular, rotational structures are common and better developed in samples LD3 and LD6. Although a periglacial origin cannot be completely ruled out, the low porosity, the absence of secondary zeolites and of mammillated vugs, deformational structures and profile settings highlight the glaciotectonic origin of rotational structures. Pressure shadows are present in sample LD6. Linear features marked by clast alignments are clearly visible in sample LD2 (Fig. 3). Thin sections were prepared on oriented samples, whose orientation is reported in Figure 3. The alignments clearly dip to the west. In sample LD1, laminated sandstones are deformed and reoriented (Fig. 4). Coal layers mark oriented planes with a westward dip direction.

µXR-CT observations reveal the heterogeneous composition of the diamictite. In sample LD5 (Fig. 5) different attenuation domains (the image is positive) are visible at the top of the core. The lighter domains are 50–500 µm in size and show a random distribution. They probably represent the doleritic clasts visible both in the hand specimen and in thin section. The very light clast at the center of the upper surface is a highly attenuated body. Its lithology is unknown. In the cut-out of the core

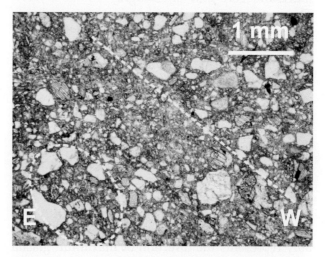

Fig. 3 Sample LD2, thin section. Clast alignment (right) and rotational structures (left) are indicated with dotted lines. E and W are indicated at the bottom of the picture.

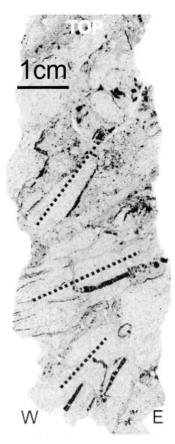

Fig. 4 Sample LD1, thin section. Linear features marked by coal alignment are indicated with dotted lines. East and west are indicated at the bottom of the picture.

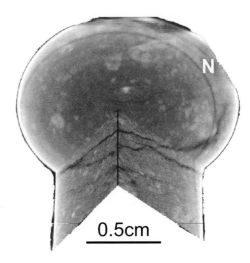

Fig. 5 Sample LD5, 3D image. Note the clasts on the top (dolerite) and the planes dipping to the west. North is indicated.

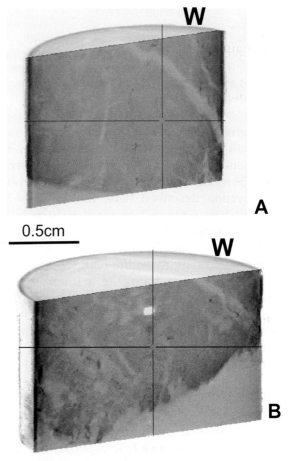

Fig. 6 Sample LD1, 3D image. Two consecutive cuts, same cut direction. Cut (B) is more central, while cut (A) is more lateral. Note the light coal layer several millimetres thick that dips to the west.

some planar structures are visible. They cut across the entire core and dip roughly 30–45° WNW. These structures mark some planes of weakness probably related to the presence of low-attenuation material concentrated along shear planes. To obtain a more precise plane orientation, the core image was rotated through rendering with 1° steps around the supposed orientation. For every step the core was cut (sequential cutting at one-degree intervals of rotation) and the dip angle was measured. Since the apparent dip angle is always less than the real maximum dip plane, rotation, cutting and measurement operations were repeated until the maximum dip angle was found.

In sample LD5 the maximum angle of 40° was measured on the cut oriented N265°, which probably represents the direction of maximum glacial stress.

In sample LD 1 (Fig. 6), representing disrupted sandstone in the bedrock, there are several different attenuation domains. These follow a planar distribution, with a coherent westward dip direction. The images are negative (Fig. 6). The light coloured, westward-dipping planes some tens of microns in thickness represent the low- attenuation coal levels also visible in thin section (Fig. 4). Through sequence rotation cutting, the maximum measured angle was 43°, with a cut direction of N273° related to the direction of maximum shear stress.

A cut in sample LD1 also reveals a very light-coloured, round domain some tens of microns in diameter (Fig. 6b). This domain has very low attenuation since the image is negative, and is probably related to a pore.

DISCUSSION

The application of µXR-CT analyses to the Ricker Hills Tillite indicates that the adopted instrument configuration and operating conditions, as well as the compositional and textural features of the investigated samples can reveal sufficiently contrasting attenuation domains. Resulting 3D images can greatly increase our ability to reconstruct the geometry of shear planes in oriented glacigenic rock samples.

Based on these preliminary studies, we are confident that similar results can be obtained in a number of onshore (e.g. Sirius Formation) or off-shore (e.g. diamictites recovered at Cape Roberts drill sites (Barrett, this volume)). These Antarctic glacigenic deposits are known to be compositionally akin to the Ricker Hills Tillite.

Micromorphological investigations performed on selected thin sections document a local E–W direction of glacier movement (Baroni and Fasano, 2006).

Comparison of micromorphological data with µX-ray tomography yields useful complementary data for interpreting the glacial stress field. Planar and linear features related to deformation, reorientation and movement resulting from shear-strain, dip uniformly westwards at various dip angles. These features indicate that the stress field is oriented E-W, like the sense of movement of the glacier responsible for till deposition. Both methods yield data that support this hypothesis, although µX-ray tomography data is more precise irrespective of the cut direction. Measurements reveal a coherent E-W direction of ice movement. A slight shift in the maximum stress direction toward the WSW is documented in the upper portion of the outcrop.

CONCLUSIONS

1 Microtextural data collected through µXR-CT investigations and those provided by conventional micromorphological studies consistently indicate a westward palaeoflow direction for the glacier responsible for the Ricker Hills Tillite.

2 Compared with micromorphology, µXR-CT yields more coherent results and better quantitative 3D data. Since structural elements visible in thin section vary according to the predetermined cut direction, µXR-CT is a better tool for investigating structural orientation.

3 The study of these materials benefits from µXR-CT for various reasons: (i) larger quantities of higher-quality microstructural and micromorphological data are acquired than would be possible through 2D analysis of just a few large-format thin sections; (ii) 3D quantitative models can be constructed; (iii) poorly consolidated specimens can be analysed; (iv) analyses are non-destructive and preserve the specimen intact.

REFERENCES

Baroni C. and Fasano F. (2006) Micromorphological evidence of warm-based glacier deposition from the Ricker Hills Tillite (Victoria Land, Antarctica). *Quatern. Sci. Rev.*, **25**, 976–992.

Capponi, G., Crispini, L., Meccheri, M., Musumeci, G., Pertusati, P.C., Baroni, C., Delisle, G. and Orsi, G. (1999) Antarctic Geological 1:250.000 map series, Mount Joyce Quadrangle (Victoria Land), Museo Nazionale dell'Antartide, Sez. Scienze della Terra, via Laterina, 8 (53100) Siena.

Ketcham R.A. and Carlson W.D. (2001) Acquisition, optimization and interpretation of X-ray computed tomographic imagery: applications to the geosciences. *Comput. and Geosci.*, **27**, 381–400.

Ketcham R.A. (2005) Three-dimensional grain fabric measurements using high-resolution X-ray computed tomography. *J. Struct. Geol.*, **27**, Issue 7, 1217–1228.

Murphy, P.C. (1986) Thin section preparation of soil and sediments. *AB Academic*, Berkhamsted, 149 pp.

van der Meer, J.J.M. (1993) Microscopic evidence of subglacial deformation. *Quatern. Sci. Rev.*, **12**, 553–587.

van der Meer, J.J.M. (1997) Particle and aggregate mobility in till: microscopic evidence of subglacial processes. *Quatern. Sci. Rev.*, **16**, 827–831.

van der Meer J.J.M. (1996). Micromorphology. In: Menzies J. (Ed.) *Glacial environments.* Vol **2**. Butterworth & Heinemann, Oxford, 335–355.

van der Meer, J.J.M., Menzies, J. and Rose, J. (2003) Subglacial till: the deforming glacier bed. *Quatern. Sci. Rev.*, **22**, 1659–1685.

van Geet M., Swennen R. and Wevers M. (2001) Towards 3-D petrography: application of microfocus computer tomography in geological science. *Comput. and Geosci.*, **27**, 1091–1099.

SOFTWARE REFERENCES

Mricro http://www.psychology.nottingham.ac.uk/staff/cr1/mricro.html

AMIDE http://amide.sourceforge.net/

The Late Ordovician glacial sedimentary system of the North Gondwana platform

JEAN-FRANCOIS GHIENNE*, DANIEL PAUL LE HERON*†, JULIEN MOREAU*, MICHAEL DENIS‡ and MAX DEYNOUX*

*Ecole et Observatoire des Sciences de la Terre, Centre de Géochimie de la surface, CNRS-UMR 7517, 1 rue Blessig, 67084 Strasbourg, France (e-mail: ghienne@illite.u-strasbg.fr)
†Institut für Geologie, Leibniz Universität Hannover, Callinstrasse 30, D-30167, Germany
‡UMR 5561, Biogéosciences, Université de Bourgogne, 6 Bd Gabriel, 21 000 Dijon, France

ABSTRACT

The Late Ordovician (Hirnantian) glaciation is examined through the North Gondwana record. This domain extended from southern high palaeo-latitudes (southeastern Mauritania, Niger) to northern lower palaeo-latitudes (Morocco, Turkey) and covered a more than 4000 km-wide section perpendicular to ice-flow lines. A major mid-Hirnantian deglaciation event subdividing the Hirnantian glaciation in two first-order cycles is recognised. As best illustrated by the glacial record in western Libya, each cycle comprises 2–3 glacial phases separated by ice-front retreats several hundreds kilometres to the south. From ice-proximal to ice-distal regions, the number of glacial surfaces differentiates (i) a continental interior with post-glacial reworking of the glacial surfaces), (ii) a glaciated continental shelf that is subdivided into inner (1–2 surfaces), middle (2–5 surfaces) and outer (a single surface related to the glacial maximum) glaciated shelves, and (iii) the non-glaciated shelf. Ice-stream-generated glacial troughs, 50–200 km in width, cross-cut these domains. These troughs are zones of preferential glacial erosion and subsequent sediment accumulation. A glacial depositional sequence, bounded by two glacial erosion surfaces, records one glacial phase. The position either within or outside a glacial trough controls the stratigraphic architecture of a glacial sequence. Glaciomarine outwash diamictites are developed at or near the maximum position of the ice-front. During ice-sheet recession, and in an ice-stream-generated trough, a relatively thin sediment cover blankets the foredeepened erosion surface. An initial rapid ice-sheet withdrawal is inferred. Marine-terminating ice fronts then evolve later into more slowly retreating, land-terminating ice fronts. In adjacent inter-stream areas where a more gradual ice-sheet recession occurred, fluvio-glacial deposits prevailed. The progradation of a delta-shelf system, coeval with fluvial aggradation, that may be locally interrupted by a period of isostatic rebound, characterises the late glacial retreat to interglacial conditions. This model should facilitate the sequence stratigraphic interpretation of Late Ordovician glacial deposits and other ancient glacial successions.

Keywords Glacial record, Hirnantian, North Africa, ice stream, sequence stratigraphy.

INTRODUCTION

The Late Ordovician glacial record comprises extensive exposures distributed over the former North Gondwana cratonic platform. The size of this domain is >1500 km from south to north (in present-day coordinates), and extended from Mauritania to Arabia. Southern regions were positioned near the ice centres while the northern part of the platform was located at lower palaeo-latitudes (Fig. 1). During the 1960s–1980s, studies in Morocco (Destombes, 1968a, Hamoumi, 1988), Algeria (Beuf et al., 1971), Libya (Klitzsch, 1981; Massa, 1988), Mauritania (Deynoux, 1980; 1985; Deynoux and Trompette, 1981) and Arabia (McLure, 1978; Vaslet, 1990; McGillivray and Husseini, 1992)

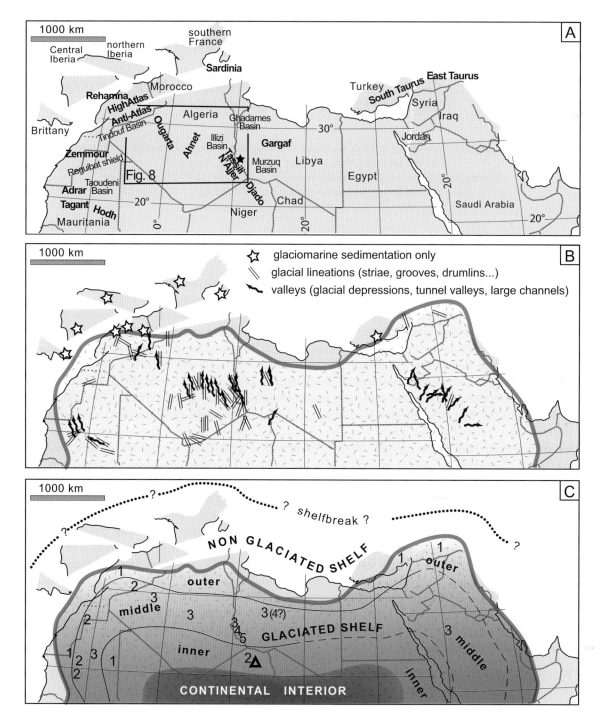

Fig. 1 The North Gondwana platform during the Hirnantian glacial event. (A) Main names cited in the text (see also Fig. 8). Names in bold indicates study areas comprising the data base. The black star locates the pre-glacial (Late Ordovician) trilobite-bearing lonestone found within the syn-glacial strata. (B) Location of areas that experienced grounded ice (glacial lineations and palaeovalleys) and those that have never been glaciated (glaciomarine sediments only). The outlined surface corresponds to the envelope of the ice fronts during the Hirnantian glacial maximum (modified from Deynoux and Ghienne, 2004). (C) Subdivisions of the glaciated platform in five domains (name in bold) based on the number of glacial surfaces preserved. The location of the shelfbreak (continental margin) is not known but was located to the north of the study area, beyond the maximum ice fronts. The Δ locates mid-Hirnantian graptolite-bearing marine facies in the Djado area, suggesting the marine limit of the associated deglaciation was located to the south of this point.

showed the extent, diversity and complexity of the syn-glacial strata. Renewed interest arose in the 1990s, when oil companies recognised the association of syn-glacial strata (reservoir rocks) with lower Silurian shales (source rocks) as one of the most significant plays in the North African Lower Palaeozoic succession (Lüning *et al.*, 2000). Recent work has included detailed field and subsurface studies conducted in Mauritania (Ghienne, 1998; Ghienne and Deynoux, 1998; Ghienne, 2003), Morocco (Ouanaimi, 1998; Sutcliffe *et al.*, 2000, 2001; Le Heron *et al.*, 2007), Algeria (Hirst *et al.*, 2002; Eschard *et al.*, 2005), Niger (Denis *et al.*, 2007), Libya (McDougall and Martin, 2000; Smart, 2000; Ghienne *et al.*, 2003; Le Heron *et al.*, 2004; El-ghali, 2005; Moreau, 2005), Jordan (Abed *et al.*, 1993; Powell *et al.*, 1994; Turner *et al.*, 2005), Saudi Arabia (Senalp and Al-Laboun, 2000) and the Horn of Africa (Ethiopia and Eritrea, Kumpulainen, 2005). In addition to these areas, studies have also been undertaken around the northern Gondwana periphery to better provide an overall understanding of the glacial record as a whole, e.g. in Sardinia (Leone *et al.*, 1995; Ghienne *et al.*, 2000) and in Turkey (Monod *et al.*, 2003).

This paper proposes a large-scale reconstruction of the North Gondwana platform in present-day North and West Africa during the Late Ordovician glaciation. The number of glacial advance-retreat events recorded from the near-polar ice centre to lower palaeo-latitudes, and the presence or absence of ice-stream tracks, are used to subdivide the platform into a number of domains characterised by a specific palaeo-glaciological evolution, particularly subglacial processes, stratigraphic architecture and depositional environments. Finally, a conceptual glacial depositional sequence that could be expected within a typical glacial advance-retreat event is depicted. Such a sequence will form the basis for a more robust stratigraphic analysis of the Late Ordovician glacial record and more generally for ancient glacial successions.

PRE-GLACIAL SETTING

No evidence for Late Ordovician glacial sediments older than Hirnantian (Fig. 2) has thus far been found in North or West Africa. However, this does not preclude the possibility that glaciers grew

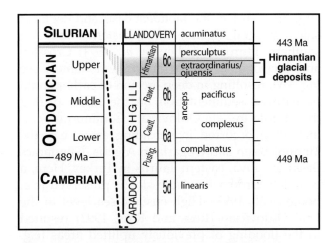

Fig. 2 Late Ordovician stratigraphic chart modified from Webby *et al.* (2004) showing the time range corresponding to Hirnantian glacial event.

elsewhere on Gondwana both before and after the Hirnantian. Saltzman and Young (2005) proposed a glacio-eustatic lowstand during the early Late Ordovician (Fig. 2) and, in South America, glacial deposits are recognized in the Silurian (Grahn and Caputo, 1992; Caputo, 1998). Ghienne (2003) suggested that the Hirnantian simply reflected a continent-wide ice maximum, during which ice sheets reached the North and West African sedimentary basins where their glacial sedimentary record could be preserved. In this paper, preglacial time is defined as the period preceding the Hirnantian, i.e. the Cambrian and Ordovician, including the main part of the Ashgill with the exception of the Hirnantian (Fig. 2). Le Heron *et al.* (2005) published a schematic lithostratigraphic summary of Hirnantian glaciogenic rocks across Africa and Arabia. A regional correlation scheme, including the names of the main preglacial and glacial formations in this huge region, is contained therein.

Geological setting

Late Ordovician glacial deposits rest on a Cambrian to Ordovician, clastic-dominated succession that has an off-shelf gradient towards the NNW (Boote *et al.*, 1998; Carr, 2002). These sediments progressively buried palaeorelief forms associated with both the late Neoproterozoic Panafrican Orogeny and Cambrian post-orogenic collapse/strike-slip basins. Early Ordovician tectonic activity uplifted large areas such as the Taoudeni Basin and Reguibat shield, Central Algeria, Ahnet, Eastern Tassili and southern Taurus (e.g. Beuf *et al.*, 1971; Crossley and McDougall, 1998) (for location names, see Fig. 1A), resulting in deep erosion of earlier strata or reduced sedimentation on the highs during the Middle to Late Ordovician. Continuous sedimentation occurred in subsiding areas, e.g. the Tindouf Basin and Ougarta in Algeria, Ghadames Basin in Libya, eastern Taurus in Turkey. Here, a complete Ordovician succession is preserved (e.g. Boote *et al.*, 1998). High eustatic sea level in the Late Ordovician (Ross and Ross, 1992) resulted in the flooding of previously uplifted areas (e.g. easternmost Turkey, Dean & Monod, 1990). In Algeria, shallow-marine Ashgill facies are known in the south of the Ougarta Range (Legrand, 1985) and Ashgill carbonates are identified in drill holes as far south as the Illizi Basin (e.g. *Oued Ahara* borehole, unpublished data). In SW Libya, some dropstones in glaciomarine facies include trilobites (located in Fig. 1A), yielding an age close to the Caradoc-Ashgill boundary (W.T. Dean, pers. comm., 2003). These fauna, which probably derived from higher southern polar palaeolatitudes, imply a southwards-directed Late Ordovician transgression, which penetrated far south onto the platform. Transgressive deposits have been difficult to recognise owing to glacially related erosion, but are herein considered as key evidence for the flooding of the main part of the North Gondwana platform prior to the Hirnantian glacial event.

Subglacial substrate

As a consequence of this pre-glacial flooding, it is clear that a large part of the glaciated areas should have been characterised by subglacial soft-sediment conditions. If Ashgill siliciclastics were not lithified in Hirnantian time, it is surprising that much older deposits were also very poorly lithified at that time, as suggested by three sets of observations: sand injections, cross-sections of palaeovalleys and a sand-dominated glacial record. Firstly, sand injections in the form of pipes or undulating dykes occur beneath glacial erosion surfaces cut in sediments as old as Early Ordovician (>30 Ma before the Hirnantian glacial event) in the western Murzuq Basin. Secondly, cross-sections

of palaeovalleys, either subglacially or fluvially formed, show low-angle margins, with slopes typically in the 1–10° range. These gradients are more compatible with erosion in soft material than in consolidated sediments. In addition, in the Ahnet area (south Algeria, Fig. 1A), palaeovalleys have a flat bottom coinciding with the basal unconformity of the Cambrian-Ordovician succession above the slightly metamorphosed late Neoproterozoic shales and sandstones. This suggests a clear contrast between the underlying metamorphosed basement and the overlying non- to poorly lithified sandy succession. Thirdly, the Late Ordovician glacial record is sand-dominated, with subordinate finer-grained, silt-dominated units. Glacial conglomerates contain clasts from Precambrian basement lithologies (granites, metasediments e.g. quartzite), but not from the underlying Cambro-Ordovician sandstones (Oujeft sandstones in Mauritania, Hamra Formation in Algeria, Ash Shabiyat or Haouaz formations in Libya), which can be recognised by abundant *Scolithos* burrows. Notwithstanding small inliers of crystalline basement or rare carbonate-cemented Ordovician strata, the evidence presented above suggests that Late Ordovician ice sheets grew and decayed on largely unconsolidated substrates.

RECOGNITION OF GLACIAL EROSION SURFACES

In the sand-dominated setting of the Late Ordovician glacial record, depths of glacial erosion range from 10 to 500 m (maximum erosion depths in Saudi Arabia, e.g. McGillivray and Husseini, 1992), and widths of glacially-cut depressions range from 10 m to >100 km. Erosional features form lineations, ovoid to elongated spoon-shaped depressions, straight to slightly sinuous channels and palaeovalleys, or basin-scale incisions. Criteria used to distinguish glacial erosion surfaces from fluvial or transgressive wave-ravinement surfaces include: large-scale morphologies, glaciotectonic structures and specific depositional features such as esker structures.

Depths of erosion greater than 100 m normally point towards subglacial processes (Fig. 3). If the erosion feature is relatively narrow (<5 km), tunnel valleys are inferred (Ghienne and Deynoux, 1998; Hirst *et al.*, 2002; Le Heron *et al.*, 2004)

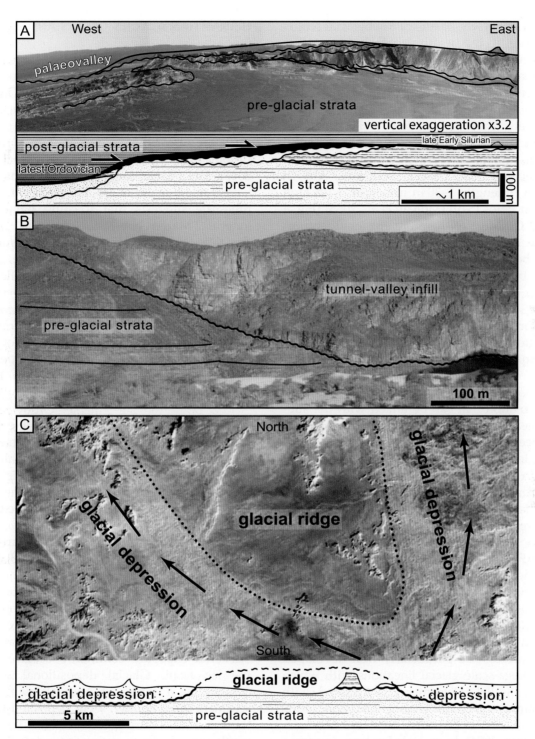

Fig. 3 Diversity in glacial erosion surfaces. (A) A wide palaeodepression within the Eastern Tassili n'Ajjer glacial trough (Ghat area, western Murzuq Basin, Libya) (view with vertical exaggeration, interpretation from field sections). Bounding surfaces of syn-glacial depositional units are gently dipping glacial erosion surfaces (wavy lines). The upper unit (in black) mainly represents a late to post glacial transgressive wedge, that reworked the latest glacial erosion surface (see also Fig. 5). The latest glacial palaeodepression was underfilled and post-glacial sediments progressively onlapped on the residual palaeotopography from the latest Ordovician to the Silurian (see also Fig. 9B). (B) An overfilled tunnel valley characterised by a narrow, steep-sided erosion surface (Adrar, Mauritania, modified from Ghienne and Deynoux, 1998). (C) Landsat image showing a glacial ridge (inverted topography, present-day lower area) with two adjacent palaeovalley-like depressions into a glacial trough (Tihemboka Arch, western Murzuq Basin).

(Fig. 3B). Conversely, basin-scale subglacial scours are inferred if the width is >50 km. In other circumstances, overdeepenings or the occurrence of erosional troughs that narrow in planform also suggest subglacial processes (Beuf et al., 1971). However, the most reliable subglacial erosion surfaces are those associated with subglacial lineations. Subglacial lineations (mega-scale glacial lineations, flow-parallel and attenuated drumlins) are characterised by parallel bedforms, <100 m to >10 km in length, 1–30 m in height, with high elongation ratios (>10) (Moreau et al., 2005) (Fig. 4 A, B and C). Most of them are erosional features cored by older sediments. Only the sediments directly beneath the sediment-ice interface are affected by intense glaciotectonic deformation (Deynoux and Ghienne, 2004, 2005).

The best outcrop-scale evidence for glacially cut erosion surfaces is preserved in the record of deformation structures formed beneath them as a combination of shear-induced and gravitational deformation structures within the sediment column. Glaciotectonic deformation includes all the deformation that can be linked with the subglacial shear zone. Structures or deposits resulting from processes at the ice-sediment interface (striae, grooves, lodgement till) are generally poorly preserved in the Late Ordovician glacial record (Deynoux and Ghienne, 2005; Le Heron et al., 2005). Intraformational deformation structures are more readily preserved and are therefore most commonly observed. Soft-sediment intraformational striated surfaces (Fig. 4F) were formed at several metres in depth beneath the ice-sediment interface (Deynoux and Ghienne, 2004; Le Heron et al., 2005). They are typically associated with intraformational grooves (Fig. 4D), drag and sheath folds (Fig. 4E), Riedel shears and water-escape structures. Intraformational striated surfaces are particularly well developed beneath mega-scale glacial lineations with which they are parallel. Deformation associated with glacial surfaces includes chaotic sandstone units ('grès bousculés' of Deynoux, 1980 and Deynoux and Ghienne, 2004), large-scale loading structures such as domes, cylindrical folds and sediment diapirs (Le Heron et al., 2005), overturned fold and thrust-and-fold belts, 5–50 m high. However, the last may also occur in association with proglacial glaciotectonic processes, such as broadly arcuate belts, or with slide-generated structures.

HIRNANTIAN GLACIES CYCLES AND PHASES

It is established that the so-called Upper Ordovician syn-glacial strata in West and North Africa are, in fact, strictly Hirnantian in age (Destombes, 1968b; Destombes et al., 1985; Paris et al., 1995, 1998; Underwood et al., 1998; Sutcliffe et al., 2000). They are also time-equivalent with a significant isotopic excursion of global extent (Brenchley et al., 2003). The Hirnantian is generally considered to have lasted less than 1 Ma, but the new Ordovician timescale suggests that it could have been as long as 2 Ma (Webby et al., 2004) (Fig. 2). Within this time-slice, a number of glacial advances and subsequent retreats occurred (Ghienne, 2003). At present, the number, as well as the significance and extent, of each of these Hirnantian glacial events is not known, but is the subject of ongoing work. However, the comparison of the stratigraphic architecture of syn-glacial successions from distinct areas gives noteworthy temporal relationships.

Areas closest to the ice centre

Glacial erosion in the areas closest to the ice centre has resulted in reworking of pre-glacial sediments, and during later glacial phases, cannibalisation of the earlier glacial sequences. In these regions, geological mapping covering a minimum representative area of about 1000 km^2 has revealed several, laterally juxtaposed glacial erosion surfaces (Fig. 5) that generate a complex stratigraphic architecture (Ghienne, 2003; Ghienne et al., 2003; Moreau, 2005).

Depositional sequences are bounded by subglacially-cut unconformities or correlative subaerial unconformities. These surfaces were formed regionally during major phases of glacial advance. Subaerial exposure commonly occurred beyond the ice front. Glacial depositional sequences, comprising a range of alluvial plain to shelf sediments (see below), were deposited on top of these surfaces during ice recession towards the south, and during ensuing interglacials. Four to five depositional sequences occur in Mauritania (Ghienne, 2003), Libya (Gargaf Uplift: Deynoux et al., 2000; Le Heron, 2004; Ghienne et al., 2003; western Murzuq Basin: Moreau, 2005; Moreau et al., 2005; Le Heron et al., 2006) and Jordan (Turner et al., 2005).

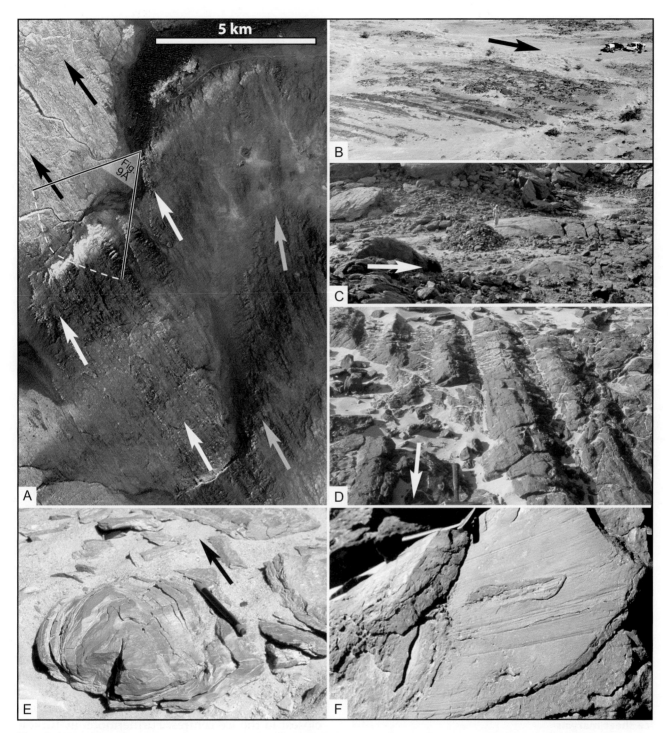

Fig. 4 Multi-scale glacial lineations associated with glacial erosion surfaces; all structures are in syn-glacial sands. Arrows show ice-flow orientation. (A) Landsat image showing three sets of ice-stream-generated mega-scale glacial lineations (Eastern Tassili n'Ajjer, Ghat area, western Murzuq Basin, Libya). (B) Glacial lineations (Tagant, Mauritania, cars for scale). (C) Asymmetrical glacial alignments shaped as a roche moutonnée (western Hodh, Mauritania, person for scale). (D) Grooves preserved within an intraformational subglacial shear zone (Ghat area, western Murzuq Basin, hammer for scale). (E) Sheath folds, the elongation of which is parallel to the shear orientation (Djado, Niger, pencil for scale). (F) Intraformational striae and ploughing structure in sandstones (Tihemboka Arch, western Murzuq Basin).

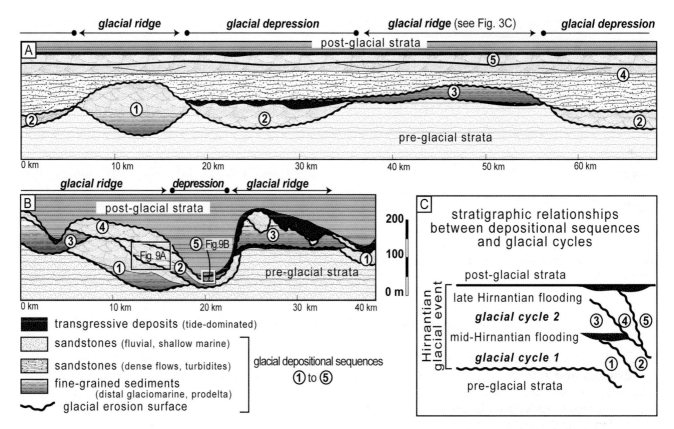

Fig. 5 Stratigraphic architecture of the syn-glacial strata in the western Murzuq Basin, Libya (profile location in Fig. 8A). Syn-glacial strata fill in palaeovalley-like depressions separated by glacial ridges. Indicated depressions and ridges were formed during the glacial maximum (fourth glacial phase). See text for more details about depositional facies. (A) Tihemboka Arch (26°10′N). (B) Eastern Tassili n'Ajjer, Ghat area (25°N). (C) Stratigraphic relationships and subdivision of the Hirnantian succession in two glacial cycles separated by the mid-Hirnantian deglaciation event, and five glacial phases of glacial advance and subsequent retreat.

In Mauritania and Libya, the most extensive ice sheet (e.g. phase 3 of Ghienne, 2003 in Mauritania; phase 4 of Moreau *et al.*, 2005 in western Libya) occurred after a major transgressive event characterised by the deposition of a tide-dominated succession (Fig. 5). This transgression is interpreted as a major phase of deglaciation within the Hirnantian.

Ice sheet margins

In ice-marginal areas, the effects of glacial erosion are less pronounced. Bounding surfaces of successive glacial depositional sequences are frequently preserved within a single vertical section (e.g. Monod *et al.*, 2003). A number of regressive-transgressive cycles are observed and interpreted

in terms of glacioeustatically-driven rhythms (Leone *et al.*, 1995). These areas also preserve evidence of three to five glacial events, as best exemplified by the glacial record of Morocco (e.g. Le Heron *et al.*, 2007) where the Hirnantian strata can be divided into two glacial successions separated by a major transgression (Fig. 6). This major transgression is interpreted to correlate with the deglaciation event associated with tidal deposits in areas closest to the ice centres. Furthermore, in Turkey, evidence for subglacial processes is found at only one stratigraphic interval (Monod *et al.*, 2003) above well-developed transgressive deposits. These regionally correlatable features indicate that a key interglacial transgression was followed by the most widespread and significant phase of ice sheet advance across North Gondwana.

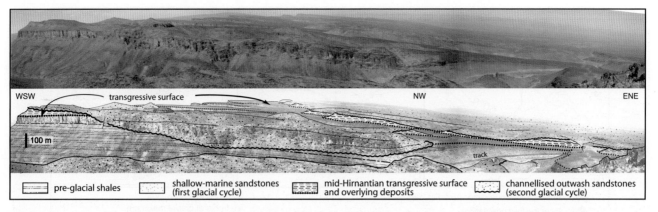

Fig. 6 The mid-Hirnantian deglaciation event in the Anti-Atlas (southern Morocco, northeast of Zagora). The glacially-related succession that mainly comprises shallow-marine and channellised outwash sandstones in contrast with pre-glacial shale-dominated sediments, is divided into two sedimentary wedges, separated by a major transgressive surface related to the mid-Hirnantian event. The panoramic view is c. 4 km in width in the foreground.

The Hirnantian glacial event: a multiphase glaciation

As glacial depositional sequences record major ice advances and retreats, they provide excellent potential for regional allostratigraphic correlation. The present understanding is that successive Hirnantian glacial events occurred within a single graptolite or chitinozoa biozone (*extraordinarius* and *elongata* Biozones respectively) (Paris *et al.*, 1995). Because no biostratigraphic subdivision is currently available, the best chronostratigraphic marker is probably the mid-Hirnantian transgression detailed above. This event of great amplitude is recorded as far south as the Djado area (northern Niger, location in Fig. 1C). Here, fauna-rich marine shales, originally considered as pre-Hirnantian strata (Legrand, 1993), and now attributed to the *extraordinarius* Biozone (P. Storch, pers. comm., 2004) were deposited between two major glacial erosion surfaces (Denis *et al.*, 2007). These data imply that the transgression advanced deep into North Africa at least within erosional troughs. This transgression is also recognised in non-glaciated area such as Sardinia (e.g. Storch & Leone, 2003).

This transgressive interval, coeval with a major deglaciation event, subdivides the Hirnantian into two first-order glacial cycles (Sutcliffe *et al.*, 2000) (Fig. 5C). Each cycle includes a limited number of glacial phases, the correlation of which throughout the North Gondwana is at present more controversial. At least two glacial phases are recognised within both the first (early Hirnantian) and the second (late Hirnantian) glacial cycles. Major ice recessions (~500 km) occurred between two successive phases, to compare with the mid-Hirnantian deglaciation associated with ice-front retreats >1000 km, possibly much more (Moreau, 2005).

Ice sheet size: the glacial maximum

Accurate reconstruction of pre-Pleistocene ice sheets, and the demonstration of synchronous, continent-wide ice fronts, is not straightforward. The size of both Permo-Carboniferous (Eyles *et al.*, 2003) and Neoproterozoic (Eyles and Januszack, 2004) ice sheets has recently been challenged, because their size may have been affected by enhanced polar wander or rates of rift propagation. Even assuming a maximum duration for the Hirnantian of 2 Ma (Webby *et al.*, 2004), and although the opening of the Palaeotethys (rifting of the Hun superterrane of Stampfli and Borel, 2002) occurred in Late Ordovician time, we argue here that these processes have not had a substantial impact on Hirnantian ice-sheet reconstructions. Synchronous growth of ice sheets, at least during the most protracted phase of glaciation, was likely from Mauritania to Turkey. This stands in contrast to Permo-Carboniferous centres of glaciation, which waxed and waned asynchronously over an interval of c. 55 Ma (Eyles *et al.*, 2003).

A reconstruction of the maximum size of the West Gondwana ice sheets, based on the occurrence of a regionally correlatable glacial erosion surface

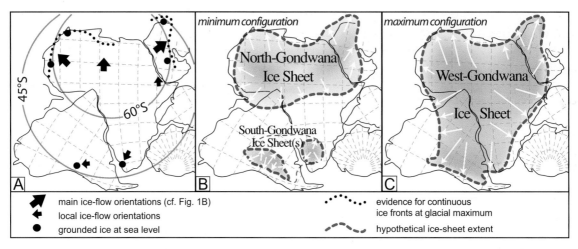

Fig. 7 The extent of the Late Ordovician glaciation. (A) Locations of identified subglacial deformation zones in the more distant areas from the ice centres. On the North Gondwana platform, these are from west to east: Tagant (western Mauritania), Rehamna (northern Morocco), Eastern Taurus (south-eastern Turkey) and Central Arabia. These four locations were glacierised at the Hirnantian glacial maximum, implying that grounded ice occurred at sea level close to the 60°S parallel. To include glacial successions from South America and South Africa (potentially of Hirnantian age) implies grounded ice occurring at sea level within the 45°S parallel. (B) The minimum-sized Hirnantian ice sheets, with a large North Gondwana Ice Sheet, and possibly penecontemporaneous subordinate ice centres in South America (linked with a pre-cordilleran setting) and South Africa, which may have coalesced together. (C) The maximum-sized Hirnantian West-Gondwana Ice Sheet, assuming fully coalescent synchronous glaciers.

following the major mid-Hirnantian interglacial, is presented in Fig. 7A. Ice-front migrations from place-to-place during the same glacial event (e.g. Boulton and Clark, 1990; Boulton *et al.*, 2001) means that ice fronts at the glacial maximum would have been best represented as an envelope rather than a continuous line. Late Ordovician ice fronts were continuous from Mauritania to western Libya. The existence of a continuous ice front between western Libya and Arabia is controversial, but likely at least during the most protracted phase of glaciation, which also affected southern Turkey (Monod *et al.*, 2003). Although Upper Ordovician glacial successions are also known in South Africa (e.g. Hiller, 1992) and South America (Caputo, 1998; Diaz-Martinez *et al.*, 2001), dating uncertainty still exists for the related glacial successions (e.g. Boucot *et al.*, 2003 for the South African example).

At the Hirnantian glacial maximum, the sedimentary record supports the concept of a continuous ice sheet across North Africa that possibly extended into Arabia, with WNW-to-NE oriented ice flow directions (Figs. 1B & 7A). A minimum-sized ice sheet at the glacial maximum had to incorporate a roughly symmetrical ice-front line, depicting a North Gondwana Ice Sheet (the

Saharan Ice Sheet of Young *et al.*, 2004), characterised by southward-flowing ice in sub-Saharan Africa (Fig. 7B). As ice flowed from south to north in Eritrea and Ethiopia (Kumpulainen, 2005; Fig. 7A), this minimum-sized ice sheet must have reached Central Africa. A maximum-sized Hirnantian ice sheet may, in addition, have overridden South America and South Africa where subglacial deformation zones have been also identified (Blignault, 1981; Martinez, 1998; Le Heron *et al.*, 2004; Deynoux and Ghienne, 2005) (Fig. 7C). The maximum-sized ice-sheet scenario envisages a huge West Gondwana Ice Sheet (Vaslet, 1990; Sutcliffe *et al.*, 2000; Ghienne, 2003) centred above Central Africa where no Upper Ordovician sediments crop out.

THE NORTH GONDWANA PLATFORM DURING THE HIRNANTIAN

This section proposes the subdivision of the North Gondwana platform, from the ice centre to lower palaeolatitudes, into several palaeogeographic domains based on the number of glacial erosion surfaces preserved. A second palaeogeographical scheme is then presented, providing a subdivision of the ice sheet perpendicular to ice-flow lines based on the occurrence of ice-stream-generated glacial troughs.

Palaeogeographic domains of the glaciated shelf

The number of glacial erosion surfaces distributed regionally define five proximal-to-distal palaeogeographic domains in North Gondwana (Fig. 1C). The continental interior corresponds to a domain that was not deglaciated until the close of the Hirnantian. The dominance of subglacial and post-glacial erosion in these areas means that the glacial record in this domain is limited and restricted to small-scale, fault-bounded intracontinental basins (Konaté *et al.*, 2003). It is conceivable that the continental interior may have been glacierised both prior and after the Hirnantian glaciation.

The inner, middle and outer domains of the glaciated continental shelf were subject to multiple phases of ice-sheet growth and decay. On the inner glaciated shelf, typified by the Niger succession (Denis *et al.*, 2007), two glacial erosion surfaces and associated overlying sequences are recorded, and, are separated by the mid-Hirnantian graptolitic shales (see above). This domain was deglaciated during the mid-Hirnantian, but remained glacierised during phases of ice-front retreat within each of the two first-order glacial cycles. Glacio-isostacy should have played a significant role in developing the stratigraphic architecture. The middle glaciated shelf corresponds to the area subjected to from 2 to 5, possibly more, phases of subglacial erosion. In this zone, encompassing the well-known successions in Mauritania, Algeria and Libya, both glacio-eustatic and glacio-isostatic processes were operative, and the most complete record of multiple advance and retreat events of the Hirnantian ice sheets is preserved. The outer glaciated shelf, including northern Morocco and Turkey, was glacierised only during the most protracted advance phase of the Hirnantian glacial maximum (Figs. 1C and 7A). Glacio-eustacy was probably more important than glacio-isostacy in these regions (e.g. Boulton, 1990). The shelf areas beyond the outer ice-sheet limit comprise a sedimentary record controlled solely by glacio-eustacy. In these areas, distal glaciomarine sediments are locally preserved (Fig. 1B; e.g. Sardinia, Brittany,

Bohemia: Robardet and Doré, 1988 and references therein; western Taurus, Turkey: Monod *et al.*, 2003; northernmost Morocco: Le Heron *et al.*, 2007).

Ice sheet configuration: the case study of Algeria

Legrand (1974, 1985) published a map of Algeria showing well-defined, 50–200 km wide, so-called sediment 'thicks' of 'upper Caradoc-Ashgill-lower Silurian' strata, separated by 150 to 300 km wide sediment 'thins'. This map is reproduced in Fig. 8A including data collected by the authors from Morocco (Anti-Atlas, Tamlelt area), Ougarta (NW Algeria) and Tassili n'Ajjer (western Murzuq Basin, Libya). It excludes smaller-scale variations, such as steep contour gradients over <10 km that could be attributed to tunnel valley incisions or fluvial incised valleys. Recent biostratigraphic work confirms that the strata outlined by Legrand (1985) are well constrained to Hirnantian syn-glacial strata (Paris *et al.*, 1995; Oulebsir and Paris, 1995; Vecoli and Le Hérissé, 2004).

It seems likely that areas with thick Hirnantian successions outline depositional lows and areas typified by thin successions designate palaeo-highs. North of the Hoggar, the data reviewed above are interpreted as a series of Late Ordovician glacially formed erosional troughs separated by plateaux or 'interfluves'. Four glacial troughs are identified across the Algerian middle glaciated shelf, from Ahnet to Tassili n'Ajjer, where the glaciogenic sediments reach >300 m in thickness. It is stressed that the isopach maps may mask the amalgamation of multiple glacial cycles in the stratigraphic record. In the interfluves, with the exception of deep but narrow palaeovalleys, the contact between pre-glacial and syn-glacial strata is stratigraphically higher.

The glacial troughs are very difficult to identify in the field as they are comparable in scale to tectonic structures (e.g. zones of preferential subsidence or post-depositional lithospheric bending). However, the easternmost glacial trough on Fig. 7A is the northward extension of a glacial trough already described in western Libya (Moreau *et al.*, 2005). This well-defined trough is >200 km in length and >80 km in width, and the depth of glacial erosion is up to 300 m, despite residual relief forms occurring on the upper surface of pre-glacial strata (Fig. 5). This trough contains several glacial surfaces

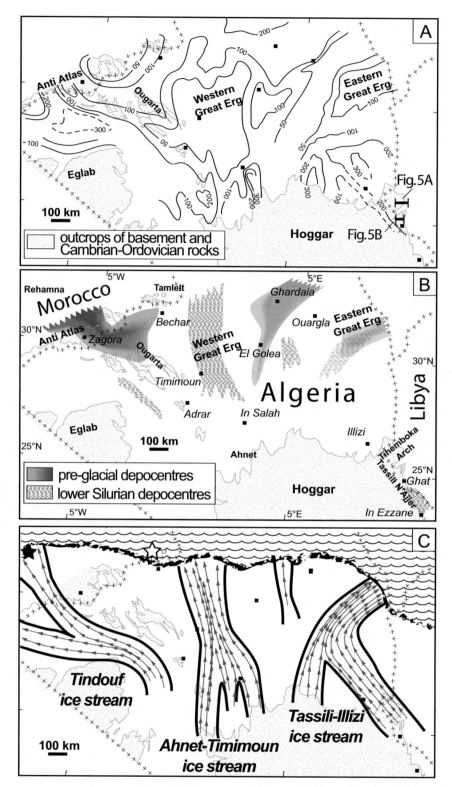

Fig. 8 Glacial troughs and ice streams in Algeria. (A) Isopach map corresponding to the syn-glacial Hirnantian strata (modified from Legrand, 1974). Erosional troughs are identified at the northern edge of the Hoggar Massif. (B) Pre-Hirnantian preferential depocentres (from Legrand, 1974 and Destombes *et al.*, 1985) and earliest to Early Silurian depocentres corresponding to depositional lows (from Klitsczh, 1981; Legrand, 1985; Lüning *et al.*, 2000). (C) Potential locations of major ice streams in Algeria as inferred from the locations of glacial erosional troughs and ice-flow orientations (cf. Fig. 1B). The black star is the northernmost glacierised area (Le Heron *et al.*, 2007); the white star indicates an area that was not glaciated (Tamlelt area, southeastern Morocco, unpublished data). Ice-stream fronts correspond to the Hirnantian glacial maximum. They are tentatively figured as shallow-marine ice fronts with no ice shelves.

characterised by mega-scale glacial lineations (Fig. 4A) (Clark, 1993; Stokes and Clark, 2001). These lineations have elongation ratios ranging from 5 to >20 and are 1–30 m in height. The lineations make up a large-scale undulating morphology on the glacial erosion surface, which is superimposed on wide elongated ridges separated by straight valley-like depressions (Figs 3C and 5). The morphology, comparable to the landforms produced by former Antarctic (e.g. Canals *et al.*, 2000; Anderson *et al.*, 2002; Evans *et al.*, 2004) or Scandinavian (Ottesen *et al.*, 2005) ice streams, defines a Late Ordovician ice-stream pathway (Moreau *et al.*, 2005). As a working hypothesis, the other glacial troughs of Algeria are interpreted as palaeo ice-stream pathways.

North of the glacial troughs, the general thinning of the sediment wedge is interpreted either as a decrease in the depth of erosion, or a decrease of sediment thickness in glacial troughs that are gradually underfilled. The latter scheme makes sense with the occurrence of elongate areas that Legrand (1985, 2003) and Lüning *et al.* (2000) have mapped, where lower Silurian organic-rich shales preferentially occur (Fig. 7B). These elongate areas are interpreted as depositional lows within which postglacial lower Silurian deposits may have accumulated, in contrast to highs that were not covered by sediments until middle Silurian time. Those depositional lows are interpreted as reflecting glacial troughs that remained underfilled after the ice-sheet retreat.

Connecting the overfilled glacial troughs of the Hoggar Massif (Fig. 8A) with underfilled depressions containing lower Silurian deposits to the north (Fig. 8B), a network of ice streams is reconstructed at regional scale (Fig. 8C). Hence, the Late Ordovician ice sheet over Algeria and western Libya was drained by a number of north-flowing ice streams separated by more stagnant ice in inter-stream areas. The shelfbreak extended a long distance from the ice fronts (Fig. 8C) and the existence of a shallow shelf (<200 m deep) may have restricted the formation of extensive frontal ice shelves.

More subtle depocentres are found northward in the Algerian outer glaciated shelf. However, they coincide spatially with pre-glacial depocentres (Fig. 8B) and are therefore not interpreted as erosional features. They cannot be considered as proxies for distal ice-stream pathways.

STRATIGRAPHIC ARCHITECTURES IN THE LATE ORDOVICIAN GLACIAL RECORD

Distinctive stratigraphic architectures faithfully reflect the position of each area on the shelf. For instance, low accommodation space and repeated 'cannibalisation' characterised the middle glaciated shelf, while restricted glacial erosion and greater accommodation space prevail on the outer glaciated shelf. This palaeogeographic subdivision of the glaciated shelf, parallel to ice flow, is coupled with a flow-transverse subdivision of the ice sheet into ice-stream troughs and inter-stream areas. That means a great deal of lateral variability should exist in the Late Ordovician glacial record. In the following section, the stratigraphic architectures of ice-stream pathways and inter-stream areas are compared for the middle and outer glaciated shelf.

Ice-stream-related depositional systems

The middle glaciated shelf: western Libya

In the Murzuq Basin (Libya), the depositional succession of the Hirnantian glacial record has been studied extensively (McDougall and Martin, 2000; Deynoux *et al.*, 2000; Sutcliffe *et al.*, 2000; Ghienne *et al.*, 2003; Le Heron, 2004; Le Heron *et al.*, 2004; El-ghali, 2005; Le Heron *et al.*, 2006). Most of the interpretations are based on field studies conducted in the Gargaf area, in the northern Murzuq Basin. In the following, we developed to some extent the glacial record of the western Murzuq Basin (Eastern Tassili n'Ajjer and Tihemboka Arch, see location in Fig. 8B). It characterises the stratigraphic architecture in ice-stream pathways on the middle glaciated shelf.

Despite partial reworking during later glacial cycles, 4–5 glacial depositional sequences are recognised, each separated by subglacial unconformities (Deynoux and Ghienne, 2004; Moreau *et al.*, 2005). The erosion depth of an individual unconformity is up to 200 m. The total thickness of the syn-glacial succession approaches 250–300 m. These sequences record the progressive infilling of glacially cut palaeovalley-like depressions. The long axis of younger depressions deviates from older examples (e.g. Smart, 2000), resulting in a complex depositional architecture (Figs 4A and 5) akin to Late Cenozoic alluvial terraces (e.g. Blum & Törnqvist, 2000).

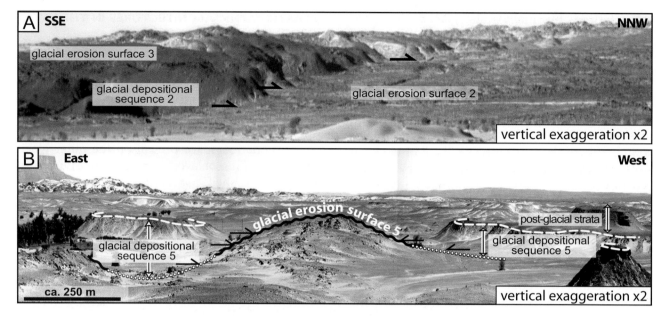

Fig. 9 Onlap relationships (black arrows) in the syn-glacial strata of the Eastern Tassili n'Ajjer (Ghat area, western Murzuq Basin). (A) A glacial depositional sequence onlaps on a glacial erosional surface with mega-scale glacial lineations (see location Fig. 4A). It is made up of fine-grained deposits at the base (distal glaciomarine, prodelta), truncated by fluvial sandstones at the top (depositional sequence 2 in Fig. 5B). (B) Glaciomarine and shelf deposits onlapping on a surface with mega-scale glacial lineations (depositional sequence 5 in Fig. 5B).

Depositional sequences 1–3. In the Eastern Tassili n'Ajjer (Ghat area), sequences 1–3 comprise coarsening upward successions, each 30–100 m thick, bounded above and below by a glacial erosion surface (Figs 5B and 9A). Sequences 1 and 2 belong to the first glacial cycle, and sequence 3 to the second glacial cycle (Fig. 5C). The lowermost 20–75 m of each sequence comprises crudely laminated micaceous sandstones and shales with occasional lonestone-bearing diamictite horizons. These are interpreted as distal glaciomarine (plume) deposits or prodelta sediments. The middle section, 5–20 m thick, rests with a sharp or rapid transitional contact with the underlying section, and comprises well sorted medium-grained to poorly sorted coarse-grained sandstones. These deposits contain current and wave ripples, trough or tabular cross-laminae and plane beds with parting lineations. They are interpreted as shallow-marine to distal alluvial plain deposits. The upper part of each sequence includes sharp-based, coarse- to very coarse-grained sandstones bearing numerous internal erosional surfaces. Cross-laminated sandstones grade laterally into fine- to medium grained sandstones bearing horizontal to low-angle lamination with

parting lineations and climbing ripple cross-lamination. These are interpreted as a series of amalgamated fluvial channels and associated over-bank sediments, deposited on a flood-dominated aggrading braid-plain.

Each depositional sequence is considered to record (i) rapid retreat of a marine-terminating ice front, followed by (ii) flooding of the glacial erosion surface, then (iii) progradation of a fluvial-delta-shelf system. A braid-plain separated the retreating, land-terminating ice margin from the shoreline. An abrupt facies change occurs within the prograding succession, either pointed out by sharp-based shallow marine deposits or by erosionally based fluvial deposits. In each of these sequences, a sea-level fall or a fluvial incision may be attributable to the effect of glacio-isostacy. Fluvial aggradation is concomitant with renewed progradation.

Depositional sequence 4. This sequence is associated with the Hirnantian glacial maximum that occurred during the second glacial cycle (see above) (Fig. 5C). In the Tihemboka Arch, a thick sandstone wedge has been deposited on the glacial erosion surface (Fig. 5A). It is overlain by a sandstone sheet, which can be followed to the south into the Eastern Tassili n'Ajjer, resting here directly on the basal glacial erosion surface (Figs 5A and B).

The sandstone wedge, 75–150 m thick, displays a coarsening and thickening-upward succession. The sandstones pass upwards from fine-grained massive sandy deposits, through a poorly stratified, medium-grained section characterised by sharp-based, horizontally laminated, graded, and de-watered sandstones, to low-angle or horizontally laminated deposits at the top. This succession is overlain by a channel-overbank system comprising cross-bedded medium- to coarse-grained sand-stones. Channel fill structures, 10–40 m in thick-

ness, 0.2–1 km in width, have transitional contact with overbank facies and any levée structures have been identified. These sandstones contain abundant subcritical climbing megaripples in 2–20 m thick beds (Figs 10A and B), associated with deep-sided cut-and-fill structures and undulating bedforms that show a steep climbing (40°) to aggradational (90°) superposition (Fig. 10C). Large glaciotectonic structures in the form of thrusted fold belts (20–40 m in height) are also preserved within these strata.

The coarsening sandstone wedge described above is capped by a medium-grained sandstone sheet, extending from the Eastern Tassili n'Ajjer

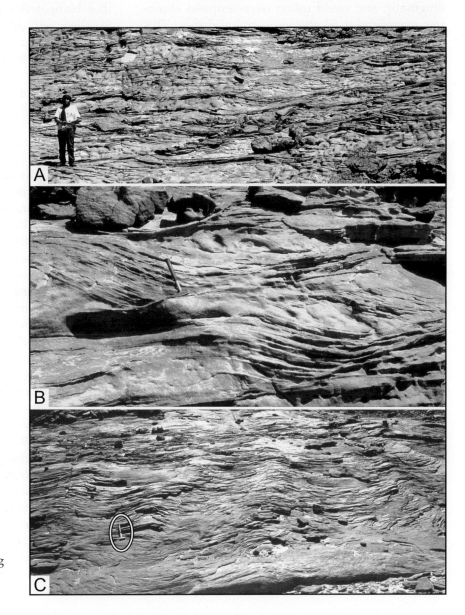

Fig. 10 Flood (possibly outburst)-dominated facies in an ice-stream-related outwash fan (glacial depositional sequence 4, Tihemboka Arch, western Murzuq Basin). (A) 2D climbing dunes composed of medium- to coarse-grained sandstones forming a 10 m-thick depositional package. (B) Close up view of facies illustrated in (A). (C) Nearly symmetrical, vertically climbing, large-scale bedforms interpreted as oversupplied climbing dunes or alternatively antidunes (encircled hammer for scale).

to the Tihemboka Arch. Its internal architecture is characterised by channel-fill structures and epsilon cross-stratification. The size of these structures increases towards the north. Horizontal lamination, climbing ripples and megaripples are ubiquitous.

Within the sandstone wedge, the climbing megaripples suggest deposition from highly concentrated sediment-laden streamflows with high rates of sediment fall-out. Such highly concentrated flows are most probably achievable by long-lived, turbulent glacial outburst floods (Russell and Arnott, 2003; Russell et al., 2003). The origin of undulating climbing to aggradational bedforms (Fig. 10C) is enigmatic, and could reflect over-supplied climbing megaripples or alternatively antidunes. This last interpretation has been proposed for correlative deposits near Djanet, Algeria (Hirst et al., 2002). Distal counterparts of this flood-dominated environment comprise poorly stratified to massive, largely dewatered sandstones deposited in environments dominated by high-density sandy turbidites. The overall sandstone wedge is interpreted as an ice-contact, flood-dominated outwash fan. The channel structures form the distributary system, which was directly linked with the subglacial drainage system. It is uncertain whether the sandstone wedge was deposited on a large submarine outwash related to a marine ice front or as part of a fan delta system related to an ice front terminating on land. In either case, its scale exceeds comparable depositional systems in the literature (Powell, 1990; Lønne, 1995; Lønne et al., 2001) by one to two orders of magnitude. The size and thickness of the sand wedge suggests that the ice-sheet retreat occurred in the study area much more slowly than for sequences 1–3. This suggests that deposition occurred during a phase of stabilisation of the ice sheet. The overlying sandstone sheet was deposited by a regional-wide sinuous to meandering, sand-dominated fluvial system when ice-front position retreated further to the south and a delta most probably developed to the north.

Depositional sequence 5. This sequence is restricted to the Ghat area (Eastern Tassili n'Ajjer) and is considered as compelling evidence for a locally developed, fifth phase of ice-sheet advance. To the north of a boundary zone corresponding to the maximum ice advance, a sandstone wedge that is similar to but thinner than that of sequence 4,

contains a channel-fill network and associated overbank deposits. To the south of the boundary zone, a glacial erosion surface bears fan-shaped drumlins with associated subglacial channel fills, both of which cross-cut older glacial lineations (Moreau et al., 2005). A thin veneer (5–20 m) of distal glaciomarine deposits caps this surface in the axis of glacially cut depressions that remained essentially underfilled after the ice sheet retreated (Figs 3A, 5B and 9B).

This depositional sequence had preserved the spatial relationship between a subglacial drainage network to the north and the proglacial outwash system to the south. The thin glaciomarine deposits that blanket the glacial surface suggest a rapid ice-sheet withdrawal. Post glacial evolution in a very last phase of Ordovician sedimentation was characterised by tide-dominated reworking processes and the deposition of a transgressive wedge on the flanks of underfilled palaeovalleys (Figs 3A and 5B).

The outer glaciated shelf: High Atlas and Meseta (Morocco)

Glacial troughs have not been found at outcrop on the outer glaciated shelf. However, a large depocentre of marine sediments has been identified in northeastern Morocco (Tazzeka Massif, Khoukhi and Hamoumi, 2001) beyond the maximum ice-front position on the non-glaciated shelf. Here, a relatively deep (200–400 m) marine fan was deposited at the continental margin (Le Heron et al., 2007). This fan system is not an analogue to the trough-mouth fans identified in front of Quaternary ice streams beyond the shelfbreak (Vorren and Laberg, 1997), within which sedimentation is dominated by very thick, poorly evolved diamictite facies. However, focused and massive sand-dominated deposition is interpreted as the signature of a sediment delivery system fed by an ice stream.

To the south, on the outer glaciated shelf, the two major Hirnantian glacial cycles are recorded, with sediments related to the first (lower) glacial cycle being somewhat more ice-distal than those of the second (upper) glacial cycle. Only the latter is discussed below. Further details are given in Le Heron et al. (2007).

In the Rehamna inlier, a thin sandy diamictite interpreted as a till is observed in close association with subglacial features such as glaciofluvial

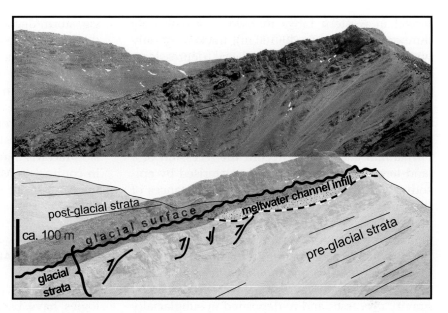

Fig. 11 The glacial succession in the High Atlas (Morocco). The following events are illustrated: (i) glaciotectonic deformation involving glaciomarine outwash fan facies, deposits of the first glacial cycle and subordinate pre-glacial strata, (ii) incision of a subglacial channel and sandy diamictite infill, (iii) truncation by a glacial erosion surface comprising streamlined bedforms (not illustrated, see Le Heron *et al.*, 2007).

sandstones, sand intrusions and a striated surface. This till horizon rest above thin outwash facies and is overlain by well developed glaciomarine diamictite facies, up to 20 m thick. In the High Atlas, c. 100 km further to the south, the sedimentary succession of the second glacial cycle begins with thick outwash facies comprising sandstones interstratified with massive to poorly stratified, sandy, clast-poor diamictites (Ouanaimi, 1998). These sediments are deformed by glaciotectonic thrusts and gravitational loads, and are truncated by 20–40 m thick, 100–300 m wide channel structures (Fig. 10). The latter are filled with stratified coarse-grained sandstones, sandy diamictites and subordinate conglomerate lenses, including boulder-sized clasts of local origin. These deposits are capped by a glacial erosion surface with subglacial streamlined bedforms overlain by a thin (<2 m) diamictite.

The Rehamna succession, probably in close proximity to the maximum ice-front position, reflects more ice-marginal conditions than the High Atlas succession (Le Heron *et al.*, 2007). Glaciomarine diamictites overlying the glacial surface in the Rehamna were probably deposited during an earlier phase of ice-front recession. This stage has immediately preceded an ice-sheet collapse that prevented further glacial sedimentation over the area, as suggested by the thin 'post-glacial' glaciomarine sediment blanket above the underlying glacial erosion surface in the High Atlas.

Combining observations from the three study areas (Tazzeka, Rehamna and High Atlas), the Moroccan succession related to the Hirnantian glacial maximum records: (i) deposition in a submarine, ice-contact outwash fan dominated by mass-flow diamictites and sandy turbidites, (ii) glaciotectonic deformation and incision by channels suggesting subglacial, focused, sediment input points, (iii) subglacial conditions characterised by fast-flowing ice indicated by streamlined bedforms on the outer glaciated shelf, whereas an ice-stream-related 'deep' marine fan developed on the non-glaciated shelf, and (iv) a rapid ice-sheet withdrawal. A depositional succession related to a glacial advance on a shallow-marine shelf, followed by the development of an ice stream and a subsequent ice-stream collapse is then apparent.

Depositional systems in inter-stream areas

The middle glaciated platform: Mauritania

Syn-glacial strata in eastern and western Mauritania (Hodh and Adrar, Fig. 1) are organised into four laterally juxtaposed and vertically superimposed depositional units (Ghienne, 2003). As in Libya, each depositional sequence rests on a glacial erosion surface and records essentially an ice-sheet recession with the influence of glacial processes rapidly disappearing up-section. Key differences with the record in western Libya include (i) a markedly thinner succession (c. 40–100 m), thickening only within palaeovalleys or subglacial depressions, (ii) widespread palaeorelief forms on the upper surface of pre-glacial strata, (iii) better developed fluvio-glacial, fluvial and delta facies, and (iv) poorly developed offshore marine facies.

In the Hodh area, syn-glacial strata of the first Hirnantian glacial cycle record the recession of a land-terminating ice margin characterised by episodic, high-frequency retreats and re-advances of the ice front, resulting in a complex depositional architecture (Deynoux, 1985; Ghienne, 2003). This architecture includes aggrading outwash sediments deposited near a stagnating ice front that is partly glaciotectonised by minor re-advances. Both tills and some glaciofluvial material were reworked by gravity processes and re-deposited in subglacially overdeepened zones (Ghienne, 1998). The final phase of ice retreat was accompanied by the incision and subsequent infilling of large subglacial or proglacial meltwater channels (Ghienne, 2003).

During the second glacial cycle, a rather thin (<50 m) succession was deposited, with the exception of tunnel-valley infills (Ghienne and Deynoux, 1998; Fig. 3B). The last depositional sequence lacks any glacial features in western Mauritania where erosion prevailed in shallow-marine environments. In the east, in the Hodh area, the sequence comprises a fining-up glaciomarine succession, characterised by thin (<10 m) submarine outwash-fan deposits at its base, truncated by nearshore sediments. This sequence indicates a significant relative sea-level fall ascribed to a late post-glacial isostatic rebound (Ghienne, 1998, 2003). It occurred, nevertheless, before the end of the Hirnantian (Paris *et al.*, 1998; Underwood *et al.*, 1998).

The outer glaciated shelf: Turkey (Taurus Mountains)

This area is characterised by laterally extensive (up to 100 km), superimposed depositional units (Monod *et al.*, 2003). Relatively clast-rich sandy diamictites, representing mass-flow or low- to high-density sandy turbidites, characterise glaciomarine outwash environments in front of a marine-terminating ice margin. The absence of large-scale channel structures suggest a limited availability of meltwater and line, rather than point, sources (e.g. grounding-line fans). Non-glacial shelf processes,

e.g. storm deposits, are also present. A rather condensed shelf sedimentation regime not influenced by glacial processes, occurred during interglacial periods.

The glacial maximum is characterised by the deposition of a thin (<0.2 m) till horizon overlying a striated pavement. Upwards, a progressively fining-up, lonestone bearing glaciomarine succession records a gradual ice-front retreat. Therefore, in contrast to Moroccan successions, no evidence for a rapid ice-sheet withdrawal is identified in Turkey.

PLATFORM-SCALE FACIES MODEL

Only rarely have sequence stratigraphic methodologies have been applied successfully in glacial settings (e.g. Proust and Deynoux, 1993; Brookfield and Martini, 1999; Powell and Cooper, 2002; El-ghali, 2005). Large-scale facies models for glacial sedimentary systems are thus still needed. When developed, they will form the basis of a more robust sequence stratigraphic approach to ancient glacial successions.

The database presented herein is regionally comprehensive (Fig. 1A) and should facilitate correlation of Late Ordovician depositional sequences across the North Gondwana platform. This correlation is complicated both by the number of palaeogeographic domains (e.g. inner, middle, outer glaciated shelves) and whether each study area lies within an ice-stream pathway or an inter-stream area. Based on the glacier dynamics inferred for each study area, a facies model illustrating the stratigraphic architecture of a glacial sequence is now proposed for two parallel S-N profiles (Fig. 12). The first set of diagrams on the left illustrates the stratigraphy within an ice-stream pathway, whereas the set on the right depicts an adjacent co-evally evolving inter-stream area. The model captures four phases in the temporal evolution of a glacial sequence, namely the glacial advance, the glacial maximum, the ice-margin recession and the interglacial minimum. The two points of maximum ice advance and ice retreat, as well as the marine limit, are not fixed palaeogeographically. They are located at different positions across the platform according to the amplitude of both the maximum and minimum ice extent associated with each glacial phase.

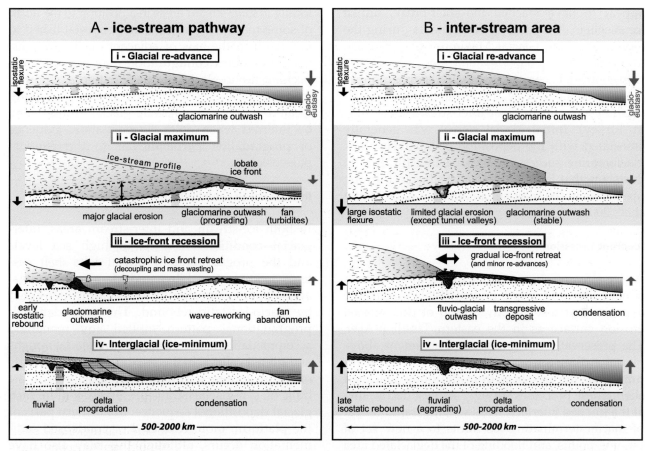

Fig. 12 An idealised glacial depositional sequence corresponding to a single glacial phase (glacial advance and subsequent retreat) in (A) ice-stream-related depositional environments (left, mainly based on the Libyan case study), and (B) inter-stream areas (right, mainly based on the Mauritanian case study, Ghienne *et al.*, 2003). (i) Glacial re-advance. (ii) Glacial maximum and initial retreat. (iii) Ice-front recession. (iv) interglacial (ice-minimum). The stratigraphic superposition of several glacial sequences (2–5), and the potential lateral migration of ice streams result in an intricate glacial record (e.g. Fig. 5). Post-glacial transgressive deposits are not figured. Vertical exaggeration: ~1000.

Glacial advance

This stage is generally poorly recorded on the glaciated platform where regressive successions are poorly developed possibly due to a rapid sea-level fall. Immediately prior to the ice maximum, glaciomarine outwash sediments are deposited in some areas such as northern Morocco. This infilling of accommodation space probably supported the ice sheet as it advanced northward on the shallow-marine platform, and was affected by concomitant fall in sea-level (e.g. Dahlgren *et al.*, 2002). It is not possible to ascertain whether ice streams had developed at this stage.

Glacial maximum and initial glacial retreat

While glaciomarine deposition occurred at the ice margin, subglacial erosion surfaces developed to the south. They are characterised by foredeepened depressions, glaciotectonic thrusts and folds, mega-scale glacial lineations and meltwater channels.

Focused sediment input-points located in front of ice-stream systems, may have been locally connected to some deeper turbiditic fans beyond a shelf edge inherited from the pre-glacial shelf architecture (e.g. in northern Morocco). This shelf edge may have controlled the maximum ice-stream advance (Le Heron *et al.*, 2007), which does not appear to have reached the shelf-break, unlike ice sheets in Antarctica or Scandinavia during the Pleistocene Late Glacial Maximum (e.g. Ottesen *et al.*, 2005).

In contrast, linear and distributed sediment-input points built small-scale, diamictite-dominated outwash systems in front of inter-stream areas (e.g. in Turkey). Initial glacial retreat was generally associated with the deposition of relatively thick glaciomarine succession above a thin till. It is at this stage that large tunnel valleys began forming up-glacier in the inter-stream areas (e.g. Le Heron *et al.*, 2004).

Ice-front recession

As ice stream retreat, only a thin sediment cover remains, comprising fine-grained distal glaciomarine deposits that rest directly on top of each glacial erosion surface as in the eastern Tassili n'Ajjer. The preservation of fine-grained sediment above glacial erosion surfaces implies the sudden withdrawal of the ice sheet from the outer glaciated shelf. These processes strongly suggest decoupling at the ice/bed interface, mass-wasting, rapid retreat of marine-terminating ice fronts to a new set of pinning-points, and flooding of the deglaciated area (Eyles and McCabe, 1989; Anderson and Thomas, 1991). This suite of processes is in agreement with the interpretation of a glacial trough as a former ice-stream pathway, characterised by foredeepened topography where intense calving may have occurred. After initial rapid retreat, a more stable and slow retreat of ice fronts produced a renewed and voluminous ice-contact, flood-dominated, outwash system such as a submarine fan or fan delta (sequence 4, Tihemboka Arch). The latter system then evolves into a delta system once a flood-dominated fluvial plain becomes established in front of the land-terminating ice margin.

In inter-stream areas, the sediments from the ice-sheet recession stage comprise the tunnel-valley basal infill (Ghienne and Deynoux, 1998; Le Heron *et al.*, 2004). Glacial recession was more gradual as no sudden glacial withdrawal is observed (Turkey). With slow continuous retreat, a land-terminating ice margin deposited braided outwash sediments grading seaward into transgressive, shallow-marine deposits such as in Mauritania. During this time, a land-terminating ice margin may have persisted locally in inter-stream areas in a position to the north of some marine ice fronts that were established to the south in the axis of adjacent glacial troughs. Such a complex configuration may explain some peculiar fluvio-glacial drainage patterns. A source in a higher inter-stream area and a depocentre located within an adjacent glacial trough may be inferred when delta systems show evidence of progradation perpendicular to the regional palaeoslope

Interglacial (ice-minimum)

In both ice-stream and inter-stream areas, interglacial conditions resulted in high sea levels and the progradation of fluvial-delta-shelf systems as in western Libya and Mauritania. Fluvial aggradation occurs, locally temporarily interrupted by glacio-isostatic rebound. This aggradational-progradational pattern infilled any remaining accommodation space in glacially cut glacial troughs and tunnel valleys. In the proximal (southern) parts of the shelf, these sediments may actually form the bulk of syn-glacial strata preserved within each glacial depositional sequence. In northern parts of the platform, more restricted or condensed sedimentation occurs, although this may also have been fed by local highlands providing an additional sediment source (Le Heron *et al.*, 2007).

CONCLUSIONS

The model given above should be regarded as a first attempt to merge numerous observations and interpretations in various parts of Northern Gondwana. The architecture of Late Ordovician glacial depositional sequences is mainly controlled by their location within or outside an ice-stream-generated trough. Ice-stream-dominated systems are characterised by up to 300 m deep incisions, a greater sediment supply typified by flood-dominated outwash-fan sedimentation, and a sedimentary record indicating rapid ice-sheet collapse. In inter-stream areas, shallower erosion surfaces are observed and sediment as well as meltwater supply are less vigorous. The general depositional features that can be portrayed for a complete cycle of glacial advance and subsequent retreat include:

1 Ice-proximal, coarse-grained glaciomarine successions are best developed at or near the maximum ice-front position and characterise the outer glaciated shelf in both ice-stream and inter-stream-dominated areas. Relatively deep marine fans occurred beyond the shelf edge in the axis of ice streams;

2 During ice-sheet recession, ice-sheet collapse in ice-stream-generated troughs resulted in a relatively thin ice-distal glaciomarine sediment cover blanketing the glacial erosion surface. Gradual ice-front recession in adjacent inter-stream areas resulted in well developed fluvio-glacial to fluvial facies;

3 Outwash systems preserved in ice-proximal segments of ice-stream troughs mark transitional conditions showing an evolution from marine-terminating to land-terminating ice fronts in association with a slowing down in the rate of ice-sheet recession;

4 Much of the Hirnantian strata in the middle glaciated shelf essentially record late glacial retreat to interglacial phases. In general, they are finer-grained than strata deposited several hundreds kilometres to the north when the ice-front position was at its maximum.

Several factors have resulted in a complicated glacial record: the stratigraphic superimposition of several, erosionally based, glacial sequences (e.g. Fig. 5B); the possibility that ice streams of successive glacial phases do not necessarily coincide spatially; and Hirnantian glacially induced reactivation of pre-existing basement faults (Ghienne *et al.*, 2003; Turner *et al.*, 2005; Denis *et al.*, 2007). The model presented herein should facilitate a better integration of outcrop and subsurface data, guiding the sequence stratigraphic interpretation of Late Ordovician glacial deposits and other ancient glacial successions. Future research looks forward to the development of a detailed chronostratigraphic framework, which will be the subject of a forthcoming paper.

ACKNOWLEDGEMENTS

This regional scale study would not have been possible without the combined support of TOTAL, ENI, NOC, REMSA, SONATRACH and TPAO for financial and logistical support necessary for both the fieldwork and writing-up phases of this work. This research was also made possible by the Centre National de la Recherche (CNRS, programme Eclipse). Daniel Paul Le Heron publishes with the support of the CASP Libya Project Consortium. The manuscript benefited from considerable improvements following the thoughtful reviews of Dag Ottesen and Risto Kumpulainen, and from the editorial support of Mike Hambrey. This is a contribution of EOST (Ecole et Observatoire des Science de la Terre), number 2006.501 – UMR 7517.

REFERENCES

Abed, M.A., Makhlouf, I.M., Amireh, B.S. and Khalil, B. (1993) Upper Ordovician glacial deposits in southern Jordan. *Episodes*, **16**, 316–328.

Anderson, J.B. and Thomas, M.A. (1991) Marine ice-sheet decoupling as a mechanism for rapid, episodic sea-level change: the record of such events and their influence on sedimentation. *Sed. Geol.*, **70**, 87–104.

Anderson, J.B., Shipp, S.S., Lowe, A.L., Wellner, J.S. and Mosola, A.B. (2002) The Antarctic Ice Sheet during the Last Glacial Maximum and its subsequent retreat history: a review. *Quatern. Sci. Rev.*, **21**, 49–70.

Beuf, S., Biju-Duval, B., De Charpal, O., Rognon P., Gariel, O. and Bennacef, A. (1971) *Les grès du Paléozoïque inférieur au Sahara.* 'Science et Technique du pétrole', **18**, Paris, 464 pp.

Blignault, H.J. (1981) Ice sheet deformation in the Table Mountain Group, Western Cape. *Annale Universiteit van Stellenbosch. Series A1 (Geology)*, **3**, 1–66.

Blum, M.D. and Törnqvist, T.E. (2000) Fluvial responses to climate and sea-level change: a review and look forward. *Sedimentology*, **47** (suppl.), 2–48.

Boote, D.R.D., Clark-Lowes, D.D. and Traut M.W. (1998) Paleozoic petroleum systems of North Africa. In: *Petroleum Geology of North Africa* (Eds D.S. Macgregor, R.T.J. Moody and D.D. Clark-Lowes), *Geol. Soc. London Spec. Publ.*, **132**, 7–68.

Boucot, A.J., Jia-Yu, R., Xu, C. and Scotese, C.R. (2003) Pre-Hirnantian Ashgill climatically warm event in the Mediterranean region. *Lethaia*, **36**, 119–131.

Boulton, G.S. (1990) Sedimentary and sea level changes during glacial cycles and their control on glaciomarine facies architecture. In: *Glacimarine environments: processes and sediments* (Eds J.A. Dowdeswell and J.D. Scourse), *Geol. Soc. London Spec. Publ.*, **53**, 15–52.

Boulton, G.S. and Clark, C.D. (1990) A highly mobile Laurentide Ice Sheet revealed by satellie images of glacial lineations. *Nature*, **346**, 613–817.

Boulton, G.S., Dongelmans, P., Punkari, M. and Broadgate M. (2001) Palaeoglaciology of an ice sheet through a glacial cycle: the European ice sheet through the Weichselian. *Quater. Sci. Rev.*, **20**, 591–625.

Brenchley, P.J., Carden, G.A., Hints, L., Kaljo, D., Marshall, J.D., Martma, T., Meidla, T. and Nolvak, J. (2003) High-resolution stable isotope stratigraphy of Upper Ordovician sequences: Constraints on the timing of bioevents and environmental changes associated with mass extinction and glaciation. *Geol. Soc. Am. Bull.*, **115**, 89–104.

Brookfield, M.E. and Martini, I.P. (1999) Facies architecture and sequence stratigraphy in glacially influenced basins: basic problems and water level/ glacier input-point controls (with an example from the Quaternary of Ontario, Canada). *Sed. Geol.*, **123**, 183–197.

Canals, M., Urgeles, R. and Calafat, A.M. (2000) Deep sea-floor evidence of past ice streams off the Antarctic Peninsula. *Geology*, **28**, 31–34.

Caputo, M.V. (1998) Ordovician-Silurian glaciations and global sea-level changes. *Bull. New York State Mus.*, **491**, 15–25.

Carr, I.D. (2002) Second-order sequence stratigraphy of the Palaeozoic of North Africa. *Mar. Petrol. Geol.*, **25**, 259–280.

Clark, C.D. (1993) Mega-scale glacial lineations and cross-cutting ice flow landforms. *Earth Surf. Proc. Land.*, **18**, 1–29.

Crossley, R. and McDougall, N. (1998) Lower Palaeozoic reservoirs of North Africa. In: *Petroleum Geology of North Africa* (Eds D.S. Macgregor, R.T.J. Moody and D.D. Clark-Lowes), *Geol. Soc. London Spec. Publ.*, **132**, 157–166.

Dahlgren, K.I.T., Vorren, T.O. and Laberg, J.S. (2002) The role of grounding-line sediment supply in ice-sheet advances and growth on continental shelves: an example from the mid-Norwegian sector of the Fennoscandian ice sheet during the Saalian and Weichselian. *Quatern. Int.*, **95–96**, 25–33.

Dean, W.T. and Monod, O. 1990. Revised stratigraphy and relationships of Lower Palaeozoic rocks, eastern Taurus Mountains, south central Turkey. *Geol. Mag.*, **127**, 333–347.

Denis, M., Buoncristiani, J.-F., Konaté, M., Ghienne, J.-F. and M. Guiraud. Hirnantian glacial and deglacial record in SW Djado Basin (NE Niger) (2007). *Geodynamica Acta*, **20**, 177–195.

Destombes, J. (1968a) Sur la présence d'une discordance générale de ravinement d'âge Ashgill supérieur dans l'Ordovicien terminal de l'Anti-Atlas (Maroc). *CR Acad. Sci. Paris*, **267**, 565–567.

Destombes, J. (1968b) Sur la nature glaciaire des sédiments du groupe du 2ᵉ Bani, Ashgill supérieur de l'Anti-Atlas, Maroc. *CR Acad. Sci. Paris*, **267**, D, 684–686.

Destombes, J., Hollart, H. and Willefert, S. (1985) Lower Paleozoic rocks of Morocco. In: *Lower Paleozoic Rocks of northwest and west Central Africa* (Ed. C.H. Holland), pp. 291–325, John Wiley, New-York.

Deynoux, M. (1980) Les formations glaciaires du Précambrien terminal et de la fin de l'Ordovicien en Afrique de l'ouest. Deux exemples de glaciation d'inlandsis sur une plate-forme stable. *Travaux des Laboratoires des Sciences de la Terre, St.Jérome, Marseille*, **B17**, 554 pp.

Deynoux, M. (1985) Terrestrial or waterlain glacial diamictites? Three case studies from the Late Precambrian and Late Ordovician glacial drifts in West Africa. *Palaeogeogr. Palaeoclimatol. Palaeoecol.*, **51**, 97–141.

Deynoux, M. and Ghienne, J.-F. (2004) Late Ordovician glacial pavements revisited – a reappraisal of the origin of striated surfaces.*Terra Nova*, **16**, 95–101.

Deynoux, M. and Ghienne, J.-F. (2005) Late Ordovician glacial pavements revisited – a reappraisal of the origin of striated surfaces. Discussion. *Terra Nova*, **17**, 488–491.

Deynoux, M. and Trompette, R. (1981) Late Ordovician tillites of the Taoudeni Basin, West Africa. In: *Earth's pre-Pleistocene glacial record* (Eds M.J. Hambrey, W.B. Harland), pp. 89–96, Cambridge University Press. Available on-line via www.aber.ac.uk/glaciology.

Deynoux, M., Ghienne, J.-F. and Manatschal, G. (2000) Stratigraphy and sedimentology of the Upper Ordovician glacially-related deposits of the western Gargaf High, Fezzan, Northern Libya. Unpublished explanatory booklet of the geological map, CNRS-EOST, Ecole et Observatoire des Sciences de la Terre, Université L. Pasteur, Strasbourg, 41 pp.

Diaz-Martinez, E., Acosta, H., Cardenas J., Carlotta, V. and Rodriguez, R. (2001) Paleozoic diamictites in the Peruvian Altiplano: evidence and tectonic implications. *J. S. Am. Earth Sci.*, **14**, 587–592.

Eschard, R., Abdallah, H., Braik, F. and Desaubliaux, G. (2005) The Lower Paleozoic succession in the Tasilli outcrops, Algeria: sedimentology and sequence Stratigraphy. *First Break*, **23**, 27–36

El-ghali, M.A.K. (2005) Depositional environments and sequence stratigraphy of paralic glacial, para-glacial and postglacial Upper Ordovician siliciclastic deposits in the Murzuq Basin, SW Libya. *Sed. Geol.*, **177**, 145–173.

Evans, J., Dowdeswell, J.A. and O'Cofaigh, C. (2004) Late Quaternary submarine bedforms and ice-sheet flow in Gerlache Strait and on the adjacent continental shelf, Antartica Peninsula. *J. Quatern. Sci.*, **19**, 397–407.

Eyles, C.H., Mory, A.J. and Eyles, N. (2003) Carboniferous–Permian facies and tectonostratigraphic successions of the glacially influenced and rifted Carnarvon Basin, western Australia. *Sed. Geol.*, **155**, 63–86.

Eyles, N. and Januszczak, N. (2004) 'Zipper-rift': a tectonic model for Neoproterozoic glaciations during the breakup of Rodinia after 750 Ma, *Earth-Sci. Rev.*, **65**, 1–73.

Eyles, N. and McCabe, A.M. (1989) The Late Devensian (<22000 BP) Irish Sea Basin: the sedimentary record of a collapsed ice sheet margin. *Quatern. Sci. Rev.*, **8**, 307–351.

Ghienne, J.-F. (1998) Modalités d'enregistrement d'une glaciation ancienne; exemple de la glaciation fini-ordovicienne sur la plate-forme nord-gondwanienne en Afrique de l'Ouest. Unpubl. PhD thesis, Université Louis Pasteur, Strasbourg, 407 pp.

Ghienne, J.-F. (2003) Late Ordovician sedimentary environments, glacial cycles, and post-glacial transgression in the Taoudeni Basin, West Africa. *Palaeogeogr., Palaeoclimatol., Palaeoecol.*, **189**, 117–145.

Ghienne, J.-F. and Deynoux, M. (1998) Large-scale channel fill structures in Late Ordovician glacial deposits in Mauritania, western Sahara. *Sed. Geol.*, **119**, 141–159.

Ghienne, J.-F., Bartier, D., Leone, F. and Loi, A. (2000) Caractérisation des horizons manganésifères de l'Ordovicien supérieur de Sardaigne: relation avec la glaciation fini-Ordovicienne. *CR Acad. Sci. Paris*, **331**, 257–264.

Ghienne, J.F., Deynoux, M., Manatschal, G. and Rubino J.L. (2003) Palaeovalleys and fault-controlled depocenters in the Late Ordovician glacial record of the Murzuq Basin (Central Libya), *CR Geosciences*, **335**, 1091–1100.

Grahn, Y. and Caputo, M.V. (1992) Early Silurian glaciations in Brazil. *Palaeogeogr. Palaeoclimatol. Palaeoecol.*, **99**, 9–15.

Hamoumi, N. (1988) La plate-forme ordovicienne du Maroc: dynamique des ensembles sédimentaires. Unpubl. PhD thesis, Université Louis Pasteur, Strasbourg, 237p.

Hiller, N. (1992) The Ordovician System in South Africa: a review. In: *Global perspectives on Ordovician geology* (Eds Webby, Laurie), 473–485 pp, Balkema, Rotterdam.

Hirst, J.P.P., Benbakir, A., Payne, D.F. and Westlake, I.R. (2002) Tunnel Valleys and Density Flow Processes in the upper Ordovician glacial succession, Illizi Basin, Algeria: influence on reservoir quality. *Mar. Petrol. Geol.*, **25**, 297–324.

Khoukhi, Y. and Hamoumi, N. (2001) L'Ordovicien de la Méséta orientale (Maroc): stratigraphie génétique-contrôle géodynamique, climatique et eustatique. *Africa Geosci. Rev.*, **8**, 289–302.

Klitzsch, E. (1981) Lower Palaeozoic rocks of Libya, Egypt, and Sudan. In: *Lower Palaeozoic of the Middle East, Eastern ans Southern Africa, and Antarctica* (Ed. C.H. Holland), pp. 131–163, John Wiley, New York.

Konaté, M, Guiraud, M., Lang, J. and Yahaya, M. (2003) Sedimentation in the Kandi extensional basin (Benin and Niger): fluvial and marine deposits related to the Late Ordovician deglaciation in West Africa. *J. Afr. Earth Sci.*, **36**, 185–206.

Kumpulainen, R.A. (2005) Ordovician glaciation in Eritrea and Ethiopia. International Conference on *Glacial Sedimentary Processes and Products*, University of Wales, Aberystwyth (22–27 august), abstract volume.

Le Heron D. (2004) The development of the Murzuq Basin, SW Libya, during the late Ordovician. Unpubl. PhD thesis, University of Wales, Aberystwyth.

Le Heron, D., Sutcliffe, O.E., Bourgig, K., Craig, J., Visentin, C. and Whittington, R. (2004) Sedimentary Architecture of Upper Ordovician Tunnel Valleys, Gargaf Arch, Libya: Implications for the Genesis of a Hydrocarbon Reservoir. *GeoArabia*, **9**, 137–160.

Le Heron, D.P., Sutcliffe, O.E., Whittington, R.J. and Craig, J. (2005) The origins of glacially related soft-sediment deformation structures in Upper Ordovician glaciogenic rocks: implication for ice sheet dynamics. *Palaeogeogr., Palaeoclimatol., Palaeoecol.*, **218**, 75–103.

Le Heron, D., Craig, J., Sutcliffe, O. and Whittington, R. (2006) Ordovician glaciogenic reservoir heterogeneity: an example from the Murzuq Basin Libya. *Mar. Petrol. Geol*, **23**, 655–677.

Le Heron, D., Ghienne, J.-F., El Houicha, M., Khoukhi, Y. and Rubino, J.-L. (2007) Maximum extent of ice sheets in Morocco during the Late Ordovician glaciation. In: *Lower Palaeozoic Palaeogeography and Palaeoclimate* (Eds Munnecke, A. and Servais, T.), *Palaeogeogr. Palaeoclimatol. Palaeoecol*, **245**, 200–226.

Legrand, P. (1974) Essai sur la paléogéographie de l'Ordovicien du sahara algérien. *Compagnie Française des Pétroles, Notes et Mémoires*, **11**, 121–138.

Legrand, P. (1985) Lower Palaeozoic rocks of Algeria. In: *Lower Palaeozoic of north-western and west-central Africa* (Ed. C.H. Holland), pp. 5–89, John Wiley, New-York.

Legrand, P. (1993) Graptolites d'âge ashgillien dans la région de Chirfa (Djado, République du Niger). *Bull. Centres Rech. Explor.-Prod. Elf Aquitaine*, **17**, 435–442.

Legrand, P. (2003) Paléogéographie du Sahara algérien à l'Ordovicien terminal et au Silurien inférieur. *Bull. Soc. Geol. Fr.*, **174**, 19–32.

Leone, F., Loi, A. and Pillola G.L. (1995) The post Sardic Ordovician sequence in SW Sardinia. In: *6th Palaeobenthos International Symposium* (Ed. A. Cherchi), *Guide book*, pp. 81–108, Cagliari University.

Lønne, I. (1995) Sedimentary facies and depositional architecture of ice-contact glaciomarine systems. *Sed. Geol.*, **98**, 13–43.

Lønne, I., Nemec, W., Blikra, L.H. and Lauritsen, T. (2001) Sedimentary architecture and dynamic stratigraphy of a marine ice-contact system. *J. Sed. Petrol.*, **71**, 922–943.

Lüning, S., Craig. J., Loydell, D.K., Storch, P. and Fitches, B. (2000) Lower Silurian 'hot shales' in North Africa and Arabia: regional distribution and depositional model. *Earth-Sci. Rev.*, **49**, 121–200.

Martinez, M. (1998) Late Ordovician glacial deposits of northwest Argentina: new evidence from the Mecoyita Formation, Santa Victoria Range. *J. Afr. Earth Sci.*, **27**, supplement 1, 136–137.

Massa, D. (1988) *Paléozoïque de Libye occidentale. Stratigraphie et Paléogéographie.* Unpublish. Thèse d'Etat, Université de Nice, 514 pp.

McDougall, N. and Martin, M. (2000) Facies models and sequence stratigraphy of Upper Ordovician outcrops in the Murzuq Basin, SW Libya. In: *Geological Exploration in Murzuq Basin* (Eds M.A. Sola and D. Worsley), pp. 223–236, Elsevier Science.

McGillivray, G.J. and Husseini, M. (1992) The Paleozoic petroleum geology of Central Saudi Arabia. *AAPG Bull.*, **76**, 1473–1490.

McLure, H.A. (1978) Early Paleozoic glaciation in Arabia. *Palaeogeogr., Palaeoclimatol., Palaeoecol.*, **25**, 315–326.

Monod, O., Kozlu, H., Ghienne, J-F., Dean, W.T., Günay, Y., Le Hérissé, A., Paris, F. and Robardet, M. (2003) Late Ordovician glaciation in southern Turkey. *Terra Nova*, **15**, 249–257.

Moreau, J. (2005) Architecture stratigraphique et dynamique des dépôts glaciaires ordoviciens du Bassin de Murzuk (Libye). Unpubl. PhD thesis, Université Louis Pasteur, Strasbourg.

Moreau, J., Ghienne, J.-F., Le Heron, D., Rubino, J.-L. and Deynoux, M. (2005) A 440 Ma old ice stream in North Africa. *Geology*, **33**, 753–756.

Ottesen, D., Dowdeswell, J.A. and Rise, L. (2005) Submarine landforms and the reconstruction of fast-flowing ice streams within a large Quaternary ice sheet: The 2500-km-long Norwegian-Svalbard margin (57°–80°N). *Geol. Soc. Am. Bull.*, **117**, 1033–1050.

Ouanaimi, H. (1998) Le passage Ordovicien-Silurien à Tizi n'Tichka (Haut-Atlas, Maroc): variations du niveau marin, *CR Acad. Sci. Paris*, **326**, 65–70.

Oulebsir, L. and Paris, F. (1995) Chitinozoaires ordoviciens du Sahara algérien: biostratigraphie et affinités paléogéographiques. *Rev. Palaeobot. Palynol.*, **86**, 49–68.

Paris, F., Elaouad-Debbaj, Z., Jaglin, J.C., Massa, D. and Oulebsir, L. (1995) Chitinozoans and Late Ordovician glacial events on Gondwana. In: *Ordovician Odyssey* (Eds J.D. Cooper, M.L. Droser and S. Finney), pp. 171–176, Short papers for the seventh international symposium on the Ordovician System, SEPM, Fullerton, California.

Paris, F., Deynoux, M. and Ghienne, J.-F. (1998) Chitinozoaires de la limite Ordovicien-Silurien en Mauritanie. *CR Acad. Sci. Paris*, **326**, 499–504.

Powell, R.D. (1990) Sedimentary processes at grounding line fans and their growth to ice-contact deltas. In: *Glacimarine environments: processes and sediments* (Eds J.A. Dowdeswell and J.D. Scourse), *Geol. Soc. London Spec. Publ.*, **53**, 53–77.

Powell, R.D., Khalil Moh'd, B. and Masri, A. (1994) Late Ordovician-Early Silurian glaciofluvial deposits preserved in palaeovalleys in South Jordan. *Sed. Geol.*, **89**, 303–314.

Proust, J.-N. and Deynoux, M. (1993) Marine to non marine sequence architecture of an intracratonic glacially related basin. Late Proterozoic of the West African platform in western Mali. In: *Earth's Glacial Record* (Eds M. Deynoux, J.MG. Miller, E.W. Domack, N. Eyles, I.J. Fairchild and G.M. Young), pp. 241–259, Cambridge University Press.

Powell, R.D. and Cooper, J.M. (2002) A glacial sequence stratigraphic model for temperate, glaciated continental shelves. In: *Glacier-Influenced Sedimentation on High-Latitude Continental Margin* (Eds J.A. Dowdeswell and C. Cofaigh), pp. 215–244, *Geol. Soc. London Spec. Publ.*, **203**.

Robardet, M. and Doré, F. (1988) The Late Ordovician diamictic formations from southwestern Europe: North-Gondwana glaciomarine deposits. *Palaeogeogr. Palaeoclimatol. Palaeoecol.*, **66**, 19–31.

Ross, J.R.P. and Ross C.A. (1992) Ordovician sea-level fluctuations. In: *Global perspectives on Ordovician geology* (Eds B.D. Webby and J.R. Laurie), pp. 327–336, Balkema, Rotterdam.

Russell, H.A.J. and Arnott, R.W.C. (2003) Hydraulic-jump and hyperconcentrated-flow deposits of a glacigenic subaqueous fan: Oak Ridges Moraines, southern Ontario, Canada. *J. Sed. Res.*, **73**, 887–905.

Russell, H.A.J., Arnott, R.W.C. and Sharpe, D.R. (2003) Evidence for rapid sedimentation in a tunnel channel, Oak Ridges Moraine, southern Ontario, Canada. *Sed. Geol.*, **160**, 33–55.

Saltzman, M.R. and Young, S.A. (2005) Long-lived glaciation in the Late Ordovician? Isotopic and sequence-stratigraphic evidence from western Laurentia. *Geology*, **33**, 109–112.

Senalp M. and Al-Laboun A. (2000) New evidence on the Late Ordovician Glaciation in Central Saudi Arabia. *Saudi Aramco Journal of Technology*, 1–40 (Spring).

Smart, J. (2000) Seismic expressions of depositional processes in the upper Ordovician succession of the Murzuq Basin, SW Libya. In: *Symposium on Geological Exploration in Murzuq Basin* (Eds M.A. Sola and D. Worsley), pp. 397–415, Elsevier, Amsterdam.

Stampfli, G.M. and Borel G.D. (2002) A tectonic model for the Paleozoic and Mesozoic constrained by dynamic plate boundaries and restored synthetic oceanic isochrones. *Earth Planet. Sci. Lett.*, 196, 17–33.

Stokes, C.R. and Clark, C.D. (2001). Palaeo-ice streams. *Quatern. Sci. Rev.*, 20, 1437–1457.

Storch, P. and Leone, F. (2003) Occurrence of the late Ordovician (Hirnantian) graptolite Normalograptus ojuensis (Koren & Mikhaylova, 1980) in south-western Sardinia, Italy. *Bol. Della Soc. Paleont. Ital.*, 42, 31–38.

Sutcliffe, O.E., Dowdeswell, J.A., Whittington, R.J., Theron, J.N. and Craig, J. (2000) Calibrating the Late Ordovician glaciation and mass extinction by the eccentricity cycles of Earth's orbit. *Geology*, 28, 967–970.

Sutcliffe, O.E., Harper, D.A.T., Aït Salem, A., Whittington, R.J. and Craig, J. (2001) The development of an atypical Hirnantia-brachiopod Fauna and the onset of glaciation in the late Ordovician of Gondwana. *Trans. Roy. Soc. Edinb. Earth Sci.*, 92, 1–14.

Underwood, C.J., Deynoux, M. and Ghienne, J.-F. (1998) High palaeolatitude recovery of graptolite faunas after the Hirnantian (top Ordovician) extinction event. *Palaeogeogr. Palaeoclimatol. Palaeoecol.*, 142, 91–105.

Turner, B.R., Makhlouf, I.M. and Armstrong, H.A. (2005) Late Ordovician (Ashgillian) glacial deposits in southern Jordan. *Sed. Geol.*, 181, 73–91.

Vaslet, D. (1990) Upper Ordovician glacial deposits in Saudi Arabia. *Episodes*, 13, 147–161.

Vecoli, M. and Le Hérissé, A. (2004) Biostratigraphy, taxonomic diversity and patterns of morphological evolution of Ordovician acritarchs (organic-walled microphytoplankton) from the northern Gondwana margin in relation to palaeoclimatic and palaeo-geographic changes. *Earth-Sci. Rev.*, 67, 267–311.

Vorren, T.O. and Laberg, J.S. (1997) Trough mouth fans – Palaeoclimate and ice-sheet monitors. *Quatern. Sci. Rev.*, 16, 865–881.

Webby, B.D., Cooper, R.A., Bergström, S.M. and Paris, F. (2004) Stratigraphic framework and time slices. In: *The Great Ordovician Diversification Event* (Eds B.D. Webby, F. Paris, M. Droser and I. Percival), pp. 41–47, Columbia University Press, New York.

Young, G.M., Minter W.E.L. and Theron, J.N. (2004) Geochemistry and palaeogeography of upper Ordovician glaciogenic sedimentary rocks in the Table Mountain Group, South Africa, *Palaeogeogr., Palaeoclimatol., Palaeoecol.*, 214, 323–345.

The Ordovician glaciation in Eritrea and Ethiopia, NE Africa

R.A. KUMPULAINEN

Department of Geology and Geochemistry, Stockholm University, 106 91 Stockholm, Sweden (e-mail: risto.kumpulainen@geo.su.se)

ABSTRACT

Ordovician (Hirnantian?) glacigenic deposits are described here for the first time from south-central Eritrea. These deposits rest on an almost peneplained Neoproterozoic basement and define, in Eritrea and Ethiopia, a depositional area measuring at least 200 km in an east-west direction and 170 km in a north-south direction. For this preliminary note, five sections through the glacigenic succession were logged in Eritrea and one in Ethiopia. Facies types are described and interpreted. An ice-proximal facies assemblage is located in the Tigray Province of northern Ethiopia, the type area of the glacigenic Edaga Arbi Beds. These proximal deposits, c. 20 m thick, are characterised by melt-out diamictites, with striated clasts, interlayered with sandstone beds displaying horizontal lamination and normal grading (sand-silt). The horizontal lamination in the section is transitional with climbing ripple beds. Ice rafted clasts in sand-granule grade are common in these sandy beds. This ice-proximal section also exhibits some, minor soft-sediment deformation, such as asymmetrically folded beds, south-dipping reverse faults and glacial grooves suggesting transport to the north. This proximal facies grades laterally into a cross-bedded arkosic sandstone, the Enticho Sandstone, which probably represents deposition on subaqueous outwash fans. Cross-beds in this sandstone dip consistently to the north also in south-central Eritrea. Glacial striae and grooves are observed on top of the Enticho Sandstone in two localities in Eritrea. These proximal facies types are overlain by a distally deposited mudstone-dominated unit, 3–40 m thick, most probably deposited from turbid overflow plumes, although it also contains ice-rafted clasts. Only this unit hosts ice-rafted clasts in Eritrea. In Eritrea it also contains some diamictites. The name Edaga Arbi Beds is adopted for this unit in Eritrea. Icebergs were probably responsible for the deposition of the diamictites in south-central Eritrea. The development of this glacigenic succession was probably related to a regular retreat of the ice margin from north to south. It is also probable, that this succession only represents one cycle of deglaciation, the last of the two (or three) recognised in other parts of North Africa. The post-glacial development is initially represented by the deposition of a probable marine dune complex migrating from north to south. Fossil evidence and trace fossils, particularly *Arthrophycus alleghaniensis* (Harlan) suggest that the age of these glacigenic deposits is Late Ordovician, probably Hirnantian.

Keywords Glaciation, Ordovician, Eritrea, Ethiopia.

INTRODUCTION

Glacial deposits have been known since the 1970s from the Tigray Province of northern Ethiopia (Dow *et al.*, 1971; Beyth, 1973). The presence of corresponding glacigenic rocks also in Eritrea was mentioned by Kumpulainen *et al.* (2002). The present paper describes for the first time these rocks, known as the Edaga Arbi Beds, and their relationship with the underlying and overlying units in Eritrea. Some complementary observations were also made in the Edaga Arbi area of northern Tigray.

These glacigenic deposits occur in a sandstone-dominated succession of the Adigrat Group (Garland 1980), which rests unconformably on a peneplained surface (Abul-Haggag, 1961) underlain by Neoproterozoic schists and pan-African

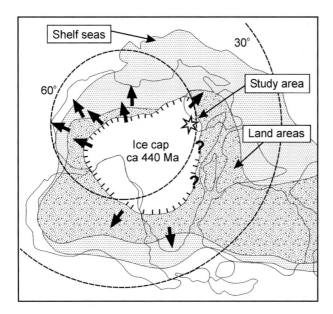

Fig. 1 Palaeogeography of Gondwana at *c*. 440 Ma, exhibiting the inferred distribution of land areas, shelf seas and the Ordovician continental ice cap. The black arrows indicate a radial ice-flow pattern. Limited amount of data is available in the east. The star indicates the location of the study area in Eritrea and Ethiopia. Modified from Vaslet (1990).

granitoid rocks that are associated with the suture zone between the East- and West-Gondwana. In the Early Palaeozoic Era, the area was clearly located within a very large continental block, allowing only a thin veneer of sediments to accumulate over a wide almost horizontal continental shelf (Fig. 1). A clastic sedimentary succession was deposited in Palaeozoic time on this flat-lying unconformity. In Ethiopia and Eritrea, this sedimentary cover is known as the Adigrat Group and is overlain by a marly limestone unit, the Antalo Limestone (Blanford, 1869; Bosellini *et al.*, 1997), containing a rich Oxfordian-Kimmeridgian marine fauna.

The lower part of the Adigrat Group is essentially non-fossiliferous. The only body fossils were described by Saxena and Assefa (1983) from the lower, glacigenic part of the clastic succession. The fossil, *Discophyllum peltatum* Hall 1847, belongs to *Siphonophorida* and is of probable Ordovician-Silurian age. Otherwise, all the other parts of the succession could have been formed any time from Cambrian to Jurassic. The recent discovery of trace-fossils particularly *Arthrophycus alleghaniensis* (Harlan), along with eight other types of traces,

from Eritrea also suggests a diagnostic Ordovician-Silurian age of the lower part of the sandstone formation resting on top of the glacigenic Edaga Arbi Beds in Eritrea (Kumpulainen *et al.*, 2006). In Ethiopia, the upper part of the possibly same sandstone exhibits Triassic dinosaur footprints (Assefa, 1987), contains Jurassic plant debris (von zur Mühlen, 1931; Mohr, 1971; Bosellini, 1989) and grades gradually into the overlying Jurassic Antalo Limestone. These conflicting palaeontological data lead to stratigraphic correlation problems. New data on stratigraphy in Eritrea has been discussed in some detail by Kumpulainen *et al.* (2006) and reviewed briefly below under the section on stratigraphy. The present paper will be limited to the Lower Palaeozoic part of the Adigrat Group.

A large amount of data has accumulated the last few decades on the Late Ordovician glaciation in North Africa (Debyser *et al.*, 1965; Beuf *et al.*, 1971; Deynoux & Trompette, 1981; Ghienne, 2003), Arabia (Vaslet, 1990) and Turkey (Monod *et al.*, 2003), and at least two major advances (Le Heron *et al.*, 2004) and maybe more than two (Ghienne, 2003) advances, separated by a recession, have been recognised in many areas. Interpretations of the extension of the ice cap have been presented by *e.g.* Hambrey, 1985; Vaslet, 1990; Sutcliffe *et al.*, 2000; Monod *et al.*, 2003; Le Heron *et al.*, 2004; Ghienne *et al.*, this volume). According to these interpretations the margin extended from Africa eastwards beyond the successions discussed here (Fig. 1).

Stratigraphy

In northern Tigray, Ethiopia (Fig. 2), the Adigrat Group is composed of three *formations* (Garland, 1980): (i) the glacigenic *Edaga Arbi Beds*, which in the Edaga Arbi and Abi Adi areas rest on the near-horizontal Neoproterozoic basement. According to Dow *et al.* (1971), the Edaga Arbi Beds are divided into four informal stratigraphic units, Units 1–4 (Fig. 3). The coarse-clastic, Unit 1 grades laterally to the west, north and east into (ii) the *Enticho Sandstone*. The dominating upper part of the Edaga Arbi Beds in Ethiopia is composed of mudstones (Units 2–4). Unit 2 contains striated dropstones, whereas units 3 and 4 are devoid of any features suggesting a glacigenic origin. Unit 3 carries trace fossils, yet to be described. The Edaga

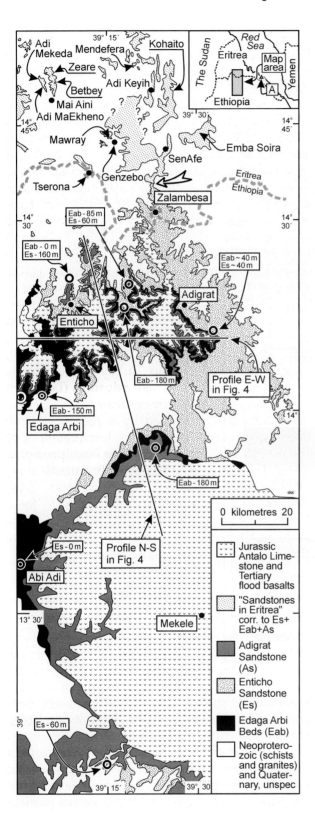

Arbi Beds and the Enticho Sandstone are both covered by (iii) the *Adigrat Sandstone*.

In the Enticho–Mekele areas (Fig. 2), the Edaga Arbi Beds form a lensoidal body, which is c. 60 km wide in the east-west direction (Fig. 4), and extends from the Enticho area to the Mekele area in the south. The Edaga Arbi Beds are limited to the west, north and east by a 'glacial type' of the Enticho Sandstone (Beyth, 1973), described as poorly sorted sandstone containing large clasts. To the east and north, the 'glacial' Enticho Sandstone grades laterally over to a 'fluvial' cross-bedded Enticho Sandstone. The facies transition from 'glacial' to 'fluvial' remains to be described later and has not been indicated in Figure 2.

According to the previous information (Dow *et al.*, 1971; Mohr, 1971; Garland, 1980), the Edaga Arbi Beds wedge out northwards, never reaching Eritrea. However, new information reveals that this glacigenic succession occurs in Eritrea (Fig. 2), although it has not been traced, so far, in outcrop across the border area between Ethiopia and Eritrea. In the Mai Aini–Adi Keyih–Genzebo areas, three distinctly different *formations* (Fig. 5) have been identified (Kumpulainen *et al.*, 2006): (i) the arkosic *Enticho Sandstone* with palaeocurrents trending approximately to the north, rests on the peneplained Neoproterozoic basement, (ii) a maroon, mudstone-dominated formation with outsized clasts, correlated with Unit 2 (Fig. 3) in the Edaga Arbi type section in Tigray, but for which the name *Edaga Arbi Beds* is applied in this paper, and (iii) the *Adigrat Sandstone formation*. The trace fossil information (Kumpulainen *et al.*, 2006) in the lower part of the Adigrat Sandstone (*sensu* Garland, 1980) suggests an Ordovician to Silurian

Fig. 2 Geological map of the area from Mekele in Ethiopia northwards to south-central Eritrea. The 'glacial and fluvial types' of the Enticho Sandstone are not distinguished on the map. The scale does not allow the distinction of formations of the Adigrat Group in the area north of the open arrow in Eritrea. In earlier literature, the name 'Eritrean sandstone' was used for these formations (Mohr, 1971). The thickness for the Edaga Arbi Beds (Eab) and Enticho Sandstone (Es), respectively, is given for a number of localities in Tigray (Beyth, 1973). The boundary between Eritrea and Ethiopia is approximate. Modified from Arkin *et al.* (1971), Garland (1980) and Hambrey (1981).

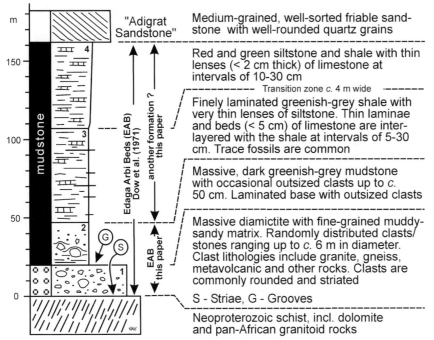

Fig. 3 Stratigraphy of the Edaga Arbi Beds in their type area, compiled from the description of Dow *et al.* (1971). Four informal units (1–4) have been identified in this succession. Unit 2 contains outsized clasts and may represent the youngest part of the glacigenic succession in this area. In contrast to the original suggestion, units 3 and 4 do not belong to the Edaga Arbi Beds, but represent younger events of sedimentation.

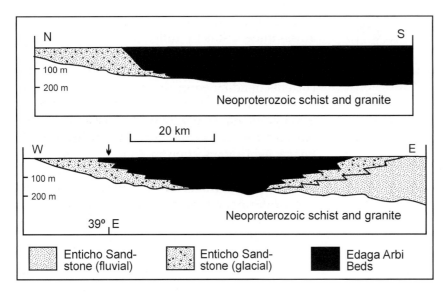

Fig. 4 Two profiles describing the facies relationships between the Edaga Arbi Beds and the Enticho Sandstone in Tigray, representing the Enticho–Adigrat area and the Enticho–Mekele area. The Edaga Arbi Beds are a lensoid body limited laterally by the 'glacial' facies of the Enticho Sandstone. The western end, left of the arrow, of the east-west profile projects outside the left margin of the map (Fig. 2). The Adigrat Sandstone and the Antalo Limestone are omitted in these profiles. Redrawn from Beyth (1973).

age for the lower part of the formation, whereas the upper part elsewhere is Triassic to Jurassic (see above). The informal *Adigrat Sandstone formation* was introduced by Kumpulainen *et al.* (2006) for the lower part of the Adigrat Sandstone (*sensu* Garland, 1980) in order to circumvent the required redefinition of the lithostratigraphical nomenclature in the study area, for which the amount of data currently available is insufficient (*cf.* Salvador, 1994). Further work may prove that two or more

sandstone formations of very different ages and characteristics instead of one Adigrat Sandstone may be present in the Horn of Africa.

The presence of trace fossils in the successions of these two areas may also be significant for stratigraphical correlations between Eritrea and Tigray. In Eritrea, a rich trace fossil fauna occurs particularly in the Adi MaEkheno Member, which rests on top of the Edaga Arbi Beds. In Tigray, trace fossils occur in Unit 3 (*sensu* Dow *et al.*, 1971) of the

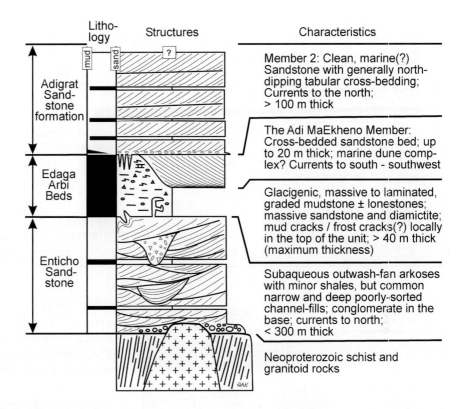

Fig. 5 The Lower Palaeozoic stratigraphy of south-central Eritrea. The Jurassic Antalo Limestone and younger units are omitted. Modified from Kumpulainen *et al.* (2006).

Edaga Arbi Beds, which does not exhibit any evidence of glacigenic origin, but rests on top Unit 2 containing dropstones. It is therefore suggested here that only Units 1 and 2 (*sensu* Dow *et al.*, 1971) belong to the Edaga Arbi Beds in Tigray, whereas Units 3 and 4 belong to another lithostratigraphical unit, *i.e.* the Adigrat Sandstone formation or constitute a new stratigraphic unit of their own.

The Adigrat Sandstone formation in Eritrea is divided into two *members*, the lower, Adi MaEkheno Member and an informal upper member, *Member 2*. The Adi MaEkheno Member is dominated by one tabular, cross-stratified sandstone bed, up to 20 m thick. The attitude of the clinoforms suggests bed-form migration towards the south or southwest. Member 2 rests with an erosional contact on the Adi MaEkheno Member and is composed of sandstone beds with cross-bedding suggesting palaeocurrents trending to the north. Hence, with their opposing palaeocurrent polarities, these two sandstone members of the Adigrat Sandstone formation are readily delimited in the outcrop. The upper boundary of Member 2 has not been studied by the present author. A schematic north-south correlation profile between the Edaga

Arbi–Enticho area and Mai Aini–Adi Keyih area is provided in Figure 6.

A similar three-fold stratigraphy was identified by Hutchinson & Engels (1970) eastwards in the Adeilo area, centred along the Eritrean coast (Fig. 2, insert map). They found a lower sandstone, a middle mudstone and an upper sandstone, and ascribed these units to the Adigrat Sandstone. New work in that area (Fig. 7) reveals that the mudstone unit carries outsized clasts up to cobble size, and is texturally and lithologically very similar to the mudstones of the Edaga Arbi Beds in south-central Eritrea (Fig. 3); this concerns also the colour of the rocks. This new information shows that the Edaga Arbi Beds extend from the Mai Aini area c. 200 km eastwards to the Adeilo area along the Eritrean coast and, indeed, further across the Red Sea to Yemen. This stratigraphy probably also extends westwards from the Mai Aini area (Zanettin *et al.*, 1999), but it has not been studied there. In the north-south direction, the glacigenic succession extends from the Mai Aini area for at least 170 km southwards to the Abi Adi area west of Mekele in Tigray, and its extension further south into other parts of Ethiopia is highly probable.

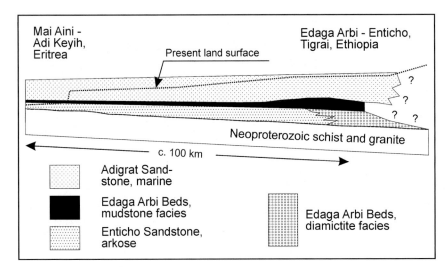

Fig. 6 A cross-section from Edaga Arbi to south-central Eritrea suggesting the presence of a three-fold stratigraphy in both areas of Ethiopia and Eritrea. The Edaga Arbi Beds grade laterally into the Enticho Sandstone. Modified from Kumpulainen *et al.* (2006).

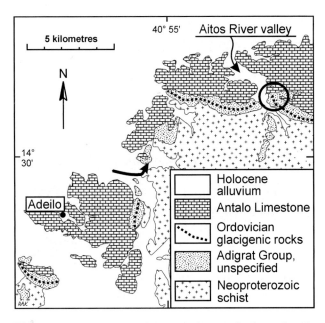

Fig. 7 Geological map of the Adeilo area (indicated with A in the insert map, Fig. 2), along the Danakil coast of Eritrea. Maroon ice-rafted debris-bearing glacigenic mudstone with sand to cobble-sized dropstones were found in the Aitos River valley (encircled). Another locality, c. 5 km NE of Adeilo, is indicated by an arrow. The extension of that mudstone unit within the unspecified Adigrat Group is inferred with a broken curve in other parts of the Adeilo area. Modified from Sagri *et al.* (1998).

with steep-sided valleys and flat valley floors (Fig. 8). The Precambrian schists exhibit an erosional, smooth hilly landscape. The top of the Precambrian sequence is close to a peneplain and particularly in Eritrea, this surface is covered by the Enticho Sandstone. The Edaga Arbi Beds are soft and easily removed by erosion. They commonly form an inclined slope covered by scree, but also occasionally providing a complete exposure across the whole Edaga Arbi Beds. The Adigrat Sandstone formation on top of the mudstones forms vertical cliffs. These circumstances provide an excellent opportunity to trace the individual formations across these regions.

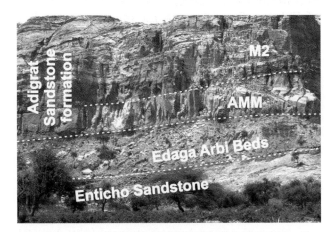

Fig. 8 The Lower Palaeozoic stratigraphy exposed in a steep hill-side north of Mawray. The various stratigraphic units are indicated. AMM – Adi MaEkheno Member, M2 – Member 2 of the Adigrat Sandstone formation.

It is worth noting that this three-fold stratigraphy, particularly in the highland areas of Eritrea and Ethiopia, produces a very impressive erosional landscape of table mountains and buttes

THE GLACIGENIC SUCCESSION

General

Glacigenic successions may be formed (*e.g.* Hambrey, 1994; Miller, 1996) in a variety of depositional settings, either (i) by melt-out or lodgement of sediment in contact with the depositing glacier or ice cap, or (ii) transport of sediment to the depositional site on land, or into lakes or seas by traction currents, gravity flows, in suspension or by ice-rafting, followed by possible subsequent redeposition.

Boundary relationships

Around Edaga Arbi and Abi Adi in Tigray, the Edaga Arbi Beds rest unconformably, with a sharp boundary, on top of the Neoproterozoic greenschist facies volcanic sedimentary rocks (e.g. Dow *et al.*, 1971). Locally, glacial striae have been reported on top of the Neoproterozoic basement. The boundary of the Edaga Arbi Beds towards the Enticho Sandstone was interpreted by Dow *et al.* (1971), Beyth (1973) and Garland (1980) as a lateral facies change.

In Eritrea, the base of the Edaga Arbi Beds is commonly sharp or gradational over a distance of a few centimetres, and they always rests on the Enticho Sandstone. Glacial striae or grooves are observed on the top of the Enticho Sandstone in two localities in Eritrea: one in the village Adi MaEkheno (Fig. 9A) with the striae oriented 45°-225° and the other in the village Zeare (Fig. 9B), where the grooves trend 320°-140°. In Adi MaEkheno, the outcrop is a striated surface of the Enticho Sandstone, but it provides no clear relationship with the overlying rocks. In Zeare, the grooved surface is covered by mudstones of the Edaga Arbi Beds. Glacial grooves are also observed on top of Unit 1 (Fig. 3) of the Edaga Arbi section in Tigray (Fig. 9C). In Adi Mekeda in Eritrea, the contact between these two units exhibits striking evidence of liquefaction. The boundary is a 'mixing zone' (Fig. 10a), about 0.5 m thick, composed of sand clasts from the underlying Enticho Sandstone and mud clasts from the overlying Edaga Arbi Beds. Irregular sandstone dykes from the underlying Enticho Sandstone cut this mixing zone and the lower 2–3 metres of the Edaga Arbi Beds.

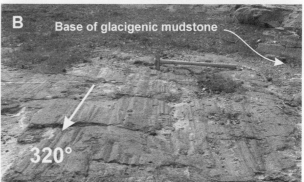

Fig. 9 Glacial striae and grooves. (A) Glacial striae pointing to 45° on top of the Enticho Sandstone, Adi MaEkheno, (B) Glacial grooves, trending 320°, on top of the Enticho Sandstone, which is overlain by the Edaga Arbi Beds (mudstone); hammer for scale, (C) Glacial grooves on top of a diamictite, with a cobble projecting out of the bed top, west of Edaga Arbi (*cf.* Fig. 11, section 1).

The upper boundary of the Edaga Arbi Beds in Tigray (Dow *et al.*, 1971; Garland, 1980), *i.e.* the top of Unit 4, has been described as an undulating unconformity, whereas the boundary between Unit 2 and 3 has not been described. The upper boundary of the Edaga Arbi Beds in Eritrea is commonly depositional and conformable, but locally

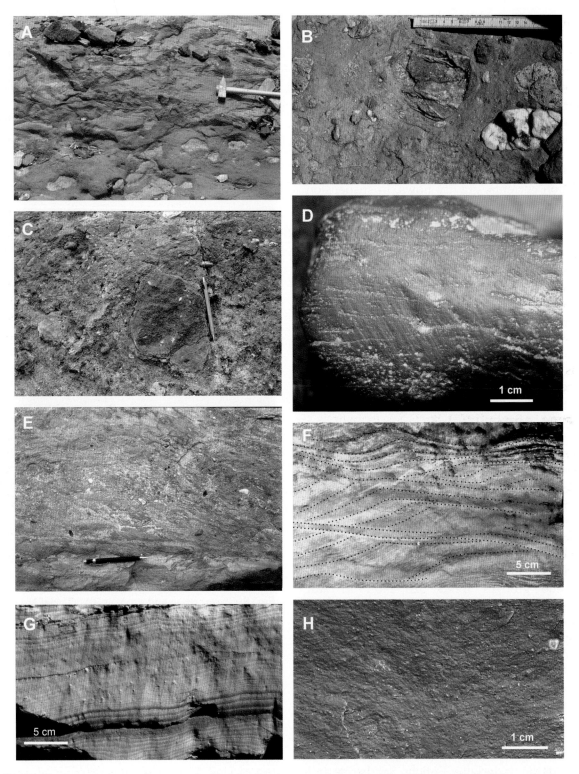

Fig. 10 (A) The boundary zone between the Enticho Sandstone and the overlying Edaga Arbi Beds is a mixing zone composed of soft-sediment deformed sand clasts from the underlying unit and mud clasts from the overlying unit; hammer shaft *c.* 25 cm; Adi Mekeda; (B) Massive matrix-supported diamictite (Dmm), west of Edaga Arbi (Fig. 2), scale increments in centimetres; (C) Facies diamictite (Dmm) carrying itself a diamictite clast, Zeare; (D) A striated clast; (E) Facies diamictite (Dms) resting on a mudstone unit in the lower part of the Edaga Arbi Beds with inclined bedding, a possible glacitectonite or a creep deposit; (F) Climbing ripples (Sr) of type-A developing into type-B, west of Edaga Arbi (Fig. 2), (G) Graded sand-silt couplets, which are slightly rippled; note the ice-rafted debris in the centre, west of Edaga Arbi (Fig. 2); (H) Massive mudstone (Mm) with ice-rafted debris up to granule grade.

erosional. In the latter case, the boundary is covered by a thin, less than 15 cm thick, conglomerate on top of the glacigenic unit. This is the case at Mawray and Mendefera in Eritrea (Fig. 2). At Mendefera, the Adi MaEkheno Member is missing, so the conglomerate rests on top of the mudcracked Edaga Arbi mudstones and the cracks penetrate c. 30–35 cm into it. However, the cracks could also be seasonal frost cracks. They are filled with material similar to that in the overlying thin granular to pebbly conglomerate.

Thickness variations

The thickness of the Edaga Arbi Beds varies from one area to another, probably reflecting changes in palaeobathymetry and depositional conditions. In northern Ethiopia, the thickness of the Edaga Arbi Beds has been estimated by Dow et al. (1971) and Garland (1980) at c. 150 to 180 m. If the new correlation (see above), based on trace fossil evidence is correct, then the Edaga Arbi Beds in Tigray may be limited to Units 1 and 2 (sensu Dow et al., 1971), and the thickness reduced to about 50 m. According to Beyth (1973), the Edaga Arbi Beds in Tigray were accumulated in a depression, being thickest along the axis of the depositional trough and wedging out laterally.

Earlier workers reported that the glacigenic succession thinned out towards the north before reaching Eritrea. Work in Eritrea (Kumpulainen et al., 2006), demonstrates that the largest thickness of the Edaga Arbi Beds, measured thus far, is about 40 m at Mendefera village. From there, it thins westwards and southwards to 10–15 m. Eastwards, in Kohayto (Fig. 3) the unit is no more than 3 m thick. Further eastwards, in the Adeilo area along the Eritrean coast, the Edaga Arbi Beds are probably again much thicker, perhaps several tens of metres, but the succession there is heavily dissected by faults precluding reliable thickness estimates.

Facies types

There are a number of important factors that influence interpretations of various facies types and the processes that are responsible for their formation (Reading, 1996). Such factors include topographical and bathymetrical conditions, circulation patterns and wave energy in marine and lacustrine settings.

Geomorphological studies in Eritrea and Ethiopia (Abul-Haggag, 1961) suggest, that the top of the Neoproterozoic schists is close to the peneplain over the area of study. The thickness of the Enticho Sandstone (up to 300 m) has been proposed to correspond with irregularities in the substratum filling of basement depressions (Mohr, 1971). No systematic morphological studies have been carried out concerning the upper surface of the Enticho Sandstone, but tracing the contact between the Enticho Sandstone and the Edaga Arbi Beds along the steep hillsides in Eritrea suggests that it is close to planar (Kumpulainen et al., 2006). Hence, the Edaga Arbi Beds were deposited on a near-horizontal surface, limiting the types and importance of transportational and depositional processes responsible for the formation of the facies assemblage of the Edaga Arbi Beds, as well as the environment in which they could have accumulated, the probable setting being a glacially influenced shallow continental shelf.

For this preliminary study six sections across the Edaga Arbi Beds were logged graphically; five sections in south-central Eritrea and one in Tigray. Logging was generally centred on a narrow, 1–2 m wide, profile perpendicular to the bedding in the succession. Occasionally, some beds were traced laterally in order to characterise the geometry of the bed. About ten different facies types were encountered and described (Table 1); they are summarised below.

Massive, matrix-supported diamictite (Dmm): This facies is a massive, poorly sorted, commonly maroon or greenish grey diamictite with sandy-muddy matrix (Fig. 10b, c). The framework clasts are commonly less than 0.5 m in diameter, but may range up to several metres. They are sub-rounded to rounded in shape and occasionally striated (Fig. 10d). The content of framework clasts varies from less than 10% to more than 30%. The lower part of a diamictite bed may exhibit deformation structures, such as flow structures and isoclinal soft-sediment folds. The bed boundary may be slightly undulating, and indeed in some localities this facies is loaded into the subjacent mudstone. In other places, it rests on laminated (Sh) and or ripple-cross bedded sandstones (Sr), which in those cases may display reverse, soft-sediment faults. The thickness of the beds of the Dmm facies may reach c. 3 m. It rests on various kinds of substratum, which

Table 1 Facies types of the Edaga Arbi Beds, Eritrea and Ethiopia

Code	Description	Interpretations
Dmm	Massive, diamictite with sandy-muddy matrix, in which stones up to metre-size occur. The content of frame-work clasts varies from less than 10% to more than 30%	Waterlain or basal melt-out diamictite
Dms	Diamictite with poor stratification. Sandy to muddy matrix, in which stones of various lithologies occur. Stone sizes vary from centimetre-size to 10–20 cm. The content of framework clasts is low	Waterlain or basal melt-out diamictite, a possible creep deposit or glacitectonite
Sm	Massive sand with possible dish structures and poorly preserved bedding	Liquifaction or fluidisation and possible redeposition of a stratified sand bed
Sr	Ripple-cross bedded and ripple-laminated sandstone. The ripples are commonly climbing	Traction currents and, in case ripples climb, high-suspension clastic input
Sh	Horizontally laminated sandstones	Lower or upper flow regime traction
Sd	Soft-sediment deformed sandstones	Sediment instability, ice-push or earthquake
SMg	Normally graded sand-mud couplets commonly less than 1 cm thick. May be combined with Sr	Possible cyclopsams, cyclopels, turbidites
Mh	Horizontally laminated mudstone	Distal supension deposit influenced by tidal currents
Mm	Massive, structureless mudstone	Distal suspension deposit from suspension plumes; rain-out from icebergs; homogenisation by liquifaction, fluidisation
IRD	Outsized clasts. Grain sizes vary from sand to boulder. Occur in facies types Sr, SMg, Mm, Sh	Ice-rafted debris melt-out from glacier base or icebergs. Difficult to recognise in coarse grained or homogenised units

occasionally may be grooved or striated. This facies is associated with Dms, Sd, Sr, Sm, Mm, Mh.

Interpretation: Three possible interpretations may be presented as to the origin of this facies. In case (i), the sequence displays essentially 'vertical' deformation, such as diapiric rise of mud and concomitant loading of the Dmm facies, and the likely interpretation of formation would be dumping of the coarse- to fine-clastic melt-out load of a tilting iceberg and the deposition of a melt-out diamictite. In case (ii), oblique deformation, such as reverse faults or asymmetric folds, is present in the substratum of the facies. Accepting these data in isolation, the original site of deposition would have been close to the grounding line of a tidewater glacier. The probable deposits could be interpreted as a waterlain till, a basal melt-out till or a debris-flow (*cf.* Eyles *et al.*, 1983). In case (iii), deposition of a diamictite and later deformation by an overriding glacier could have produced the oblique deformation structures, but recognised as facies Dms (below).

Poorly stratified and poorly sorted diamictite (Dms): This facies is a poorly sorted and poorly stratified, maroon to grey rock with a sandy to muddy matrix, in which clasts of various lithologies occur in an aligned fashion. These rocks are commonly folded (Fig. 10e) and display flow structures. Clast sizes vary from granule to pebble. The clasts are commonly rounded to well-rounded. The content of framework clasts is low. These units may display reverse grading, normal grading or no grading. In the lower part of some diamictite beds, the framework clasts (less than 10 cm in diameter) are rotated and together define an overturned fold. Bed thicknesses range up to 1–2 m. This facies may be associated with Dmm, Sr, Sm, Mm, Mh, SMg.

Interpretation: The most likely origin is as a waterlain or basal melt-out till with subsequent reworking by traction or gravity (*cf.* Eyles *et al.*, 1983), but may also be a creep deposit or a glacitectonite (Isbell *et al.*, 2001).

Massive sandstone (Sm): This sandy facies contains dish structures and poorly developed lamination. The grain size is medium to coarse sand and the grains are sub-angular to rounded. The content of matrix is very low. The lower bed boundary may display loading structures. The thickness of beds may reach 4 m. This facies is associated with facies Mh, Mm and Sm.

Interpretation: The probable mode of deposition was liquefaction or fluidisation and possibly redeposition of an originally stratified, subaqueous sandstone bed(s) or a rapidly deposited sandflow or other density-flow unit (*cf.* Middleton & Hampton, 1973).

Ripple-cross laminated sandstone (Sr): Ripple-cross laminated sandstone forms bedsets up to more than a metre in thickness or, less commonly, thin single beds. Climbing ripples of both type-A- and type-B (*sensu* Ashley *et al.*, 1982) are common (Fig. 10f) and may develop in the section from one end member to the other, particularly in the succession in Tigray. Successive cross-beds indicate unimodal as well as opposing transport polarities. This facies is associated with Dmm, Dms, Sh, SMg and commonly contains outsized clasts.

Interpretation: Ripple-cross bedding is formed by traction current in the lower flow regime (*cf.* Middleton & Southard, 1984). Formation of climbing ripples depends on the interplay between clastic input and current velocity. At high clastic input and low current velocity, deposition takes place also on the stoss side of a ripple (*cf.* Ashley *et al.*, 1982) and may cause a steep climbing angle (type-B). Increasing current velocity and low clastic input produces asymmetrical ripples with lower climbing angle (type-A). Climbing ripples are common in ice-proximal subaqueous sites, where icebergs provided the clastic contribution to the ice-rafted debris (Miller, 1996; Ó Cofaigh & Dowdeswell, 2001).

Horizontally laminated sandstone (Sh): Only a few thin laminae, from less than a millimetre to beds a few centimetres thick of horizontally laminated sandstones, are recorded in the measured sections. The grain sizes are medium to coarse sand. The boundaries with other facies are commonly sharp and abrupt. No grading is observed in this facies. This facies is associated with Mm, Mh, SMg and is transitional to SMg.

Interpretation: Horizontally laminated sandstones may be deposited from traction currents in the lower and upper flow regime (*cf.* Middleton & Southard, 1984). In a subaqueous, glacigenic setting, very thin sand laminae may form when sandy material rains-out from an iceberg and is deposited under the influence of sea currents (Ó Cofaigh & Dowdeswell, 2001).

Soft-sediment deformed sandstone (Sd): Some sandstone beds may display deformation features, such as symmetric or asymmetric folds, and normal or reverse faults. The original depositional features, particularly lamination are still partly preserved. This facies is associated with Sh, Mm, SMg, Dms and Dmm.

Interpretation: The most likely explanations are sediment instability resulting from deposition on over-steepened depositional slope, excess pore pressure within the sand bed induced either by ice-push, seismic or tectonic activity or glacial loading (*cf.* Middleton & Hampton, 1973; Allen, 1984)

Normally graded sand-mud couplets (SMg): Normally graded sand-mud couplets in this area are commonly less than 1 cm thick and form units up to c. 0.5 m thick. A set of normally graded laminae may develop into climbing ripples, where the individual laminae are also normally graded (Fig. 10g) and contain outsized clasts commonly in the sand-granule grade. This facies is associated with facies Mh, Mm, Sr, Gms.

Interpretation: The normally graded sand-mud couplets may be glacial cyclopsams and cyclopels, deposited from overflow plumes, or thin turbidites (*cf.* Eyles *et al.*, 1983; Mackiewicz *et al.*, 1984; Cowan & Powell, 1990; Ashley 1995; Ó Cofaigh & Dowdeswell, 2001).

Horizontally laminated mudstone (Mh): Horizontally laminated mudstone facies comprises mud interbedded with thin laminae and beds of fine sandstone. It forms beds from a few centimetres up to c. 4 m in thickness, that extend laterally for several metres or more. Boundaries to beds of other facies types, Mm, SMg, Sh, Sm, are commonly sharp and horizontal and only occasionally deformed. This facies contains outsized clasts.

Interpretation: The responsible mechanism for formation of horizontally laminated mudstone is traction currents in lower flow regime (*cf.* Middleton & Southard, 1984) or, in a subaqueous glacial setting, deposition from turbid overflow plumes (Ó Cofaigh & Dowdeswell, 2001). The outsized clasts represent ice-rafted debris.

Massive, structureless mudstone (Mm): Massive, structureless mudstone is an important component, particularly in the Eritrean successions of the Edaga Arbi Beds. Some of these massive mudstones contain outsized clasts (Fig. 10h), commonly of sand or granules, whereas pebbles or larger clasts are less common. The outsized clasts are commonly concentrated in certain, laterally persistent intervals of varying thickness and varying levels within this facies. The thickness of massive mudstone units may range up to c. 15 m. No bioturbation has been observed in this facies, which is commonly intercalated with facies Mh, Sh and Sd; bed boundaries are sharp or diffuse.

Interpretation: The massive character of this facies may be interpreted in three different ways, by (i) rain out of fine-clastic material from icebergs (*cf.* Miall, 1983; Ó Cofaigh & Dowdeswell, 2001), (ii) homogenisation due to liquefaction or fluidisation of mud (Middleton & Hampton, 1973) or (iii) bioturbation (Singer & Anderson, 1984). Some of the massive sandy mudstones may be fine-grained diamictites (Ó Cofaigh & Dowdeswell, 2001). No trace fossils have been observed in this unit so far, hence, bioturbation is an unlikely interpretation for its formation.

Outsized clasts: This is not a facies type of its own, but it contributes to the clastic component of other facies types. Outsized clasts (Fig. 10g) are common in the Edaga Arbi Beds in Eritrea and Ethiopia and they vary from sand to boulder, but the dominating fractions are sand to small pebbles. Outsized clasts, which also may display striae, have been observed in facies types Sr, SMg, Mh and Mm.

Interpretation: Outsized clasts are commonly, but not exclusively deposited as glacier ice melts and releases the trapped clastic particles, which may be rain-out in a variety of settings and environ-

ments (Miller, 1996). Alternatively, and particularly in terrestrial settings, larger clasts may roll, or be transported by wind into low-energy environments. Generally, the outsized clasts may be recognised in clearly finer-grained deposits. In the present case, the outsized clasts are interpreted as ice-rafted debris and the code IRD (ice-rafted debris) is therefore applied here. Striated ice-rafted clasts are typical of glacilacustrine and glacimarine environments.

Studied sections

Interpretations of the various facies types encountered in the study area are presented above. The selected sections through the Edaga Arbi Beds are described below with respect to the various facies types and the individual stratigraphies that they exhibit. The interpretation of the depositional setting of the Edaga Arbi Beds in the study area is then discussed, based on the information from the various sections collectively.

The Edaga Arbi section. The Edaga Arbi Beds have their type section near the village Edaga Arbi in Tigray, northern Ethiopia (Fig. 2) and the section was described by Dow *et al.* (1971). Depending on the interpretation outlined above, the Edaga Arbi Beds reach c. 50 m, as proposed in this paper, or 180 m in thickness, and may be divided into two or four informal subunits (Fig. 3). The lowermost unit, Unit 1, rests on a striated Neoproterozoic basement surface and has been described (Dow *et al.*, 1971) as massive diamictite containing clasts up to c. 6 m in diameter, some of the clasts being striated.

For this paper, a section of the lowermost 20 m of the Edaga Arbi Beds, is described from a river gorge c. 4 km west of the type section (Fig. 11, section 1; 14° 02' 20" N, 38° 59' 15" E). This section probably corresponds to Unit 1 *sensu* Dow *et al.* (1971). Here, the succession is deposited in a minor north-south trending valley, a few metres deep. This section apparently deviates from that described by Dow *et al.* (1971), since it comprises nearly equal proportions of diamictite and sand-dominated beds, not only 'massive diamictite'. Many of the sand-dominated beds display ripple-cross bedding, including type-A and type-B climbing ripples (Sr), *sensu* Ashley *et al.* (1982), which in a section may develop from one end-member to the other (Fig. 10f). Normally-graded varve-like laminae or

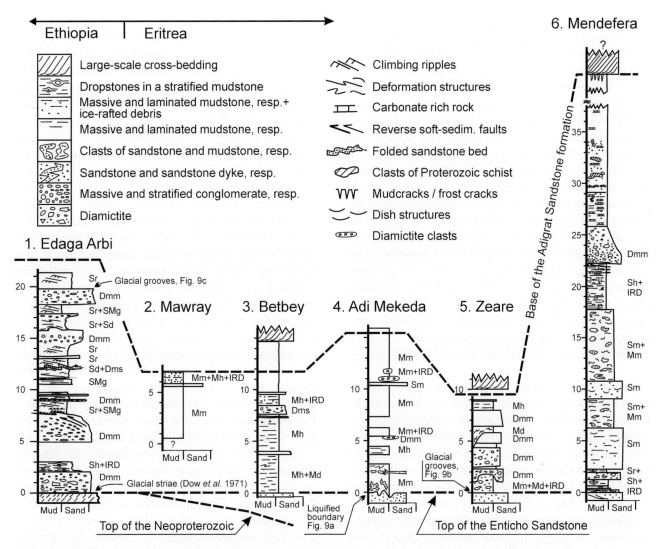

Fig. 11 Six logged sections, five from Eritrea and one west of Edaga Arbi in Tigray, Ethiopia, where only the lowermost part of the glacigenic unit was studied. For facies codes see the text. Scale increments are in metres. For locations, see Fig. 2.

cyclopsams (SMg, Fig. 10g) are also common. In a vertical section, the graded laminae may gradually develop into climbing ripples, where the individual laminae retain normal grading from sand to silt. Occasionally, these facies types display soft-sediment deformation. Several of the sand-dominated units also contain outsized clasts. Four diamictite beds (Dmm, Dms), each about 2 m thick and three thinner diamictite beds (Dmm, Dms), less than 0.5 m thick are interlayered with the sandy beds in this section. Three of the thicker beds have a gradual, transitional boundary with either the overlying or underlying sand-dominated

unit. The uppermost diamictite exhibits a sharp lower and upper boundary. The top of this uppermost diamictite bed displays parallel grooves trending approximately north-south. Clast lithologies in this upper diamictite bed include black marble, white quartzite, black schist, mica schist, chlorite schist, quartz, metabasalt, metarhyolite, quartz-K-feldspar-porphyry, K-feldspar-granite. Some of the diamictite beds or their boundaries display minor soft-sediment folds. Additionally, south-dipping, reverse soft-sediment faults, are encountered particularly on top of the second, from bottom, diamictite bed.

The Mawray section (14° 42′ 43″ N, 39° 15′ 38″ E). In the Mawray area (Fig. 11, section 2), the Edaga Arbi Beds is about 20 m thick. The section on the hillside just north of the village of Mawray exposes c. 6.5 m of the upper part of the formation. The lower part of the exposed section is dominated by massive mudstone (Mm), which is overlain by a structureless sandstone bed (Sm), 15 cm thick. The uppermost metre of the Edaga Arbi unit is composed of sandy to granular mudstone (Sm + Sh + IRD), interrupted by a thin, 10 cm, and distinctly laminated mudstone bed (Mh). This glacigenic unit is overlain by a mud-clast dominated conglomerate, 10 cm thick, in the base of mudstones of the Adi MaEkheno Member of the Adigrat Sandstone formation. The Adi MaEkheno Member displays here poorly preserved trace fossils similar to *Palaeophycus tubularis* (Hall, 1847). The contact between the Edaga Arbi Beds and the Adigrat Sandstone formation is conformable, but may be slightly erosional.

The Betbey section (14° 51′ 13″ N, 39° 07′ 01″ E). The Betbey section on the hillside next to the old village Betbey exhibits a complete section through the Edaga Arbi Beds, which here attain the thickness of 10 m (Fig. 11) and rest with a sharp yet conformable contact upon the Enticho Sandstone. The upper boundary is a diffuse transition into trace-fossil-bearing sandy, mudstone of the Adi MaEkheno Member of the Adigrat Sandstone formation. The succession is characterised by reddish, laminated mudstone (Mh), which in some intervals displays normal grading (SMg) and contain occasional laminae in sand grade with sharp boundaries (Sh). Beds of massive sandstone (Sm; each <10 cm thick) are interbedded with the mudstone. Some sandstone beds, (e.g. at 7–8 m on the section), are lenticular and partly undulating. A few isolated outsized clasts are scattered throughout the section, but are particularly concentrated in a 1 m interval at about 8 m on the logged section (Fig. 11, section 3), where the largest clast is 18 cm in diameter. The uppermost sandstone bed, which displays a normal coarse-tail grading, is covered by trace-fossil-bearing mudstone (*c.* 3 m thick) where the carbonate component increases upwards. The section ends in cross-bedded friable sandstones with a rich *Cruziana*-type, trace fossil fauna (Kumpulainen *et al.*, 2006).

The Adi Mekeda section (14° 53′ 14″ N, 39° 03′ 22″ E). The Adi Mekeda village exposes an almost complete section through the Edaga Arbi Beds (Fig. 11, section 4). The unit rests on the Enticho Sandstone, and it is overlain by nodular mudstones of the Adi MaEkheno Member of the Adigrat Sandstone formation. The top of the Enticho Sandstone and the base of the Edaga Arbi unit are both liquefied and form together a mixing zone (Sm + Mm), which is c. 0.5 m thick, and composed of clasts from both sedimentary units (Fig. 10a). Additionally, sandstone dykes cut this mixing zone and also intrude into the apparently liquefied mudstone (Mm). At about 2 m above the base, a probable sandstone bed is also liquefied (Sm) and forms a sill-like unit of irregular thickness, 5–50 cm, and displays minor sandstone dykes in the base and the top of the liquified bed. The Adi Mekeda section is dominated by massive mudstone (Mm), although a 1 m thick interval at c. 4 m on the section exhibits a horizontal lamination (Mh). Other interbeds are a single discontinuous, massive sandstone bed (Sm), about 5 cm thick, in the upper part of the section. Sand-grade ice-rafted debris is particularly common in two intervals, which are about 1 m thick each, one at 6 m and the other 11–12 m. A few clasts of diamictite are also observed.

Zeare section (14° 53′ 30″ N, 39° 06′ 25″ E). The Edaga Arbi Beds are exposed almost continuously for at least 200 m along the hillside in the Zeare village. The measured section, is located *c.* 50 m north of the church, and exposes an almost complete sequence, about 9 m thick, through the Edaga Arbi Beds (Fig. 11, section 5). Mudstone of this glacigenic unit rest conformably, with a sharp contact on the presumably glacier-grooved top of the Enticho Sandstone. Apart from the grooves, no other deformation is observed in the underlying sandstone. The upper boundary of the Edaga Arbi Beds is hidden in an unexposed gap, perhaps a metre wide. The unit is overlain by friable, cross-bedded sandstones of the Adigrat Sandstone formation, which carry a rich trace-fossil fauna. The basal c. 2 m of the Edaga Arbi Beds along the hillside is a mudstone-dominated part of the section. This lower mudstone unit may be sub-divided into beds with characteristically different features. It begins with a laminated mudstone (Mh), c. 25 cm thick, which, except for the distinctly-laminated lowermost 5 cm, contains sandy ice-rafted debris. This basal unit is overlain by a massive claystone (Mm) without ice-rafted debris, containing irregularly shaped clasts of mudstone with ice-rafted debris (Fig. 12A). The next bed, about 10 cm thick, is a laminated mudstone with distinct sand laminae (Sh), whereas the uppermost 20–30 cm are again rich in mud clasts and characterised by some soft-sediment deformation. A mixed diamictite-dominated succession with laterally discontinuous diamictite beds (Dmm) rests on top of this basal, mudstone-dominated part of the Zeare section. The diamictite beds are

separated by laminated and partly soft-sediment deformed mudstone, which rise diapirically up into, and wrap around, the lower parts of the loading diamictite pillows. The diamictite has a muddy-sandy matrix, in which stones of granule to cobble size are distributed. Stone lithologies include granitoid rocks, schist and also diamictite (Fig. 10c).

Mendefera section (14° 54′ 25″ N, 39° 19′ 20″ E). The hillside north of the village of Mendefera north of Adi Keyih exposes a section, about 40 m thick, through the Edaga Arbi Beds. The lower boundary of the unit towards the underlying Enticho Sandstone is sharp and conformable. The upper boundary of the unit is sharp and mudcracked. The unit is overlain by the basal conglomerate, c. 10 cm thick, of Member 2 of the Adigrat Sandstone formation. Laminated, ice-rafted debris-bearing mudstones (Mh + IRD) form the lowermost part of this glacigenic unit. Otherwise the lowermost 18 m are dominated by beds of homogenised sandstone (Sm), 1.5 m and 4 m thick, respectively, and deformed muddy sandstone which forms units 2.5 m and 7 m thick, respectively. The sandstone beds display diffuse lamination and dish structures; loading structures occur in the base of the beds. The muddy sandstone units are pervasively soft-sediment deformed and contain a mixture of irregular clasts of mudstone and sandstone, and additionally are cut by a network of thin sandstone dykes (Fig. 12B). The proportions between sand and mud vary through these two deformed units. The upper half of the section is dominated by laminated mudstone (Mh). The lower 4 m of these upper mudstones are laminated, contain some ice-rafted debris and display only minor soft-sediment deformation. A diamictite bed (Dmm), c. 3–4 m thick, rests on the mudstones at c. 22 m on the section. This diamictite has a sandy matrix and a low content of pebbly outsized clasts. It exhibits normal grading and a southeast-dipping fissility. The lower part of the mudstone above the diamictite is laminated (Mh) and contains some carbonate; the carbonate content increases upwards and the mudstone becomes apparently massive in the upper part of the unit. The top of the mudstone is cut by a network of mud cracks filled by material in sand and granule grade, resembling the material in the basal conglomerate bed of Member 2 of the Adigrat Sandstone formation.

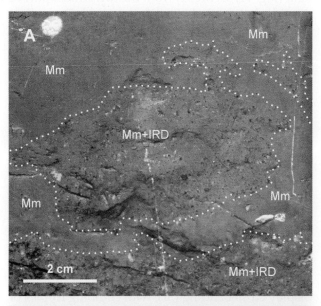

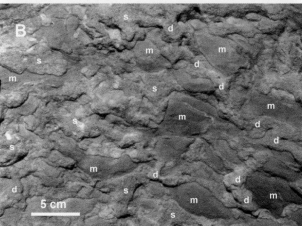

Fig. 12 (A) A soft-sediment deformed massive sandy mudstone (Mm) clast including ice-rafted debris in a massive claystone, Zeare; (B) Pervasively soft-sediment deformed sandstone-mudstone; m – mud clast, s – sand clast, d – sandstone dyke, Mendefera (Fig. 11, section 6.)

DISCUSSION

Pre-glacial morphology

Eritrea and Ethiopia were part of the pan-African orogenic mountain chain in early Cambrian time and developed to a large peneplained landscape in Late Ordovician time. The reason for this peneplanation is uncertain. It could have been related to earlier events of the Ordovician glaciation. Variations in the thickness of the Enticho Sandstone, in the range of 0–300 m, have been proposed (Mohr, 1971) to indicate filling of basement depressions by those sandstones, but those variations could also depend on depositional conditions (Fig. 4) and need be assessed later.

In other parts of North Africa that were covered by the Late Ordovician glaciers, sinuous channels, several tens of kilometres wide, a few tens of metres deep were cut in the subjacent rocks and soft sediments. Much of the later glacial deposits were accumulated in these channels (Le Heron *et al.*, 2004; Ghienne *et al.*, this volume). Channel directions coupled with information on deformation structures together with glacial striae and grooves in North Africa are indicated in Figure 13. The present study area in Eritrea and Ethiopia has a limited extension, and the succession, particularly that in south-central Eritrea, can be accommodated in a glacial channel of dimensions similar to those in other parts of northern Africa. A channel-like feature in which the Edaga Arbi Beds were deposited was described by Beyth (1973) in Tigray

(Fig. 4). It is not an erosional valley, however, but depends on localized deposition of the Edaga Arbi Beds and successive accumulation of traction-current deposits, i.e. the Enticho Sandstone. Thus far, the data do not support the presence of large glacial valleys in the Horn of Africa.

Depositional environments

A proximal depositional setting is inferred for the succession in the Edaga Arbi–Enticho areas of Tigray (Dow *et al.*, 1971; Garland, 1980). Similarly, the new data (Fig. 11, section 1) from the lowermost part of the Edaga Arbi Beds (i.e. Unit 1, *sensu* Dow *et al.*, 1971; Fig. 3), west of the Edaga Arbi type section, supports this interpretation. Here, Unit 1 contains several coarse-clastic diamictite beds, which probably were deposited sub-aqueously as melt-out diamictite (Dmm) released from the ice-margin, and were partly reworked by traction currents, particularly melt-water currents from subaqueous ice tunnels or by gravity transport (Dms). The diamictites are interbedded with sandstone beds composed of horizontally laminated sand (Sh), which may be graded (SMg) and form cyclopsams of sand-silt (Fig. 10g). These facies types were probably deposited from turbid overflow plumes under the influence of tidal currents (Ó Cofaigh & Dowdeswell, 2001). The graded laminae could also be turbidites. The sandstone beds additionally display cross-bedding (Sr), with climbing ripples of types A and B (*sensu* Ashley *et al.*, 1982) developing in one case (Fig. 10f) in a vertical section, suggesting

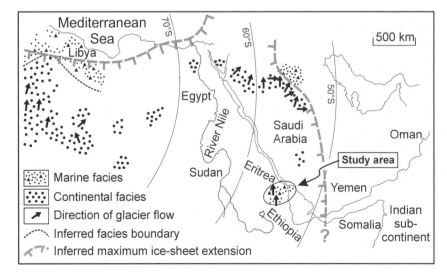

Fig. 13 Distribution of the known occurrences of Upper Ordovician glacigenic deposits in North Africa, Arabia and those in Eritrea and Ethiopia. Compiled from Vaslet (1990), Klitzsch and Wycisk (1999) and Monod *et al.* (2003).

in this case waning current velocity. The observed characteristics support the interpretation that the studied section through the lower part of the Edaga Arbi Beds in Tigray was deposited in a subaqueous environment near a tidewater glacier margin.

The succession comprising Unit 1 displays soft-sediment deformation, such as asymmetric north-verging folds, south-dipping reverse faults and glacial grooves trending to the north on top of the uppermost diamictite bed of the logged section (Fig. 9C, Fig. 11). As this section was deposited in a south to north-trending minor valley, a general glacier transport to the north is inferred. However, all these deformation structures are considered to be minor, and most sedimentary structures are well preserved (Fig. 10F & G), which is unlikely to be consistent with an ice-contact setting, unless the margin was retreating and exerted little or no stress on the deposits along the ice margin. The deformation structures could have been formed by the collapse of the ice-margin or from icebergs. In the event of prograding ice, much more severe types of deformation, such as large-scale folds and associated large faults, could have been expected, but such features have not been observed. However, this does not preclude their presence in other parts of Tigray and Ethiopia.

The evidence presented above suggests a subaqueous setting for Unit 1 of the Edaga Arbi Beds (Dow *et al.*, 1971). As this coarse-clastic facies appears to grade laterally into the Enticho Sandstone, which is consistent with models of ice-marginal sedimentation (Hambrey, 1994; Miller, 1996; Le Heron *et al.*, 2004), then the Enticho Sandstone must also be of subaqueous origin and glacigenic, rather than fluvial as suggested previously. The likely depositional site would have been subaqueous outwash fans along the retreating ice margin, where the melt-out debris was continually reworked by melt-water currents, obliterating the massive sediments. Melt-out diamictites were first preserved in Tigray as the ice had already receded from Eritrea. This hypothesis needs be tested by future work.

In Tigray, Unit 1 is overlain by ice-rafted debris-bearing mudstones of Unit 2 (Fig. 3). According to the earlier descriptions (Dow *et al.*, 1971), Unit 2 in Tigray resembles the mudstone-dominated glacigenic succession in Eritrea. These two successions most probably constitute a mudstone unit extending across the entire study area, covering the coarse-clastic glacial facies in Tigray and the Enticho Sandstone in Eritrea. This extensive mudstone blanket is dominated by two facies types: (i) distinctly laminated mudstones (Mh) and (ii) massive mudstones (Mm). Beds of these facies types occur in various proportions in all sections, commonly exhibiting sharp bed boundaries, which are conformable and sub-horizontal. The mudstones commonly host ice-rafted debris, from sand to granule, rarely to boulder size, and also diamictite clasts, whereas other intervals are devoid of ice-rafted debris. In a glacially-influenced environment, mudstones are deposited sub-aqueously, distal to the contemporaneous ice margin, *e.g.* from turbid overflow plumes (Mh) or by melt-out from icebergs (Mm). Alternatively, the massive mudstone (Fig. 10H) could have been deposited first as a laminated mudstone and then homogenised as a result of a glacier over-riding, or from an earthquake chock.

Diamictite is a minor facies component in the Eritrean succession and has, thus far, been encounters in three localities. Two of them are the Zeare and Mendefera sections (Fig. 10C; Fig. 11), but they display different characteristics. In Zeare, the diamictites are down-loaded into the underlying partly laminated, partly reworked mudstone, while the mudstone beds locally diapirically intrude the diamictite. No oblique deformation is observed here, and the likely interpretation is that the diamictite beds were dumped from an iceberg that may have grounded close to the measured section. The diamictite in the Mendefera section, which here occurs (at c. 22–26 m in the section) on top of an obviously soft-sediment deformed lower part, displays a clear south-dipping preferred clast orientation, suggesting either ice push or slow creep on a depositional slope. In the third locality, half way between Zeare and Adi Keyih, the diamictite rests, as in the Zeare section, on a lower, but here massive, mudstone-dominated unit. The contact between these facies types is sharp and on the large scale planar, but in detail slightly undulating with the wavelength less than 15 cm and the amplitude less than 5 cm. The lower part of the diamictite bed displays (Fig. 10E) rotated framework clasts (less than 10 cm in diameter) that together define an overturned fold. Again, this could be the product of ice push or creep.

Cross-bedded sandstones are a minor component in the Eritrean succession, whereas two beds of massive sand (Sm) are important components in the lower part of the Mendefera section. These two beds represent a rapid input of fairly clean sand into the depositional system probably by a gravity-flow mechanism or by homogenisation of a sequence of sand beds. These sand beds occur in the similarly soft-sediment deformed (Fig. 12B) mixed, sand-mud facies (Sm + Mm) that was originally interlayered thinly bedded sand and mud.

The distribution of the various facies types in the Eritrean glacigenic succession presents a problem, since the coarse-clastic and less organised facies types (Dmm, Dms, Sm) occur in the northernmost known successions (Fig. 2), that are interpreted here as the most distal, as opposed to proximal in Tigray. The presence of the coarse-clastic and less organised facies types (Dmm, Dms, Sm) in the distal succession in south-central Eritrea may be interpreted in two different ways: (i) as initial ice retreat towards the south, followed by a glacier growth in the north before the final retreat back towards the south or (ii) deposition by dumping from icebergs. In the first case, the growing glacier would flow over earlier deposits and most likely produce a compacted basal till (*cf.* Evans *et al.*, 2004). A rapid glacier advance in the Horn of Africa would be consistent with the interpretations presented for some other successions of the Late Ordovician glacial record of North Africa (Le Heron *et al.*, 2004; Ghienne *et al.*, this volume).

Evidence that is in conflict with the hypothesis of a second glacier advance and over-riding of the entire Eritrean glacigenic succession derives from the measured sections. They, indeed, display various degrees of soft-sediment deformation, yet in most sections distintcly laminated mudstone forms the lowermost beds (e.g. Betbey, Zeare and Mendefera), or that they are also interlayered with beds of massive, sandy mudstone (e.g. Fig. 10H) exhibiting very little deformation. Neither basal till nor other features such as large-scale isoclinal folding of the beds have been recognised in these sections. The most deformed section is that at Mendefera, whereas Mawray and Betbey are least deformed. It is perhaps more likely that the coarse-clastic facies were transported to the distal environment, e.g. Zeare, by icebergs. Likewise, the soft-sediment deformation, including that at

Mendefera, was produced by grounded icebergs. The diamictite in the middle of the section (Fig. 11) maybe also have been deposited from grounded icebergs.

Development of the glacigenic succession

Interpretation of the development of the glacigenic sedimentary succession in Eritrea and Ethiopia relies on the following evidence: (i) an ice-proximal facies in Tigray, which (ii) grades laterally over to the traction-current deposited Enticho Sandstone, exhibiting (iii) cross-bedding with consistently north-dipping foresets, (iv) palaeocurrents in Tigray indicate transport to a general north.

Assuming retreat of a single ice-margin from north to south, whilst still in Eritrea, the margin produced melt-out diamictite, which were continually reworked by jet currents from subglacial, subaqueous melt-water tunnels. The melt-water tunnels also issued clastic debris into the sedimentary system, forming underflow-fans (the Enticho Sandstone). The ice-margin migrated to Tigray, perhaps becoming stationary there for some time and releasing the coarse-clastic deposits that, more than in other parts of the succession, display features indicating an ice-margin. Fine-clastic sediments accumulated from turbid overflow plumes and produced successively a mud-dominated blanket (Edaga Arbi Beds in Eritrea) on top of the older, gravelly and sandy deposits. Icebergs were discharged into this water body, rafting clastic material, single clasts and diamictite clasts as well as larger quantities of debris to form beds of melt-out diamictite in the distal environment. Locally, the icebergs probably came into contact with the soft deposits causing local deformation in the older non-lithified sedimentary succession. The depositional environment is depicted in a cartoon in Figure 14.

Post-glacial development

The sections studied and earlier work by Kumpulainen *et al.* (2006) also provide useful information about sedimentation spanning the glacial and post-glacial times. Interestingly, whilst the glacier in the south fed the basin from south to north, the subsequent clastic input in post-glacial times was from approximately north to south and

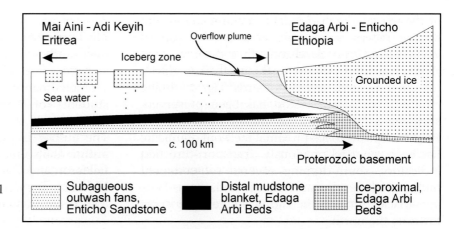

Fig. 14 A schematic profile of glacial depositional environment in south-central Eritrea and northern Tigray.

deposited on top of the glacigenic succession as the large-scale cross-bedded, Adi MaEkheno Member, which is probably a sub-marine dune complex. The source area for that northern material is not known. Conditions for life improved greatly, so that a *Cruziana* type trace-fossil fauna was established, and is now preserved in the Adi MaEkheno Member, maybe also in Member 2 of the Adigrat Sandstone formation.

The sub-marine dune complex migrated southwards; it reached as far as the Mendefera area, where the thickest known glacigenic succession in south-central Eritrea forms a depositional and bathymetric high. Engulfing first that high, the dune complex migrated further south to at least the Genzebo area, a total distance of more than 40 km. It is possible that the mudstones of Unit 3 (and maybe also Unit 4 of the Edaga Arbi Beds, *sensu* Dow *et al.*, 1971) constitute the distal facies of the Adi MaEkheno Member in Tigray. Further work is required to assess these relationships. The Adi Maekheno Member wedges out towards the Mendefera depositional high. Otherwise, the lower boundary of the Adi MaEkheno Member is comformable, whereas its upper surface is a disconformity overlain by a local thin conglomerate in the base of Member 2, suggesting a possibly minor erosional event before deposition of that member.

At the start of deposition of Member 2, the transport polarity again reversed towards the north and tabularly cross-bedded sandstone beds accumulated on top of the minor disconformity. It is interesting to note that Member 2 rests on top of the Edaga Arbi Beds in the Mendefera high, where the muddy

top of the high emerged for a certain time, and being subject to some erosion and desiccation, giving rise to a mud-crack network. The mud cracks were then filled by material similar to that in the thin conglomerate at the base of Member 2, indicating that this member began accumulating before the Edaga Arbi Beds were lithified. Alternatively, the mud cracks may represent seasonal frost cracks (*cf.* Berg & Black, 1966; Péwé, 1966; Washburn, 1970). Frost action may also have influenced the solid rocks, but in that case the cracks would probably be straight, because if lithified, the mudstone would have split into blocks. The cracks are not straight, however, but undulating and irregular, suggesting that they were formed in a non-lithified mud. Also frost action requires that the top of the Edaga Arbi Beds had emerged above the waterline.

CONCLUSIONS

A glacigenic succession has been located in Eritrea and Ethiopia within the greatest limit of the Ordovician glaciation in the Horn of Africa and Arabia. The succession rests on a nearly horizontal nonconformity, underlain by Neoproterozoic low-grade metamorphosed volcanic sedimentary and intrusive rocks. This succession exhibits facies assemblages that indicate sedimentation in both proximal and distal glacial environments, and was probably formed during one cycle of deglaciation towards the end of the Ordovician Period. The glacigenic succession began accumulating in Eritrea with the deposition of cross-bedded, subaqueous underflow outwash fans, whilst at the same time most of the massive melt-out debis was reworked and became incorporated in the cross-bedded succession, the Enticho Sandstone that displays consistent transport polarity to the north. The glacier margin retreated slowly towards Tigray, where an ice-proximal succession has been preserved, and where the lateral facies change from traction-current deposits to ice-proximal melt-out diamictite has been reported previously. Transport-direction indicators, south-dipping reverse soft-sediment faults, north-verging asymmetric folds, cross-beds and glacial grooves in Tigray suggest transport towards the north.

In the distal setting, mud-dominated sediment was deposited from turbid overflow plumes under the influence of currents. Massive and laminated mudstone was formed. Icebergs rafted clastic debris into this distal setting, depositing clastic material either as single grains or as aggregated (diamictite clasts) and also as beds of diamictite. Locally, icebergs disturbed the already deposited non-lithified muddy sediments.

In other parts of North Africa and Arabia, two ice sheet advances have been recognised. The present interpretation for the succession in Eritrea and Ethiopia assumes only one retreat and no second advance, although it is possible that the area experienced two ice sheet advances, the second advance removing the deposits of the first retreat (interglacial). In that case, the deposits in Eritrea and Ethiopia would represent the second retreat.

Post-glacial deposition in the area is represented by a large-scale cross-stratified bed (Adi MaEkheno Member; c. 20 m thick), a probable sub-marine dune complex. The south- or southwest-dipping clinoforms suggest that this complex has its source area in the north and it migrated to the south or southwest. This unit rests conformably on the glacigenic succession although minor erosion is locally observed. Fossil evidence, particularly the trace fossil *Arthrophycus alleghaniensis* (Harlan) suggests that the glacigenic unit is of Late Ordovician, possibly Hirnantian, age.

ACKNOWLEDGEMENTS

This study was financed by the University of Asmara, Eritrea; Uppsala University, Sweden; the Swedish International Development Cooperation Agency and Stockholm University, Sweden. The ideas presented here were gained from field visits and discussions with E. Ferrow, S. Ghirmay, T. Kreuser, A. Kumar, J. Shoshani, M. Teklay, B. Woldehaimanot, B. Zerai and the field course students (of January 1998). R. Bussert borrowed his copy of the geological map sheet Mekele. The author is grateful for all this support. The author is also indebted to the two reviewers, D.P. LeHeron and O.E. Sutcliffe and the volume editor M.J. Hambrey, who all provided many, useful comments on an earlier version of the manuscript.

REFERENCES

Abul-Haggag, Y. (1961) *A contribution to the physiography of northern Ethiopia*. University of London, The Athlone Press, 153 pp.

Allen, J.R.L. (1984) *Sedimentary structures: their characteristics and physical basis*. Elsevier, Amsterdam, 663 pp.

Arkin, Y., Beyth, M., Dow, D.B., Levitte, D., Haile, T. and Hailu, T. (1971) Mekele map sheet, ND 37–11, 1:250 000, Geol. Surv. Ethiopia (preliminary version).

Ashley, G.M. (1995) Glaciolacustrine environments. In: *Modern Glacial Environments: Processes, Dynamics and Sediments* (Ed. J. Menzies), pp. 417–444. Butterworth-Heinemann. Oxford.

Ashley, G.M., Southard, J.B. and Boothroyd, J.C. (1982) Deposition of climbing-ripple beds: a flume simulation. *Sedimentology*, **29**, 67–79.

Assefa, G. (1987) First record of Triassic Dinosaur footprints from the lower sandstone unit of Kersa area, eastern Ethiopia. *Riv. Ital. di Paleontol. Stratigr.*, **93**, 171–180.

Berg, T.E. and Black, R.F. (1966) Preliminary measurements of growth of non-sorted polygons, Victoria Land, Antarctica. *Am. Geophys. Union Publ.*, **1418**, 61–108.

Beuf, S., Biju-Duval, B., De Charpal, O., Rognon, P., Gabriel, O. and Bannacef, A. (1971) Les grès du Paléozoïque inférieur au Sahara. Sédimentation et discontinuités, évolution structurale d'un craton. *Sci. Techniq. Pétrole*, **18**, Paris, 464 pp.

Beyth, M. (1973) Correlation of Paleozoic-Mesozoic Sediments in Northern Yemen and Tigre, Northern Ethiopia. *AAPG Bull.*, **57** (for 1972), 2440–2446.

Blanford, W.T. (1869) On the geology of a portion of Abyssinia. *Q. J. Geol. Soc. London*, **25**, 401–406.

Bosellini, A. (1989) The continental margins of Somalia: their structural evolution and sequence stratigraphy. *Mem. Sci. Geol.*, **41**, 373–458.

Bosellini, A., Russo, A., Fantozzi, P.L., Assefa, G. and Solomon, T. (1997) The Mesozoic succession of the Mekele Outlier (Tigre Province, Ethiopia). *Sci. Geol. Mem.*, **49**, 96–116.

Cowan, E.A. and Powell, R.D. (1990) Suspended sediment transport and deposition of cyclically interlaminated sediment in a temperate fjord, Alaska, USA. In: *Glacimarine Environments: Processes and Sediments* (Eds J.A. Dowdeswell and J.D. Scourse), *Geol. Soc. London Spec. Publ.*, **53**, 75–89.

Debyser, J., De Charpal, V. and Meralet, O. (1965) Sur le caractère glaciaire de la sédimentation de l'Unité IV au Sahara Central. *C. R. Acad. Sci., Paris*, **261**, 5575–5576.

Deynoux, M. and Trompette, R. (1981) Late Ordovician tillites of the Taodeni Basin, West Africa. In: *Earth's pre-Pleistocene glacial record* (Eds M.J. Hambrey and W.B. Harland), pp. 89–96. Cambridge Univ. Press. Cambridge.

Dow, D.B., Beyth, M. and T. Hailu, (1971) Palaeozoic glacial rocks recently discovered in northern Ethiopia. *Geol. Mag.*, **108**, 53–59.

Evans, J., Pudsey, C.J., Ó Cofaigh, C., Morris, P. and Domack, E. (2004) Late Quaternary glacial history, flow dynamics and sedimentation along the eastern margin of the Antarctic Peninsula Ice Sheet. *Quatern. Sci. Rev.*, **24**, 741–774.

Eyles, N., Eyles, C.H. and Miall, A.D. (1983) Lithofacies types and vertical profile models; an alternative approach to the description and environmental interpretation of glacial diamict and diamictite sequences. *Sedimentology*, **30**, 393–410.

Garland, C.R. (1980) Geology of the Adigrat Area. *Geol. Surv. Ethiopia, Mem.*, **1**, 1–51.

Ghienne, J.-F. (2003) Late Ordovician sedimentary environments, glacial cycles, and post-glacial transgression in the Taoudeni Basin, West Africa. *Palaeogeogr. Palaeoclimatol. Palaeoecol.*, **189**, 117–145.

Ghienne, J.-F., Heron, D., Moreau, J. and Deynoux, M. (this volume) The Late Ordovician glacial sedimentary system of the North Gondwana platform.

Hall, J. (1847) *Paleontology of New York*, vol. **1**. C. Van Benthuysen, Albany, 338 p.

Hambrey, M.J. (1981) Palaeozoic tillites in northern Ethiopia. In: *Earth's pre-Pleistocene glacial record* (Eds M.J. Hambrey and W.B. Harland), pp. 38–40. Cambridge Univ. Press. Cambridge.

Hambrey, M.J. (1985) The Late Ordovician–Early Silurian glacial period. *Palaeogeogr. Palaeoclimatol. Palaeoecol.*, **51**, 273–289.

Hambrey, M.J. (1994) *Glacial environments*. UCL Press Ltd. London, 296 pp.

Harlan, R. (1831) Description of an extinct species of fossil vegetable, of the family Fucoides. *J. Acad. Nat. Sci. Philadelphia*, **6** (for 1830), 289–295.

Hutchinson, R.W. and Engels, G.G. (1970) Tectonic significance of regional geology and evaporite lithofacies in Northeasten Ethiopia. *Phil. Trans. Roy. Soc. London*, **A 267**, 313–329.

Isbell, J.L., Miller, M.F., Babcock, L.E. and Hasiotis, S.T. (2001) Ice-marginal environment and ecosystem prior to initial advance of the late Palaeozoic ice sheet in the Mount Butters area of the central Transantarctic Mountains, Antarctica. *Sedimentology*, **48** (5), 953–970.

Klitzsch, E. und Wycisk, P. (1999) Beckenentwicklung und Sedimentationsprozesse in kratonalen Bereichen Nordost-Afrikas im Phanerozoikum. In: *Nordost-Afrika: Strukturen und Ressourcen: Ergebnisse aus dem Sonderforschungsbereich 'Geowissenschaftliche Probleme in ariden und semiariden Gebieten'*. (Eds E. Klitzsch and U. Thorweihe), pp. 61–108. Wiley-VCH Verlag, Weinheim.

Kumpulainen, R.A., Kumar, A. and Woldehaimanot, B. (2002) The Horn of Africa in the cross road of two glaciations; new evidence from Eritrea and Ethiopia. 16th Int. Sed. Congr., Johannesburg. Abstr., 204–205.

Kumpulainen, R.A., Uchman, A., Woldehaimanot, B., Kreuser, T. and Ghirmay, S. (2006) Trace fossil evidence from the Adigrat Sandstone for an Ordovician glaciation in Eritrea, NE Africa. *J. Afr. Earth Sci.*, **45**, 408–420.

Le Heron, D., Sutcliffe, O., Bourgig, K., Craig, J., Visentin, C. and Whittington, R. (2004) Sedimentary architecture of Upper Ordovician tunnel valleys, Gargaf Arch, Libya: Implications for the genesis of a hydrocarbon reservoir. *GeoArabia*, **9** (2), 137–160.

Mackiewicz, N.E., Powell, R.D., Carlson, P.R. and Molnia, B.F. (1984) Interlaminated ice-proximal glacimarine sediments in Muir inlet, Alaska. *Mar. Geol.*, **57**, 113–147.

Miall, A.D. (1983) Glaciomarine sedimentation in the Gowganda Formation (Huronian), Northern Ontario. *J. Sed. Petrol.*, **53**, 477–491.

Middleton, G.V. and Hampton, M.A. (1973) Sediment gravity flows: mechanics of flow and deposition. In: *Turbidites and Deep-Water Sedimentation* (Eds D.J. Stanley and D.J.P. Swift). *SEPM Short Course*, **1**, 1–38.

Middleton, G.V. and Southard, J.B. (1984) Mechanics of sediment movement. *SEPM Short Course* **3**, 1–401.

Miller, J.M.G. (1996) Glacial sediments. In: *Sedimentary Environments: Processes, Facies and Stratigraphy* (Ed. H.G. Reading), pp. 454–484. Blackwell. Oxford.

Mohr, P. (1971) *The Geology of Ethiopia*. Haile Selassie I University Press. Addis Ababa. 268 pp.

Monod, O., Kozlu, H., Ghienne, J.-F., Dean, W.T., Günay, Y., Le Hérissé, A., Paris, F. and Robardet, M. (2003) Late Ordovician glaciation in southern Turkey. *Terra Nova*, **15**, 249–257.

Ó Cofaigh, C. and Dowdeswell, J.A. (2001) Laminated sediments in glacimarine environments: diagnostic criteria for their interpretation. *Quatern. Sci. Rev.*, **20**, 1411–1436.

Péwé, T.L. (1966) Ice-wedges in Alaska – classification, distribution, and climate. *Nat. Acad. Sci., Nat. Res. Council Can.*, **1287**, 76–81.

Reading, H.G. (1996) *Sedimentary Environments: Processes, Facies and Stratigraphy*. Blackwell. Oxford. 688 pp.

Sagri, M., Abbate, E., Azzaroli, A., Balestrieri, M.L., Benvenuti, M., Bruni, P., Fazzuoli, M., Ficcarelli, G., Marcucci, M., Papini, M., Pavia, G., Reale, V., Rook, L. and Tecle, T.M. (1998) New data on the Jurassic and Neogene sedimentation in the Danakil Horst and Northern Afar Depression, Eritrea. *Mém. Mus. natn. Hist. Nat.*, **177**, 193–214.

Salvador, A. (1994) *International stratigraphic guide*. 2nd Ed. Int. Union. Geol. Sci. and Geol. Soc. Am., Boulder, Colorado. USA. 214 pp.

Saxena, G.N. and Assefa, G. (1983) New evidence on the age of the glacial rocks of northern Ethiopia. *Geol. Mag.*, **120**, 549–554.

Singer, J.K. and Anderson, J.B. (1984) Use of total grain-size distribution to define bed erosion and transport for poorly sorted sediment undergoing simulated bioturbation. *Mar. Geol.*, **57**, 335–359.

Sutcliffe, O.E., Dowdeswell, J.A., Whittington, R.J., Theron, J.N. and Craig, J. (2000) Calibrating the Late Ordovician glaciation and mass extinction by the eccentricity cycles of Earth's orbit. *Geology*, **28**, 967–970.

Vaslet, D. (1990) Upper Ordovician glacial deposits in Saudi Arabia. *Episodes*, **13**, 147–161.

von zur Mühlen, L. (1931) Zur Geologie der Gegend von Harrar und Deder in Ostabessinien. *Z. Deut. Geol. Ges.*, **83**, 625–634.

Washburn, A.L. (1970) An approach to genetic classification of patterned ground. *Acta Geogr. Lodz.*, **24**, 437–446.

Zanettin, B., Bellieni, G., Justin-Visentin, E. and Haile, T. (1999) The volcanic rocks of the Eritrean Plateau; stratigraphy and evolution. *Acta Vulcanol.*, **11**, 183–193.

Neoproterozoic glaciated basins: a critical review of the Snowball Earth hypothesis by comparison with Phanerozoic glaciations

J.L. ETIENNE*‡, P.A. ALLEN*†, R. RIEU*§ and E. LE GUERROUÉ*▲

*Geological Institute, Department of Earth Sciences, ETH-Zentrum, Haldenbachstrasse 44, CH-8092 Zürich, Switzerland
†Department of Earth Science and Engineering, Imperial College London, South Kensington Campus, London SW7 2AS, UK
‡Neftex Petroleum Consultants Ltd, 115BD Milton Park, Abingdon, Oxfordshire, OX14 4SA, UK (e-mail: james.etienne@neftex.com)
§Repsol YPF, Exploration & Production, Calle Orense 34, Planta 3a, 28020 Madrid, Spain
▲Structure et Propriétés de la Matière, Géosciences Rennes, 263 Avenue du General

ABSTRACT

The Neoproterozoic is widely considered to have experienced some of the most severe climatic perturbations recorded in Earth history, with extensive glaciations often referred to as 'Snowball Earth' events. The Snowball Earth and competing hypotheses seek to explain a wide range of geological data on Neoproterozoic pre-, syn- and post-glacial successions including glacial sedimentology, chemostratigraphy, palaeoceanography, geochronology, palaeomagnetism and palaeogeography, geodynamics, tectonics, palaeontology and palaeobiogeochemistry. However, our understanding of the Phanerozoic and particularly the Cenozoic and contemporary glacial geological record is often relatively neglected when evaluating the evidence for apparent severe and prolonged periods of globally synchronous glaciation. This paper presents a review of the available geological data for Neoproterozoic glacial successions in the light of what we know about the Cenozoic and recent glacial record. Most Neoproterozoic successions are shown to exhibit spatial and temporal variability, with sediment stacking patterns and facies associations indicative of dynamic ice masses. These characteristics are typical of sedimentary sequences deposited along glaciated continental margins throughout Earth history, without the need for global synchroneity or necessarily severe climatic excursions. Although recurrent very widespread glaciation is envisaged in the Neoproterozoic, the presence of analogous glacigenic and interglacial successions in the Neoproterozoic and Cenozoic suggest the operation of a similar set of processes across a similar range of depositional environments. Consequently, an unambiguous sedimentary record of hydrological shutdown during a prolonged global glaciation appears to be lacking. This indicates either a preservational bias in Neoproterozoic successions of the advance and recessional stages of glacial epochs, or the occurrence of dynamic, non-global glaciations during the Neoproterozoic.

Keywords Glacial sedimentation, Neoproterozoic, Snowball Earth

INTRODUCTION

Earth's Neoproterozoic glacial record retains a number of features that are apparently contradictory to our understanding of Phanerozoic glaciation, the foremost of which are palaeomagnetic data suggesting marine-terminating glaciers at low latitudes and the association of ^{13}C-depleted carbonates above and locally below glacigenic successions across the globe (Sumner et al., 1987; Schmidt et al., 1991; Schmidt & Williams, 1995; Williams, 1996; Hoffman et al., 1998a, b; Sohl et al., 1999; Schrag et al., 2002; Halverson et al., 2004; Fig. 1a). Several radical models have been proposed in order to explain these observations, including long-lived, but rapidly terminated (10^4

Fig. 1 Glacially influenced Neoproterozoic deposits: (A) contact between Fiq diamictite and Hadash cap carbonate, Wadi Bhani Kharus, Jabal Akhdar, (B) stratified diamictite lithofacies, Ayn Fm. (formerly Mirbat Sandstone Fm., Lower member), Dhofar; (C) glacially striated clast in diamictite lithofacies of the Blaini Fm. near Dadahu; (D) dropstone in glaciolacustrine deposits, Ayn Fm., Dhofar; (E) wave-rippled deposits in the Fiq member, Ghadir Manqil Fm., Wadi Sahtan, Jabal Akhdar; (F) Gilbert-type delta foresets in the Ayn Fm. in Wadi Anushar, Dhofar. All photographs Sultanate of Oman, except (C) from the Lesser Himalaya, northern India. Hammer for scale in (A) and (B) is 30 cm in length. Lens cap in (C) 3.5 cm diameter; lens cap in (D) to (F), 5 cm diameter.

to 10^6 years) global glaciations (the Snowball Earth hypothesis; Kirschvink, 1992; Hoffman *et al.*, 1998a, b; Hoffman & Schrag, 2002), high obliquity (Williams & Schmidt, 2004) and progressive rifting during the breakup of the Rodinia super-continent (the 'Zipper Rift' model of Eyles and Januszczak, 2004). The high obliquity theory suffers since a mechanism is required in order to reduce obliquity (Williams & Schmidt, 2004), while the Snowball Earth and Zipper Rift models argue the genetic nature of diamictite facies and their palaeo-climatic significance. Although the Zipper Rift model concedes that low solar luminosity may have allowed glaciation at lower latitudes than during the Phanerozoic, the quality of the palaeo-magnetic evidence for tropical or equatorial

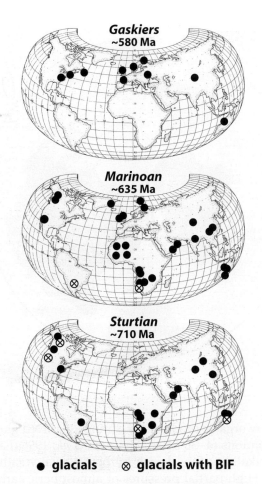

• glacials   ⊗ glacials with BIF

Fig. 2 Modern distribution of Sturtian (~710 Ma), Marinoan (~635 Ma) and Gaskiers (~580 Ma) successions. Crossed circles indicate association of Banded Iron Formations (BIF). Reproduced from Hoffman (2005) with permission of the Geological Society of South Africa. Copyright (2005) Geological Society of South Africa.

glaciation at low altitudes is seriously questioned (Eyles & Januszczak, 2004). Despite their differences, all models agree on at least local glacial activity, typically subdivided temporally into the Sturtian (c. 700 Ma), Marinoan (c. 635 Ma) and Gaskiers (c. 580 Ma) glacial epochs (Fig. 2). Historical reviews on the development of Proterozoic climate models can be found in Hoffman and Schrag (2002) and Eyles and Januszczak (2004).

While advances have been made in geochronology and the chemostratigraphic characteristics of overlying 'cap' carbonate sequences, relatively little detailed work has been undertaken on the glacial facies assemblages in the context of the Snowball Earth hypothesis (with notable exceptions; Arnaud & Eyles, 2002a, b; Allen *et al.*,

2004; Benn & Prave, 2006; Dobrzinski & Bahlburg, *in Press*; Rieu *et al.*, unpublished data). Since the style of Neoproterozoic glaciation remains to be rigorously tested, our understanding of the Phanerozoic, and particularly the Cenozoic glacial record, provides the most appropriate means by which to test depositional models and identify the dominant glaciological processes which operated in Neoproterozoic basins. This paper aims to review sedimentological data that provide insights into the palaeoglaciological characteristics of Neoproterozoic glaciation. A better understanding of climatic variability during this time is highly desirable, since it is here that the first fossil evidence for metazoan life is recorded (Narbonne *et al.*, 1994), and because the Snowball hypothesis challenges our knowledge of the boundary conditions of Earth climate.

The Palaeomagnetic Record

While most Neoproterozoic glacial successions lack reliable palaeomagnetic constraints, the predominance of low-latitude palaeopoles is notable (Evans, 2000) and is atypical by Phanerozoic standards (Evans, 2003). However, relatively few samples have conclusively passed syn-sedimentary fold tests to demonstrate natural remanent magnetization (NRM), and the age of magnetization acquisition is often open to debate. The issue is complicated further given the evidence for a significant (~30%) octopole component in the Proterozoic Earth's magnetic field (e.g. Kent & Smethurst, 1998), although Williams and Schmidt (2004) argue that this would be insufficient to make moderate latitude palaeopoles appear equatorial. Our understanding of Neoproterozoic palaeogeography is limited for the time between 720 Ma and 600 Ma (Meert & Powell, 2001), and competing high and low latitude models for Laurentia have different implications for testing both the high obliquity and Snowball Earth hypotheses (Meert & Torsvik, 2004). Nevertheless, good palaeogeographical models have been developed for 750 Ma and 580 Ma which may provide palaeolatitudinal constraints on some of the older Neoproterozoic basins, and those that are equivalent to the Gaskiers epoch (Fig. 3). Given the problems in resolving the timing of 'Sturtian' glacial events (Halverson, 2005; see below), a detailed palaeogeographical model for the end of the Marinoan glacial epoch at

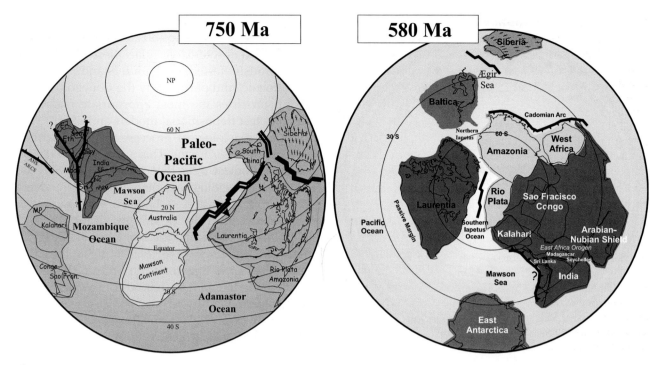

Fig. 3 Palaeogeographical reconstructions for 750 Ma and 580 Ma. Reproduced from Meert and Torsvik (2003) with permission of Elsevier.

635 Ma (Hoffmann *et al.*, 2004; Condon *et al.*, 2005) is now much required. A comprehensive review of Neoproterozoic palaeomagnetic data may be found in Evans (2000).

Chemostratigraphy

One of the key characteristics of Proterozoic glacial successions is their association with ^{13}C depleted carbonates. However, the exact nature of the relationship between ^{13}C depleted carbonates and glaciation is a subject of debate. Negative $\delta^{13}C$ anomalies occur in preglacial stratigraphic units in Svalbard (Halverson *et al.*, 2004), Namibia (Halverson *et al.*, 2002), Ethiopia (Miller *et al.*, 2003), Canada (Hoffman & Schrag, 2002), Australia (McKirdy *et al.*, 2001) and Scotland (Brasier & Shields, 2000). Schrag *et al.* (2002) proposed that the low-latitude position of continental landmasses during this time would have led to more efficient burial of organic carbon and the development of large methane reservoirs, the slow release of which resulted in the negative $\delta^{13}C$ anomalies observed in preglacial carbonates. Negative $\delta^{13}C$ values observed in postglacial cap carbonates are interpreted differently, and are considered to be a result of low organic productivity during long-lived global glaciations (4–30 Myrs; Hoffman *et al.*, 1998a) and an alkalinity flux driven by post-glacial weathering. High partial pressures of atmospheric carbon dioxide are required to initiate deglaciation in the Snowball Earth model, and are achieved by volcanic outgassing of CO_2 (Hoffman *et al.*, 1998a; Hoffman & Schrag, 2002). Some geochemical evidence in support of elevated pCO_2 has been presented for the Marinoan glacial succession in Namibia, although the associated $\delta^{13}C$ excursion may not be fully accounted for (Kasemann *et al.*, 2005).

Overall the linkage between Proterozoic glaciation and carbon cycling remains unclear. For example, geochemical analysis of the Ediacaran Shuram Formation of the Nafun Group in Oman shows a pronounced negative (–12‰ $\delta^{13}C_{carb}$) excursion which persists through hundreds of metres of stratigraphy (Le Guerroué *et al.*, 2006; Fig. 4). This perturbation is greater in magnitude and more long lived than those recorded in cap carbonate sequences, and highlights that large negative $\delta^{13}C$ shifts do not require glaciation as a precursory condition. However, it is interesting to note that as with cap carbonates, the Shuram excursion is associated with a transgressive cycle.

Fig. 4 Composite δ¹³C record for the Neoproterozoic modified from the Halverson (2005) version 2 reconstruction (calibrated by lithostratigraphic correlation of the Petrovbreen Member diamictites (Svalbard) with the Chuos Fm. in Namibia), with revised calibration for the Ediacaran period following Le Guerroué *et al.* (2006a, b). A simplified Phanerozoic δ¹³C record is included for comparison after Jacobsen and Kaufman (1997) and Hayes *et al.* (1999). Note the difference in scale between Proterozoic and Phanerozoic composite records. Data for Canada, Svalbard and Namibia after Halverson (2005), Oman (Le Guerroué *et al.*, 2006a, b; Burns & Matter, 1993; McCarron, 2000; Cozzi and Al-Siyabi, 2004), and China (Condon *et al.*, 2005).

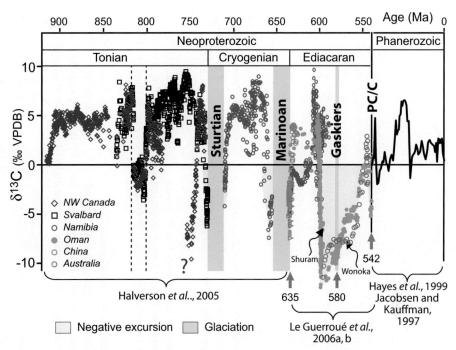

Although local extreme (−41‰ δ¹³C_carb) isotope values occur within some cap carbonate sequences, such values have been interpreted as a result of methane clathrate release (e.g. Jiang *et al.*, 2003; see also Kennedy *et al.*, 2001). Studies of geologically recent submarine continental slope failures (past 45 ka) are co-incident with ice-core records of elevated atmospheric methane, and highlight the linkage between submarine mass movement processes and release of methane hydrates over glacial-interglacial transitions (Maslin *et al.*, 2004). Since the Phanerozoic record illustrates a relationship between deglaciation and gas hydrate release, similar mechanisms are likely to have caused local, short-lived perturbations of the carbon cycle in the Neoproterozoic.

PALAEOENVIRONMENTAL DISCRIMINATION OF DIAMICTITE FACIES

In order to critically evaluate and interpret sediments of glacigenic origin in the geological record, an understanding of contemporary processes of glacial and glacially influenced sedimentation is required. Knowledge of the range and distinctiveness of depositional systems, processes, resulting lithofacies assemblages and sediment-landform

associations is essential for accurate palaeoenvironmental reconstruction. However, glacial deposystems tend to be complex and in some cases, even basic tasks such as distinguishing marine from terrestrial glacigenic sediments can be challenging (e.g. Lowe *et al.*, 2001, and references therein). This is particularly the case in the absence of palaeoclimatologically significant faunal assemblages. Perhaps one of the key issues in many Neoproterozoic successions is determining to what degree the evidence favours a glacial origin. Eyles and Januszczak (2004) argued that many 'tillites' of former workers are in fact tectonically induced mass flow deposits, many of which do not contain a demonstrable glaciclastic debris content (but see Table 1). Similar arguments have been provoked over many Neoproterozoic diamictite-bearing formations (e.g. Schermerhorn, 1974; Bhatia & Kanwar, 1975; Eyles, 1992; Eyles & Januszczak, 2004, and references therein), which is unsurprising since sediment gravity flows are a common component of glacimarine environments (e.g. Benn & Evans, 1998; Eyles *et al.*, 2001; Laberg & Vorren, 2000; Elverhøi *et al.*, 2002; Taylor *et al.*, 2002; Nygård *et al.*, 2002; Ó Cofaigh *et al.*, 2004). Indeed, stacked debris flows are thought to be the principal building blocks of large glacial fan systems in the polar North Atlantic (Dowdeswell *et al.*,

Table 1 Sedimentological characteristics of some Neoproterozoic diamictite-bearing successions. Note that a significant literature is represented in the Hambrey and Harland (1981) volume on Earth's Pre-Pleistocene glacial history; the reader is referred to this for age constraints (including biostratigraphy) at the time of publication, and interpretations for the genesis of individual successions. This volume is freely available in PDF format at http://www.aber.ac.uk/glaciology. Some examples are provided of formation thicknesses, but it should be noted that many successions show considerable lateral thickness variations

Stratigraphic unit		Characteristics	References
AUSTRALIA			
Louisa Downs Gp.	Egan Fm. 80 m	Diamictite, dolomite, limestone, sandstone, arkose, conglomerate, shale and siltstone. Well preserved striated pavement occurs in the Mount Ramsay area. Striations indicate ice flow from the North. Capped by a laminated flaggy pink dolomite with red shale interbeds	Coats & Preiss (1980) Corkeron et al. (2001) Griffin et al. (1998)
Kuniandi Gp.	Landrigan Tillite 210 m	2 polished and striated pavements occur beneath the Landrigan tillite; east–west oriented striae; pluck marks indicate ice flow from the east; polished and striated boulders; diamictites are poorly sorted and lack stratification; clasts mainly extra-basinal	Roberts et al. (1972) Coats & Preiss (1980) Yeats & Muhling (1977)
Duerdin Gp. (E. Kimberleys)	Moonlight Valley Tillite 200 m	Diamictites with massive red siltstone matrix; contain abundant striated and faceted clasts; overlain by lonestone-bearing shales. Diamictites in the Osmond Range overlie a striated pavement with crescentic structures – ice flow from the northeast. Capped by a pink finely laminated dolomite with red shale partings and passes gradationally upwards into the Ranford Fm.	Coats & Preiss (1980) Dow & Gemuts (1967) Blake et al. (1998) Dunster et al. (2000)
	Fargoo Tillite >100 m	Diamictites contain abundant polished, striated and faceted clasts	
Mount House Gp. (Kimberleys)	Walsh Tillite	Abundant striated clasts, erosively based (unconformable), clast imbrication possibly indicative of lodgement till	Coats & Preiss (1980)
Umberatana Gp., Yudnamutana Subgroup (Adelaide Rift Basin)	Bolla Bollana Tillite 204 m	Massive diamictite; shale, siltstone, mudstone, pebbly calcareous diamictite, quartzite and arkose. Rare striated and faceted clasts, including extra-basinal errratics	Coats (1981) Preiss (2000) Krieg et al. (1991)
	Wilyerpa Fm. 584 m	Laminated mudstone, siltstones, arenites, grits and minor diamictites; gradationally overlies and local intertongues with the Appila Tillite	Preiss (2000) McKirdy et al. (2001) Krieg et al. (1991)
	Pualco Tillite <3300 m	Calcitic diamictite, orthoquartzite, siltstone, sandstone and limestone. Rare faceted and striated clasts, but locally abundant in the Central Flinders Ranges (Daily & Forbes, 1969)	Daily & Forbes (1969) Coats (1981) Young & Gostin (1991)

Unit	Description	References
Elatina Fm. (100–500 m)	Sandstone with local lenses of diamictite, interpreted as glacifluvial outwash; some evidence of ice-contact deformation. Abundant faceted and striated clasts (Mawson, 1949; Dalgarno & Johnson, 1964)	Mawson (1949) Dalgarno & Johnson (1964) Preiss et al. (1998) Preiss (2000) Lemon & Reid (1998) Lemon & Williams (1998) Schmidt & Williams (1995) Krieg et al. (1991) Coats (1973) Parkin (1969)
Sturt (300 m)/ Appila Tillite Bibliando, Hanborough, Merinjina (650–1500 m) and Calthorinna tillites (650 m)	Muddy and sandy diamictites. The Appila Tillite (~170 m thick) consists of pebbly diamictite and conglomerate with a disconformable base and contains abundant striated and faceted clasts. The Sturt Tillite contains rare faceted clasts (Sprigg, 1942). A number of other diamictite units occur at similar stratigraphic levels including the Bibliando, Hanborough, Merinjina and Calthorinna tillites. The Hansborough tillite contains rare faceted clasts; the Merinjina Tillite contains abundant striated clasts. Further details on these may be found in Coats (1981)	Coats (1981) and references therein Dunn et al. (1971) Sprigg (1942) http://www.ga.gov.au/oracle/stratnames_info.jsp
Umberatana Gp., Yerelina Subgroup Pepuarta ~280 m Mount Curtis Tillite ~90 m	Pepuarta: massive pebbly-cobble siltstone with lenticular sandstone bodies. Mount Curtis: predominantly dolomitic silty diamictite. Both units contain abundant faceted and striated clasts (Coats, 1981 and references therein).	Coats (1981) and references therein Preiss (2000)
Olympic Fm. 36 m (Halls Creek Gp.)	Cross-bedded arkosic sandstone with conglomerate lenses; poorly sorted sandstones with shale interbeds, conglomerates, diamictites and carbonate beds. Considerable lateral variation in facies associations; diamictites interpreted by Lindsay (1989) as mass flow deposits; channelised conglomerates; striated and faceted clasts common (Wells, 1981)	Wells et al. (1967) Wells (1981) Lindsay (1989) Freeman et al. (1991)
Amadeus Basin Areyonga Fm. (230 m)	Diamictites, with intercalated sandstone, conglomerate and carbonate interbeds, shales and siltstones; diamictites are 1–40 m thick; diamictites towards the top of the formation show weak stratification and contain conglomerate pellets, abundant sandstone lenses; intra- and extra-basinal clast lithologies; boulder pavements widely distributed throughout the formation; rare (<1%) striated and faceted clasts. Other units of diamictite containing striated and faceted clasts are known from elsewhere in the Amadeus Basin, and occur in the Boord Fm. and the Inindia beds (Wells, 1981)	Wells (1981) Lindsay (1989) Prichard & Quinlan (1962) redefined at http://www.ga.gov.au/oracle/stratnames_info.jsp
Boord Fm.	Sandstone, siltstone, carbonate, diamictite and calcilutite, calcarenite; faceted clasts are common, with occasional examples of striated clasts	Wells (1981)

Table 1 (*cont'd*)

Stratigraphic unit	Characteristics	References
Ngalia Basin		
Naburula Fm. (8 m)	Shale, siltstone, diamictite and dolomite. Common striated and faceted clasts	Wells (1981) http://www.ga.gov.au/oracle /stratnames_info.jsp
Mount Doreen Fm. (340 m)	Shale, diamictite and dolomite. Striated and faceted quartzite clasts	Wells (1976) Wells (1981)
Georgina Basin		
Mount Cornish Fm. (680 m), Yardida Tillite (650–2900 m) and Mount Stuart Fm.	Siltstone, shale, diamictite associated with arkose and dolomite. Frequent faceted and striated clasts are known from diamictites in both the Mount Cornish Fm. and the Yardida Tillite; lensoid bodies of diamictite are also known from the Mount Stuart Formation which crops out in the Georgina Basin, and locally in outliers between the Ngalia and Georgina basins.	Walter (1981) Wells (1981) Kruse et al. (2002)
Grassy Gp. (**Tasmania**)		
Cottons Brecca	Diamictite, volcaniclastic sandstones; till pellets and dropstones in interbedded laminites, wide range of clast lithologies; capped by a pale pinkish-grey laminated Cumberland Creek Dolostone	Jago (1974) Calver & Walter (2000) Calver et al. (2004)
Togaria Gp. (NW **Tasmania**)		
Croles Hill Diamictite <250 m	Diamictite, rare laminated mudstone and siltstone interbeds containing dropstones; no known cap carbonate unit	Calver et al. (2004)
AFRICA		
Morocco, Anti-Atlas; Siroua Series		
Tiddiline Fm. Anzi Fm.	Conglomerates, turbiditic greywackes, laminated siltstone, diamictite and feldspathic sandstones. Greywackes contain possible dropstones; no faceted or striated clasts observed	Leblanc (1981)
Central **Sierra Leone**; Rokel River Gp.		
Tibai Mbr., Tabe Fm.	Diamictites (including paraconglomerates and orthoconglomeratic facies), medium-coarse sandstones, laminated siltstone, fine sandstone. Slump folds and contorted bedding indicate mass-movement processes. Bedded facies contain lonestones	Tucker & Reid (1981)
Algeria, western Hoggar		
Série Pourprée	Conglomerates, diamictites, lonestone-bearing claystones, arkosic sandstones, pelites, siltstones, greywackes and breccias. Clasts in the Adafar glacigenic beds are striated, faceted and bear pressure marks; glacifluvial facies in the Ouallen-In Semmen Gp. also contain striated clasts; 'varved' claystones at Ouallen contain small dropstones.	Caby & Fabre (1981)
W. **Uganda**		
Bunyoro Series	Rare faceted, 'scratched' exotic pebbles in diamictites, associated with rhythmically laminated argillites interpreted as varvites	Davies (1939) Bjørlykke (1981)
Northern **Zaïre**		
Lower Lenda Tillite, Itari district	Poorly sorted pebbly mudstone containing striated cobbles	Cahen (1954)
Lower **Zaïre**		
Niari Tillite	Striated and faceted clasts in diamictites locally abundant	Cahen & Lepersonne (1967) Cahen & Lepersonne (1981)

Region / Group	Unit	Description	References
Lower **Zaïre**	Bamba Mt. tillite	Rhythmically laminated ('varvite') containing sparsely distributed pebbles. Faceted clasts and clasts bearing percussion marks have been reported from equivalent strata in Angola (Schermerhorn & Stanton, 1963; Kröner & Correia, 1973), but are interpreted as non-glacial.	Cahen & Lepersonne (1967) Cahen & Lepersonne (1981)
NW **Angola,** Schisto-calcaire Gp	'Upper Tilloid Formation' <200 m	Lithofacies: diamictite, mudstone, greywacke, conglomerate, breccia, quartzite and limestones; striated clasts reported, but their significance debated	Schermerhorn & Stanton (1963) Schermerhorn (1981)
NW **Angola,** Haut Shiloango Gp	'Lower Tilloid Formation' <500 m	Lithofacies: diamictites, conglomerate, breccia, arkosic and calcareous quartzite, greywacke, mudstone, limestone. Exotic (extra-basinal clasts) reported from tilloids (diamictites?).	Schermerhorn & Stanton (1963) Schermerhorn (1981)
Urungwe District, **Zimbabwe**	Msukwi River tillite ~105 m	Diamictites overlain by laminated shales; clasts in the diamictite are faceted and bear striae	Bond (1981)
Namibia; Otavi Gp.	Chuos southwestern congo	Dropstone intervals in laminated hemipelagic sediment. Dropstones interpreted by Hoffman et al., 1998 and Condon et al., (2002) as rafted by glacier ice; Eyles & Januszczak (2004) contest the presence of convincing dropstone structures, and interpret lonestone-bearing units as debrites and highlight previous sedimentological investigations which identify predominantly mass flow facies, with no indication of glaciclastic debris (Schermerhorn, (1974, 1975; Martin et al., 1985)	Schermerhorn (1974, 1975) Martin et al. (1985) Hoffman et al. (1998) Condon et al. (2002) Eyles & Januszczak (2004)
Namibia; Swakop Gp.	Ghaub congo	Dropstone intervals in laminated hemipelagic sediment. Dropstones interpreted by Hoffman et al., 1998 and Condon et al., (2002) as rafted by glacier ice; Eyles & Januszczak (2004) contest the presence of convincing dropstone structures, and interpret lonestone-bearing units as debrites.	Condon et al. (2002) Eyles & Januszczak (2004)
Namibia; Kuibis sub-group	Blaubeker Fm. >500 m	Diamictite, conglomerate, quartzite, shale; diamictites contain abundant striated and faceted clasts	Kröner (1981)
SW Namibia; Gariep Basin	Numees	Massive diamictite; rhythmically laminated pelites with erratic dropstones; thin iron formation. Numerous faceted and striated clasts reported (De Villers & Söhnge, 1959), although Kröner, (1981) argues many are probably tectonic features, a glaciclastic debris component is accepted.	Rogers (1916) Martin (1965) Kröner (1981)
	Kaigas	Diamictite, arkose, greywackes; dropstones; striated and faceted clasts. Sorting, graded bedding and conglomerate channel-fills led Kröner (1975) to suggest a fluviatile origin.	Kröner (1981) Von Veh (1993)
NW Zambia; Kundelungu Basin	Grand Conglomerat <500 m	Diamictites containing striated and faceted clasts; associated facies variably interpreted as ground moraine, glacilacustrine, glacifluvial sediments and glacially influenced marine deposits dominated by mass flow facies.	Gray (1930) Cahen (1954) Robert (1956) Cahen & Lepersonne (1967) Binda & Van Eden (1972) Cahen & Lepersonne (1981)
	Petit Conglomerat	Striated pebble-sized clasts in diamictites	Cahen (1954)

Table 1 (cont'd)

Stratigraphic unit	Characteristics	References
Mauritania, Mali, E. Senegal, Guinea Jbéliat Fm./'Triad'	Facies: terrestrial glacigenic tillites, deltaic, lacustrine (or marine), and fluviatile facies. Tillites contain striated clasts, lenticular pockets of sandstone and conglomerate which are folded and fractured; rare stratified tillites. Striated pavements, roches moutonnées, glacitectonic structures including step fractures and folds and breccias in preglacial substrate. Polygonal structures associated with sandstone wedges occur at the top of the glacial sequence, always beneath the cap dolostones. These are interpreted as periglacial features. 3–5% striated clasts	Trompette (1973) Deynoux (1978) Deynoux & Trompette (1981) Deynoux (1982) Deynoux (1985)
N. Ethiopia; Tambien Gp. Matheos Fm. ~1000 m	Carbonate, slate and pebbly slate ('diamictite'). The pebbly slate contains striated and polished cobble-sized clasts	Miller et al. (2003)
NORTH AMERICA		
Canada; Newfoundland; Conception Gp. (4 km thick) Gaskiers Fm.	Massive and crudely stratified tabular units of diamictite, interbedded with and overlain by turbidites. Striated and faceted clasts occur. Associated volcanics suggest glaciation on a volcanic cone of a complex island arc (Eyles, 1990).	Eyles (1990)
Canada; Windermere Supergroup Mount Vreeland <1200 m	Massive to crudely bedded diamictite and sandstones; dropstones provide evidence for iceberg rafting. Overlain by a thin laminated grey cap dolostone.	Hein & McMechan (1994) Ross et al. (1995)
Toby Fm. <2500 m	Diamictite, conglomerate, breccia, pelite, carbonate and sandstones	Aalto (1971, 1981) Eisbacher (1981) Ross et al. (1995)
Canada; NW Territories; Rapitan Gp. Sayunei (>500 m) Shezal (<300 m)	Diamictites (<250 m thick) associated with turbidites bearing dropstones. Stratified diamictite occurs with mudstone interbeds. Scoured and polished pavement beneath lowermost diamictite units. Rare striated clasts. Extrabasinal (erratic) clasts in laminated siltstones, till pellets and dropstones. The Sayunei Formation is predominantly siliciclastic rhythmite; where well developed, the Shezal Formation contains reasonably abundant striated clasts, with thin diamictite sheets separated by shale, siltstone and sandstone beds. Clusters of outsize clasts locally occur which may represent iceberg dump-features.	Yeo (1981) Eisbacher (1985) Young (1995)
NW Canada; McKenzie Mountains Icebrook Formation, Stelfox Member	Dropstones, till pellets, angular quartz grains, rare striated clasts. Diamictites interpreted as glacimarine, interbedded with laminated mudstone and siltstones and associated with slope deposits and olistostromes (Aitken, 1991).	Aitken, (1991)
U.S.A.; Kingston Peak Fm. Surprise Mbr. Wildrose Mbr.	Argillite, diamictite, siltstone, sandstone, conglomerates. Small proportion of striated and faceted clasts occurs. Dropstone intervals in laminated hemipelagic sediment	Miller et al. (1981) Condon et al. (2002) Lund et al. (2003)
U.S.A.; Idaho; Windermere Supergroup Edwardsburg Fm. 700–1200 m	Quartzite, diamictites with deformed subangular to subrounded clasts (no dropstones or faceted clasts recorded), volcaniclastics including sandstones and conglomerates. Link et al., (1994) interpret a proximal to distal glacimarine succession, but no definitive evidence presented of glaciclastic debris	Link et al. (1994)

Location	Formation	Description	References
U.S.A.; SE Idaho; Pocatello Fm.	Scout Mountain Mbr.,	Diamictites containing striated clasts; local iron-rich laminites. The succession is capped by a pink dolomite.	Link et al. (1994) Fanning & Link (2004)
U.S.A.; Virginia	Mechum River Fm. (>400 m)	Mudstones, boulder conglomerates, rhythmites, diamictites; discontinuous lenses of arkose sandstones (possible meltwater channel-fills). Rare striated clasts, but apparent lack of dropstones.	Bailey & Peters (1998)
U.S.A.; Southwest Virginia	Konnarock Fm. (formerly Mount Rogers Fm.)	Diamictite, laminated pebbly mudstone; dropstones; striated and faceted clasts 'exceedingly' rare, but do occur	Schwab (1981)
U.S.A.; N. Carolina	Grandfather Mountain Fm.	Laminated pebbly mudstone; widely dispersed dropstones	Schwab (1981)
U.S.A.; Roxbury Conglomerate, Boston Basin	Squantum Tillite Mbr. ~215 m	Diamictite, conglomerate, feldspathic sandstone, laminated (rhythmically) mudstone and siltstones. Rare striated and faceted clasts have been reported, but are considered by some to be of tectonic origin. Rip-up clasts support a mass flow origin for at least part of the succession. Further literature may be found in Rehmer (1961) and Eyles & Januszczak (2004).	Dott (1961) Rehmer (1981) Eyles & Januszczak (2004)
U.S.A.; Utah	Mineral Fork Fm. ~900 m	Conglomerate, sandstone, siltstone, shale, diamictites; diamictites contain rip-up clasts, silty lenses and irregular slumped tops, thought to represent mass flow or ice contact phenomena, however, striated, polished pavement on the underlying Big Cottonwood Fm., and associated whaleback forms, rare faceted and striated clasts reported from diamictites, till pellets; little evidence for major uplift or tectonism during deposition. Interpreted as continental to marine or fully marine glacially-influenced succession	Ojakangas & Matsch (1980) Christie-Blick (1982) Young (2002)
Greenland; Tillite Gp.	Storeelv Fm.	Clast fabrics in diamictites consistent with lodgement till characteristics. Striated clasts in diamictites; Two striated levels, one at base of Storeelv, and 52 m above base of the formation. Facies include basal tillites, waterlain tillites, debris flow facies, ice-proximal and distal glacimarine deposits, rhythmites (proximal and distal facies). Till pellets locally occur. Far-travelled erratics from extra-basinal sources. Lithologies: diamictite, sandstone and conglomerates. Rare sandstone wedges and downfolds interpreted as periglacial features. Capped by laminated orange dolostone associated with grey shaly dolomitic mudstones of the Canyon Fm.	Hambrey (1988) Moncrieff (1989a, b) Moncrieff & Hambrey (1988, 1990) Hambrey et al. (1989) Hambrey & Spencer (1987) Fairchild & Hambrey (1995)
	Arena Fm.	Lithologies: sandstone and dolomitic mudstones; wave ripples towards base indicate open marine conditions; local dropstones.	
	Ulvesø Fm.	Possible Aeolian sandstone facies at base of formation; periglacial thermal contraction and load structures above the youngest diamictite; evaporitic facies include non-ferroan dolomite pseudomorphs after gypsum including alabastrine gypsum and gypsum laths; cubic halite pseudomorphs also occur – 30 m above the base of the formation and at the top of the Arena Fm. which divides the Ulvesø and Storeelv Fms. Lithologies: diamictite, sandstone, conglomerate and mudstone. Sandstone wedges and downfolds with polygonal arrangement in plan view at the top of the formation, interpreted as periglacial features. Interpreted as transitional low level terrestrial to glacimarine depositional environments.	

Table 1 (*cont'd*)

Stratigraphic unit	Characteristics	References
SOUTH AMERICA		
Brazil São Francisco Supergroup — Jequitaí Fm. (Macaúbas Gp)	Massive diamictites dominant in the western part of the Macaúbas Gp., but in the east are associated with quartzites, phyllites, siltstones, greenschists and laminites. Faceted and striated clasts are locally abundant. Overlain by the Bebedouro cap carbonate	Rocha-Campos & Hasui (1981)
Brazil; Jaganá Gp. (1400 m thick) — Puga Fm.	Diamictite, conglomerate, shale and sandstone. Diamictites contain 1–2% faceted clasts and occasional striated clasts interpreted as terrestrial glacial deposits which grade laterally into more distal glacimarine facies in the Paraguay-Araguaia geosyncline Overlain by cap carbonate of the Araras Fm. Diamictites are also known of the Jacadigo Group where they are associated with banded iron formations similar to those of the Rapitan Group (Gaucher et al., 2003).	Rocha-Campos & Hasui (1981) de Almeida (1964a) de Almeida (1964b) Alvarenga et al. (2004) Nogueira et al. (2003) Gaucher et al. (2003)
EUROPE		
Scotland; Dalradian Supergroup — Port Askaig Formation	Diamictites, conglomerates, sandstones, siltstones. Faceted clasts occur, but no striated clasts have been reported. A mass flow origin is indicated for many of the diamictite sheets that occur in the formation (Arnaud & Eyles, 2002b)	Spencer (1985) Arnaud & Eyles (2002b)
Scotland — Kinlochlaggan boulder bed	'Boulder bed' associated with feldspathic quartzite and massive and bedded quartzitic psammites. Dropstones reported, but no striated clasts.	Treagus (1981)
Scotland, Southern Highland Group — Loch na Cille boulder bed	Interpretations differ considerably from volcanic hyaloclastic breccia (Gower, 1977) to glacial (Prave, 1999). The beds contain extra-basinal clasts which may support a glacial influence.	Condon & Prave (2000); Prave (1999)
France; Upper Brioverian Supergroup — Granville Fm.	Pebbly mud diamictites associated with thick sequences of Brioverian turbidites and volcaniclastics. Interpreted as non-glacial debris flows.	Eyles (1990)
Norway; Vestertana Gp. — Mortensnes Fm. 10–60 m	Laminated mudstones bearing dropstones, structureless tillite (?diamictite) contains extra-basinal erratic clasts and rare striated and faceted clasts.	Edwards & Føyn (1981)
— Smalfjord Fm.	Rare striated and faceted clasts, a single striated pavement at Bigganjargga; diamictites interpreted as subaqueous mass flow deposits.	Edwards & Føyn (1981) Arnaud & Eyles (2002a)
S. Norway, Hedmark Group — Moelv Fm.	Massive and stratified diamictites occur in association with conglomerates, sandstones, laminated mudstones containing dropstones and rare striated and faceted clasts. Interpreted as basal tillite transitional with glacimarine sediments with iceberg rafting.	Bjørlykke & Nystuen (1981)

Region	Formation	Description	References
Sweden; Swedish Caledonides	Sito Tillite	Diamictite associated with siltstone and dolomite. No striated clasts have been reported from this unit, and a non-glacial origin is favoured in the absence of dropstones.	Strømberg (1981)
	Långmarkberg Fm.	Laminated siltstones bearing abundant lonestones, massive and stratified (with siltstone and sandstone interbeds) sandy diamictite facies bearing striated clasts.	Thelander (1981)
	Lillfjället Fm.	Massive diamictite, laminated siltstones, dolomitic sandstone and weakly stratified diamictite. No striated or faceted clasts have been reported, but rare dropstones occur.	Kumpulainen (1981)
Southern **Sweden**	Lilla Hals Boulder Bed ~240 m	Bedded diamictites associated with feldspathic and arkosic sandstone, laminated shales and mudstones. Faceted clasts occur, but no striated examples reported. Interpreted as submarine debris flows by Vidal & Bylund (1981).	Vidal & Bylund (1981)
Russia; Rybachiy Peninsula	Motka tilloids	Sandstone, conglomerate, tilloids (?diamictites), breccia, mudstones interpreted as mass flow (slump, slide, turbidites) deposits deposited in a non-glacial setting.	Chumakov (1981)
Russia; S. and Middle Urals	Tolparovo Fm. 600–650 m	Sandstones interbedded with mixtite, gritstone, conglomerate and argillite.	Maslov (2000)
	Kurgashlya Fm. 160–200 m	Mixtites intercalated with sandstone, massive and thinly-bedded siltstones, gritstone, conglomerate, breccia and dolomite interpreted as distal glacimarine sediments reworked by density currents. Mixtites contain very rare striated and faceted clasts.	Chumakov (1981) Chumakov (1998)
	Koiva Fm.	Shale, siltstone, carbonates with localised mixtites and volcanics	Maslov (2000)
	Tany Fm. 360–800 m	Mixtite, sandstones and shales with subordinate carbonate and volcanics. Mixtites contain extra-basinal clasts and dropstones in laminated hemipelagic sediments (Chumakov, 1996). Possible lateral equivalents include the Vil'va Fm. which comprises weakly stratified mixtites (Maslov, 2000), although metamorphism and tectonism are regarded to have destroyed original clast surface features, such that striae have not been observed (Chumakov, 1981).	Chumakov (1981) Chumakov (1996) Maslov (2000)
Belarus	Vilchitsy Fm.	Sandstone, diamictite, rhythmically laminated siltstones containing lonestones, sandy diamictites containing striated clasts.	Chumakov (1981)
Russia; N. Urals; Polyudov Ridge	Churochnaya tillite	Shale, sandstone, brecciated dolomite, diamictite, chert. Diamictites contain extra-basinal clasts and striated clasts.	Chumakov (1981)
Svalbard; Polarisbreen Gp. (1075 m thick)	Wilsonbreen Fm. >200 m	Dolomitic diamictite, discontinuous sandstone and conglomerate bodies, dolostone, limestone, dolomitic shale and rhythmites. Dolostone underlying the Wilsonbreen Formation is frost-wedged. The Gropbreen Mbr. (73 m) diamictites contain <15% striated clasts. Overlain by cap carbonate of the Dracoisen Fm. Sequence interpreted as temperate glacimarine and terrestrial environment.	Hambrey (1988, 1989) Hambrey et al. (1981) Fairchild et al. (1989) Harland et al. (1993) Fairchild & Hambrey (1984, 1995) Harland (1997) Halverson et al. (2004)
	Elbobreen Fm. >400 m	Dolomitic diamictite, dolomitic conglomerate, rhythmites, shale and silty homogenous dolostone. The Petrovbreen Mbr. diamictites contain <15% striated clasts and shows considerable lateral thickness variations.	

Table 1 (*cont'd*)

Stratigraphic unit	Characteristics	References
ASIA		
Oman; Abu Mahara Gp.		
Fiq Mbr. (Ghadir Manquil Fm.)	Striated clasts, rare dropstones, massive diamictite sheets; overlain by cap dolostone of the Hadash Fm.	Rabu (1988) Kellerhals & Matter (2003)
Ghubrah Fm.	Massive diamictites with mixed clast composition; occasional striated clasts.	Allen et al. (2005)
Oman; Mirbat Gp. (formerly Mirbat Sandstone Fm.)		
Shareef Fm.	Diamictites and laminated lonestone-bearing muds; abundant polished and striated clasts.	Leather et al. (2002) Rieu et al. (*In Review*)
Ayn Fm.	Abundant striated and faceted clasts, dropstones in stratified diamictite facies. Glacifluvial facies in Gilbert-type deltas.	This study
India; Baliana Gp.		
Blaini Fm. <300 m	Abundant striated and faceted clasts; massive and weakly stratified diamictite sheets and rhythmically laminated mudstones and siltstones; rare sandstones. Overlain by a laminated pink or grey microcrystalline dolomite with red shale interbeds.	Bhatia & Kanwar (1975) and references therein Jain & Varadaraj (1978) Bhatia & Prasad (1981) Brookfield (1987) This study
China; Yangtze Platform, South China		
Nantuo Fm. ~210 m	Major facies types include lodgement tillites, glacifluvial and proglacial subaqueous deposits. Lithofacies: stratified and massive clast-rich (30–50% clasts) diamictites, laminated siltstones bearing lonestones and argillo-arenaceous rocks containing <10% clasts. Discontinuous sandstone and conglomerate channel fills occur in the diamictites. Abundant dropstones, striated and faceted clasts.	Songnian et al. (1985) Songnian & Lesheng (1987) Dobrzinski et al. (2004) Dobrzinski & Bahlburg (*In Press*)
Chang'an Fm. <3700 m	Argillaceous pebbly sandstone, slate, sandstone and pebbly sandy mudstone. Striated clasts also occur.	Yuelun et al. (1981)
Central **China**; E. Qinling Range, Henan Province,		
Luoquan Fm. (204 m)	Massive, bedded and weakly bedded diamictites, conglomerates, pebbly sandstones and rhythmites containing up to 15% dropstones. Diamictites contain abundant striated clasts, and locally overlie striated pavements with p-forms and friction cracks.	Baode et al. (1986)
Northwest **China**; Tarim Block,		
Beiyixi Fm. Altungal Fm. Tereeken Fm. Hangelchaok Fm.	Diamictites, sandstones, conglomerates, rhythmically laminated slates (varvites?) containing dropstones. A number of different facies types are recognised including tillites, turbidites, glacilacustrine, glacimarine and glacifluvial facies associations (Zhenjia & Jianxin, 1985). General indications in favour of glaciation include glacially polished bedrock surfaces, pebbles bearing striations, grooves, abrasion pits and cracks, dropstones in laminated hemipelagic sediment. Permafrost features also occur.	Zhenjia & Jianxin (1985)

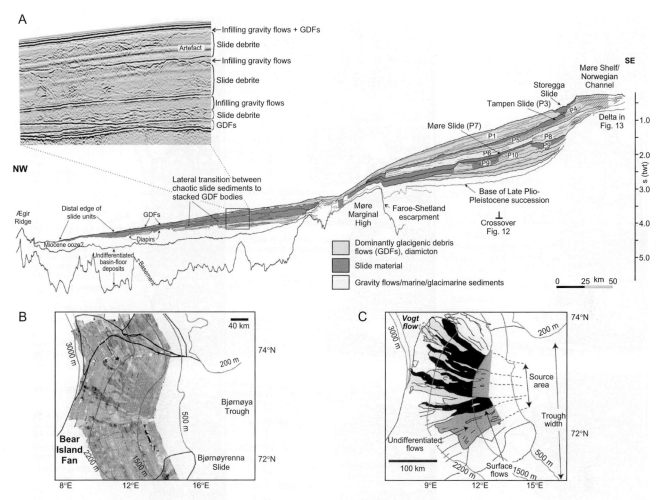

Fig. 5 Submarine glacial fan systems in the polar North Atlantic. (A) Geoseismic section across the North Sea Fan. P1 to P10 are Late Plio–Pleistocene seismic sequences. GDFs: Glacigenic debris flows. Reprinted from Sejrup *et al.* (2004) with permission from Elsevier. (B) GLORIA side-scar sonar imagery of the Bear Island Fan, Barents Sea, and (C) location of subsurface (light grey), surface (black) and undifferentiated (dark grey) glacigenic debris flows. Reproduced from Taylor *et al.* (2002) with permission of the Geological Society, London.

1996; Fig. 5). Examples include the Storfjorden and Isfjorden fans on the continental margin of the Barents Sea and to the west of Svalbard, the Scoresby Sund Fan off the east coast of Greenland and the massive (~350,000 km³) Bear Island Fan, which covers much of the western part of the Barents Sea (Dowdeswell *et al.*, 1996). Redistribution of glacigenic sediment is also a characteristic feature of grounding-line fan assemblages developed on continental shelves (e.g. Lønne, 1995; Powell & Cooper, 2002; Figs 6, 7). In order to derive some palaeoglaciological data for Neoproterozoic successions, it is relevant therefore to examine some basic characteristics of glacial debris entrainment,

transport and sedimentation processes as a means for palaeoenvironmental discrimination.

Subglacial debris entrainment

The entrainment of subglacial debris is dependent on a number of factors, principally the nature of the geological substrate and glacier thermal regime (Kirkbride, 1995; Table 2). Polythermal glaciers (for example on Brøggerhalvøya, Svalbard) tend to be effective erosional agents (e.g. Boulton, 1970; Sugden, 1978), as these ice masses have considerable areas where basal melting and re-freezing of meltwater occurs. Such areas are dependent on the

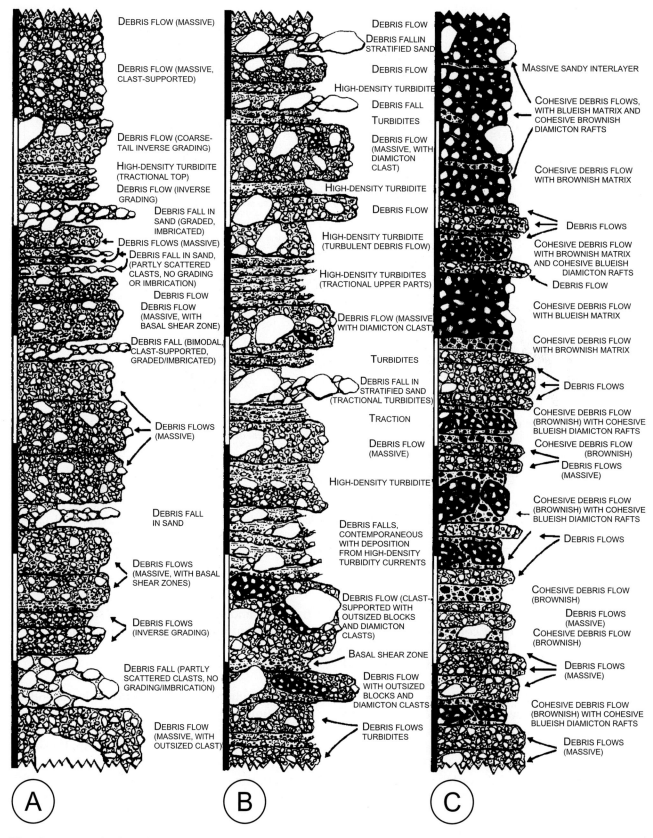

Fig. 6 Stratigraphic logs from the mid-upper part of an ice-contact submarine fan system at Storsand, Oslofjorden, southern Norway. (A) Cohesionless debrites interbedded with turbidites and debris-fall gravel; (B) debrites containing blocks of ice-rafted diamicton; (C) cohesive debrites. Vertical scale bars are metres. Reprinted from Lønne (1995) with permission from Elsevier.

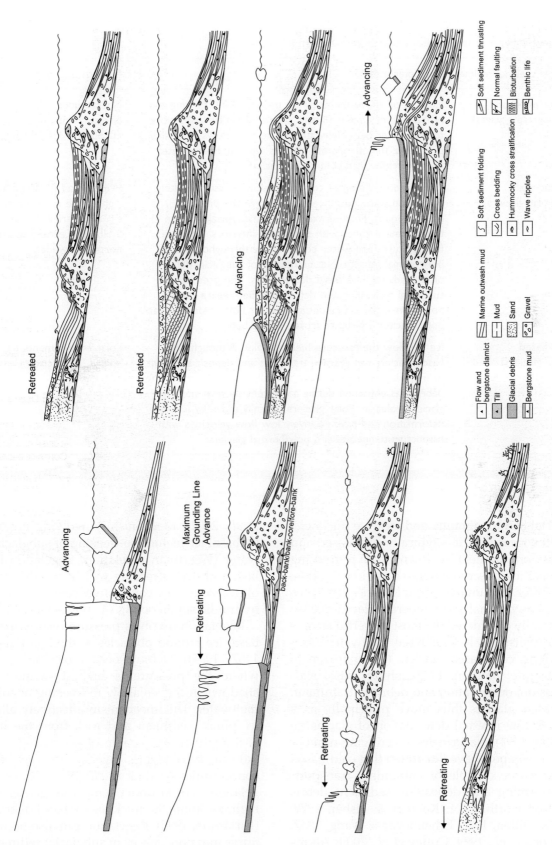

Fig. 7 Sequence stratigraphic models for glaciomarine depositional successions. Reproduced from Powell and Cooper (2002) with permission of the Geological Society, London.

Table 2 Glacier characteristics under different thermal regimes

Glacier thermal regime	Characteristics	Modern distribution
Warm/wet-based (Temperate glaciers)	Ice is at or above the pressure melting point. Melting and re-freezing of basal ice keeps subglacial debris loads close to the glacier bed. Meltwater at the bed allows basal sliding processes, thus temperate glaciers typically have greater flow velocities than cold-based glaciers.	Mid-latitudes, Alpine environments, e.g. European Alps, Iceland.
Polythermal (Subpolar glaciers)	Variable distribution of cold ice (below pressure melting point) and warm basal ice (at or above the pressure melting point). Polythermal glaciers are typically frozen in their terminal zones with temperate interiors. Both ice temperatures and water content vary throughout polythermal glaciers. Net adfreezing of subglacial debris often leads to thick basal debris loads. The presence of subglacial meltwater and deformable bed means that polythermal glaciers can attain greater flow velocities than polar glaciers which are frozen at the bed.	Widespread in Arctic and high Alpine environments, e.g. Brøggerhalvøya in Svalbard; Kebnekaise massif, northern Sweden.
Cold-based (Polar glaciers)	Ice is below the pressure melting point. Although cold-based glaciers can entrain, transport and deposit sediment debris, basal debris loads are typically very low, unless the debris was entrained during a different thermal stage in glacier evolution. Polar glaciers typically move by internal deformation and have relatively low flow velocities with respect to temperate and polythermal glaciers.	Polar environments e.g. Dry Valleys, Antarctica.

subglacial bed topography and local ice thickness; where the ice is thin, it is typically cold-based and promotes entrainment by downstream re-freezing and regulation of sediment into basal ice (Boulton, 1979; Hutter & Olunloyo, 1981; Menzies, 1981; see below). Polythermal and temperate glacial masses are generally considered the most powerful erosive agents, whilst the lack of basal meltwater associated with cold-based glaciers (being frozen to the substrate; Boulton, 1972) inhibits basal sliding, erosion, deformation and debris entrainment. Cold-based glaciers thus move principally as a result of slow internal deformation of glacier ice (Paterson, 1994). Nevertheless, a growing number of research papers have illustrated that cold-based ice masses *are* capable of entraining, transporting, deforming and depositing sediment debris (e.g. Holdsworth, 1974; Koerner & Fisher, 1979; Chinn & Dillon, 1987; Echelmeyer & Wang, 1987; Fitzsimons *et al.*, 1999; Cuffey *et al.*, 2000; Atkins

et al., 2002). Mechanisms involving entrainment by basal ice include re-freezing of subglacial meltwater (Weertman regelation), regulation into the bed, and net adfreezing which includes processes of freeze-on by conductive cooling and glacio-hydraulic supercooling of subglacial meltwater.

Regelation involves pressure-related melting of basal ice around obstacles at the bed (Weertman, 1957, 1964). Melting occurs on the stoss face, where the pressure is highest, while pressure shadows in the lee allow re-freezing of subglacial meltwater. This mechanism effectively allows ice to 'pluck' sediment and rock from the bed, but is thought to be capable of producing only thin (<0.1 m) basal debris layers, with low sediment concentrations (Alley *et al.*, 1997). Repetitive regelation, typical under temperate glacier thermal regimes, limits the thickness of basal debris layers (Kirkbride, 1995). Regelation can also occur *downwards* into pore spaces in subglacial sediment, and is considered a much more effective entrainment mechanism, capable of generating thick basal debris layers (e.g. Alley *et al.*, 1997; Iverson, 2000). Debris contents vary depending on the substrate, but can be very high (Boulton, 1970; Harris & Bothamley, 1984). This mechanism is similar to the basal freeze-on model of Christoffersen and Tulaczyk (2003), which involves the formation of a segregated ice layer as pore water accretes to the glacier sole. The freezing front then migrates downwards into the substrate, allowing effective entrainment of large volumes of debris (Christoffersen, 2003).

Net adfreezing occurs where freezing dominates over melting at the bed, allowing considerable volumes of debris-rich basal ice to form. This is particularly effective when meltwater flows into areas of the bed that are cold-based (Weertman, 1961; Hubbard & Sharp, 1989; Hubbard, 1991), and thus explains why polythermal glaciers are important transporters of large volumes of sediment debris (Elverhøi *et al.*, 1998). Freeze-on by conductive cooling occurs as a result of changing basal thermal conditions (Alley *et al.*, 1997). The thick ice-sheets envisaged in Snowball Earth models are likely to have been effective at insulating the bed, and thus resisted cold-based glacial conditions; however, where ice was thin over highs in the subglacial topography (typical of dispersal centres) cold-based conditions probably existed (*cf.* Kleman *et al.*, 1997; Kleman & Hättestrand, 1999). Changes in basal thermal conditions may be initiated over time by surface cooling, or increased accumulation leading to downward advection of cold surface ice (Alley *et al.*, 1997). This mechanism is likely to be very important in the long-term evolution of ice sheets, because of the resulting changes in dynamics. Indeed, Christoffersen (2003) has suggested that basal freeze-on is a viable mechanism for the shutdown of ice streams observed in West Antarctica. Changes in basal thermal conditions over time can therefore radically affect the overall flow dynamics of ice masses. Dependent on the duration of glaciation, surface temperature conditions and degree of precipitation, it is likely that these processes would operate in a Snowball Earth glaciation. Given the important role played by sea ice dynamics in numerical climate models that simulate Snowball Earth conditions (e.g. Warren *et al.*, 2002; Goodman & Pierrehumbert, 2003;

Lewis *et al.*, 2003), a consideration of terrestrial ice sheet dynamics is also required.

Net freeze-on by glaciohydraulic supercooling occurs when subglacial meltwater is supercooled as it rises from overdeepenings at the bed. Thick debris-rich layers can be accreted on to the glacier sole (e.g. Lawson *et al.*, 1996; Strasser *et al.*, 1996). This mechanism of entrainment is likely to be important over a range of spatial scales, for temperate glaciers and ice sheets that have complex basal topography, and particularly in Snowball Earth scenarios.

Particle roundness and other shape parameters

Numerous different processes operate in the transfer of sediment debris through the glacier system, and, in some cases involve large-scale reorganisation of debris between supraglacial, englacial and subglacial transport zones. Active *versus* passive transport (Boulton, 1978) relates to the relative histories of material in subglacial transport compared with supraglacial and englacially transported debris. Supraglacial material is considered to be passively transported, with little modification to particle shape by physical processes other than local reworking by surface meltwater, freeze-fracturing, and, in some instances, aeolian ventifact formation. Since the majority of supraglacial debris is derived from rockfall and talus, the debris is often very angular or angular in clast roundness. Glaciers with considerable supraglacial debris loads are most common in areas of high relief (e.g. in the Himalaya; Benn *et al.*, 2001; Benn & Owen, 2002; the Southern Alps, New Zealand; Hambrey & Ehrmann, 2004), particularly where there is a strong coupling between mountain slope and glacier transport systems (Kirkbride, 1995), although supraglacial debris can also accumulate on ice shelves which fringe mountainous terrain (Evans & Ó Cofaigh, 2003). By contrast, debris incorporated in subglacial transport displays a wide variety of particle sizes and shapes, with surface features resulting from abrasive wear. This material is said to be *actively* transported and is significantly modified during subglacial transport.

Actively and passively transported debris are often analysed in terms of their constituent particle roundness (following Powers, 1953) and other aspects of particle shape. One of the more recent

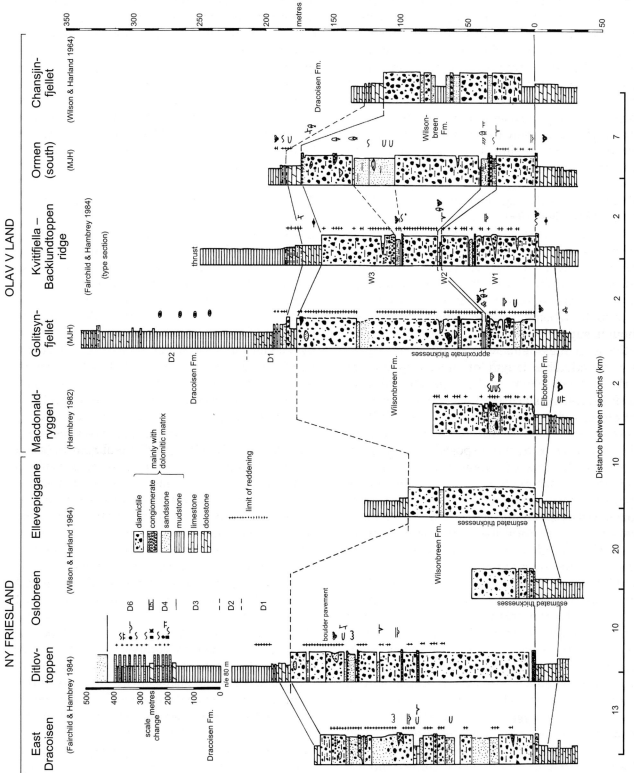

Fig. 8 Lithostratigraphic logs and correlation panel for the Neoproterozoic Wilsonbreen Fm., Olav V Land and Ny Friesland, Svalbard. Reproduced from Harland *et al.,* 1993 with permission of Norsk Polarinstitutt.

approaches employed involves the use of the RA/C_{40} index (the percentage of angular and very angular clasts plotted against the percentage of clasts with a c/a axial ratio ≤0.4), following the methods outlined by Benn and Ballantyne (1993, 1994). This technique can provide good discrimination between different glacigenic lithofacies in Arctic environments (Bennett *et al.*, 1997), and is particularly effective in separating supraglacial (passively transported) from subglacial (actively transported) debris. Other techniques for evaluating clast shape such as the use of maximum projection sphericity and oblate-prolate indices formerly used to distinguish beach and fluvial gravels (e.g. Dobkins & Folk, 1970; Stratten, 1974; Gale, 1990) have limited applicability since modern glaciofluvial gravels have similar values to beach sediments (Etienne, 2004).

From a palaeoenvironmental perspective, the presence or absence of supraglacially derived angular debris is significant, as it provides a means to evaluate whether or not nunataks existed, although supraglacial debris buried by primary stratification may remain in englacial transport for a considerable length of time. Supraglacial debris is most likely to be deposited during glacier recession, since surface ablation leads to melt-out of sediment buried in primary stratification (e.g. Glasser & Hambrey, 2001), and mechanical weathering of freshly exposed rock is likely to increase the flux of supraglacial debris. The dominance of subangular to subrounded clasts (typical of many subglacial tills) in the Neoproterozoic Wilsonbreen and Petrovbreen diamictites (Svalbard; Fig. 8) indicates that nunataks were not important debris sources, although angular debris has been reported locally from the former (e.g. Hambrey, 1983; Fairchild & Hambrey, 1984). Diamictites containing angular gravel-sized clasts are also known from many other Neoproterozoic successions (Hambrey & Harland, 1981 and references therein), including those in Oman (Ayn Formation), North India (Blaini Formation), Brazil (Puga Formation; Gaucher *et al.*, 2003) and Africa (Kundelungu Basin; Cahen, 1963).

Subglacial facies

Subglacial deposits (tills) are typically poorly sorted (often diamicton lithofacies), with poly-modal particle-size distributions (Boulton, 1978) and include a wide variety of particle sizes and shapes. Diamicton, and its equivalent term for lithified rocks 'diamictite' are non-genetic terms introduced by Flint *et al.* (1960a, b) for poorly sorted deposits comprising sand and/or larger (gravel-sized) particles in a muddy matrix. Since this time, the terminology for poorly sorted sediments has moved towards quantitative classifications for diamicts which provide more specific textural information (e.g. Moncrieff 1989; Hambrey, 1994; Table 3). The Udden-Wentworth particle-size scale has also been expanded allowing better description of very coarse grained deposits (Blair & McPherson, 1999). Gravel clast lithologies and heavy mineral fractions reflect the geology of the glacierised catchment (e.g. Dewez & Geurts, 1996; Lee *et al.*, 2002), and clasts which are faceted or bear striations, crescentic gouges or chattermarks are characteristic of subglacial transportation (e.g. Agassiz, 1838; Chamberlin, 1888; Hambrey, 1994; Miller, 1996; Benn & Evans, 1998; Fig. 1c). Glacial striae are easily distinguished from tectonic features such as slickenlines which tend to be more regular and are often associated with mineralization. Fine-grained calcareous precipitates may occur on large clasts as a result of solute precipitation from subglacial water films in response to localised variations in basal ice-contact pressures (Weertman, 1957; Hallet, 1979; Hubbard & Sharp 1993), although such features may be difficult to distinguish in carbonate-cemented diamictites typical of many Neoproterozoic successions (e.g. in Namibia, Scotland). It is worth noting that some studies have indicated that modern subglacially precipitated carbonates are isotopically depleted in $\delta^{13}C$ (Souchez & Lemmens, 1985; Aharon, 1988).

Alignment of gravel clast a-axes in tills can reflect glacier palaeoflow; and, as such, three-dimensional clast macrofabrics are commonly used for palaeo-environmental reconstruction. Well developed a-axial clast macrofabrics are often supportive of till deposition by either lodgement or meltout processes (Boulton, 1970; Dreimanis, 1988; Hambrey, 1994; Ham & Mickelson, 1994; Menzies & Shilts, 1996); however, fast-flowing debris, flow- or 'deformation' tills can produce similar fabrics. Clast orientation data may be analysed using principal direction analysis, or in terms of overall fabric shape, based on ratios between eigenvalues for the measured population (Dowdeswell *et al.*, 1985). Fabrics which

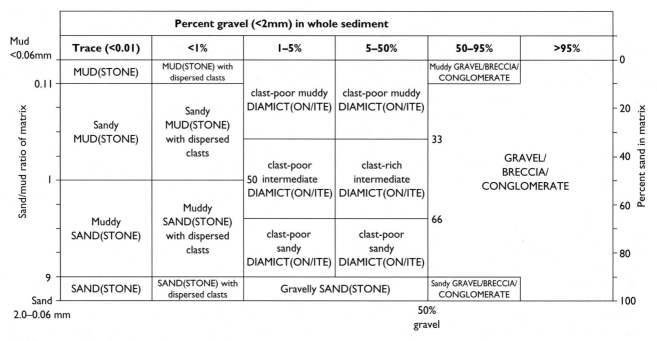

Table 3 Classification for poorly sorted sediments, based on Moncrieff (1989) and modified by Hambrey and Glasser (2003)

have low isotropy and moderate to high elongation values are thought to be characteristic of lodgement till (Dowdeswell *et al.*, 1985; Benn & Evans, 1998), and shallow upglacier clast imbrications develop in some lodgement tills (Dowdeswell & Sharp, 1986; Krüger, 1994). However, genetic discrimination between massive till facies is often difficult and interpretations based *solely* on macrofabric analyses are considered tenuous (e.g. Bennett *et al.*, 1999). Doubts have also been raised regarding the reliability of fabric strength in determining depositional process, with problems attributed to sampling effects (Benn & Ringrose, 2001). In the Neoproterozoic record, additional factors have to be considered in clast fabric and shape analyses including post-depositional compaction, growth of diagenetic cements and, in tectonically deformed terrain (particularly orogenic belts), metamorphism, flattening, stretching or clast re-orientation resulting from shearing and pressure solution. These factors inhibit the use of clast fabric analyses in many successions, including (but not limited to) the Blaini Formation in North India, the Tambien Group in NE Africa (Beyth *et al.*, 2003), the Toby and Edwardsburg Formations in Idaho, USA (Aalto, 1981; Lund *et al.*, 2003) and parts of the Port

Askaig Formation in the UK. This technique has been applied to the little-deformed Petrovbreen diamictites in Svalbard, East Greenland and those of the Jbéliat Formation in Mauritania where clast fabrics are similar to waterlain and lodgement tills (Deynoux, 1985; Harland *et al.*, 1993).

Given complications in clast fabric analyses, other factors need to be taken into account. Association with other lithofacies, particularly with regard to their structural deformation features, is important. For example, tills deposited as a result of lodgement or meltout often overlie deformed sediments, as large shear stresses can be generated by the overriding ice mass (Boulton, 1996). Conversely, deformation concentrated *within* subglacial till may act as a buffer, reducing the intensity of substrate deformation. Layered or stratified deformation tills may develop (as described from Breiðamerkurjökull in Iceland; Boulton, 1979; Boulton & Hindmarsh, 1987; Benn & Evans, 1996), which overlie nondeformed soft-sediment. The character of these subglacial deposits reflects variations in incremental strain (Boulton, 1996), and is inherently related to subglacial porewater pressure (e.g. Hiemstra & van der Meer, 1997). The gross-scale architecture of basal till sheets, particularly thickness and spatial distribution, is also considered to be closely linked to subglacial drainage conditions, and is partly dependent on the character of pre-existing substrate (e.g. Kjaer *et al.*, 2003). However, it is likely that mechanisms such as deformation and melt-out are part of a continuum of basal processes that vary over spatial and temporal scales. Features traditionally thought to be diagnostic of deformation at the bed, such as clast pavements and high porosity, are also thought to develop during passive melt-out (e.g. Mickelson *et al.*, 1992; Ronnert & Mickelson, 1992), although examples of pavements with planed-off tops are unlikely to develop by this process. Discrimination between deformation and melt-out till is considerable importance, since they imply different subglacial conditions at the time of sedimentation and the resultant dynamics of the ice mass in question (Benn, 1995).

Gross-scale stratigraphic architecture, associated lithofacies or subglacial erosional features such as striated pavements, meltwater channels (including Nye channels and p-forms) and glacial lineations including flutes, drumlins and roches moutonnées all provide additional information on the likely depositional setting of diamictite facies. For example, striated pavements are known beneath numerous successions in Australia (e.g. Egan, Landrigan and Moonlight Valley tillites; Coats & Preiss, 1980), Greenland (beneath the Storeelv Formation; Hambrey & Spencer, 1987), Mauritania (Jbéliat Formation; Deynoux & Trompette, 1981), Norway (Smalfjord Formation; Arnaud & Eyles, 2002a), Brazil (Macaubas Megasequence; Isotta *et al.*, 1969) and India (Blaini Formation; this study) where they provide solid evidence for grounded ice. Roches moutonnées and other whaleback forms are also known from the Jbéliat (Deynoux & Trompette, 1981) and Mineral Fork Formations (Ojakangas & Matsch, 1980), and may indicate relatively thin ice, since subglacial quarrying tends to be enhanced by cavities at the bed (Benn & Evans, 1998).

Glacitectonic structures

Much recent research on modern and Pleistocene glacigenic sediments has highlighted the importance of using soft sediment (glacitectonic) or thermal contraction deformation structures to provide additional information for determining sediment transport mode and deposition (Fig. 9). These studies have improved our understanding of processes of moraine formation (Bennett *et al.*, 1996a, b, 1998; Hambrey *et al.*, 1997), sediment stacking patterns in proglacial environments (Hambrey & Huddart, 1995; Hart & Boulton, 1991) and palaeo-environmental interpretations of diamicton and diamictite lithofacies (e.g. Menzies & Maltman, 1992; Rijsdijk *et al.*, 1999, 2001; Maltman *et al.*, 2000; Menzies, 2000; van der Wateren *et al.*, 2000; Khatwa & Tulaczyk, 2001; Lachniet *et al.*, 2001; van der Meer *et al.*, 2003). Such structures are only occasionally recognized or utilized for palaeoenvironmental reconstruction in older geological terrains (e.g. Le Heron *et al.*, 2005), where they are often overprinted by diagenetic, metamorphic or regional tectonic events. Those working on microstructures in diamicton lithofacies have been primarily concerned with the identification of features which act as clues towards sediment genesis, and the ability to differentiate deposits resulting from primary subglacial deposition, re-distribution of sediment by subaerial debris flows or accumulation of periglacial slope deposits (e.g. Harris, 1998; Lachniet *et al.*, 2001). However, few diagnostic criteria can be presented since terrestrial debris flow deposits commonly display similar microstructural features to subglacially deformed till (Khatwa & Tulaczyk, 2001). The presence of fractured grains (observed in thin-sections) is considered by some as evidence of glacitectonism (Hiemstra & van der Meer, 1997), although terrestrially deposited tills can be reworked by paraglacial slope processes with little or no modification of gravel clast shapes, imbrication, texture, consolidation or granulometry (Curry & Ballantyne, 1999). Thus structurally similar deposits can form as a result of different depositional processes. Differentiating subglacial deposits from other poorly sorted facies types is problematic, particularly with regard to mass flow deposits such as debris flows, or massive sediments containing 'outsized' gravel lonestones, such as glacimarine or glacilacustrine sediments. Overcompaction or stratification are not considered sufficiently critical for discounting a subglacial origin, and Kluiving *et al.*, (1999) advocate sorting and the presence of dropstones as key criteria. Dropstones are defined as outsized clasts in finely stratified or laminated sediment, which deform and truncate underlying laminae (Hambrey & Harland, 1981). Isolated clasts are probably the best

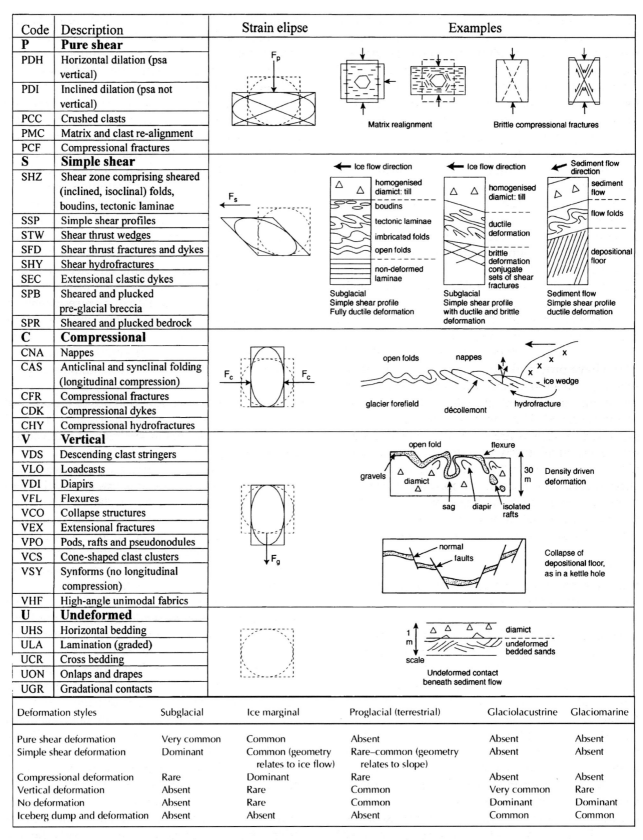

Fig. 9 Soft sediment deformation structures and codes for description of macroscopic glaciotectonic structures. Strain ellipses indicate deformation style under different stress regimes; forces Fp, Fs, Fc and Fg refer to pure shear, simple shear, compressional and gravitational, respectively. From McCarroll and Rijsdijk (2003). Copyright (2003) John Wiley & Sons, Limited. Reproduced with permission.

indicators, since debrite clasts can also deform and truncate underlying laminae. For Proterozoic sediments, it is worth noting that perennial sea and lake ice would also have been capable of entraining and rafting gravel-sized material (see Smith, 2000). Pebble-sized clasts are generally considered the best proxy for ice-rafted debris (Grobe, 1987; Andrews et al., 1997, Smith & Andrews, 2000).

Other features that are important for palaeoenvironmental discrimination include macroscopic deformation structures, which are often associated with ice-marginal sediment assemblages. These include faults (normal, reverse, low-angle thrust faults and shear zones; Croot, 1988), folds (Hart & Boulton, 1991) and more chaotic structures resulting from subglacial deformation of poorly consolidated deformable substrate (Benn & Evans, 1996; Boulton, 1996), drumlinisation (Hart, 1995a, b), density inversion in saturated depositional sequences (e.g. Rijsdijk, 2001) and injection of clastic dykes (Le Heron & Etienne, 2005). McCarroll and Rijsdijk (2003) have provided a detailed account of macroscopic glacitectonic structures and their relative dominance in different glacially influenced environments (Fig. 9). These features complement glaciclastic debris such as dropstones, striated clasts and till pellets highlighted by Eyles and Januszczak (2004) as key criteria for identifying glacial influences on sedimentation in Neoproterozoic successions, and may provide a means to distinguish primary subglacial facies from sediment redistributed by subaqueous debris flow processes.

THICKNESSES OF GLACIALLY INFLUENCED MARINE SUCCESSIONS

Eyles and Januszczak (2004) argued that less than 100 m of stratigraphic section is deposited over a typical 'glacial cycle,' emphasizing the much thicker preserved successions of Neoproterozoic glacially influenced strata. While this statement may be true for terrestrial successions, or those deposited over short-term glacial/interglacial cycles, sediment yields in basins for glacial epochs spanning tens of millions of years far exceed this.

With the exception of DSDP/ODP (Deep Sea Drilling Programme/Ocean Drilling Programme), CIROS (Cenozoic Investigations of the Ross Sea) and CRP (Cape Roberts Project) recovery, much of what we know about glacimarine sedimentation is limited to short gravity core and geophysical investigations. Nonetheless, AMS (Accelerated Mass Spectrometry) ^{14}C dates provide important constraints on Pleistocene and Holocene sedimentation rates in glacially influenced basins. Predictably, sediment accumulation rates vary both spatially and temporally. For example, sedimentation rates across the North Atlantic region over the past 3 Myrs varied from 0.02 to 0.1 m ka^{-1}, and 0.12 m ka^{-1} on the Norwegian continental margin (Heinrich et al., 2002 and references therein). Similar figures have been presented for shallow, distal areas of the Barents Sea (0.03 m ka^{-1}; Elverhøi et al., 1989). Spatial and temporal variability is exhibited by Quaternary sedimentation rates for the Reykjanes Ridge between Heinrich events 1 and 2 (0.09 to 0.15 m ka^{-1}) and events 3 and 4 (0.12 to 0.22 m ka^{-1}; Moros et al., 2002) and from Pleistocene and Holocene data on Kejser Franz Joseph Fjord (Greenland) and the adjacent continental margin (Evans et al., 2002). Here Evans et al., (2002) reported sedimentation rates of 0.3 m ka^{-1} on the upper continental slope and 0.16 m ka^{-1} on the mid-lower slope during the glacial maxima, with deglacial fluxes in the order of 0.51–0.79 m ka^{-1} (mid-lower slope), and 1.11 m ka^{-1} (in the fjord and inner slope). High sedimentation rates are also known from other fjords in Greenland including Nansen fjord where proximal sedimentation rates are calculated at 1.8 m ka^{-1} and 1.3 m kyr^{-1} in distal areas, and accumulation rates of 0.1 to 0.3 m kyr^{-1} in Scoresby Sund (Dowdeswell et al., 2000). However, these examples are relatively low by comparison with Kongsfjorden in Svalbard, where present-day annual sedimentation rates are ~70 mm yr^{-1} (Elverhøi et al., 1998).

Based on the above examples, if we assume a modest average sedimentation rate of 0.05 m/kyr over a 30 million year period (the upper predicted limit for the duration of a Snowball Earth event; Hoffman et al., 1998a), it is possible to generate ~1.5 km of non-compacted stratigraphy. Given the higher sedimentation rates recorded in ice-proximal regions, and the vast volumes of sediment deposited in grounding zone wedges (cf. Shipp et al., 2002), considerably thicker sequences may be preserved in glacially influenced marine basins, particularly those associated with temperate or polythermal ice masses (see Elverhøi et al., 1998). However, it should be noted that in the long-term, the thickness of sedimentary successions is ultimately controlled by tectonically generated accommodation space.

Glacial and non-glacially influenced strata deposited over the past 800,000 years on the south-western part of the Barents Sea Shelf are known to approach 150 m in thickness (Rafaelsen *et al.*, 2002), Miocene to Quaternary strata in the Polar North Atlantic exceed 1 km in thickness (Thiede *et al.*, 1998), Miocene and Pliocene deposits of the Pagodroma Group in Antarctica are ~300 m thick (Hambrey & McKelvey, 2000) and Miocene to Pleistocene deposits of the Yakataga Formation along the southern continental margin of Alaska are ~7 km thick (Zellers & Lagoe, 1992). In the lower part, the Yakataga Formation consists primarily of debrites and turbidites, passing up into glaci-marine diamictites, sandstones and mudstones (Eyles & Lagoe, 1990; Zellers & Lagoe, 1992), and in this respect bears similarity to many Neoproterozoic glacially influenced facies associations. For example, 1.5 km of glacially influenced Eocene-Pleistocene strata including mudstones, sandstones, diamictites and conglomerates have been proven in core from the Victoria Land basin in the East Antarctic rift system (Taviani & Beu, 2003; Figs 10, 11b). Cyclical controls on basin sedimentation are

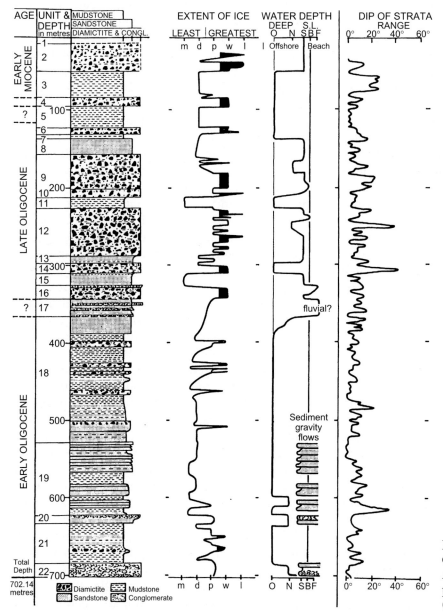

Fig. 10 Lithostratigraphy of the CIROS 1 drillcore, McMurdo Sound, Antarctica. Reproduced with permission from Hambrey *et al.* (1989b).

clear from IRD (ice rafted debris) proxy data and sedimentary facies, while spatial variability in thickness and stratigraphic preservation are evident from condensed sequences preserved over basement highs and extensive disconformities across the basin-fill (Hambrey *et al.*, 2002). These Cenozoic successions show that marine glacially influenced strata can achieve thicknesses comparable to the Neoproterozoic in similar depositional settings over similar timescales. The Victoria Land basin preserves both rift-related volcanics and huge volumes of glacially influenced strata, and in this respect is an excellent analogue for the Neoproterozoic Abu Mahara rift basin in North Oman which contains pillowed basalts of the Saqlah Formation, beneath the c. 1.5 km thick glacially influenced Fiq Member of the Ghadir Manqil Formation (Leather *et al.*, 2002; Allen *et al.*, 2004; Fig. 11a). Some characteristic features regarding the preservation potential of glaciomarine successions across different basin settings (craton, shelf, slope and deep basin environments) are dealt with in Brookfield (1994). For example, the most complete sequences tend to be found in actively subsiding basins in basin slope settings as illustrated in Figure 12.

SEDIMENTARY SEQUENCES

Glacimarine facies assemblages comprise a wide spectrum of stacking patterns, including progradational, aggradational and retrogradational packages (Powell & Cooper, 2002; Fig. 7). Distinct processes associated with advance, maxima and recessional glacial stages permit the construction of sequence stratigraphic models for glacially influenced marine basins which may be applied to Neoproterozoic basin-fill successions (Powell & Cooper, 2002). Although grounding line fan systems can be stratigraphically complex (Powell *et al.*, 2000), sediment accumulations on continental margins are remarkably consistent, with a stratigraphic architecture dominated by prograding clinoforms overlain by well-defined topsets (Eyles *et al.*, 2001). This seismic architecture has been recognised from glaciated continental margins around Antarctica (Fig. 13; Hambrey *et al.*, 1991; Eyles *et al.*, 2001; Escutia *et al.*, 2005), Greenland (Vanneste *et al.*, 1995; Solheim *et al.*, 1998), Norway (Vorren *et al.*, 1984; Saettem *et al.*, 1992), Canada (Hiscott &

Aksu, 1994) and Alaska (Powell & Cooper, 2002). Analogous sediment-stacking patterns also characterize many Neoproterozoic successions, including the Chang'an-Nantuo sequence in south China (Jiang *et al.*, 2003), the Umberatana Group in the Adelaide Rift basin in Australia (Young & Gostin, 1991; McKirdy *et al.*, 2001), parts of the Windermere Supergroup in Canada (Eisbacher, 1985) and the Smalfjord Formation in Norway (Fig. 14; Arnaud & Eyles 2002a; but see also Edwards and Føyn, 1981). In the Antarctica Peninsular, upper Miocene to Quaternary strata are dominated by turbidites, diamictites interbedded with lonestone-bearing muds and subglacially cannibalized marine sediments (Eyles *et al.*, 2001). This lithofacies association is characteristic of many Neoproterozoic successions, particularly those where sediment gravity flows comprise a significant component (e.g. Fiq Formation, Oman, Allen *et al.*, 2004; Smalfjord Formation, Arnaud & Eyles, 2002a; Port Askaig Formation, Arnaud & Eyles, 2002b; Mineral Fork Formation, Young, 2002; Table 1).

DURATION AND CYCLICITY OF GLACIAL EPOCHS

It is widely accepted that Cenozoic glacial-interglacial climatic transitions were driven largely as a result of orbital forcing (Milankovitch cycles), although several other factors are known to be important, including changes in atmospheric levels of radiative gases, creation of topography resulting from rift flank uplift or continental collision and changes in oceanic circulation (e.g. Broecker *et al.*, 1988; Haug *et al.*, 2001; Hiscott *et al.*, 2001; Smith & Pickering, 2003; Piotrowski *et al.*, 2005). Yet, the exact nature of the relationship between insolation, ice volume, carbon cycling and thermohaline circulation remains to be firmly established, particularly since some millennial scale climatic oscillations may have been triggered by changes in oceanic circulation (Piotrowski *et al.*, 2005), glaciation may have been initiated by global warming (Kukla & Gavin, 2005), and significant ice volume may actually be required to amplify weak insolation in order for deglaciation to occur (Parrenin & Paillard, 2003). On a simple level, we know that these short-term glacial/interglacial transitions are superimposed on higher order cycles

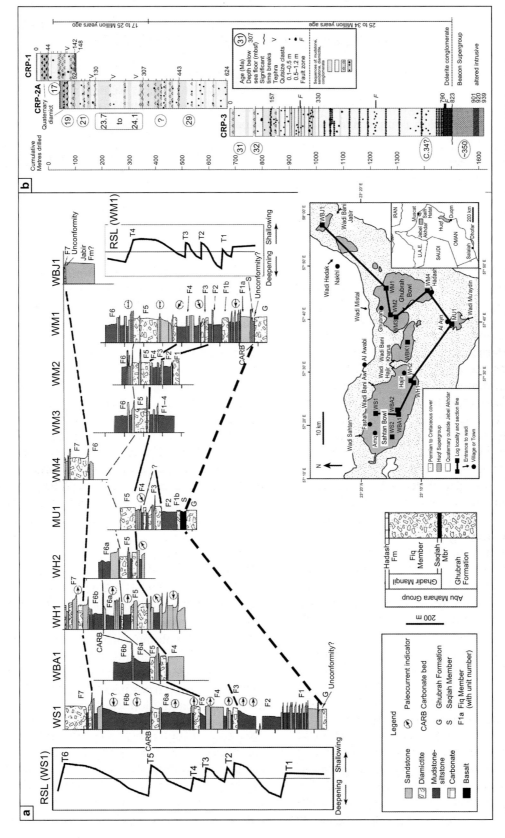

Fig. 11 Glacially influenced rift basin-fill successions (a) the Neoproterozoic Fiq Member of the Ghadir Manqil Formation, Jebel Akhdar, Sultanate of Oman. Reproduced from Leather *et al.* (2002) with permission of the Geological Society of America; (b) Cenozoic CRP (Cape Roberts Project) core stratigraphy for the Victoria Land Basin, Antarctica. Reprinted from Taviani and Beu (2003) with permission from Elsevier.

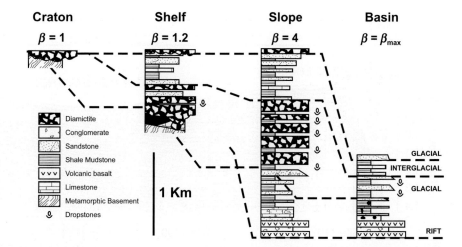

Fig. 12 Simplified facies models for glacial deposits across craton, shelf, slope and deep basin environments. Re-drawn from Brookfield (1994). Reproduced with permission from Elsevier.

of glaciation, represented in the Cenozoic by the onset of southern hemisphere glaciation at least as far back as the latest Eocene (Hambrey *et al.*, 2002; Taviani & Beu, 2003), in high latitude settings during the Miocene and Pliocene, and mid-latitudes during the Quaternary (Ehlers & Gibbard, 2003). It is worth noting the recognition of orbital cycles in glacimarine sediments around the Oligocene-Miocene boundary (Naish *et al.*, 2001). Climate models suggest that ice sheets were also sensitive to orbital changes during the Late Ordovician glaciation of Gondwana (Poussart *et al.*, 1999), but longer term glacial cycles (5–7 Myr) identified from Permo-Carboniferous glacial deposits in the Karoo Basin do not coincide with known orbital cycles, and Scheffler *et al.* (2003) suggested that changes in global temperature gradients or atmospheric-oceanic circulation were important factors.

In contrast to the Phanerozoic glacial record, which is relatively well constrained by radiometric and biostratigraphic markers, the timing of Neoproterozoic glaciations remains poorly understood. Independent attempts to distinguish the number of Neoproterozoic glaciations have involved lithostratigraphic correlation (Zhenjia & Jianxin, 1985), calibration of carbon and strontium isotope curves (Kaufman *et al.*, 1997; Halverson, 2005) and cladistic analysis (Kennedy *et al.*, 1998), with estimations ranging between 2, 3, 4 or possibly 5 glacial episodes. Over the past decade, and particularly in the last five years, a number of new radiometric ages have been published that provide firmer controls on the timing and duration of Neoproterozoic glacial epochs. New U-Pb

and Re-Os ages from sections in Africa, the United States and Australia challenge the conventional Sturtian-Marinoan subdivision typically applied to Neoproterozoic glacial successions; however, undisputed evidence for glacial influences on sedimentation is yet to be demonstrated for some of these successions (Tables 1, 4). At the present time, the best constrained diamictite-bearing successions include the Gaskiers Formation (~580 Ma; Bowring *et al.*, 2003), the Squantum Tillite (570–589 Ma; Thompson & Bowring, 2000; Thompson *et al.*, 2000), the Nantuo Formation (635–667 Ma; Condon *et al.*, 2005), the Scout Mountain Member (662–714 Ma; Fanning & Link 2004), the Kaigas Formation (735–777 Ma; Frimmel *et al.*, 1996, 2002) and the Grand Conglomerat of the Kundelungu Basin (730–770 Ma; Key *et al.*, 2001). Syn-depositional ages further constrain sedimentation between 730–735 Ma (Key *et al.*, 2001), 739–713 Ma (Brasier *et al.*, 2000), 702–705 Ma (Tollo & Hutson, 1996), 678–692 Ma (Lund *et al.*, 2003) and 634.3–636.7 Ma (Hoffman *et al.*, 2004). Some details of the sedimentology of these sections can be found in Table 1. Table 4 provides some published radiometric age constraints for different Neoproterozoic diamictite-bearing successions, variably interpreted as glacial in origin. Many of the ages should be approached with caution, particularly where there are conflicts dependent on different dissolution techniques (e.g. the Olympic Formation is constrained by competing [187]Re-[187]Os maximum ages of 592 ± 14 Ma; Schaefer & Burgess (2003) and 658 ± 5.5 Ma; Kendall & Creaser, 2004) or the full details of the isotopic data remain to be published.

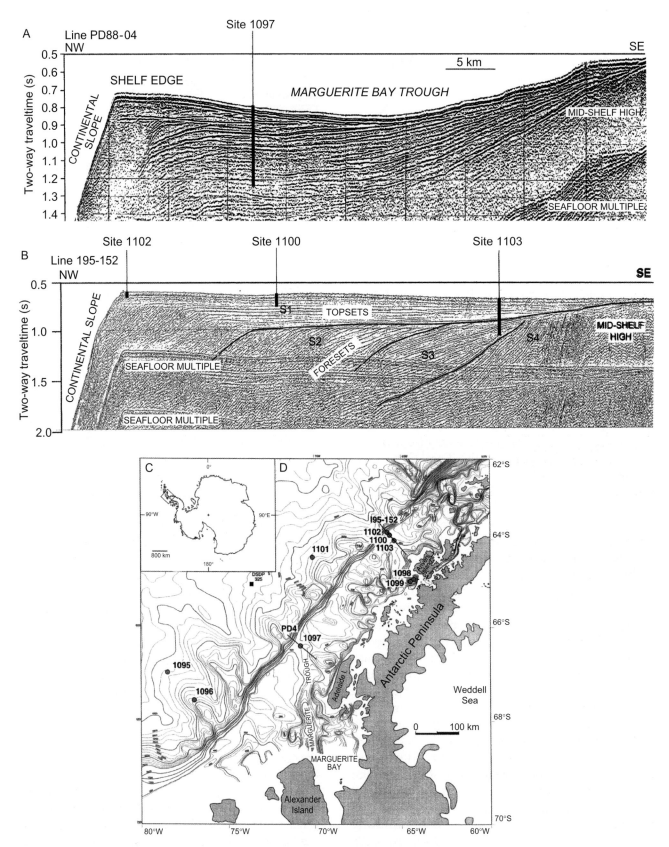

Fig. 13 Acoustic stratigraphy of glacially influenced marine deposits on the Antarctic peninsula; (A) ODP leg 178 site 1097; (B) ODP leg 178 site 1103; (C) and (D) location of drill sites. (A) and (B) after Bart and Anderson (1995) and Barker *et al.* (1998). Reprinted from Eyles (2002) with permission from Elsevier.

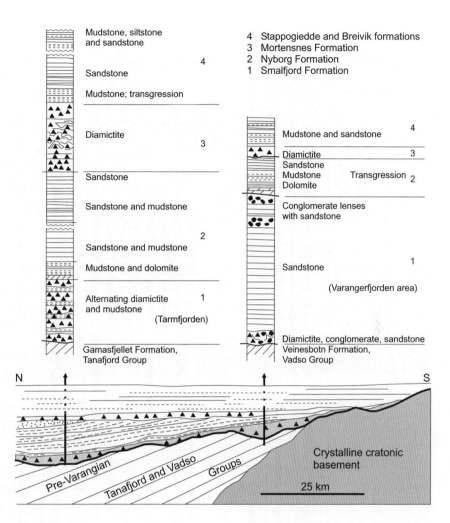

Fig. 14 Stratigraphy and geometry of the Smalfjord Formation in east Finnmark. Redrawn after Arnaud and Eyles (2002a) modification of Banks *et al.* (1971) and Nystuen (1985). Copyright (2002) Blackwell Publishing, reproduced with permission.

Estimations of glacial epoch durations have been derived from palaeomagnetic reversal studies (Sohl *et al.*, 1999), thermal subsidence modeling (Hoffman *et al.*, 1998a) and accumulation rates of inter-planetary dust particles (Bodiselitsch *et al.*, 2005), all of which indicate long-lived glacial activity. While many successions lie within the predicted range of Snowball Earth timescales, none are unusual in their longevity. It is accepted, for example, that the Permo-Carboniferous glacial epoch lasted ~90 Myrs from the Tournaisian (Early Carboniferous) until the Roadian (Mid-Permian; Crowell, 1978), and the Pleistocene (2.6 Ma; Jansen & Sjøholm, 1991; Larsen *et al.*, 1994; Ehlers & Gibbard, 2003) is merely the latest expression of Cenozoic glaciation initiated at least as far back as the late-Eocene (~35 Ma) in Antarctica (e.g. Hambrey *et al.*, 2002; Escutia *et al.*, 2005), and the Miocene in Alaska, the Polar North Atlantic,

southern South America and New Zealand (Thiede *et al.*, 1998; Ehlers & Gibbard, 2003). At the present time, geochronological constraints are too poor to constrain any shorter-term cyclicity that may be represented in Precambrian successions. However, depositional cyclicity may be used to infer palaeoclimatic conditions. This approach has been adopted by Leather (2000), Leather *et al.* (2002) and Allen *et al.* (2004), for the Fiq Member of the Ghadir Manqil Formation in North Oman. Seven gross depositional cycles are recognised which are interpreted to reflect glacial-interglacial transitions (Allen *et al.*, 2004). These cycles are also characterized by variations in CIA (Chemical Index of Alteration) values, which are thought to be indicative of climatic changes from cold/arid to warm/wet conditions (Rieu *et al.*, unpublished data). CIA may thus be used as a proxy tool for evaluating Neoproterozoic

Table 4 Radiometric age constraints on Neoproterozoic diamictite-bearing successions. We consider the U-Pb zircon ages as most reliable since any open-system behaviour may be evaluated by comparison of the ^{238}U-^{206}Pb and ^{235}U-^{207}Pb parent-daughter isotope systems (see Bowring & Schmitz, 2003). For information on individual radiometric ages and error bars we advise the reader to refer to the original source references

Stratigraphic Unit	Radiometric age (Ma)	Isotope system	Material dated	Significance of date	Source reference
AUSTRALIA					
Wilyerpa Fm. (including underlying Appila & Pualco Tillites), Elatina Fm. & Pepuarta Tillite.	802 ± 10	SHRIMP ^{238}U-^{206}Pb	Rook Tuff in the Callana Gp.	Max. age for the Elatina Fm.	Fanning et al. (1986)
	~800	^{147}Sm-^{143}Nd	Mafic dykes intruding the Pandurra Fm. Stuart Shelf	Max. age for the Elatina Fm.	Zhao et al. (1994)
	777 ± 7	SHRIMP ^{238}U-^{206}Pb		Max. age for the Elatina Fm.	Walter et al. (2000)
	750 ± 53	^{87}Rb-^{87}Sr	Boucat volcanics/Rhynie sandstone rhyolite; zircon	Max. age for the Elatina Fm?	Preiss (1987)
	690 ± 21	^{87}Rb-^{87}Sr		Max. age for the Elatina Fm?	Jenkins & Cooper (1998)
	657 ± 17	^{238}U-^{206}Pb	Tapley Hill Fm.	Max. age for Elatina Fm.	Ireland et al. (1998)
	601 ± 68	^{87}Rb-^{87}Sr	Enamora Shale and Trezona Fm.	Min. age for Elatina Fm?	Preiss (1987)
	526 ± 4	SHRIMP ^{238}U-^{206}Pb	Marino Arkose underlying Elatina; detrital zircon Brachina Fm. overlying the Nuccaleena Fm. cap dolomite Thin tuff in Early Cambrian Heatherdale Shale; zircon	Min. age for Elatina Fm.	Cooper et al. (1992)
Areyonga and Olympic Fm.	897 ± 9	^{87}Rb-^{87}Sr	Stuart Dyke Swarm	Max. age for the Areyonga Fm.	Marjoribanks & Black (1974) Black et al. (1980)
	592 ± 14	^{187}Re-^{187}Os	Black shale underlying Olympic Fm., overlies Areyonga Fm.	Min. age for Areyonga Fm.; Max. age for Olympic Fm.	Schaefer & Burgess (2003)
Sturt Diamictite	777 ± 7	^{238}U-^{206}Pb zircon	Boucat volcanics	Possible max. age for Sturt?	Walter et al. (2000)
	724 ± 40	^{87}Rb-^{87}Sr	Postglacial Yudnapinna beds	Min. age for the Sturt tillite	Preiss (1987) Drexel et al. (1993)
Cottons Breccia	579 ± 16	^{147}Sm-^{143}Nd	Bold Head and Shower Droplet Volcanics overlying the Yarra Creek Shale and Cottons Breccia	Min. age for Cottons Breccia	Calver et al. (2004)
	574.7 ± 3	SHRIMP ^{238}U-^{206}Pb zircon	Intermediate sills (Grimes Intrusive Suite) intruded into Cottons Breccia		Meffre et al. (2004)
Moonlight Valley and Fargoo Tillites	672 ± 70	^{87}Rb-^{87}Sr	Ranford Fm. shales	Max. age for Moonlight valley and Fargoo tillites	Coats & Preiss (1980)

Location / Formation	Age	Method	Rock unit	Interpretation	Reference
Croles Hill Diamictite	582.1 ± 4.1	SHRIMP ^{238}U–^{206}Pb zircon	Rhyodacite flow unit underlying partly glacigenic Croles Hill Diamictite	Croles Hill Diamictite interpreted as stratigraphically equivalent to Cottons Breccia	Calver et al. (2004)
AFRICA					
Chuos & Ghaub Fms. **Namibia**	758.5 ± 3.5	^{238}U–^{206}Pb zircon	Ombombo Sub-group	Max. ages for Chuos and Ghaub	Hoffmann & Prave (1996); Hoffman et al. (1998)
	756 ± 2	^{238}U–^{206}Pb zircon	Ombombo Sub-group	Max. ages for Chuos and Ghaub	Buchwaldt et al. (1999); Hoffman et al. (1998)
	746 ± 2	^{238}U–^{206}Pb zircon	Ash in underlying Naawpoort Fm. (Nosib Gp.)	Max. ages for Chuos and Ghaub	Buchwaldt et al. (1999); Hoffman et al. (1998)
	635.5 ± 1.2	^{238}U–^{206}Pb zircon	Ash in uppermost Ghaub Fm.	Syn-depositional age for Ghaub; Max. age for Chuos	Hoffmann et al. (2004)
	538 ± 12	^{40}K–^{40}Ar	Mulden Gp. fines	Min. age constraint for the Ghaub	Clauer & Kröner (1979)
	537 ± 7	^{87}Rb–^{87}Sr	Mulden Gp. fines	Min. age constraint for the Ghaub	Clauer & Kröner (1979)
	534 ± 7	^{238}U–^{206}Pb zircon	Syn-tectonic syenogranites (Damara orogeny)	Min. age constraint for the Ghaub	Briqueu et al. (1980)
	508 ± 2	^{238}U–^{206}Pb monazite	Post-tectonic sheeted leucogranites	Min. age constraint for the Ghaub	Briqueu et al. (1980)
Blaubeker Fm. **Namibia**	545 ± 1	^{238}U–^{206}Pb zircon	Overlying Spitskopf Fm.	Min. age constraint for the Blaubeker	Grotzinger et al. (1995)
	543.3 ± 1	^{238}U–^{206}Pb zircon	Overlying Spitskopf Fm.	Min. age constraint for the Blaubeker	Grotzinger et al. (1995)
	539.4 ± 1	^{238}U–^{206}Pb zircon	Overlying Nomtsas Fm.	Min. age constraint for the Blaubeker	Grotzinger et al. (1995)
Kaigas & Numees Fm. **Namibia**	781 + 34/−31	^{87}Rb–^{87}Sr	Lekkersing granite basement	Max. age for Kaigas and Numees	Allsopp et al. (1979) recalculated in: Frimmel & Frank (1998)
	771 ± 6	^{238}U–^{206}Pb zircon	Granite of Richtersveld Igneous Complex	Max. age for Kaigas and Numees	Frimmel et al. (2001)
	741 ± 6	^{207}Pb–^{206}Pb zircon	Rosh Pinah Rhyolite	Min. age for Kaigas; Max. age for Numees	Frimmel et al. (1996)
	717 ± 11	^{87}Rb–^{87}Sr	Gannakouriep Suite Mafic Dyke	Min. age for Kaigas; Max. age for Numees	Frimmel et al. (2002); Reid et al. (1991)
	542 ± 4	^{40}K–^{40}Ar	Metamorphism of the Gannakouriep mafic dyke	Min. age for Numees	Onstott et al. (1986)
	521 + 24/−20	^{238}U–^{206}Pb zircon	Post-tectonic alkaline intrusive Bremen complex	Min. age for Numees	Allsopp et al. (1979) recalculated in: Frimmel & Frank (1998)
Grand Conglomerat, Kundelungu Basin **Zambia**	765 ± 5	SHRIMP ^{238}U–^{206}Pb	Mwashia Gp. Lavas; zircon	Max. age of Grand Conglomerat	Key et al. (2001)
	763 ± 6	SHRIMP ^{238}U–^{206}Pb	Mwashia Gp. Lavas; zircon	Max. age of Grand Conglomerat	Key et al. (2001)
	735 ± 5	SHRIMP ^{238}U–^{206}Pb	Altered volcanic pods in contact with glacial strata; zircon	Min./syn-depositional age of Grand Conglomerat	Key et al. (2001)

Table 4 (*cont'd*)

Stratigraphic Unit	Radiometric age (Ma)	Isotope system	Material dated	Significance of date	Source reference
Jbéliat Fm. **Mauritania**	632 ± 13	^{87}Rb-^{87}Sr	Detrital smectite grains in Jbéliat Fm. diamictites	Max. age for Jbéliat Fm.	Clauer & Deynoux (1987)
	595 ± 43	^{87}Rb-^{87}Sr	Fine micas	Min. age for Jbéliat Fm.	Clauer (1976) Clauer et al. (1982)
	633.8 ± 0.5	^{40}Ar-^{39}Ar	Muscovite in Kara Nappe	Min. age constraint for Jbéliat Fm.	Attoh et al. (1997)
	608.1 ± 1.2	^{40}Ar-^{39}Ar	Muscovite in quartz schist of basal Atacora Nappe	Min. age constraint for Jbéliat Fm.	Attoh et al. (1997)
Matheos Fm. **Ethiopia**	854 ± 3	^{207}Pb-^{206}Pb zircon	Low-grade metavolcanics	Max. age based on a tentative lithostratigraphic correlation between the Tsaliet Gp. (underlying Tambien Gp.) and a sequence in neighbouring Eritrea.	Teklay (1997) Miller et al. (2003)
	~800	^{207}Pb-^{206}Pb zircon	Bizen Domain metavolcanics	Max. age based on correlation between Bizen	Teklay (1997)
	796	^{207}Pb-^{206}Pb zircon	Ghedem Domain paraschists and orthogneisses underlying the Bizen Domain metavolcanics	Domain metavolcanics in Eritrea and Tsaliet metavolcanics in N. Ethiopia.	Beyth et al. (2003) Teklay (1997) Beyth et al. (2003)
	720–800	^{147}Sm-^{143}Nd and ^{87}Rb-^{87}Sr	Units similar to the Tsaliet Gp in the west, intruded by syn-tectonic granodiorites	Max. age for Matheos Fm.	Tadesse et al. (2000)
	~630	^{207}Pb-^{206}Pb zircon	Granitoids in Lehazin, E. Eritrea	Min. age of Matheos	Teklay (1997) Beyth et al. (2003)
	613.4 ± 0.9	^{207}Pb-^{206}Pb age	Post-D1 tectonic deformation intrusions	Min. age of Matheos	Miller et al. (2003)
	606.0 ± 0.9	^{207}Pb-^{206}Pb age	Post-D1 tectonic deformation intrusions	Min. age of Matheos	Miller et al. (2003)
	545 ± 24	Th-U-total Pb on zircon	Post-tectonic granite intrusion	Min. age of Matheos	Tadesse et al. (1997) Beyth et al. (2003)
NORTH AMERICA					
Toby Fm. **Canada**	762–728	^{238}U-^{206}Pb zircon and ^{147}Sm-^{143}Nd	Granitic and volcanic rocks which unconformably underlie, or are intruded into the base of the succession	Max. age for the Windermere Supergroup	Ross et al. (1995) Ross & Villeneuve (1997)
	634 ± 57	^{187}Re-^{187}Os	Post-glacial chlorite-grade black shale of Upper Old Fort Point Fm.	Min. age for Toby Fm.	Kendall et al. (2004)
	607.8 ± 4.7	^{187}Re-^{187}Os	Volcanics unconformably overlying Windermere Supergroup	Min. age for Toby Fm.	Kendall et al. (2004)
	569.6 ± 5.3	^{238}U-^{206}Pb zircon		Min. age for Toby Fm.	Colpron et al. (2002)

Location/Formation	Age	Method	Rock unit	Interpretation	Reference
Sayunei, Shezal & Icebrook Fm. **Canada**	755 ± 18	^{238}U–^{206}Pb zircon	Leucogranite dropstone in Sayunei Formation	Max. age for Rapitan Gp.	Ross & Villeneuve (1997)
Gaskiers Fm. **Canada**	631 ± 2	^{238}U–^{206}Pb zircon	Volcanics of the Harbour Main Group	Max. age for Gaskiers	Krogh et al. (1988)
	622.6 + 2.3/−2.0	^{238}U–^{206}Pb zircon	Volcanics of the Harbour Main Group	Max. age for Gaskiers	Krogh et al. (1988)
	606 + 3.7/−2.9	^{238}U–^{206}Pb zircon	Volcanics of the Harbour Main Group	Max. age for Gaskiers	Krogh et al. (1988)
	604 +4/−3	^{238}U–^{206}Pb zircon	Volcanics of the Harbour Main Group	Max. age for Gaskiers	Myrow & Kaufman (1999)
	580	^{238}U–^{206}Pb zircon	Ash beds lying below, within and above glacigenic deposits	Timing and duration of glaciation to <1 Ma	Bowring et al. (2003)
	565 ± 3	^{238}U–^{206}Pb zircon	Uppermost Conception Group	Min. age for Gaskiers	Dunning pers. comm. in Benus (1988)
Musgravetown Gp. **Canada**	620 ± 1	^{238}U–^{206}Pb zircon	Underlying Love Cove Group volcanics	Max. age constraint on Musgravetown Gp.	Dec et al. (1992)
	~610	^{238}U–^{206}Pb zircon	Thin ash bed in underlying Connecting Point Group	Max. age constraint on Musgravetown Gp.	O'Brien et al. (1992)
Roxbury Conglomerate; Squantum Tillite Mbr. **Massachusetts U.S.A.**	610 ± 2.2	^{238}U–^{206}Pb zircon	Porphyritic granophyre within the diamictite	Max. age of Squantum tillite	Thompson et al. (2000a)
	606 ± 3.7	^{238}U–^{206}Pb zircon	Welded tuff clast within the diamictite	Max. age of Squantum tillite	Thompson et al. (2000a)
	601 ± 3.7	^{238}U–^{206}Pb zircon	Crystal-poor tuff	Max. age of Squantum tillite	Thompson et al. (2000a)
	595 ± 2	^{238}U–^{206}Pb zircon	Welded tuff clast within the tillite	Max. age of Squantum tillite	Thompson & Bowring (2000)
	587 ± 2	^{207}Pb–^{206}Pb	Vesicular basaltic andesite	Max. age of Squantum tillite	Thompson et al. (2000b)
	570	^{238}U–^{206}Pb zircon	Ash bed in overlying Cambridge Argillite.	Min. age of Squantum tillite	Thompson & Bowring (2000)
Edwardsburg Fm. **Idaho, U.S.A.**	685 ± 7	SHRIMP ^{238}U–^{206}Pb	Volcanic rocks interbedded with glacigenic facies of 1st Windermere glaciation	Approximate age for timing of glaciation, dependant on the likely duration of volcanic activity	Lund et al. (2003)
	684 ± 4	SHRIMP ^{238}U–^{206}Pb	Windermere glaciation = Rapitan? Zircon		
Scout Mountain Member, Pocatello Fm. **Idaho, U.S.A.**	717 ± 4	SHRIMP ^{238}U–^{206}Pb	Porphyritic rhyolite clast; zircon	Max. age for Scout Mountain Mbr.	Fanning & Link (2004)
	709 ± 5	SHRIMP ^{238}U–^{206}Pb	Epiclastic plagioclase-phyric tuff breccia immediately below	Max. age for Scout Mountain Mbr.	Fanning & Link (2004)
	667 ± 5	SHRIMP ^{238}U–^{206}Pb	Scout Mountain Member; zircon simple igneous zircon population from reworked fallout tuff bed 20 m above uppermost diamictite and cap carbonate? and immediately below a second cap carbonate; zircon	Min. age for glacial Scout Mountain Mbr. (lithostratigraphically correlative to Edwardsburg Fm.)	Fanning & Link (2004)
Mechum River Fm.	580 ± 7	^{40}Ar–^{39}Ar	Browns Hole Formation extrusive volcanics	Min. age for Pocatello Fm.	Christie-Blick & Levy (1989)
	729	^{238}U–^{206}Pb zircon	Robertson River granitoid;	Max. age of Mechum River Fm.	Tollo & Aleinikoff (1992) Tollo & Hutson (1996)
Virginia, U.S.A	702–705	^{238}U–^{206}Pb zircon	Robertson River Igneous Suite; 2 Perialkaline units of the Battle Mountain volcanic centre	Syn-depositional age based on interpreted timing of rhyolite eruption of Mechum River	Bailey & Peters (1998)

Table 4 (*cont'd*)

Stratigraphic Unit	Radiometric age (Ma)	Isotope system	Material dated	Significance of date	Source reference
Konnarock Fm. (formerly Mount Rogers Fm.) **Virginia, U.S.A**	758 ± 12	^{238}U-^{206}Pb zircon	Mount Rogers volcanics	Max. age of the Konnarock Fm.	Aleinikoff et al. (1995)
SOUTH AMERICA					
Jequitaí Fm. **Brazil**	900	^{238}U-^{206}Pb zircon	Detrital zircons in Jequitaí Fm. diamictites	Max. age of Jequitaí Fm.	Buchwaldt et al. (1999); Pedrosa-Soares et al. (2000); Pimentel & Fuck (1992)
	740 ± 22	^{207}Pb-^{206}Pb	'Cap' carbonates of the Bambuí Group	Min. age of Jequitaí Fm.	Babinski & Kaufman (2003)
EUROPE					
Port Askaig Fm. Kinlochlaggan and Loch na Cille boulder beds **Scotland, U.K.**	806	^{238}U-^{206}Pb monazite	Shear zone truncating the base of the Grampian Group	Max. age for Port Askaig Fm.	Noble et al. (1996)
	601 ± 4	^{238}U-^{206}Pb zircon	Tayvallich Volcanic Fm.	Max. age for Loch na Cille; Min. age for Port Askaig	Dempster et al. (2002)
	595 ± 4	^{238}U-^{206}Pb zircon	Tayvallich Volcanic Fm.; submarine keratophyre	Min. age for Port Askaig	Halliday et al. (1989)
	590 ± 2	^{238}U-^{206}Pb zircon	Ben Vuiruch Granite	Min. age for Dalradian block (and thus glacial successions in Dalradian)	Dempster et al. (2002)
	470 ± 9	^{238}U-^{206}Pb zircon	Gabbros of Insch and Morven-Cabrach in Aberdeenshire	Min. age for Dalradian block (and thus glacial successions in Dalradian)	Dempster et al. (2002)
Smalfjord & Mortensnes Fm., Moelv Tillite **Norway**	654 ± 7	^{87}Rb-^{87}Sr	Argillite of the Nyborg Fm.	Min. age for Smalfjord	Roberts et al. (1997)
	630	^{87}Rb-^{87}Sr	Fine mica; shale	Age of Smalfjord glacial sequence	Gorokhov et al. (2001)
	560	^{87}Rb-^{87}Sr	Fine mica; shale	Age of 'Mortensnes' glacial sequence	Gorokhov et al. (2001)
	807 ± 19	^{87}Rb-^{87}Sr	Underlying Klubbnasen Fm.	Max. age for Smalfjord	Sturt et al. (1975)
	612 ± 18	^{87}Rb-^{87}Sr	Ekre Fm.	Min. age for Moelv tillite	Siedlecka & Roberts (1992); Sokolov (1998)
Petrovbreen & Gropbreen Mbrs. **Svalbard**	950	^{238}U-^{206}Pb zircon	Detrital zircons in the Veteranen group in NE Svalbard and zircon in sub-Veteranen Gp. granites in Nordauslandet	Max. ages for Petrovbreen and Gropbreen mbrs.	Larianov et al. (1998)
	780	^{87}Rb-^{87}Sr	Middle Grusdievbreen Fm.	Max. ages for Petrovbreen and Gropbreen mbrs.	Jacobsen pers. comm. in Halverson et al. (2004)

ASIA

Formation / Region	Age (Ma)	Method	Rock / Sample	Notes	Reference
Ghubrah & Fiq Fms. N. Oman					
	~825	^{238}U-^{206}Pb zircon	Halfayn Fm.	Max. age for Ghubrah and Fiq	Leather (2001)
	~800	^{238}U-^{206}Pb zircon	Crystalline basement in the El Jebal El Akdhar mountains, north Oman	Max. age for Ghubrah and Fiq	Leather (2001)
	~780	^{238}U-^{206}Pb zircon	Pan-African crystalline basement	Max. age for Ghubrah and Fiq	Platel et al. (1992); Kramers & Frei (1992)
	723 + 16/–10	^{238}U-^{206}Pb zircon	Tuffaceous ash near top of Ghubrah Formation	Syn-depositional age for uppermost Ghubrah	Brasier et al. (2000)
	544.5 ± 3.3	^{238}U-^{206}Pb zircon	Ignimbrites in the Fara Fm.	Min. age for underlying Nafun and Abu Mahara Gps.	Brasier et al. (2000)
	542 ± 0.6	^{238}U-^{206}Pb zircon	Ash in subsurface lower part of the Ara Gp.	Min. age for underlying Nafun and Abu Mahara Gps.	Amthor et al. (2003)
	542.6 ± 0.3	^{238}U-^{206}Pb zircon	Ash in subsurface lower part of the Ara Gp.	Min. age for underlying Nafun and Abu Mahara Gps.	Amthor et al. (2003)
Blaini Fm. N. India					
	525 ± 8	^{238}U-^{206}Pb zircon	Detrital zircons from basal Tal Group	Minimum age for Blaini Fm.	Myrow et al. (2003)
Chang'an, Tiesiao & Nantuo Fm. South China					
	819 ± 7	SHRIMP ^{238}U-^{206}Pb	Granite; zircon	Max. age for Chang'an, Tiesiao and Nantuo successions	Ma et al. (1984)
	809 ± 16	SHRIMP ^{238}U-^{206}Pb	Tuffaceous bed; zircon	Max. age for Chang'an, Tiesiao and Nantuo successions	Yin et al. (2003)
	761 ± 8	^{207}Pb-^{206}Pb	Gabbro intruding Sanmenjie Fm.	Max. age for Chang'an, Tiesiao and Nantuo successions	Ge et al. (2001)
	758 ± 23	SHRIMP ^{238}U-^{206}Pb	Dieshuihe Fm; zircon	Max. age for Chang'an, Tiesiao and Nantuo successions	Yin et al. (2003)
	748 ± 12	SHRIMP ^{238}U-^{206}Pb	Volcanic ash; zircon	Max. age for Chang'an, Tiesiao and Nantuo successions	Ma et al. (1984)
	663 ± 4	^{238}U-^{206}Pb zircon	Tuffaceous bed	Max. age for Nantuo Fm.; Min. age for Chang'an & Tiesiao Fms.	Zhou et al. (2004)
	635.4 ± 1.3	^{207}Pb-^{206}Pb	Volcanic ash	Min. age for Nantuo Fm.	Condon et al. (2005)
	635.23 ± 0.57	^{238}U-^{206}Pb zircon	Volcanic ash	Min. age for Nantuo Fm.	Condon et al. (2005)
	632.50 ± 0.48	^{238}U-^{206}Pb zircon	Volcanic ash	Min. age for Nantuo Fm.	Condon et al. (2005)
	632.4 ± 1.3	^{207}Pb-^{206}Pb	Volcanic ash	Min. age for Nantuo Fm.	Condon et al. (2005)
	599.3 ± 4.2	^{207}Pb-^{206}Pb	Doushantuo phosphorites	Min. age for Nantuo Fm.	Barfod et al. (2002)
	598 ± 26	^{207}Pb-^{206}Pb	Doushantuo phosphorites	Min. age for Nantuo Fm.	Chen et al. (2004)
	584 ± 26	^{176}Lu-^{177}Hf	Doushantuo phosphorites	Min. age for Nantuo Fm.	Barfod et al. (2002)
	576 ± 14	^{207}Pb-^{206}Pb	Doushantuo phosphorites	Min. age for Nantuo Fm.	Chen et al. (2004)
	551.07 ± 0.61	^{238}U-^{206}Pb zircon	Volcanic ash	Min. age for Nantuo Fm.	Condon et al. (2005)
	550.55 ± 0.75	^{207}Pb-^{206}Pb	Volcanic ash	Min. age for Nantuo Fm.	Condon et al. (2005)
	538.2 ± 1.5	SHRIMP ^{238}U-^{206}Pb	Tuff	Min. age for Nantuo Fm.	Jenkins et al. (2002)
Beiyixi, Altungal, Tereeken & Hangelchaok Fm. North China					
	755 ± 15	SHRIMP ^{238}U-^{206}Pb	Volcanics	Max. age for Beiyixi, Altungal, Tereeken & Hangelchaok glacial successions	Xu et al. (2005)

climate change (*cf.* Young, 2002), as it has for the Permo-Carboniferous glaciation of Gondwana (Scheffler *et al.*, 2003), although care is required in order to evaluate the effects of diagenesis on pre-burial geochemical composition, sorting effects and changes in sediment provenance over time.

NEOPROTEROZOIC PALAEOGLACIOLOGY

Relatively few detailed sedimentological studies have been undertaken on Neoproterozoic diamictite-bearing successions in the context of Snowball Earth theory, although a wealth of existing literature is available (e.g. Edwards, 1975; Hambrey & Harland, 1981 and references therein; Fairchild & Hambrey, 1984; Spencer, 1985; Hambrey & Spencer, 1987; Moncrieff, 1988; Eyles, 1990; Harland *et al.*, 1993; Arnaud & Eyles, 2002a, b; Allen *et al.*, 2004). Recent investigations have been undertaken on successions in the British Isles (Port Askaig Formation) and in Norway (Smalfjord Formation) which highlight the importance of sediment gravity flow deposits with minor glaciclastic debris components (Arnaud & Eyles, 2002a, b; alternative interpretations of these successions can be found in Edwards (1975), Edwards & Føyn (1981) and Spencer (1985)). Eyles and Januszczak (2004) argued that sediment gravity flow deposits preserved in many Neoproterozoic basins reflect tectonic instability associated with rift-related break-up of the Rodinia supercontinent. However, since sediment gravity flows are a common (if not dominant) process in the accumulation of Quaternary shelf-break fan systems (e.g. Dowdeswell *et al.*, 1996; Powell & Cooper, 2002), distinguishing mass flow diamictites deposited purely as a function of tectonics *versus* glacially transported debris is complicated. Subaqueous mass flow deposits containing a glaciclastic debris component on passive margins are likely to result directly from continental glaciation, but where glaciers were nucleated on uplifted rift basin margins, the picture is less clear, and a combination of both tectonic and climatic controls is likely (*cf.* Allen *et al.*, 2004; Eyles & Januszczak, 2004). Many Neoproterozoic glacial successions fall into the latter category, where a rift to post-rift transition is considered likely, including the Blaini Formation in India (Kumar & Brookfield, 1987), the Chang'an and Nantuo

Formations in south China (Jiang *et al.*, 2003), the Fiq in Oman (Allen *et al.*, 2004) and the Rapitan Group in the North American Cordillera (Young, 1995). Sediment redistribution during postglacial eustatic recovery is also problematic, since successions may be significantly reworked during large-scale submarine failures (e.g. Maslin *et al.*, 2004). However, not all glacially influenced successions are dominated by sediment gravity flows. The Ayn Formation (formerly the Lower Member of the Mirbat Sandstone Formation), in Dhofar, south Oman records a terrestrial to marginal marine succession characterized by Gilbert-type glaciofluvial delta systems, which are laterally associated with more distal stratified glacimarine diamictites, containing abundant dropstones and ubiquitous glacially polished and striated clasts, indicative of temperate or polythermal glacial conditions (Figs 1, 15).

Generally speaking, the huge volumes of debris preserved in Neoproterozoic basins comprising a significant glaciclastic debris component in passive margin settings points towards extensive continental ice sheets with either temperate or polythermal basal characteristics. Temperate or polythermal conditions are also indicated from striated and polished clasts, subglacial pavements and plucked bedrock surfaces including roches moutonnées and whaleback forms (Coats & Preiss, 1980; Ojakangas & Matsch, 1980; Deynoux & Trompette, 1981) and welded subglacial breccia deposits which have been used to infer stick-slip basal behaviour (Bestmann *et al.*, 2006). However, the degree of synchroneity between the accumulation of these successions is limited given the current radiometric age constraints (Table 4), and the long-term thermal response of these ice masses to Neoproterozoic climate change is difficult to evaluate. Thus although a demonstrable record of polythermal or temperate glacial systems exists, cold-based ice sheets may also have persisted for considerable periods of time. Suggestions that Ulvesø (Greenland) and Petrovbreen diamictites of Svalbard reflect cold-based glaciation because of ineffective bedrock quarrying (Halverson *et al.*, 2004) are at odds with the volumes of debris associated with these successions (notably the Ulvesø Formation). In the Quaternary record, subglacial debris loads are commonly dominated by local basin lithologies, and the absence of basement

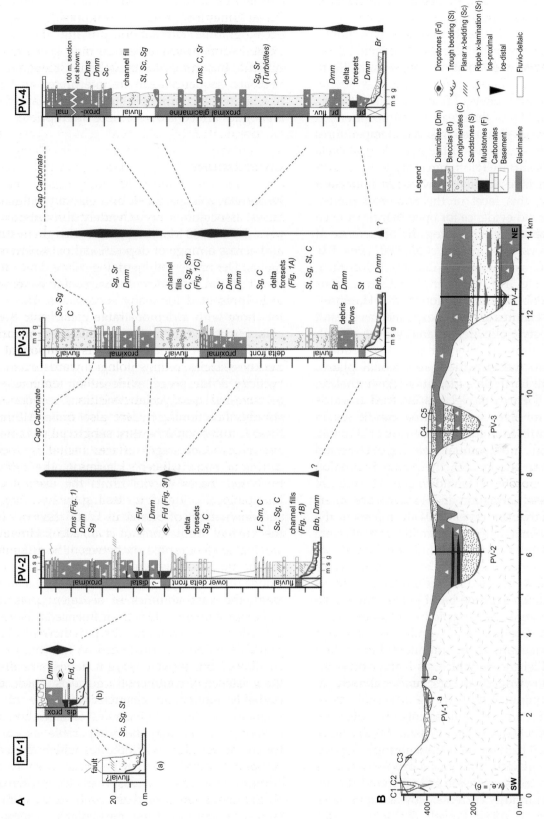

Fig. 15 Lithostratigraphy and interpretive correlation panel of Neoproterozoic valley-fill deposits of the Ayn Fm., Dhofar, south Oman; (A) palaeovalley fills. Section PV-4 modified after Kellerhals & Matter (2003); (B) lateral correlation of palaeovalley fills and elevation of basement, based on logged sections and field mapping. Modified from Rieu *et al.* (2006).

clasts does not necessarily negate temperate or polythermal glaciation.

HYDROLOGICAL SHUTDOWN

One of the central tenets of Snowball Earth theory is that decoupling of the ocean and atmosphere, in combination with plummeting surface temperatures would lead to shutdown of the hydrological cycle at the Earth's surface (Hoffman *et al.*, 1998a). Sedimentological investigations have been important in contesting this facet of the Snowball model, with widespread evidence for open marine or even terrestrial conditions, including the occurrence of dropstone horizons (Condon *et al.*, 2002; Fig. 1d), wave-rippled sandstones (Williams, 1996; Allen *et al.*, 2004; Fig. 1E), and periglacial involutions (Hambrey & Spencer, 1987; Moncrieff & Hambrey, 1990). These features are easily explained as a result of Cenozoic-style climate variability over glacial-interglacial transitions, although they could equally be applied to the growth or recessional phases of a Snowball Earth-type glaciation. Nevertheless, the Snowball Earth model has evolved towards one which involves open marine conditions in 'oases' resulting from the early demise of sikussak ice (shorefast multi-annual sea ice; Halverson *et al.*, 2004). In this model, the Arena Formation (Greenland) and MacDonaldryggen Member of the Elbobreen Formation in Svalbard are interpreted by Halverson *et al.*, (2004) to represent the Snowball maxima, when continental ice sheets were effectively landlocked by shorefast sikussak ice and laminites were deposited by density currents resulting from heavy brine formation. This is difficult to defend for the basal part of the Arena Formation where wave-rippled sandstones occur (Hambrey & Spencer, 1987), and is inconsistent with our understanding of laminite sedimentation during the Pleistocene which, as Halverson *et al.*, (2004) acknowledged, occurs predominantly in interglacial periods (Orheim & Elverhøi, 1981; Dowdeswell *et al.*, 1998; but see also Dowdeswell *et al.*, 2000; O'Grady & Syvitski, 2002). Since laminites may be deposited from a range of glacially and non-glacially influenced processes (e.g. as tidal rhythmites, suspension fallout from buoyant plumes and turbidity currents; Stow & Piper, 1984; Pickering *et al.*, 1986; Powell & Molnia, 1989;

Cowan & Powell, 1990; Cowan *et al.*, 1997, 1998), and dense brines form a natural component of modern oceanographic circulation, proving the former existence of extensive sea ice may be extremely difficult. In summary, although the Greenland and Svalbard sequences can be interpreted in terms of a Snowball-compatible succession, Phanerozoic analogues are considered more appropriate.

CONCLUSIONS

Neoproterozoic glacigenic and glacially influenced facies associations occur widely throughout the period 780–580 Ma, in passive margin settings and across a range of depositional palaeoenvironments. Definitive sedimentological evidence for a direct glacial influence on sedimentation remains to be presented for many successions. However, for those with a demonstrable glaciclastic debris component, sedimentological data are compatible with Phanerozoic analogues in terms of succession thickness, facies, sedimentological architecture and cyclicity. Widespread evidence for open marine or terrestrial periglacial conditions testifies to a strongly functioning hydrological cycle. Although these features could be attributed to pulsed growth and recessional stages of continental ice sheets during a Snowball-type glaciation, the recently proposed 'oases' model limits the use of sedimentological evidence for testing the hydrological shutdown facet of Snowball Earth theory. Given the current limitations of available radiometric and palaeogeographic databases, it is difficult to demonstrate globally synchronous glaciation, or the extent to which glaciation occurred. On a simple level, the vast volumes of sediment preserved in Neoproterozoic glacially influenced basins are consistent with temperate or polythermal glaciation. Glaciation was nucleated on locally uplifted rift flanks, but passive margin deposits testify to the existence of continental ice sheets which were probably similar in character to those of the Phanerozoic. The relative roles of glaciation and tectonic activity may be inseparable as causes for the accumulation of debrites which are a key component of the glacimarine sedimentary record. Although repeated widespread Neoproterozoic glaciations are envisaged, the evaluation of climate change during this time, particularly in terms of

Snowball Earth theory, requires further enhancement of the existing sedimentological, geochronological and palaeomagnetic datasets.

ACKNOWLEDGEMENTS

J.L. Etienne and P.A. Allen acknowledge financial support from Schweizerischer Nationalfonds Grant 103502. R. Rieu and E. Le Guerroué are supported by ETH postgraduate studentships. We are most grateful to Joachim Amthor and colleagues at PDO, Manuele Faccenda, Matthias Papp, Gion Kuper, Dhiraj Banerjee (Delhi University), Sumit Ghosh and Rafique Islam (Wadia Institute of Himalayan Geology) for logistical support and discussion during fieldwork in Oman and North India. We also thank Paul Hoffman (Harvard), Galen Halverson (University Paul Sabatier, Toulouse), Martin Kennedy (University of California, Riverside) and Andrea Cozzi (ETH-Zürich) for ongoing and insightful discussion. The constructive comments of referees Mike Hambrey and Ian Fairchild are gratefully acknowledged.

REFERENCES

Aalto, R.K. (1971) Glacial marine sedimentation and stratigraphy of the Toby Conglomerate (Upper Proterozoic), southeastern British Columbia, northwestern Idaho and northeastern Washington. *Can. J. Earth Sci.*, **8**, 753–787.

Aalto, K.R. (1981) The Late Precambrian Toby Formation of British Columbia, Idaho and Washington. In: *Earth's Pre-Pleistocene Glacial Record* (Eds M.J. Hambrey and W.B. Harland), pp. 731–735. Cambridge University Press, Cambridge.

Agassiz, L. (1838) On the polished and striated surfaces of the rocks which form the beds of glaciers in the Alps. *Proc. Geol. Soc. London.*, **3**, 321–322.

Aharon, P. (1988) Oxygen, carbon and U-series isotopes of aragonites from Vestfold Hills, Antarctica: Clues to geochemical processes in subglacial environments. *Geochim. Cosmochim. Acta*, **52**, 2321–2331.

Aitken, J.D. (1991) Two late Proterozoic glaciations, Mackenzie Mountains, northwestern Canada. *Geology*, **19**, 445–448.

Aleinikoff, J.N., Zartman R.E., Walter, M. Rankin D.W., Lyttle P.T. and Burton W.C. (1995) U-Pb ages of metarhyolites of the Catoctin and Mount Rogers Formations, central and southern Appalachians:

Evidence for two pulses of Iapetan rifting. *Am. J. Sci.*, **295**, 428–454.

Allen, P., Leather, J. and Brasier, M. (2004) The Neoproterozoic Fiq glaciation and its aftermath, Huqf Supergroup of Oman. *Basin Res.*, **160**, 507–534.

Alley, R.B., Cuffey, K.M., Evenson, E.B., Strasser, J.C., Lawson, D.E. and Larson, G.J. (1997) How glaciers entrain and transport basal sediment: physical constraints. *Quatern. Sci. Rev.*, **16**, 1017–1038.

Allsopp, H.L., Köstlin, E.O., Welke, H.J., Burger, A.J., Kröner, A. and Blignault, H.J. (1979) Rb–Sr and U–Pb geochronology of Late Precambrian–Early Paleozoic igneous activity in the Richtersveld (South Africa) and southern South West Africa. *Trans. Geol. Soc. S. Afr.*, **82**, 185–204.

Alvarenga, C.J.S., Santos, R.V. and Dantas, E.L. (2004) C–O–Sr isotopic stratigraphy of cap carbonates overlying Marinoan-age glacial diamictites in the Paraguay Belt, Brazil. *Precambrian Res.*, **131**, 1–21.

Amthor, J.E., Grotzinger, J.P., Schröder, S., Bowring, S.A., Ramezani, J., Martin, M.W. and Matter, A. (2003) Extinction of Cloudina and Namacalathus at the Precambrian-Cambrian boundary in Oman. *Geology*, **31**, 431–434.

Andrews, J.T., Smith, L.M., Preston, R., Cooper, T. and Jennings, A.E. (1997) Holocene patterns of ice-rafted detritus (IRD) in cores from the East Greenland shelf. *J. Quatern. Sci.*, **12**, 1–13.

Arnaud, E. and Eyles, C.H. (2002a) Glacial influence on Neoproterozoic sedimentation: the Smalfjord Formation, Northern Norway. *Sedimentology*, **49**, 765–788.

Arnaud, E. and Eyles, C.H. (2002b) Catastrophic mass failure of a Neoproterozoic glacially influenced continental margin, the Great Breccia, Port Askaig Formation, Scotland. *Sed. Geol.*, **151**, 313–333.

Atkins, C.B., Barrett, P.J. and Hicock, S.R. (2002) Cold glaciers erode and deposit: evidence from Allan Hills, Antarctica. *Geology*, **30**, 659–662.

Attoh, K. Dallmeyer, R.D. and Affaton, P. (1997) Chronology of nappe assembly in the Pan-African Dahomeyide Orogen, West Africa; evidence from (super 40) Ar/(super 39) Ar mineral ages. *Precambrian Res.*, **82**, 153–171.

Babinski, M. and Kaufman, A.J. (2003) First direct dating of a Neoproterozoic post-glacial cap carbonate. In: *South American Symposium on Isotope Geology 4 Short Papers*, **1**, pp. 321–323.

Bailey, C.M. and Peters, S.E. (1998) Glacially influenced sedimentation in the late Neoproterozoic Mechum River Formation, Blue Ridge province, Virginia. *Geology*, **26**, 623–626.

Banks, N.L., Edwards, M.B., Geddes, W.P., Hobday, D.K. and Reading, H.G. (1971) Late Precambrian and

Cambro-Ordovician sedimentation in East Finnmark. *Nor. Geol. Unders.*, **269**, 197–236.

Baode, G., Ruitang, W., Hambrey, M.J. and Wuchen, G. (1986) Glacial sediments and erosional pavements near the Cambrian-Precambrian boundary in western Henan Province, China. *J. Geol. Soc. London*, **143**, 311–323.

Barfod, G.H., Albarède, F., Knoll, A.H., Xiao, S., Télouk, P., Frei, R. and Baker, J. (2002) New Lu–Hf and Pb–Pb age constraints on the earliest animal fossils. *Earth Planet. Sci. Lett.*, **201**, 203–212.

Barker, P.F., Camerlenghi, A. and Acton, G.D. (1998) Antarctic glacial history and sea-level change. ODP, Preliminary report No. 78. Ocean Drilling Program, Texas A & M University, College Station, TX 77845–9547, USA.

Bart, P.J. and Anderson, J.B. (1995) Seismic record of glacial events affecting the Pacific margin of the northwestern Antarctic Peninsula. In: *Geology and Seismic Stratigraphy of the Antarctic Margin* (Eds A.K. Cooper, P.F. Barker and G. Brancolini) *Antarct. Res. Ser.*, **68**, 75–96.

Benn, D.I. (1995) Fabric signature of till deformation, Breiðamerkurjökull, Iceland. *Sedimentology*, **42**, 735–747.

Benn, D.I. and Ballantyne, C.K. (1993) The description and representation of particle shape. *Earth Surf. Proc. Land.*, **18**, 665–672.

Benn, D.I. and Ballantyne, C.K. (1994) Reconstructing the transport history of glacigenic sediments: a new approach based on the co-variance of clast form indices: *Sed. Geol.*, **91**, 215–227.

Benn, D.I. and Evans, D.J.A. (1998) Glaciers and Glaciation. Arnold, London, 734 pp.

Benn, D.I. and Owen, L.A. (2002) Himalayan glacial sedimentary environments: a framework for reconstructing and dating the former extent of glaciers in high mountains. *Quatern. Int.*, **97–8**, 3–25.

Benn, D.I. and Ringrose, T.J. (2001) Random variation of fabric eigenvalues: Implications for the use of A-axis fabric data to differentiate till facies. *Earth Surf. Proc. Land.*, **26**, 295–306.

Benn, D.I., Wiseman, S. and Hands, K.A. (2001) Growth and drainage of supraglacial lakes on debris-mantled Ngozumpa Glacier, Khumbu Himal, Nepal. *J. Glaciol.*, **47**, 626–638.

Benn, D.I. and Prave, A. (2006) Subglacial and pro-glacial glacitectonic deformation in the Neoproterozoic Port Askaig Formation, Scotland. *Geomorphology*, **75**, 266–280.

Bennett, M.R., Hambrey, M.J., Huddart, D. and Ghienne, J.F. (1996a) The formation of a geometrical ridge network by the surge-type glacier Kongsvegen, Svalbard. *J. Quatern. Sci.*, **11**, 437–449.

Bennett, M.R., Huddart, D., Hambrey, M.J. and Ghienne, J.F. (1996b) Moraine development at the high-arctic valley glacier Pedersenbreen, Svalbard. *Geogr. Ann.*, **78 A**, 209–222.

Bennett, M.R., Hambrey, M.J. and Huddart, D. (1997) Modification of clast shape in high-arctic environments. *J. Sed. Res.*, **67**, 550–559.

Bennett, M.R., Hambrey, M.J., Huddart, D. and Glasser, N.F. (1998) Glacial thrusting and moraine-mound formation in Svalbard and Britain: the example of Coire a'Cheud-chnoic (Valley of Hundred Hills), Torridon Scotland. In: Owen, L.A. (Ed.) Mountain Glaciation. *Quatern. Proc.*, **6**, 17–34.

Bennett, M.R., Waller, R.I., Glasser, N.F., Hambrey, M.J. and Huddart, D. (1999) Glacigenic clast fabrics: genetic fingerprint or wishful thinking? *J. Quatern. Sci.*, **14**, 125–135.

Benus, A.P. (1988) Sedimentological context of a deep-water Ediacaran fauna (Mistaken Point Formation, Avalon Zone, Eastern Newfoundland). In: *Trace Fossils, Small Shelly Fossils, and the Precambrian-Cambrian Boundary* (Eds E. Landing, G. Narbonne and P. Myrow). *N.Y. State Mus. Bull.*, **463**, 8–9.

Bestmann, M., Rice, A.H.N., Langenhorst, F., Grasemann, B. and Heidelbach, F. (2006) Subglacial bedrock welding associated with glacial earthquakes. J. Geol. Soc. London, **163** (3), 417–420.

Beyth, M., Avigad, D., Wetzel, H.-U., Matthews, A. and Berhe, S.M. (2003) Crustal exhumation and indications for Snowball Earth in the East African Orogen: north Ethiopia and east Eritrea. *Precambrian Res.*, **123**, 187–201.

Bhatia, M.R. and Prasad, A.K. (1981) Evolution of Late Paleozoic glacial marine sedimentation in the Simla Hills, Lesser Himalaya, India. *N. Jb. Geol. Paläeontol. Mh.*, **5**, 267–288.

Bhatia, S.B. and Kanwar, R.C. (1975) Blaini and Related Formations. *Indian Geol. Assoc. Spec. Iss.*, **8**, 279 pp.

Binda, P.L. and Van Eden, J.G. (1972) Sedimentological evidence on the origin of the Precambrian Great onglomerate (Kundelungu Tillite), Zambia. *Palaeogeogr., Palaeoclimatol., Palaeoecol.*, **12**, 151–168.

Bjørlykke, K. (1981) Late Precambrian tillites of the Bunyoro Series, western Uganda. In: *Earth's Pre-Pleistocene Glacial Record* (Eds M.J. Hambrey and W.B. Harland), pp. 151–152. Cambridge University Press, Cambridge.

Bjørlykke, K. and Nystuen, J.P. (1981) Late Precambrian tillites of South Norway. In: *Earth's Pre-Pleistocene Glacial Record* (Eds M.J. Hambrey and W.B. Harland), pp. 624–628. Cambridge University Press, Cambridge.

Black, L.P., Shaw, R.D. and Offe, L.A. (1980) The age of the Stuart Dyke Swarm and its bearing on the onset of Late Precambrian sedimentation in central Australia. *J. Geol. Soc. Aust.*, **27**, 151–155.

Blake, D.H., Tyler, I.M., Griffin, T.J., Sheppard, S., Thorne, A.M. and Warren, R.G. (1998) Geology of the Halls Creek 1:100 000 sheet area (4461), Western Australia, Australian Geological Survey Organisation, 1v, 36 pp.

Blair, T.C. and McPherson, J.G. (1999) Grain-size and textural classification of coarse sedimentary particles. *J. Sed. Res.*, **69**, 6–19.

Bodiselitsch, B., Koeberl, C., Master, S. and Reimold, W.U. (2005) Estimating Duration and Intensity of Neoproterozoic Snowball Glaciations from Ir Anomalies. *Science*, **308**, 239–242.

Bond, G. (1981) A possible Late Precambrian tillite from the Urungwe District, Zimbabwe. In: *Earth's Pre-Pleistocene Glacial Record* (Eds M.J. Hambrey and W.B. Harland), pp. 178–179. Cambridge University Press, Cambridge.

Boulton, G.S. and Hindmarsh, R.C.A. (1987) Sediment deformation beneath glaciers: rheology and sedimentological consequences. *J. Sed. Res.*, **92**, **B9**, 9059–9082.

Boulton, G.S. (1970) On the deposition of subglacial and melt-out tills at the margins of certain Svalbard glaciers. *J. Glaciol.*, **9**, 231–245.

Boulton, G.S. (1972) The role of thermal regime in glacial sedimentation. In: *Polar Geomorphology* (Eds R.J. Price and D.E. Sugden). *Institute of British Geographers, Special Publication*, **4**, 1–19.

Boulton, G.S. (1978) Boulder shapes and grain-size distribution of debris as indicators of transport paths through a glacier and till genesis. *Sedimentology*, **25**, 773–799.

Boulton, G.S. (1979) Processes of glacier erosion on different substrata. *J. Glaciol.*, **23**, 15–38.

Boulton, G.S. (1996) Theory of glacial erosion, transport and deposition as a consequence of subglacial sediment deformation. *J. Glaciol.*, **140**, 43–62.

Bowring, S.A. and Schmitz, M.D. (2003) High-precision U-Pb zircon geochronology and the stratigraphic record. *Rev. Mineral. Geochem.*, **53**, 305–326.

Bowring, S., Myrow, P., Landing, E., Ramezani, J. and Grotzinger, J. (2003) Geochronological constraints on terminal Proterozoic events and the rise of the Metazoans. *Geophys. Res. Abstr.*, **5**, 13219.

Brasier, M.D. and Shields, G. (2000) Neoproterozoic chemostratigraphy and correlation of the Port Askaig glaciation, Dalradian Supergroup of Scotland. *J. Geol. Soc. London*, **157**, 909–914.

Brasier, M.D., McCarron, G., Tucker, R., Leather, J., Allen, P. and Shields, G. (2000) New U-Pb zircon dates for the Neoproterozoic Ghubrah glaciation and for the top of the Huqf Supergroup, Oman. *Geology*, **28**, 175–178.

Briqueu, L., Lancelot, J.R., Valois, J.-P. and Walgenwitz, F. (1980) Géochronologie U-Pb et genèse d'un type de mineralisation uranifère: Les alaskites de Goanikontès (Namibie) et leur encaissant. *Centr. Rech. Explor. Prod. Elf, Aquitaine Bull.*, **4**, 759–811.

Broecker, W.S., Andree, M., Wolfli, W., Oeschger, H., Bonani, G., Kennett, J.P. and Peteet, D. (1988) The chronology of the last deglaciation: Implications to the cause of the Younger Dryas event. *Palaeoceanog.*, **3**, 1–19.

Brookfield, M. (1987) Lithostratigraphic correlation of Blaini Formation (late Proterozoic, Lesser Himalaya, India) with other late Proterozoic tillite sequences. *Geol. Rund.*, **76**, 477–484.

Brookfield, M. (1994) Problems in applying preservation, facies and sequence models to Sinian (Neoproterozoic) glacial sequences in Australia and Asia. *Precambrian Res.*, **70**, 113–143.

Buchwaldt, R., Toulkeridis, T., Babinski, M., Noce, C.M., Martins Neto, M. and Hercos, C.M. (1999) Age determination and age related Provenance Analysis of the Proterozoic glaciation in central eastern Brazil. *An. Acad. Ci.*, **71(3)**, 527–548.

Burns, S.J. and Matter, A. (1993) Carbon isotopic record of the latest Proterozoic from Oman. *Eclogae Geol. Helv.*, **86**, 595–607.

Caby, R. and Fabre, J. (1981) Late Proterozoic to Early Palaeozoic diamictites, tillites and associated glacigenic sediments in the Serie Pourpree of western Hoggar, Algeria. In: *Earth's Pre-Pleistocene Glacial Record* (Eds M.J. Hambrey and W.B. Harland), pp. 140–145. Cambridge University Press, Cambridge.

Cahen, L. (1954) *Geologie du Congo belge*. H. Vaillant-Carmanne, Liege.

Cahen, L. (1963) Glaciations anciennes et dérive des continents. *Annls. Soc. Geol. Bélg.*, **86 B**, 79-B 84.

Cahen, L. and Lepersonne, J. (1976) Les mixtites du Bas-Zaire: mise au point interimaire. *Rapp. a. Dept. Geol. Miner. Mus. R. Afr. Centr.*, 1975. 33–57.

Cahen, L. and Lepersonne, J. (1981) Proterozoic diamictites of Lower Zaire, In: *Earth's Pre-Pleistocene Glacial Record* (Eds M.J. Hambrey and W.B. Harland), pp. 153–157. Cambridge University Press, Cambridge.

Calver, C.R. (2000) Isotope stratigraphy of the Ediacaran (Neoproterozoic III) of the Adelaide Rift Complex, Australia, and the overprint of water column stratification. *Precambrian Res.*, **100**, 121–150.

Calver, C.R. and Walter, M.R. (2000) The late Neoproterozoic Grassy Group of King Island, Tasmania: Correlation and palaeogeographic significance. *Precambrian Res.*, **100**, 299–312.

Calver, C.R., Black, L.P., Everard, J.L. and Seymour, D.B. (2004) U-Pb zircon age constraints on late Neoproterozoic glaciation in Tasmania. *Geology*, **32**, 893–896.

Chamberlin, T.C. (1888) The rock scourings of the great ice invasions. *U.S. Geol. Surv. Ann. Rep.*, **7**, 155–248.

Chen, D.F., Dong, W.Q., Zhu, B.Q. and Chen, X.P. (2004) Pb–Pb ages of Neoproterozoic Doushantuo phosphorites in South China: constraints on early metazoan evolution and glaciation events. *Precambrian Res.*, **132**, 123–132.

Chinn, T.J.H. and Dillon, A. (1987) Observations on a debris-covered polar glacier 'Whisky Glacier' James Ross Island, Antarctic Peninsula, Antarctica. *J. Glaciol.*, **33**, 300–310.

Christie-Blick, N. (1982) Upper Proterozoic (Eocambrian) Mineral Fork Tillite of Utah: A continental glacial and glaciomarine sequence: Discussion. *Geol. Soc. Am. Bull.*, **93**, 184–186.

Christie-Blick, N. and Levy, M. (1989) Stratigraphic and tectonic framework of Upper Proterozoic and Cambrian rocks in the western United States. In: *Late Proterozoic and Cambrian Tectonics, Sedimentation, and Record of Metazoan Radiation in the Western United States, Fieldtrip Guidebook T331* (Eds N. Christie-Blick and M. Levy), pp. 7–23. 28th International Geological Congress. American Geophysical Union.

Christoffersen, P. (2003) *Thermodynamics of basal freeze-on: subglacial property changes and ice sheet response.* Ph.D. thesis, Technical University of Denmark, pp. 92.

Christoffersen, P. and Tulaczyk, S. (2003) Response of subglacial sediments to basal freeze-on: I. Theory and comparison to observations from beneath the West Antarctic Ice Sheet. *J. Geophys. Res., B*, **108**, 4.

Chumakov, N.M. (1981a) Late Precambrian tilloids of the Rybachiy Peninsula, U.S.S.R. In: *Earth's Pre-Pleistocene Glacial Record* (Eds M.J. Hambrey and W.B. Harland), pp. 602–605. Cambridge University Press, Cambridge.

Chumakov, N.M. (1981b) Late Precambrian glacial deposits of the Vilchitsy Formation of western regions of the U.S.S.R. In: *Earth's Pre-Pleistocene Glacial Record* (Eds M.J. Hambrey and W.B. Harland), pp. 655–659. Cambridge University Press, Cambridge.

Chumakov, N.M. (1981c) Late Precambrian Churochnya tillites of the Polyudov Ridge, U.S.S.R. In: *Earth's Pre-Pleistocene Glacial Record* (Eds M.J. Hambrey and W.B. Harland), pp. 666–669. Cambridge University Press, Cambridge.

Chumakov, N.M. (1996) Tillites and Tilloids on the Western Slope of the Middle Urals (Upper Riphean and Vendian–Early Paleozoic Sections: Guidebook of Geological Trips, All-Russia Conference on the Vendian–Early Paleozoic Paleogeography), Ekaterinburg: *Inst. Geol. Geofiz., Ural. Otd., Ross. Akad. Nauk*, 74–82.

Chumakov, N.M. (1998) Reference Section of Vendian Glacial Deposits in the Southern Urals (Kurgashlya Formation of the Krivoi Luk Graben), (The Urals: Fundamental Problems of the Geodynamics and Stratigraphy), pp. 138–153. Moscow: Nauka.

Clauer, N. (1976) Géochimie isotopique du strontium des milieux sédimentaires. Application à la geochronology de la couverture du craton ouest-africain. *Mém. Sci. Géol. Strasbourg*, **45**, 256.

Clauer, N. and Deynoux, M. (1987) New Information on the Probable Isotopic Age of the Late Proterozoic Glaciation in West Africa. *Precambrian Res.*, **37**, 89–94.

Clauer, N. and Kröner, A. (1979) Strontium and Argon Isotopic Homogenization of Pelitic Sediments during Low-Grade Regional Metamorphism: The Pan-African Upper Damara Sequence of Northern Namibia (South West Africa). *Earth Planet. Sci. Lett.*, **43**, 117–131.

Clauer, N., Caby, R., Jeannette, D. and Trompette, R. (1982) Geochronology of sedimentary and metasedimentary Precambrian rocks of the West African Craton. *Precambrian Res.*, **18**, 53–71.

Coats, R.P. (1973) Copley, South Australia, 1:250 000 explanatory notes. Sheet SH/54–09, 1st edition. *Geol. Surv. S. Aust.*, 1v, 38 pp.

Coats, R.P. (1981) Late Proterozoic (Adelaidean) tillites of the Adelaide Geosyncline. In: *Earth's Pre-Pleistocene Glacial Record* (Eds M.J. Hambrey and W.B. Harland), pp. 537–548. Cambridge University Press, Cambridge.

Coats, R.P. and Preiss, W.V. (1980) Stratigraphic and geochronological reinterpretation of late Proterozoic glaciogenic sequences in the Kimberley region, Western Australia. *Precambrian Res.*, **13**, 181–208.

Colpron, M., Logan, J.M. and Mortensen, J.K. (2002) U–Pb zircon age constraint for late Neoproterozoic rifting and initiation of the lower Paleozoic passive margin of western Laurentia, *Can. J. Earth Sci.*, **39**, 133–143.

Condon, D.J. and Prave, A.R. (2000) Two from Donegal: Neoproterozoic glacial episodes on the northeast margin of Laurentia. *Geology*, **28**, 951–954.

Condon, D.J., Prave, A.R. and Benn, D.I. (2002) Neoproterozoic glacial rain-out intervals: Observations and Implications. *Geology*, **30**, 35–38.

Condon, D. Zhu, M., Bowring, S., Wang, W., Yang, A. and Jin, Y. (2005) U-Pb Ages from the Neoproterozoic Doushantuo Formation, China. *Science*, **308**, 95–98.

Cooper, J.A., Jenkins, R.J.F., Compston, W. and Williams, I.S. (1992) Ion-probe zircon dating of a mid-Early Cambrian tuff in South Australia. *J. Geol. Soc. London.*, **149**, 185–192.

Corkeron, M.L. and George, A.D. (2001) Glacial incursion on a Neoproterozoic carbonate platform in the Kimberley region, Australia. *Geol. Soc. Am. Bull.*, **113**, 1121–1132.

Corsetti, F.A. and Kaufman, A.J. (2003) Stratigraphic investigations of carbon isotope anomalies and Neoproterozoic ice ages in Death Valley, California. *Geol. Soc. Am. Bull.*, **115**, 916–932.

Cowan, E.A. and Powell, R.D. (1990) Suspended sediment transport and deposition of cyclically interlaminated sediment in a temperate glacial fjord, Alaska, U.S.A. In: *Glacimarine Environments: Processes and Sediments* (Eds J.A. Dowdeswell and J.D. Scourse) pp. 75–89. *Geol. Soc. Spec. Publ.*, **53**.

Cowan, E.A., Cai, J., Powell, R.D., Clark, J.D. and Pitcher, J.N. (1997) Temperate Glacimarine Varves: an example from Disenchantment Bay, Southern Alaska. *J. Sed. Res.*, **67**, 536–549.

Cowan, E.A., Cai, J., Powell, R.D., Seramur, K.C. and Spurgeon, V.L. (1998) Modern tidal rhythmites deposited in a deep-water estuary. *Geo-Marine Letters*, **18**, 40–48.

Cozzi, A. and Al-Siyabi, H.A. (2004) Sedimentology and play potential of the late Neoproterozoic Buah Carbonates of Oman. *GeoArabia*, **9**, 11–36.

Croot, D.G. (1988) Morphological, structural and mechanical analysis of neoglacial ice-pushed ridges in Iceland. In: *Glaciotectonics, Forms and Processes* (Ed. D.G. Croot), pp. 33–47. A.A. Balkema, Rotterdam.

Crowell, J.C. (1978) Gondwana glaciation, cyclothems, continental positioning, and climate change. *Am. J. Sci.*, **278**, 1345–1372.

Cuffey, K.M., Conway, H., Gades, A.M., Hallet, B., Lorrain, R., Severinghaus, J.P., Steig, E.J., Vaughn, B. and White, J.W.C. (2000) Entrainment at cold glacier beds. *Geology*, **28**, 351–354.

Curry, A.M. and Ballantyne, C.K. (1999) Paraglacial modification of glacigenic sediment. *Geogr. Ann.*, **81A**, 409–419.

Daily, B. and Forbes, B.G. (1969) Notes on the Proterozoic and Cambrian, southern and central Flinders Ranges, South Australia. In: *Geological Excursions Handbook* (Ed. B. Daily), 23–30. ANZAAS, Section 3.

Dalgarno, C.D. and Johnson, J.E. (1964) Glacials of the Marinoan Series. *Quart. Geol. Notes geol. Surv. S. Austr.*, **11**, 3–4.

Davies, K.A. (1939) The glacial sediments of Bunyoro, N.W. Uganda. *Bull. Geol. Surv. Uganda*, **3**, 20–37.

de Almeida, F.F.M. (1964a) Geologia do centro-oeste matogrossense. *Brazil Minist. Minas Energ. Dep. Nac. Prod. Miner. Bol. Div. Geol. Mineral.*, **215**, 1–137.

de Almeida, F.F.M. (1964b) Glaciação Eocambriana em Mato Grosso. *Brazil Minist. Minas Energ. Dep. Nac. Prod. Mineral Notas Prelim. Est.*, **117**, 1–10.

Dec, T., O'Brien, S.J. and Knight, I. (1992) Late Precambrian volcanoclastic deposits of the Avalonian Eastport Basin (Newfoundland Appalachians): petrofacies, detrital clinopyroxene geochemistry

and plate tectonic implications. *Precambrian Res.*, **59**, 243–262.

Dempster, T.J., Rogers, G., Tanner, P.W.G., Bluck, B.J., Muir, R.J., Redwood, S.D., Ireland, T.R. and Paterson, B.A. (2002) Timing of deposition, orogenesis and glaciation within the Dalradian rocks of Scotland: constraints from U–Pb zircon ages. *J. Geol. Soc. London.*, **159**, 83–94.

Dewez, V. and Geurts, M.A. (1996) Multivariate mineralogical analysis of sediments from the Upper Wisconsinan of the southwestern Yukon Territory. *Can. J. Earth Sci.*, **33**, 42–45.

Deynoux, M. (1982) Periglacial polygonal structures and sand wedges in the late Precambrian glacial formations of the Taoudeni Basin in Adrar of Mauretania (West Africa). *Palaeogeogr., Palaeoclimatol., Palaeoecol.*, **39**, 55–70.

Deynoux, M. (1985) Terrestrial of waterlain glacial diamictites? Three case studies from the late Precambrian and Late Ordovician glacial drifts in West Africa. *Palaeogeogr., Palaeoclimatol., Palaeoecol.*, **51**, 97–141.

Deynoux, M. and Trompette, R. (1981) Late Precambrian tillites of the Taoudeni Basin, West Africa. In: *Earth's Pre-Pleistocene Glacial Record* (Eds M.J. Hambrey and W.B. Harland), pp. 89–96. Cambridge University Press, Cambridge.

Dobkins, J.E. and Folk, R.L. (1970) Shape development on Tahiti-Nui. *J. Sed. Petrol.*, **40**, 1167–1203.

Dobrzinski, N. and Bahlburg, H. (*in Press*). Sedimentology and environmental significance of the Cryogenian successions of the Yangtze platform, South China block. *Palaeogeogr., Palaeoclimatol., Palaeoecol.*,

Dobrzinski, N., Bahlburg, H., Strauss, H. and Zhang, Q. (2004) Geochemical climate proxies applied to the Neoproterozoic glacial succession on the Yangtze Platform, South China. In: *The extreme Proterozoic: Geology, Geochemistry and Climate* (Eds Jenkins, G.S., McMenamin, M.A.S., McKay, C.P. and Sohl, L.). – *AGU monograph series*, **146**, 13–32.

Domack, E.W. and Lawson, D.E. (1985) Pebble fabric in an ice rafted diamicton. *J. Geol.*, **93**, 577–591.

Dott, R.H. (1961) Squantum 'Tillite', Massachusetts; evidence of glaciation or subaqueous mass movements? *Geol. Soc. Am. Bull.*, **72**, 1289–1306.

Dow, D.B. and Gemuts, I. (1967) Dixon Range, W.A., 1:250 000. *Geol. Set., Bur. Miner. Resour. Geol. Geophys., Anst., Explan. Notes*, SE/52–6.

Dowdeswell, J.A. and Sharp, M. (1986) Characterization of pebble fabrics in modern terrestrial glacigenic sediments. *Sedimentology*, **33**, 699–710.

Dowdeswell, J.A., Hambrey, M.J. and Wu, R. (1985) A comparison of clast fabric and shape in late Precambrian and modern glacigenic sediments. *J. Sed. Petrol.*, **55**, 691–704.

Dowdeswell, J.A., Uenzelmann-Neben, G., Whittington, R.J. and Marienfeld, P. (1994) The Late Quaternary sedimentary record in Scoresby Sund, East Greenland. *Boreas*, **23**, 294–310.

Dowdeswell, J.A., Kenyon, N.H., Elverhøi, A., Laberg, J.S., Hollender, F-J, Mienert, J. and Siegert, M.J. (1996) Large-scale sedimentation on the glacier-influenced Polar North Atlantic margins: Long-range side-scan sonar evidence. *Geophys. Res. Lett.*, **23**, 3535–3538.

Dowdeswell, J.A., Elverhøi, A. and Spielhagen, R. (1998) Glacimarine sedimentary processes and facies on the polar North Atlantic Margins. *Quatern. Sci. Rev.*, **17**, 243–272.

Dowdeswell, J.A., Whittington, R.J., Jennings, A.E., Andrews, J.T., Mackensen, A. and Marienfeld, P. (2000) An origin for laminiated glacimarine sediments through sea-ice build-up and suppressed iceberg rafting. *Sedimentology*, **47**, 557–576.

Dreimanis, A. (1988) Tills: Their genetic terminology and classification. In: *Genetic Classification of Glacigenic Deposits* (Eds R.P. Goldthwait and C.L. Matsch), pp. 17–84. A.A. Balkema, Rotterdam.

Drexel, J.F., Preiss, W.V. and Parker, A.J. (1993) *The Geology of South Australia, Vol. 1. The Precambrian. Geol. Surv. S. Aust. Bull.*, **54**, 242 pp.

Dunn, P.R., Thomson, B.P. and Rankama, K. (1971) Late Precambrian glaciation in Australia as a stratigraphic boundary. *Nature*, **231**, 498–502.

Dunster, J.N., Beier, P.R., Burgess, J.M. and Cutovinos, A. (2000) Auvergne, Northern Territory, 1:250 000 geological Map Series, Sheet SD 52–15 – Explanatory Notes, Northern Territory Geological Survey. 1:250 000 geological map series. Explanatory notes., 1v, 34 pp.

Echelmeyer, K. and Wang, Z. (1987) Direct observation of basal sliding and deformation of basal drift at sub-freezing temperatures. *J. Glaciol.*, **33**, 83–98.

Edwards, M.B. and Føyn, S. (1981) Late Precambrian tillites in Finnmark, North Norway. In: *Earth's Pre-Pleistocene Glacial Record* (Eds M.J. Hambrey and W.B. Harland), pp. 606–609. Cambridge University Press, Cambridge.

Ehlers, J. and Gibbard, P.L. (2003) Extent and chronology of glaciations. *Quatern. Sci. Rev.*, **22**, 1561–1568.

Eisbacher, G.E. (1981) Sedimentary tectonics and glacial record in the Windermere Supergroup, Mackenzie Mountains, Northwest Territories. *Geol. Surv. Can., Pap.*, 80–27, 40 pp.

Eisbacher, G.H. (1985) Late Proterozoic rifting, glacial sedimentation, and sedimentary cycles in the light of Windermere deposition, Western Canada. *Palaeogeogr., Palaeoclimatol., Palaeoecol.*, **51**, 231–254.

Elverhøi, A., Pfirman, S.L., Solheim, A. and Larssen, B.B. (1989) Glaciomarine sedimentation in epicontinental seas exemplified by the Northern Barents Sea. *Mar. Geol.*, **85**, 225–250.

Elverhøi, A., Hooke, R. LeB. and Solheim, A. (1998) Late Cenozoic erosion and sediment yield from the Svalbard-Barents Sea region: implications for understanding erosion of glacierised basins. *Quatern. Sci. Rev.*, **17**, 209–241.

Elverhøi, A., de Blasio, F.V., Butt, F.A., Issler, D., Harbitz, C., Engvik, L., Solheim, A. and Marr, J. (2002) Submarine mass-wasting on glacially-influenced continental slopes: processes and dynamics. In: *Glacier-Influenced Sedimentation on High-Latitude Continental Margins* (Eds J.A. Dowdeswell and C.O. Cofaigh), Geol. *Soc. London Spec. Publ.*, **203**, 73–87.

Escutia, C., De Santis, L., Donda, F., Dunbar, R.B., Cooper, A.K., Brancolini, G. and Eittreim, S.L. (2005) Cenozoic ice sheet history from East Antarctic Wilkes Land continental margin sediments. *Glob. Planet. Change.*, **45**, 51–81.

Etienne, J.L. (2004) *Quaternary glacigenic sedimentation along the Welsh margin of the Irish Sea basin.* Unpublished Ph.D. thesis, University of Wales, Aberystwyth.

Evans, D.A.D. (2000) Stratigraphic, geochronological, and paleomagnetic constraints upon the Neoproterozoic climatic paradoxes. *Am. J. Sci.*, **300**, 347–443.

Evans, J. and Ó Cofaigh C. (2003) Supraglacial debris along the front of the Larsen-A Ice Shelf, Antarctic Peninsula. *Antarctic Sci.*, **15(4)**, 503–506.

Evans, J., Dowdeswell, J.A., Grobe, H., Niessen, F., Stein, R., Hubberten, H.-W. and Whittington, R.J. (2002) Late Quaternary sedimentation in Kejser Franz Joseph Fjord and the continental margin of East Greenland. In: *Glacier-Influenced Sedimentation on High-Latitude Continental Margins* (Eds J.A. Dowdeswell and C.O. Cofaigh), Geol. *Soc. London Spec. Publ.*, **203**, 149–179.

Eyles, C.H. and Lagoe, M.B. (1998) Slump-generated megachannels in the Pliocene-Pleistocene glaciomarine Yakataga Formation, Gulf of Alaska. *Geol. Soc. Am. Bull.*, **110**, 395–408.

Eyles, N. (1990) Marine debris flows: Late Precambrian 'tillites' of the Avalonian-Cadomian orogenic belt. *Palaeogeogr., Palaeoclimatol., Palaeoecol.*, **79**, 73–98.

Eyles, N., Daniels, J., Osterman, L.E. and Januszczak, N. (2001) Ocean Drilling Program Leg 178 (Antarctic Peninsula): sedimentology of glacially influenced continental margin topsets and foresets. *Mar. Geol.*, **178**, 135–156.

Eyles, N. and Januszczak, N. (2004) 'Zipper-Rift': a tectonic model for Neoproterozoic glaciations during the breakup of Rodinia after 750 Ma. *Earth Sci. Rev.*, **65**, 1–73.

Fairchild, I.J. and Hambrey, M.J. (1984) The Vendian succession of north-eastern Spitsbergen: petrogenesis of a dolomite-tillite association. *Precambrian Res.*, **26**, 111–167.

Fairchild, I.J., Hambrey, M.J., Spiro, B. and Jefferson, T.H. (1989) Late Proterozoic glacial carbonates in northeast Spitsbergen: new insights into the carbonate-tillite association. *Geol. Mag.*, **126**, 469–490.

Fairchild, I.J. and Hambrey, M.J. (1995) Vendian basin evolution in East Greenland and NE Svalbard. *Precambrian Res.*, **73**, 217–233.

Fanning, C.M. and Link, P.K. (2004) U-Pb SHRIMP ages of Neoproterozoic (Sturtian) glaciogenic Pocatello Formation, southeastern Idaho. *Geology*, **32**, 881–884.

Fanning, C.M., Ludwig, K.R., Forbes, B.G. and Preiss, W.V. (1986) Single and multiple grain U–Pb zircon analysis for the early Adelaidean Rook Tuff, Willouran Ranges. *S. Aust. Abstr. Geol. Soc. Aust.*, **15**, 71–72.

Fitzsimons, S.J., McManus, K.J. and Lorrain R.D. (1999) Structure and strength of basal ice and substrate of a dry based glacier: evidence for substrate deformation at subfreezing temperatures. *Ann. Glaciol.*, **28**, 236–240.

Freeman, M.J., Oaks, R.Q. Jr. and Shaw, R.D. (1991) Stratigraphy of the Late Proterozoic Gaylad Sandstone, northeastern Amadeus Basin, and recognition of an underlying regional unconformity. In: *Geological and geophysical studies in the Amadeus Basin, central Australia* (Eds Korsch and Kennard). *Bur. Mineral Resour., Aust., Bull.*, **236**, 137–154.

Frimmel, H.E., Foelling, P.G. and Eriksson, P.G. (2002) Neoproterozoic tectonic and climatic evolution recorded in the Gariep Belt, Namibia and South Africa. *Basin Res.*, **14**, 55–67.

Frimmel, H.E. and Frank, W. (1998) Neoproterozoic tectono-thermal evolution of the Gariep Belt and its basement, Namibia and South Africa. *Precambrian Res.*, **90**, 1–28.

Frimmel, H.E., Kloetzli, U.S. and Siegfried, P.R. (1996) New Pb-Pb single zircon age constraints on the timing of Neoproterozoic glaciation and continental break-up in Namibia. *J. Geol.*, **104**, 459–469.

Frimmel, H.E., Zartman, R.E. and Späth, A. (2001) Dating Neoproterozoic continental break-up in the Richtersveld Igneous complex, South Africa. *J. Geol.*, **109**, 493–508.

Gale, S.J. (1990) The shape of beach gravels. *J. Sed. Petrol.*, **60**, 787–789.

Gaucher, C., Boggiani, P.C., Sprechmann, P., Sial, A.N. and Fairchild, T. (2003) Integrated correlation of the Vendian to Cambrian Arroyo del Soldado and Corumbá Groups (Uruguay and Brazil): palaeogeographic, palaeoclimatic and palaeobiologic implications. *Precambrian Res.*, **120**, 241–278.

Ge, W.C., Li, X.H., Li, Z.X. and Zhou, H.W. (2001) Mafic intrusions in Longsheng area: age and its geological implications. *Chin. J. Geol.*, **36**, 112–118.

Glasser, N.F. and Hambrey, M.J. (2001) Styles of sedimentation beneath Svalbard valley glaciers under changing dynamic and thermal regimes. *J. Geol. Soc.*, **158**, 697–707.

Goodman, J.C. and Pierrehumbert, R.T. (2003) Glacial flow of floating marine ice in 'Snowball Earth'. *J. Geophys. Res.*, **108** (C10), 3308.

Gorokhov, I.M., Siedlecka, A., Roberts, D., Melnikov, N.N. and Turchenko, T.L. (2001) Rb–Sr dating of diagenetic illite in Neoproterozoic shales, Varanger Peninsula, northern Norway. *Geol. Mag.*, **138**, 541–562.

Gower, P.J. (1977) The Dalradian rocks of the west coast of the Tayvallich peninsula. *Scottish Journal of Geology*, **13**, 125–33.

Gray, A. (1930) The correlation of the ore-bearing sediments of the Katanga and Rhodesian Copperbelt. *Econ. Geol.*, **25**, 783–801.

Griffin, T.J., Tyler, I.M., Orth, K. and Sheppard, S. (1998) *Geology of the Angelo* 1:100 000 Sheet. *Geol. Surv. W. Aust.* Explanatory Notes., 1v, 27 pp.

Grobe, H. (1987) A simple method for the determination of ice-rafted debris in sediment cores. *Polarforschung*, **57**, 123–126.

Grotzinger, J.P., Bowring, S.A., Saylor, B.Z. and Kaufman, A.J. (1995) Biostratigraphic and geochronologic constraints on early animal evolution. *Science*, **270**, 598–604.

Hallet, B. (1979) Subglacial regelation water film. *J. Glaciol.*, **23**, 321–334.

Halliday, A.N., Graham, C.M., Aftalion, M. and Dymoke, P. (1989) The depositional age of the Dalradian Supergroup: U-Pb and Sm-Nd isotopic studies of the Tayvallich Volcanics, Scotland. *J. Geol. Soc. London.*, **146**, 3–6.

Halverson, G.P. (2005) A Neoproterozoic chronology. In: S. Xiao (Editor), Neoproterozoic Geobiology. *Kluwer Academic Publishers, Delft, Netherlands*.

Halverson, G.P., Hoffman, P.F., Schrag, D.P. and Kaufman, A.J. (2002) A major perturbation of the carbon cycle before the Ghaub glaciation (Neoproterozoic) in Namibia: prelude to snowball Earth? *Geochem., Geophys., Geosyst.*, **3**.

Halverson, G.P., Maloof, A.C. and Hoffman, P.F. (2004) The Marinoan glaciation (Neoproterozoic) in northeast Svalbard. *Basin Res.*, **16**, 297–324.

Halverson, G.P., Hoffman, P.F., Schrag, D.P., Maloof, A.C. and Rice, A.H.N. (2005) Towards a Neoproterozoic composite carbon isotope record. *Geol. Soc. Am. Bull.*, **117**, 1181–1207.

Ham, N.R. and Mickelson, D.M. (1994) Basal till fabric and deposition at Burroughs Glacier, Glacier Bay, Alaska. *Geol. Soc. Am. Bull.*, **106**, 1552–1559.

Hambrey, M.J. (1983) Correlation of Late Proterozoic tillites in the North Atlantic region and Europe. *Geol. Mag.*, **120**, 209–232.

Hambrey, M.J. (1988) Late Proterozoic Stratigraphy of the Barents Shelf. In: *Geological Evolution of the Barents Shelf Region* (Eds W.B. Harland and E.K. Dowdeswell), pp. 49–72. Graham & Trotman, London, U.K.

Hambrey, M.J. (1989) The Late Proterozoic sedimentary record of East Greenland: its place in understanding the evolution of the Caledonide Orogen. In: *The Caledonide Geology of Scandinavia* (Ed. R.A. Gayer), pp. 257–262. Graham & Trotman, London, U.K.

Hambrey, M.J. (1994) *Glacial Environments*. UCL Press, London. viii, 296 pp.

Hambrey, M.J. and Glasser, N.F. (2003) Glacial sediments: processes, environments and facies. In: *Encyclopedia of Sediments and Sedimentary Rocks* (Ed. G.V. Middleton). Dordrecht: Kluwer, 316–331.

Hambrey, M.J. and Harland, W.B. (1981) *Earth's Pre-Pleistocene Glacial Record*. Cambridge University Press, Cambridge, 1004 pp.

Hambrey, M.J., Harland, W.B. and Waddams, P. (1981) Late Precambrian tillite of Svalbard. In: *Earth's Pre-Pleistocene Glacial Record* (Eds M.J. Hambrey and W.B. Harland), pp. 592–601. Cambridge University Press, Cambridge.

Hambrey, M.J. and Spencer, A.M. (1987) *Late Precambrian glaciation of central East Greenland*. Meddelelser om Grønland, *Geoscience*, **19**, 50 pp.

Hambrey, M.J., Peel, J.S. and Smith, M.P. (1989a) Upper Proterozoic and Lower Palaeozoic strata in northern East Greenland. *Grønlands Geologiske Undersøgelse Rapport.*, **145**, 103–108.

Hambrey, M.J., Barrett, P.J., Hall, K.J. and Robinson, P.H. (1989b) Stratigraphy. In: *Antarctic Cenozoic history from the CIROS-1 drillhole, McMurdo Sound, Antarctica* (Ed. P.J. Barrett). DSIR Bulletin 245, 23–48, Wellington, New Zealand.

Hambrey, M.J. and Huddart, D. (1995) Englacial and proglacial glaciotectonic processes at the snout of a thermally complex glacier in Svalbard. *J. Quatern. Sci.*, **10**, 313–326.

Hambrey, M.J., Huddart, D., Bennett, M.R. and Glasser, N.F. (1997) Genesis of 'hummocky moraines' by thrusting in glacier ice: evidence from Svalbard and Britain. *J. Geol. Soc. London*, **154**, 623–632.

Hambrey, M.J. and McKelvey, B. (2000) Neogene fjordal sedimentation on the western margin of the Lambert Graben, East Antarctica. *Sedimentology*, **47**, 577–607.

Hambrey, M.J., Barrett, P.J. and Powell, R.D. (2002) Late Oligocene and early Miocene glaciomarine sedimentation in the SW Ross Sea, Antarctica: the record from offshore drilling. In: *Glacier-Influenced Sedimentation on High-Latitude Continental Margins* (Eds J.A. Dowdeswell and C. O Cofaigh), *Geol. Soc. Lond. Spec. Publ.*, **203**, 105–128.

Harland, W.B., Hambrey, M.J. and Waddams, P. (1993) Vendian Geology of Svalbard. *Nor. Polarinst. Skr.*, **193**, Oslo. 150 pp.

Harland, W.B. (1997) The Geology of Svalbard. *Geol. Soc. London Mem.*, **17**, 521 pp.

Harris, C. (1998) The micromorphology of paraglacial and periglacial slope deposits: a case study from Morfa Bychan, west Wales, UK. *J. Quatern. Sci.*, **13**, 73–84.

Harris, C. and Bothamley, K. (1984) Englacial deltaic sediments as evidence for basal freezing and marginal shearing, Leirbreen, southern Norway. *J. Glaciol.*, **30**, 30–34.

Hart, J.K. (1995a) Drumlin formation in southern Anglesey and Arvon, northwest Wales. *J. Quatern. Sci.*, **10**, 3–14.

Hart, J.K. (1995b) Drumlin formation in southern Anglesey and Arvon, northwest Wales – Reply. *J. Quatern. Sci.*, **10**, p. 399.

Hart, J.K. and Boulton, G.S. (1991) The interrelation of glaciotectonic and glaciodepositional processes within the glacial environment. *Quatern. Sci. Rev.*, **10**, 335–350.

Haug, G.H., Tiedemann, R., Zahn, R. and Ravelo, A.C. (2001) Role of Panama uplift on oceanic freshwater balance. *Geology*, **29**, 207–210.

Hein, F.J. and McMechan, M.E. (1994) Proterozoic-Lower Cambrian strata of the Western Canada Sedimentary Basin. In: *Geological Atlas of the western Canada Sedimentary Basin* (Eds G.D. Mossop and I. Shetsen), pp. 57–68. Canadian Soc. Petrol. Geol. Calgary.

Heinrich, R., Baumann, K.-H., Huber, R. and Meggers, H. (2002) Carbonate preservation records of the past 3 Myr in the Norwegian-Greenland Sea and the northern North Atlantic: implications for the history of NADW production. *Mar. Geol.*, **184**, 17–39.

Hiemstra, J.F. and van der Meer, J.J.M. (1997) Pore-water controlled grain fracturing as an indicator for subglacial shearing in tills. *J. Glaciol*, **43**, 446–454.

Hiscott, R.N. and Aksu, A.E. (1994) Submarine debris flows and continental slope evolution in the front of Quaternary ice sheets. *Bull. Am. Assoc. Petrol. Geol.*, **78**, 445–460.

Hiscott, R.N., Aksu, A.E., Mudie, P.J. and Parsons, D.F. (2001) A 340,000 year record of ice rafting, palaeoclimatic fluctuations, and slef-crossing glacial advances in the southwestern Labrador Sea. *Glob. Planet. Change*, **28**, 227–240.

Hoffman, P.F. (2005) 28th DeBeers Alex. Du Toit Memorial Lecture, 2004. On Cryogenian (Neoproterozoic) ice-sheet dynamics and the limitations of the glacial sedimentary record. *S. Afr. J. Geol.*, **108**, 557–578

Hoffman, P.F. and Schrag, D.P. (2002) The snowball Earth hypothesis: testing the limits of global change. *Terra Nova*, **14**, 129–155.

Hoffman, P.F., Kaufman, A.J., Halverson, G.P. and Schrag, D.P. (1998a) A Neoproterozoic snowball Earth. *Science*, **281**, 1342–1346.

Hoffman, P.F., Kaufman, A.J. and Halverson, G.P. (1998b) Comings and goings of global glaciations in a Neoproterozoic tropical platform in Namibia. *Geol. Soc. Am. Today*, **8**, 1–9.

Hoffmann, K.-H. and Prave, A.R. (1996) A preliminary note on a revised subdivision and regional correlation of the Otavi Group based on glaciogenic diamictites and associated cap dolostones. *Commun. Geol. Surv. Namibia*, **11**, 81–86.

Hoffmann, K.-H., Condon, D.J. Bowring, S.A. and Crowley, J.L. (2004) U-Pb zircon date from the Neoproterozoic Ghaub Formation, Namibia: Constraints on Marinoan glaciation. *Geology*, **32**, 817–820.

Holdsworth, G. (1974) Meserve Glacier, Wright Valley, Antarctica: Part 1. Basal Processes. *Ohio State University, Institute of Polar Studies Report*, **37**, 104 pp.

Hubbard, B. (1991) Freezing-rate effects on the physical characteristics of basal ice formed by net adfreezing. *J. Glaciol.*, **37**, 339–347.

Hubbard, B. and Sharp, M.J. (1989) Basal ice formation and deformation: a review. *Progress in Physical Geography*, **13**, 529–558.

Hubbard, B. and Sharp, M. (1993) Weertman regelation, multiple refreezing events and the isotopic evolution of the basal ice layer. *J. Glaciol.*, **39**, 275–291.

Hutter, K. and Olunloyo, V.O.S. (1981) Basal stress condition due to abrupt changes in boundary conditions: A cause for high till concentration at the bottom of a glacier. *Ann. Glaciol.*, **2**, 29–33.

Ireland, T.R., Flöttmann, T., Fanning, C.M., Gibson, G.M. and Preiss, W.V. (1998) Development of the Early Paleozoic Pacific Margin of Gondwana from detrital zircon ages across the Delamerian Orogen. *Geology*, **26**, 243–246.

Isotta, C.A.L., Rocha-Campos, A.C. and Yoshida, R. (1969) Striated pavement of the the Upper-Precambrian glaciation in Brazil. *Nature*, **222**, 467–468.

Iverson, N.R. (2000) Sediment entrainment by a soft-bedded glacier: a model based on regelation into the bed. *Earth Surf. Proc. Land.*, **25**, 881–893.

Jago, J.B. (1974) The origin of the Cottons Breccia, King Island, Tasmania. *R. Soc. South Austr. Trans.*, **98**, 13–28.

Jain, A.K. and Varadaraj, N. (1978) Stratigraphy and provenance of Late Palaeozoic diamictites in parts of Garhwal Lesser Himalaya, India. *Geol. Rund.*, **67**, 49–72.

Jansen, E. and Sjøholm, J. (1991) Reconstruction of glaciation over the past 6 Myr from ice-borne deposits in the Norwegian Sea. *Nature*, **349**, 600–603.

Jenkins, R.J.F., Cooper, J.A. and Compston, W. (2002) Age and biostratigraphy of Early Cambrian tuffs from SE Australia and southern China. *J. Geol. Soc. Lond.*, **159**, 645–658.

Jenkins, R.J.F. and Cooper, J.A. (1998) Rb–Sr whole-rock dating of the transition between the Bunyeroo and Wonoka Formations. In: *The Ediacaran in South Australia: Proposal and Field Guide supporting GSSP Position 'C' at Wearing Dolomite, Flinders Ranges* (Eds R.J.F. Jenkins, D.M. McKirdy and C. Nedin), pp. 11–15. IUGS Working Group on the Terminal Proterozoic System, University of Adelaide.

Jiang, G., Christie-Blick, N., Kaufman, A.J., Banerjee, D.M. and Rai, V. (2002) Sequence stratigraphy of the Neoproterozoic Infra Krol Formation and Krol Group, Lesser Himalaya, India. *J. Sed. Res.*, **72**, 524–542.

Jiang, G., Kennedy, M.J. and Christie-Blick, N. (2003a) Stable isotopic evidence for methane seeps in Neoproterozoic postglacial cap carbonates. *Nature*, **426**, 822–826.

Jiang, G., Sohl, L.E. and Christie-Blick, N. (2003b) Neoproterozoic stratigraphic comparison of the Lesser Himalaya (India) and Yangtze block (south China): Paleogeographic implications. *Geology*, **31**, 917–920.

Kasemann, S.A., Hawkesworth, C.J., Prave, A.R., Fallick, A.E. and Pearson, P.N. (2005) Boron and calcium isotope composition in Neoproterozoic carbonate rocks from Namibia: evidence for extreme environmental change. *Earth Planet. Sci. Lett.*, **231**, 73–86.

Kaufman, A.J., Knoll, A.H. and Narbonne, G.M. (1997) Isotopes, ice ages, and terminal Proterozoic earth history. *Proc. Natl. Acad. Sci. USA*, **94**, 6600–6605.

Kellerhals, P. and Matter, A. (2003) Facies analysis of a glaciomarine sequence, the Neoproterozoic Mirbat Sandstone formation, Sultanate of Oman. *Eco. Geolog. Helvet.*, **96**, 49–50.

Kendall, B.S., Creaser, R.A., Ross, G.M. and Selby, D. (2004) Constraints on the timing of Marinoan 'Snowball Earth' glaciation by ^{187}Re–^{187}Os dating of a Neoproterozoic, post-glacial black shale in Western Canada. *Earth Planet. Sci. Lett.*, **222**, 729–740.

Kendall, B.S. and Creaser, R.A. (2004) Re-Os depositional age of Neoproterozoic Aralka Formation (Amadeus basin, Australia) revisited. *Geol. Soc. Am. Abstr. Prog.*, **36**, 459.

Kennedy, M.J., Runneger, B., Prave, A.R., Hoffman, K.-H. and Arthur, M.A. (1998) Two or four Neoproterozoic glaciations? *Geology*, **26**, 1059–1063.

Kennedy, M.J., Christie-Blick, N. and Sohl, L.E. (2001) Are Proterozoic cap carbonates and isotopic excursions a record of gas hydrate destabilization following Earth's coldest intervals? *Geology*, **29**, 443–446.

Kent, D.V. and Smethurst, M.A. (1998) Shallow bias of paleomagnetic inclinations in the Paleozoic and Precambrian. *Earth Planet. Sci. Lett.*, **160**, 391–402.

Key, R.M., Liyungu, A.K., Njamu, F.M., Somwe, V., Banda, J., Mosley, P.N. and Armstrong, R.A. (2001) The western arm of the Lufilian Arc in NW Zambia and its potential for copper mineralization. *J. Afr. Earth Sci.*, **33**, 503–528.

Khatwa, A. and Tulaczyk, S. (2001) Microstructural interpretations of modern and Pleistocene subglacially deformed sediments: the relative role of parent material and subglacial processes. *J. Quatern. Sci.*, **16**, 507–517.

Kirkbride, M.P. (1995) Processes of transportation. In: *Modern Glacial Environments, Processes, Dynamics and Sediments (Glacial Environments: Volume I)* (Ed. J. Menzies), pp. 261–292. Butterworth-Heinemann, Oxford.

Kirschvink, J.L. (1992) Late Proterozoic low latitude glaciation: the Snowball Earth. In: *The Proterozoic Biosphere: A Multidisciplinary Study* (Eds J.W. Schopf and C. Klein), pp. 51–52. Cambridge University Press, Cambridge.

Kjaer, K.H., Kruger, J. and van der Meer, J.J.M. (2003) What causes till thickness to change over distance? Answers from MΩrdalsjökull, Iceland. *Quatern. Sci. Rev.*, **22**, 1687–1700.

Kleman, J. and Hättestrand, C. (1999) Frozen-bed Fennoscandian and Laurentide ice sheets during the Last Glacial Maximum. *Nature*, **402**, 63–66.

Kleman, J., Hättestrand, C., Borgström, I. and Stroeven, A. (1997) Fennoscandian palaeoglaciology reconstructed using a glacial geological inversion model. *J. Glaciol.*, **43**, 283–300.

Kluiving, S.J., Bartek, L.R. and van der Wateren, F.M. (1999) Multi-scale analyses of subglacial and glaciomarine deposits from the Ross Sea continental shelf, Antarctica. *Ann. Glaciol.*, **28**, 90–96.

Koerner, R.M. and Fisher, D.A. (1979) Discontinuous flow, ice texture, and dirt content in the basal debris layers of the Devon Island ice cap. *J. Glaciol.*, **23**, 209–221.

Kramers, J.D. and Frei, R. (1992) Report on age determination carried out on rocks from the Mirbat region, south Oman. Unpublished Report, Mineralogy-Petrography Institute, University of Bern.

Krieg, G.W., Rogers, P.A., Callen, R.A., Freeman, P.J., Alley, N.F. and Forbes, B.G. (1991) *Curdimurka, South Australia, 1:250 000 Sheet Explanatory Notes. Geol. Surv. S. Aust.*, 1v, 60 pp.

Krogh, T.E., Strong, D.F., O'Brien, S.J. and Papezik, V.S. (1988) Precise U-Pb zircon dates from the Avalon Terrane in Newfoundland. *Can. J. Earth Sci.*, **25**, 442–453.

Kröner, A. (1981) Late Precambrian diamictites of South Africa and Namibia. In: *Earth's Pre-Pleistocene Glacial Record* (Eds M.J. Hambrey and W.B. Harland), pp. 167–177. Cambridge University Press, Cambridge.

Kröner, A. and Correia, H. (1973) Further evidence for glaciogenic origin of Late Precambrian mixtites in Angola. *Nature*, **246**, 115–117.

Krüger, J. (1994) Glacial processes, sediments, landforms, and stratigraphy in the terminus region of Myrdalsjökull, Iceland. *Folia Geographica Danica*, **21**, 1–233.

Kruse, P.D., Brakel, A.T., Dunster, J.N. and Duffett, M.L. (2002) Tobermory, Northern Territory 1:250 000 geological map series. Explanatory Notes Sheet SF 53–12, Northern Territory Geological Survey, 58 pp.

Kukla, G. and Gavin, J. (2005) Did glacials start with global warming? *Quatern. Sci, Rev.*, **24**, 1547–1557.

Kumar, R. and Brookfield, M.E. (1987) Sedimentary environments of the Simla Group (Upper Precambrian), Lesser Himalaya, and their palaeotectonic significance. *Sed. Geol.*, **52**, 27–43.

Kumpulainen, R. (1981) The Late Precambrian Lillfjället Formation in the southern Swedish Caledonides. In: *Earth's Pre-Pleistocene Glacial Record* (Eds M.J. Hambrey and W.B. Harland), pp. 620–623. Cambridge University Press, Cambridge.

Laberg, J.S. and Vorren, T.O. (2000) Flow behaviour of the submarine glacigenic debris flows on the Bear Island Trough Mouth Fan, western Barents Sea. *Sedimentology*, **47**, 1105–1117.

Lachniet, M.S., Larson, G.J., Lawson, D.E., Evenson, E.B. and Alley, R.B. (2001) Microstructures of sediment flow deposits and subglacial sediments: a comparison. *Boreas*, **30**, 254–262.

Larianov, A., Gee, D.G., Tebenkov, A.M. and Witt-Nillson, P. (1998) Detrital zircon ages from the Planetfjella Group of the Mosselhalvøya Nappe, NE Spitsbergen, Svalbard. In: *International Conference on Arctic Margins, III, Celle, Germany, Abstr.*, 109–110.

Larsen, H.C., Saunders, A.D., Clift, P.D., Beget, J., Wei, W., Spezzaferri, S. and ODP Leg 152 Scientific Party. (1994) Seven million years of glaciation in Greenland. *Science*, **264**, 952–955.

Lawson, D.E., Evenson, E.B., Strasser, J.C., Alley, R.B. and Larson, G.J. (1996) Subglacial supercooling, ice accretion, and sediment entrainment at the Matanuska Glacier, Alaska. *Geol. Soc. Am. Abstr. Prog.*, **28**.

Le Guerroué, E., Allen, P.A., Cozzi, A. and Fanning, M. (2006) 50 Million Year Duration negative carbon isotopic excursion in the Ediacaran ocean. *Terra Nova*, **18**, 147–153.

Le Heron, D.P. and Etienne, J.L. (2005) A complex subglacial clastic dyke swarm, Sólheimajökull, southern Iceland. *Sed. Geol.*, **181** (1–2), 25–37.

Le Heron, D.P., Sutcliffe, O.E., Whittington, R.J. and Craig, J. (2005) The origins of glacially related soft-sediment deformation structures in Upper Ordovician glaciogenic rocks: implication for ice-sheet dynamics. *Palaeogeogr., Palaeoclimatol., Palaeoecol.*, **218**, 75–103.

Leather, J. (2001) *Sedimentology, chemostratigraphy and geochronology of the lower Huqf Supergroup, Oman.* Unpublished PhD Thesis, Trinity College, Dublin. 227 pp.

Leather, J., Allen, P.A., Brasier, M.D. and Cozzi, A. (2002) Neoproterozoic snowball earth under scrutiny: evidence from the Fiq glaciation of Oman. *Geology*, **30**, 891–894.

Leblanc, M. (1981) The Late Precambrian Tiddiline Tilloid of the Anti-Atlas, Morocco. In: *Earth's Pre-Pleistocene Glacial Record* (Eds M.J. Hambrey and W.B. Harland), pp. 120–122. Cambridge University Press, Cambridge.

Lee, J.R., Rose, J., Riding, J.B., Moorlock, B.S.P. and Hamblin, R.J.O. (2002) Testing the case for a Middle Pleistocene Scandinavian glaciation in Eastern England: evidence for a Scottish ice source for tills within the Corton Formation of East Anglia, U.K. *Boreas*, **31**, 345–355.

Lemon, N.M. and Reid, P.W. (1998) The Yaltipena Formation of the central Flinders Ranges. *MESA J.*, **8**, 37–39.

Lemon, N.M. and Williams, G.E. (1998) Flinders Ranges field trip, 16–22 June 1998. *Adjunct Field Guide. IUGS Working Group on the Terminal Proterozoic System.* The University of Adelaide (unpublished).

Lewis, J.P., Weaver, A.J., Johnston, S.T. and Eby, M. (2003) The Neoproterozoic 'Snowball Earth': dynamic sea ice over a quiescent ocean. *Paleoceanography*, **18** (4).

Lindsay, J.F. (1989) Depositional controls on glacial facies associations in a basinal setting, Late Proterozoic, Amadeus Basin, Central Australia. *Palaeogeogr., Palaeoclimatol., Palaeoecol.*, **73**, 205–232.

Link, P.K., Miller, J.M.G. and Christie-Blick, N. (1994) Glacial-marine facies in a continental rift environment: Neoproterozoic rocks of the western United States Cordillera. In: *International Geological Correlation Project 260: Earth's glacial record* (Eds M. Deynoux, J.M.B. Miller, E.W. Domack, N. Eyles, I.J. Fairchild and G.M. Young), pp. 29–46. Cambridge University Press, London.

Lønne, I. (1995) Sedimentary facies and depositional architecture of ice-contact glaciomarine systems. *Sed. Geol.*, **98**, 13–43.

Lowe, J.J., McCarroll, D., Knight, J. and Rijsdijk, K. (2001) The glaciation of the Irish Sea basin. *J. Quatern. Sci.*, **16** (5).

Lund, K., Aleinikoff, J.N., Evans, K.V. and Fanning, C.M. (2003) SHRIMP U-Pb geochronology of Neoproterozoic Windermere Supergroup, central Idaho: Implications for rifting of western Laurentia and synchroneity of Sturtian glacial deposits. *Geol. Soc. Am. Bull.*, **115**, 349–372.

Ma, G.G., Lee, H. and Zhang, Z. (1984) An investigation of the age limits of the Sinian System in South China. Bull. *Yichang Inst. Geol. Miner. Res., Chinese Acad. Geol. Sci.*, **8**, 1–29.

Maltman, A.J., Hubbard, B. and Hambrey, M.J. (2000) Deformation *of Glacial Materials. Geol. Soc. London, Spec. Publ.*, **176**, pp. 344.

Marjoribanks, R.W. and Black, L.P. (1974) Geology and geochronology of the Arunta Complex, north of Ormiston George, central Australia. *J. Geol. Soc. Aust.*, **21**, 291–300.

Martin, H. (1965) The Precambrian geology of South West Africa and Namaqualand. *Precambrian Res. Unit, Univ. Cape Town, Bull.*, **4**, 1–177.

Martin, H., Porada, H. and Walliser, O.H. (1985) Mixtite deposits of the Damara sequence, Namibia: problem of interpretation. *Palaeogeogr. Palaeoclimatol. Palaeoecol.*, **51**, 159–196.

Maslin, M., Owen, M., Day, S. and Long, D. (2004) Linking continental-slope failures and climate change: Testing the clathrate gun hypothesis. *Geology*, **32**, 53–56.

Maslov, A.V. (2000) Some Specific Features of Early Vendian Sedimentation in the Southern and Middle Urals. *Lithology and Mineral Resources*, **35**, 556–570. (Translated from Russian).

Mawson, E. (1949) The Elatina Glaciation. A third recurrence of glaciation evidenced in the Adelaide System. *Trans. R. Soc. S. Austr.*, **73**, 117–121.

McCarroll, D. and Rijsdijk, K.F. (2003) Deformation styles as a key for interpreting glacial depositional environments. *J. Quatern. Sci.*, **18**, 473–489.

McCarron, G.M.E. (2000) *The sedimentology and chemostratigraphy of the Nafun Group, Huqf Supergroup, Oman.* Unpublished Ph.D. thesis, Oxford University, 175 pp.

McKirdy, D.M., Burgess, J.M., Lemon, N.M., Yu, X., Cooper, A.M., Gostin, V.A., Jenkins, R.J.F. and Both, R.A. (2001) A chemostratigraphic overview of the late Cryogenian interglacial sequence in the Adelaide Fold-Thrust Belt, South Australia. *Precambrian Res.*, **106**, 149–186.

Meert, J.G. and Powell, C. McA. (2001) Introduction to the special volume on the assembly and breakup of Rodinia. *Precambrian Res.*, **110**, 1–8.

Meert, J.G. and Torsvik, T.H. (2004) Paleomagnetic Constraints on Neoproterozoic 'Snowball Earth' Continental Reconstructions. In: *The Extreme Proterozoic: Geology, Geochemistry, and Climate* (Eds G.S., Jenkins, M.A.S. McMenamin, C.P. McKay and L. Sohl), pp. 5–11. Geophysical Monograph Series 146, A.G.U.

Meffre, S., Direen, N.G., Crawford, A.J. and Kamenetsky, V. (2004) Mafic volcanics on King Island, Tasmania: Evidence for plume-triggered breakup in East Gondwana at around 579 Ma. *Precambrian Res.*, **135**, 177–191.

Menzies, J. (1981) Freezing fronts and their possible influence upon processes of subglacial erosion and deposition. *Ann. Glaciol.*, **2**, 52–56.

Menzies, J. (1988) Microstructures within subglacial diamictons. In: *Relief and Deposits of Present-day and Pleistocene Glaciation of the Northern Hemisphere – Selected problems* (Ed. A. Kostrzewski), pp. 153–166. Adam Mickiewicz University Press, Geography Series no. 58. Poznan, Poland.

Menzies, J. (2000) Micromorphological analyses of microfabrics and microstructures indicative of deformation processes in glacial sediments. In: *Deformation of Glacial Materials* (Eds A.J. Maltman, B. Hubbard and M.J. Hambrey), *Geol. Soc. London, Spec. Publ.*, **176**, 245–258.

Menzies, J. and Maltman, A.J. (1992) Microstructures in diamictons – evidence of subglacial bed conditions. *Geomorphology*, **6**, 27–40.

Menzies, J. and Shilts, W.W. (1996) Subglacial Environments. In: *Past Glacial Environments: Sediments, Forms and Techniques* (Ed. J. Menzies), pp. 15–136. Butterworth-Heinemann, Oxford.

Menzies, J., Zaniewski, K. and Dreger, D. (1997) Evidence, from microstructures, of deformable bed conditions within drumlins, Chimney Bluffs, New York State. *Sed. Geol.*, **111**, 161–175.

Mickelson, D.M., Ham N.R. and Ronnert, L. (1992) Striated Clast Pavements – Products Of Deforming Subglacial Sediment – Comment. *Geology*, **20**, 285–285.

Miller, J.M.G. (1996) Glacial Sediments. In: *Sedimentary Environments: Processes, Facies and Stratigraphy* (Ed. H.G. Reading), pp. 454–484. Oxford; Blackwell Science.

Miller, J.M.G., Wright, L.A. and Troxel, B.W. (1981) The Late Precambrian Kingston Peak Formation, Death Valley region, Cailfornia. In: *Earth's Pre-Pleistocene Glacial Record* (Eds M.J. Hambrey and W.B. Harland), pp. 45–748. Cambridge University Press, Cambridge.

Miller, N.R., Alene, M., Sacchi, R., Stern, R.J., Conti, A., Kröner, A. and Zuppi, G. (2003) Significance of the Tambien Group (Tigrai, N. Ethiopia) for Snowball Earth events in the Arabian–Nubian Shield. *Precambrian Res.*, **121**, 263–283.

Moncrieff, A.C.M. (1988) *The Vendian Stratigraphy and Sedimentology of East Greenland.* Unpublished Ph.D. thesis, Cambridge University.

Moncrieff, A.C.M. (1989a) Classification of poorly sorted sedimentary rocks. *Sed. Geol.*, **65**, 191–194.

Moncrieff, A.M. (1989b) The Tillite Group and related rocks of East Greenland: implications for Late Proterozoic palaeogeography. In: *The Caledonide Geology of Scandinavia* (Ed. R.A. Gayer). Graham and Trotman, London, 285–297.

Moncrieff, A.C.M. and Hambrey, M.J. (1988) Glacial erosional pavements in the Late Precambrian Tillite Group, central East Greenland. *Palaeogeog., Palaeoclimatol. Palaeoecol.*, **65**, 183–200.

Moncrieff, A.C.M. and Hambrey, M.J. (1990) Marginal-marine glacial sedimentation in the late Precambrian succession of East Greenland. In: *Glacimarine environments: processes and sediments* (Eds J.A. Dowdeswell and J.D. Scourse), *Geol. Soc. London Spec. Publ.*, **53**, 387–410.

Moros, M., Kuijpers, A., Snowball, I., Lassen, S., Bäckstöm, D., Gingele, F. and McManus, J. (2002) Were glacial iceberg surges in the North Atlantic triggered by climate warming? *Mar. Geol.*, **192**, 393–417.

Myrow, P.M. and Kaufman, A.J. (1999) A newly discovered cap carbonate above Varanger-age glacial deposits in Newfoundland. *J. Sed. Res.*, **69**, 784–793.

Myrow, P.M., Hughes, N.C., Paulsen, T.S., Williams, I.S., Parcha, S.K., Thompson, K.R., Bowring, S.A., Peng, S.-C. and Ahluwalia, A.D. (2003) Integrated tectonostratigraphic analysis of the Himalaya and implications for its tectonic reconstruction. *Earth Planet. Sci. Lett.*, **212**, 433–441.

Naish T.R., Woolfe, K.J., Barrett, P.J., Wilson, G.S., Atkins, C., Bohaty, S.M., Bucker, C.J., Claps, M., Davey, F.J., Dunbar, G.B., Dunn, A.G., Fielding, C.R., Florindo, F., Hannah, M.J., Harwood, D.M., Henrys, S.A., Krissek, L.A., Lavelle, M., van Der Meer, J., McIntosh, W.C., Niessen, F., Passchier, S., Powell, R.D., Roberts, A.P., Sagnotti, L., Scherer, R.P., Strong, C.P., Talarico, F., Verosub, K.L., Villa, G., Watkins, D.K., Webb, P.N. and Wonik, T. (2001) Orbitally induced oscillations in the East Antarctic Ice Sheet at the Oligocene-Miocene boundary. *Nature*, **413**, 719–723.

Narbonne, G.M., Kaufman, A.J. and Knoll, A.H. (1994) Integrated chemostratigraphy and biostratigraphy of the Windermere Supergroup, northwestern Canada: Implications for Neoproterozoic correlations and the early evolution of animals. *Geol. Soc. Am. Bull.*, **106**, 1281–1292.

Noble, S.R., Hyslop, E.K. and Highton, A.J. (1996) High-precision U-Pb monazite geochronology of the c. 806 Ma Grampian shear zone and the implications for the evolution of the Central Highlands of Scotland. *J. Geol. Soc. Lond.*, **153**, 511–514.

Nogueira, A.C.R., Riccomini, C., Sial, A.N., Moura, C.A.V. and Fairchild, T.R. (2003) Soft-sediment deformation at the base of the Neoproterozoic Puga cap carbonate (southwestern Amazon craton, Brazil): Confirmation of rapid icehouse to greenhouse transition in snowball Earth. *Geology*, **31**, 613–616.

Nygård, A., Sejrup, H.P., Haflidason, H. and King, E.L. (2002) Geometry and genesis of glacigenic debris flows on the North Sea Fan: TOBI imagery and deep-tow boomer evidence. *Mar. Geol.*, **188**, 15–33.

Nygård, A., Sejrup, H.P., Haflidason, H. and Bryn, P. (2005) The glacial North Sea Fan, southern Norwegian Margin: architecture and evolution from the upper continental slope to the deep-sea basin. *Mar. Petrol. Geol.*, **22**, 71–84.

Nystuen, J.P. (1985) Facies and preservation of glaciogenic sequences from the Varanger ice age in Scandinavia and other parts of the North Atlantic Region. *Palaeogeogr. Palaeoclimatol. Palaeoecol.*, **51**, 209–229.

Ó Cofaigh, C., Dowdeswell, J.A., Evans, J., Kenyon, N.H., Taylor, J., Mienert, J. and Wilken, M. (2004) Timing and significance of glacially influenced mass-wasting in the submarine channels of the Greenland Basin. *Mar. Geol.*, **207**, 39–54.

O'Brien, S.J., Tucker, R.D., Dunning, G.R. and O'Driscoll, C.F. (1992) Four-fold subdivision of the late Precambrian magmatic record of the Avalon Zone area (east Newfoundland): nature and significance. *Geol. Assoc. Can. Abstr. Prog.*, **17**, A85.

O'Grady, D.B. and Syvitski, J.P.M. (2002) Large-scale morphology of Arctic continental slopes: the influence of sediment delivery on slope form. In: *Glacier-Influenced Sedimentation on High-Latitude Continental Margins* (Eds J.A. Dowdeswell and C.O. Cofaigh), Geol. Soc. London Spec. Publ., **203**, 11–31.

Ojakangas, R.W. and Matsch, C.L. (1980) Upper Precambrian (Eocambrian) Mineral Fork Tillite of Utah: A continental glacial and glaciomarine sequence. *Geol. Soc. Am. Bull.*, **91**, 495–501.

Onstott, T.C., Hargraves, R.B. and Reid, D.L. (1986) Constraints on the tectonic evolution of the Namaqua Province; III, Palaeomagnetic and (super 40úAr/ (super 39úAr results from the Gannakouriep dyke swarm. *Trans. Geol. Soc. S. Afr.*, **89**, 171–183.

Orheim, O. and Elverhøi, A. (1981) Model for submarine glacial deposition. *Ann. Glaciol.*, **2**, 123–128.

Parkin, L.W. (1969) Handbook of South Australian Geology. *Geol. Surv. S. Austr.* 268 pp.

Parrenin, F. and Paillard, D. (2003) Amplitude and phase of glacial cycles from a conceptual model. *Earth Planet. Sci. Lett.*, **214**, 243–250.

Paterson, W.S.B. (1994) *The Physics of Glaciers*. Pergamon Press, Oxford, 480 pp.

Pedrosa-Soares, A.C., Cordani, U.G. and Nutman, A. (2000) Constraining the age of Neoproterozoic glaciation in eastern Brazil: first U–Pb (SHRIMP) data of detrital zircons. *Rev. Bras. Geoc.*, **30**, 58–61.

Pickering, K.T., Stow, D.A.V., Watson, M.P. and Hiscott, R.N. (1986) Deep water facies, processes and models: a review and classification scheme for modern and ancient sediments. *Earth-Sci. Rev.*, **23**, 75–174.

Pimentel, M.M. and Fuck, R.A. (1992) Neoproterozoic crustal accretion in central Brazil. *Geology*, **20**, 375–379.

Piotrowski, A.M., Goldstein, S.L., Hemming, S.R. and Fairbanks, R.G. (2005) Temporal Relationships of Carbon Cycling and Ocean Circulation at Glacial Boundaries, *Science*, **307**, no. 5717, 1933–1938.

Platel, J.P., Roger, J., Peters, T., Mercolli, I., Kramers, J.D. and Le Metour, J. (1992) Geological map of Salalah NE 40–09 1:250,000 with explanatory notes. Directorate General of Minerals, Oman Ministry of Petroleum and Minerals.

Poussart, P.F., Weaver, A.J. and Barnes, C.R. (1999) Late Ordovician glaciation under high atmospheric CO_2: A coupled model analysis: *Paleoceanography*, **14**, 542–558.

Powell, R.D. and Molnia, B.F. (1989) Glacimarine sedimentary processes, facies and morphology of the south-southeast Alaska shelf and fjords. *Mar. Geol.*, **85**, 359–390.

Powell R.D. and Cooper, J.M. (2002) A glacial sequence stratigraphic model for temperate, glaciated continental shelves. In: *Glacier-Influenced Sedimentation on High-Latitude Continental Margins* (Eds J.A. Dowdeswell and C.O. Cofaigh), Geol. Soc. London Spec. Publ., **203**, 215–244.

Powell, R.D., Krissek, L.A. and van der Meer, J.J.M. (2000) Preliminary depositional environmental analysis of CRP-2/2A, Victoria Land Basin, Antarctica: Palaeogeographical and palaeoclimatic inferences. *Terra Antarctica*, **7**, 313–322.

Powers, M.C. (1953) A new roundness scale for sedimentary particles. *J. Sed. Petrol.*, **23**, 117–119.

Prave, A.R. (1999a) Two diamictites, two cap carbonates, two $\delta^{13}C$ excursions, two rifts: Te Neoproterozoic Kingstone Peak Formation, Death Valley, Cailfornia. *Geology*, **27**, 339–342.

Prave, A.R. (1999b) The Neoproterozoic Dalradian Supergroup of Scotland: an alternative hypothesis. *Geol. Mag.*, **136**, 609–617.

Preiss, W.V. (1987) The Adelaide Geosyncline—late Proterozoic stratigraphy, sedimentation, palaeontology and tectonics. *S. Australian Geol. Surv. Bull.*, **53**, 438 pp.

Preiss, W.V. (2000) The Adelaide Geosyncline of South Australia and its significance in Neoproterozoic continental reconstruction. *Precambrian Res.*, **100**, 21–63.

Preiss, W.V., Dyson, I.A., Reid, P.W. and Cowley, W.M. (1998) Revision of lithostratigraphic classification of the Umberatana Group. *MESA J.*, **9**, 36–42.

Prichard, C.E. and Quinlan, T. (1962) The geology of the southern half of the Hermannsburg 1:250 000 Sheet. *Bur. Mineral. Resour. Geol. Geophys. Rep.*, **61**.

Rabu, D. (1988) Géologie de l'autochtone des montagnes d'Oman, la fenêtre du Jabal Akhdar. Thèse Doct. ès-Sciences de l'Université Pierre et Marie Curie, Paris 6, and documents BRGM, **130**.

Rafaelsen, B., Andreassen, K., Kuilman, L.W., Lebesbye, E., Hogstad, K. and Midtbø, M. (2002) Geomorphology of buried glacigenic horizons in the Barents Sea from three-dimensional seismic data. In: *Glacier-Influenced Sedimentation on High-Latitude Continental Margins* (Eds J.A. Dowdeswell and C.O. Cofaigh), Geol. *Soc. London Spec. Publ.*, **203**, 259–275.

Rehmer, J. (1981) The Squantum tilloids member of the Roxbury Conglomerate of Boston, Massachusetts. In: *Earth's Pre-Pleistocene Glacial Record* (Eds M.J. Hambrey and W.B. Harland), pp. 756–759. Cambridge University Press, Cambridge.

Reid, D.L., Ransome, I.G.D., Onstott, T.C. and Adams, C.J. (1991) Time of Emplacement and Metamorphism Of Late Precambrian Mafic Dykes Associated with the Pan-African Gariep Orogeny, Southern Africa: Implications for the Age of the Nama Group. *J. Afr. Earth Sci.*, **13**, 531–541.

Rijsdijk, K., Owen, G., Warren, W.P., McCarroll, D. and van der Meer, J.J.M. (1999) Clastic dykes in consolidated tills: evidence from Killiney Bay, eastern Ireland. *Sed. Geol.*, **129**, 111–126.

Rijsdijk, K.F. (2001) Density-driven deformation structures in glacigenic consolidated diamicts: examples from Traeth y Mwnt, Cardiganshire, Wales, UK. *J. Sed. Res.*, **71**, 122–135.

Robert, M. (1956) *Géologie et Géographie du Katanga.* Hayez, Bruxelles, 620 pp.

Roberts, D. *et al.* (1998) Rb-Sr dating of illite fractions from Neoproterozoic shales on Varanger Peninsula, North Norway. *Nor. Geol. Unders. Bull.*, **433**, 24–25.

Roberts, H.G., Gemuts, I. and Halligan, R. (1972) Adalaidean and Cambrian stratigraphy of the Mount Ramsay, 1:250 000 sheet area, Limberley Region, Western Australia. *Rep. Bur. Miner. Resour. Geol. Geophys., Aust.*, **150**.

Rocha-Campos, A.C. and Hasui, Y. (1981a) Tillites of the Macaúbas Group (Proterozoic) in central Minas Gerais and southern Bahia, Brazil. In: *Earth's Pre-Pleistocene Glacial Record* (Eds M.J. Hambrey and W.B. Harland), pp. 933–938. Cambridge University Press, Cambridge.

Rocha-Campos, A.C. and Hasui, Y. (1981b) Late Precambrian Jangada Group and Puga Formation of central western Brazil. In: *Earth's Pre-Pleistocene Glacial Record* (Eds M.J. Hambrey and W.B. Harland), pp. 916–919. Cambridge University Press, Cambridge.

Rogers, A.W. (1916) The geology of part of Namaqualand. *Trans. Geol. Soc. S. Afr.*, **18**, 72–101.

Ronnert, L. and Mickelson, D.M. 1992. High porosity of basal till at Burroughs glacier, southeastern Alaska. *Geology*, **20**, 849–852.

Ross, G.M. and Villeneuve, M.E. (1997) U–Pb geochronology of stranger stones in Neoproterozoic diamictites, Canadian Cordillera: implications for provenance and ages of deposition. *Radiogenic Age and Isotopic Studies: Report* **10**, Geol. Surv. Can. Curr. Res. 1997-F, 141–155.

Ross, G.M., Bloch, J.D. and Krouse, H.R. (1995) Neoproterozoic strata of the southern Canadian Cordillera and the isotopic evolution of seawater sulfate. *Precambrian Res.*, **73**, 71–99.

Saettem, J., Pool, D.A.R., Eilingsen, L. and Sejrup, H.P. (1992) Glacial geology of outer Björnöyrenna, Southwestern Barents Sea. *Mar. Geol.*, **103**, 15–51.

Schaefer, B.F. and Burgess, J.M. (2003) Re-Os isotopic age constraints on deposition in the Neoproterozoic Amadeus Basin: implications for the 'Snowball Earth' *J. Geol. Soc. Lond.*, **160**, 825–828.

Scheffler K., Hoernes, S. and Schwark, L. (2003) Global changes during Carboniferous-Permian glaciation of Gondwana; linking polar and equatorial climate evolution by geochemical proxies. *Geology.*, **31**, 605–608.

Schermerhorn, L.J.G. (1974) Late Precambrian mixtites: glacial and/or nonglacial? *Am. J. Sci.*, **274**, 673–824.

Schermerhorn, L.J.G. (1975) Tectonic framework of Late Precambrian supposed glacials. In: *Ice Ages: Ancient and Modern* (Eds A.E. Wright and F. Moseley), pp. 241–274. Geol. J. Spec. Publ.

Schermerhorn, L.J.G. (1981) Late Precambrian tilloids of northwest Angola. In: *Earth's Pre-Pleistocene Glacial Record* (Eds M.J. Hambrey and W.B. Harland), pp. 158–161. Cambridge University Press, Cambridge.

Schermerhorn, L.J.G. and Stanton, W.I. (1963) Tilloids in the west Congo geosyncline. *Q. Jl. Geol. Soc. Lond.*, **119**, 201–241.

Schmidt, P.W. and Williams, G.E. (1995) The Neoproterozoic climatic paradox: equatorial paleolatitude for Marinoan glaciation near sea level in South Australia. *Earth Planet. Sci. Lett.*, **134**, 107–124.

Schmidt, P.W., Williams, G.E. and Embleton, B.J.J. (1991) Low palaeolatitudes of Late Proterozoic glaciation: early timing of remanence in haematite of the Elatina Formation, South Australia. *Earth Planet. Sci. Lett.*, **105**, 355–367.

Schrag, D.P., Berner, R.A., Hoffman, P.F. and Halverson, G.P. (2002) On the initiation of a snowball Earth. *Geochem., Geophys., Geosyst.*, **3**, No. 6.

Schwab, F.L. (1981) Late Precambrian tillites of the Appalachians. In: *Earth's Pre-Pleistocene Glacial Record* (Eds M.J. Hambrey and W.B. Harland), pp. 751–755. Cambridge University Press, Cambridge.

Sejrup, H.P., Haflidasona, H., Hjelstuena, B.O., Nygård, A., Brynb, P. and Lienb, R. (2004) Pleistocene development of the SE Nordic Seas margin. *Mar. Geol.*, **213**, 169–200.

Shipp, S.S., Wellner, J.S. and Anderson, J.B. (2002) Retreat signature of a polar ice stream: sub-glacial geomorphic features and sediments from the Ross Sea, Antarctica. In: *Glacier-Influenced Sedimentation on High-Latitude Continental Margins* (Eds J.A. Dowdeswell and C.O. Cofaigh), Geol. *Soc. London Spec. Publ.*, **203**, 277–303.

Siedlecka, A. and Roberts, D. (1992) The bedrock geology of Varanger Peninsula, Finnmark, north Norway: an excursion guide. *Nor. Geol. Unders. Spec. Publ.*, **5**, 45 pp.

Smith, A.G. and Pickering, K.T. (2003) Oceanic gateways as a critical factor to initiate icehouse Earth. *J. Geol. Soc. Lond.*, **160**, 337–340.

Smith, I.R. (2000) Diamictic sediments within high Arctic lake sediment cores: evidence for lake ice rafting along the lateral glacial margin. *Sedimentology*, **47**, 1157–1179.

Smith, L.M. and Andrews, J.T. (2000) Sediment characteristics in iceberg dominated fjords, Kangerlussuaq region, East Greenland. *Sed. Geol.*, **130**, 11–25.

Sohl, L.E., Christie-Blick, N. and Kent, D.V. (1999) Paleomagnetic polarity reversals in Marinoan (ca. 600 Ma) glacial deposits of Australia: implications for the duration of low-latitude glaciations in Neoproterozoic time. *Geol. Soc. Am. Bull.*, **111**, 1120–1139.

Sokolov, B.S. (1998) Ocherki stanovleniya venda. KMK Ltd., Moscow.

Solheim, A., Faleide, J.I., Andersen, E.S., Elverhøi, A., Forsberg, C.F., Vanneste, K., Uenzelmann-Neben, G. and Channell, J.E.T. 1998. Late Cenozoic seismic stratigraphy and glacial geological development of the East Greenland and Svalbard-Barents Sea continental margins. *Quatern. Sci. Rev.*, **17**, 155–184.

Songnian, L., Guogan, M., Zhenjia, G. and Weixing, L. (1985) Sinian ice ages and glacial sedimentary facies-areas in China. *Precambrian Res.*, **29**, 53–63.

Songnian, L. and Zhenjia, G. (1987) Characteristics of the Sinian glaciogenic rocks of the Shennongjia region, Hubei Province, China. *Precambrian Res.*, **36**, 127–142.

Souchez, R.A. and Lemmens, M. (1985) Subglacial carbonate deposition: an isotopic study of a present-day case. *Palaeogeogr., Palaeoclimatol., Palaeoecol.*, **51**, 357–364.

Spencer, A.M. (1985) Mechanisms and environments of deposition of Late Precambrian geosynclinal tillites: Scotland and East Greenland. *Palaeogeogr., Palaeoclimatol., Palaeoecol.*, **51**, 143–157.

Sprigg, R.C. (1942) Geology of the Eden-Moana Fault Block. *Trans. R. Soc. S. Austr.*, **66**, 185–214.

Stow, D.A.V. and Piper, D.J.W. (1984) Deep-water fine-grained sediment: facies models. In: *Fine-grained sediment: deep water processes and facies* (Eds D.A.V. Stow and D.J.W. Piper), pp. 611–646. Blackwell Scientific, Oxford.

Strasser, J.C., Lawson, D.E., Larson, G.C., Evenson, E.B. and Alley, R.B. (1996) Preliminary results of tritium analyses in basal ice, Matanuska Glacier, Alaska, USA: evidence for subglacial ice accretion. *Ann. Glaciol.*, **22**, 126–133.

Stratten, T. (1974) Notes on the application of shape parameters to differentiate between beach and river deposits in southern Africa. *Trans. Geol. Soc. S. Afr.*, **77**, 59–64.

Strømberg, A.G.B. (1981) The Late Precambrian Sito tillite and the Vakkejokk breccia in the northern Swedish Caledonides. In: *Earth's Pre-Pleistocene Glacial Record* (Eds M.J. Hambrey and W.B. Harland), pp. 611–614. Cambridge University Press, Cambridge.

Sturt, B.A., Pringle, I.R. and Roberts, D. (1975) Caledonian nappe sequence of Finnmark, northern Norway, and the timing of the orogenic deformation and metamorphism. *Geol. Soc. Am. Bull.*, **86**, 710–718.

Sugden, D.E. (1978) Glacial erosion by the Laurentide Ice Sheet. *J. Glaciol.*, **20**, 367–391.

Sumner, D.Y., Kirschvink, J.L. and Runnegar, B.N. (1987) Soft-sediment paleomagnetic fold tests of late Precambrian glaciogenic sediments. *Eos*, **68**, 1251.

Tadesse, T., Hoshino, M., Suzuki, K. and Iisumi, S. (2000) Sm–Nd, Rb–Sr and Th–U–Pb zircon ages of syn- and post-tectonic granitoids from the Axum area of northern Ethiopia. *J. Afr. Earth Sci.*, **30**, 313–327.

Tadesse, T., Suzuki, K. and Hoshino, M. (1997) Chemical Th–U total Pb isochron age of zircon from the Mareb Granite in northern Ethiopia. *J. Earth Planet. Sci. Negoya Univ.*, **44**, 21–27.

Taviani, M. and Beu, A.G. (2003) The palaeoclimatic significance of Cenozoic marine macrofossil assemblages from Cape Roberts Project drillholes, McMurdo Sound, Victoria Land Basin, East Antarctica. *Palaeogeogr., Palaeoclimatol., Palaeoecol.*, **198**, 131–143.

Taylor, J., Dowdeswell, J.A., Kenyon, N.H. and Ó Cofaigh, C. (2002) Late Quaternary architecture of trough-mouth fans: debris flows and suspended sediments on the Norwegian margin. In: *Glacier-Influenced Sedimentation on High-Latitude Continental Margins* (Eds J.A. Dowdeswell and C.O. Cofaigh), *Geol. Soc. London Spec. Publ.*, **203**, 55–71.

Teklay, M. (1997) Petrology, geochemistry and geochronology of Neoproterozoic magmatic arc rocks from Eritrea: implications for crustal evolution in the southern Nubian Shield, Eritrea. *Dept. Mines Mem.* **1**, 125 pp.

Thelander, T. (1981) Late Precambrian (?) Långmarkberg Formation in the central Swedish Caledonides. In: *Earth's Pre-Pleistocene Glacial Record* (Eds M.J. Hambrey and W.B. Harland), pp. 615–619. Cambridge University Press, Cambridge.

Thiede, J., Winkler, A., Wolf-Welling, T., Eldholm, O., Myhre, A.M., Baumann, K.-H., Henrich, R. and Stein, R. (1998) Late Cenozoic history of the polar North Atlantic: results from ocean drilling. *Quatern. Sci. Rev.*, **17**, 185–208.

Thompson, M.D. and Bowring, S.A. (2000) Age of the Squantum 'tillite', Boston Basin, Massachusetts: U–Pb zircon constraints on terminal Neoproterozoic glaciation. *Am. J. Sci.*, **300**, 630–655.

Thompson, M.D., Keefe, K.L.D., Martin, M.W. and Bowring, S.A. (2000) Refined U-P zircon age of terminal Neoproterozoic Varanger (?) glaciation in the Boston Basin, eastern Massachusetts. *Geol. Soc. Am. Abstr. Prog.*, **32**, 375–376.

Thompson, M.D., Keefe, K.L.D., Martin, M.W. and Bowring, S.A. (2000) Maximum depositional age of the Neoproterozoic Squantum 'Tillite', Boston Basin, Massachusetts; new U-Pb zircon age constraint on Varanger glaciation. *Geol. Soc. Am. Abstr. Prog.*, **32**, 78.

Tollo, R.P. and Aleinikoff, J.M. (1992) Age and compositional relations of the Robertson River Igneous Suite, Blue Ridge Province, Virginia: Implications for the nature of Laurentian rifting. *Geol. Soc. Am. Abstr. Prog.*, **24**, A365.

Tollo, R.P. and Hutson, F.E. (1996) 700 Ma age for the Mechum River Formation, Blue Ridge province, Virginia: A unique time constraint on pre-Iapetan rifting of Laurentia. *Geology*, **24**, 59–62.

Treagus, J.E. (1981) The Lower Dalradian Kinlochlaggan Boulder Bed, central Scotland. In: *Earth's Pre-Pleistocene Glacial Record* (Eds M.J. Hambrey and W.B. Harland), pp. 637–639. Cambridge University Press, Cambridge.

Trompette, R. (1973) Le Precambrien supérieur et le Paléozoi'que inférieur de l'Adrar de Mauritanie (bordure oceidentale du basin de Taoudeni, Afrique de l'Ouest). Un exemple de sédimentation de craton. Etude stratigraphique et sédimentologique. *Tray. Lab. Sci. Terre St-Jr.*, Marseille, B, **7**, 702 pp.

Tucker, M.E. and Reid, P.C. (1981) Late Precambrian glacial sediments, Sierra Leone. In: *Earth's Pre-Pleistocene Glacial Record* (Eds M.J. Hambrey and W.B. Harland), pp. 132–133. Cambridge University Press, Cambridge.

van der Meer, J.J.M., Menzies, J. and Rose, J. (2003) Subglacial till: the deforming glacier bed. *Quatern. Sci. Rev.*, **22**, 1659–1685.

van der Wateren, F.M., Kluiving, S.J. and Bartek, L.R. (2000) Kinematic indicators of subglacial shearing. In: Maltman, A.J., Hubbard, B. and Hambrey, M.J. (eds). *Deformation of Glacial Materials. Geol. Soc. London Spec. Publ.*, **176**, 259–278.

Vanneste, K., Uenzelmann, G. and Miller, H. (1995) Seismic evidence for long-term history of glaciation on central East Greenland shelf south of Scoresby Sund. *Geo-Marine Letters*, **15**, 63–70.

Vidal, B. and Bylund, G. (1981) Late Precambrian boulder beds in the Visingsö Beds, southern Sweden. In: *Earth's Pre-Pleistocene Glacial Record* (Eds M.J. Hambrey and W.B. Harland), pp. 629–631. Cambridge University Press, Cambridge.

Von Veh, M.W. (1993) The stratigraphy and structural evolution of the Late Proterozoic Gariep Belt in the Sendelingsdrif-Annisfontein area, northwestern Cape Province. *Precambrian Res. Unit, Univ. Cape Town, Bull.*, **38**, 1–174.

Vorren, T.O., Hald, M. and Thomsen, E. (1984) Quaternary sediments and environments on the continental shelf off northern Norway. *Mar. Geol.*, **57**, 229–257.

Walter, M.R. (1981) Late Proterozoic tillites of the southwestern Georgina Basin, Australia. In: *Earth's Pre-Pleistocene Glacial Record* (Eds M.J. Hambrey and W.B. Harland), pp. 525–527. Cambridge University Press, Cambridge.

Walter, M.R., Veevers, J.J., Calver, C.R., Gorjan, P. and Hill, A.C. (2000) Dating the 840–544 Ma Neoproterozoic interval by isotopes of strontium, carbon, and sulfur in seawater, and some interpretive models. *Precambrian Res.*, **100**, 371–433.

Warren, S.G., Brandt, R.E., Grenfell, T.C. and McKay, C.P. (2002) Snowball Earth: ice thickness on the tropical ocean. *J. Geophys. Res.*, **107**(C10), 3167.

Weertman, J. (1957) On the sliding of glaciers. *J. Glaciol.*, **3**, 33–38.

Weertman, J. (1961) Mechanism for the formation of inner moraines found near the edge of cold ice caps and ice sheets. *J. Glaciol.*, **3**, 965–978.

Weertman, J. (1964) The theory of glacier sliding. *J. Glaciol.*, **5**, 287–303.

Wells, A.T. (1976) Ngalia Basin. In: *Economic Geology of Australia and Papua New Guinea 3. Petroleum* (Eds R.B. Leslie, H.J. Evans, and C.L. Knight, C.L.), pp. 226–230. Aus. IMM. Monograph Series, **7**.

Wells, A.T. (1981) Late Proterozoic diamictites of the Amadeus and Ngalia Basins, central Australia. In: *Earth's Pre-Pleistocene Glacial Record* (Eds M.J. Hambrey and W.B. Harland), pp. 515–524. Cambridge University Press, Cambridge.

Wells, A.T., Ranford, L.C., Stewart, A.J., Cook, P.J. and Shaw, R.D. (1967) The geology of the north-eastern part of the Amadeus Basin, Northern Territory. *Bur. Miner. Resour. Geol. Geophys., Rep.*, **113**, 97 pp.

Williams, G.E. (1996) Soft-sediment deformation structures from the Marinoan glacial succession, Adelaide foldbelt: implications for the paleolatitude of late Neoproterozoic glaciation. *Sediment. Geol.*, **106**, 165–175.

Williams, G.E. and Schmidt, P.W. (2004) Neoproterozoic Glaciation: Reconciling Low Paleolatitudes and the Geologic Record. In: *The Extreme Proterozoic: Geology, Geochemistry, and Climate. Geophysical Monograph Series,* **146**. A.G.U. 145–159.

Xu, B., Jian, P. Zheng, H., Zou, H., Zhang, L. and Liu, D. (2005) U–Pb zircon geochronology and geochemistry of Neoproterozoic volcanic rocks in the Tarim Block of northwest China: implications for the breakup of Rodinia supercontinent and Neoproterozoic glaciations. *Precambrian Res.*, **136**, 107–123.

Yeates, A.N. and Muhling, P.C. (1977) Explanatory Notes on the Mount Bannerman 1:250 000 Geological Sheet. *Bur. Miner. Resour. Geol. Geophys. and Geol. Surv. W. Austr. Rep.*, 24 pp.

Yeo, G.M. (1981) The Late Proterozoic Rapitan glaciation in the northern Cordillera. In: *Proterozoic basins of Canada* (Ed. F.H.A. Campbell). *Geol. Surv. Canada Paper*, 81–10, 25–46.

Yin, C., Liu, D., Gao, L., Wang, Z., Xing, Y., Jian, P. and Shi, Y. (2003) Lower boundary age of the Nanhua System and the Gucheng glacial stage; evidence from SHRIMP II dating. *Chinese Sci. Bull.*, **48**, 1657–1662.

Young, G.M. (1995) Are Neoproterozoic glacial deposits preserved on the margins of Laurentia related to the fragmentation of two supercontinents? *Geology*, **23**, 153–156.

Young, G.M. (2002) Geochemical investigation of a Neoproterozoic glacial unit: The Mineral Fork Formation in the Wasatch Range, Utah. *Geol. Soc. Am. Bull.*, **114**, 387–399.

Young, G.M. and Gostin, V.A. (1991) Late Proterozoic (Sturtian) succession of the North Flinders Basin, South Australia: an example of temperate glaciation in an active rift setting. In: Glacial Marine Sedimentation: Paleoclimatic Significance (Eds J.B. Anderson and G. Ashley), pp. 207–223. *Geol. Soc. Am. Spec. Publ. Pap.*, **261**.

Yuelun, W., Songnian, L., Zhengjia, G., Weixing, L. and Guogan, M. (1981) Sinian tillites of China. In: *Earth's Pre-Pleistocene Glacial Record* (Eds M.J. Hambrey and W.B. Harland), pp. 386–401. Cambridge University Press, Cambridge.

Zellers, S.D. and Lagoe, M.B. (1992) Stratigraphic and seismic analyses of offshore Yakataga Formation sections, northeastern Gulf of Alaska. In: *Proceedings of the International Conference on Arctic Margins, Anchorage.* OCS Study. MMS 94–0040, 111–116.

Zhao, J., McCulloch, M.T. and Korsch, R.J. (1994) Characterisation of a plume-related approximately 800 Ma magmatic event and its implications for basin formation in central-southern Australia. *Earth Planet. Sci. Lett.*, **121**, 349–367.

Zhenjia, G. and Jianxin, Q. (1985) Sinian glacial deposits in Xinjiang, Northwest China. *Precambrian Res.*, **29**, 143–147.

Zhou, C., Tucker, R., Xiao, S., Peng, Z., Yuan, X. and Chen, Z. (2004) New constraints on the ages of Neoproterozoic glaciations in south China. *Geology*, **32**, 437–440.

Index

Note: page numbers in *italics* refer to figures, those in **bold** refer to tables